WITHDRAWN

About the Cover
Technical Writing: A Practical Approach, Fifth Edition
William Sanborn Pfeiffer

The Japanese artist Ando Hiroshige (1797–1858) designed the print on the cover of this book, entitled "Fuji From the Mountains of Izu." Besides being a beautiful piece of art, it was chosen for three other reasons.

First, the print suggests the book's focus on both Western and non-Western culture—evident in all five editions since its initial publication in 1991. A concerted effort has been made to weave case studies and examples from around the world into the fabric of the book, to give students an international context for their study.

Second, Japanese artists like Hiroshige created woodblock prints using the same basic collaborative skills employed by today's technical writers. This degree of cooperation, common for centuries in Asian work and play, also has become an integral part of the workplace in the West.

Third, the piece on the cover embodies an important principle of all good writing—simplicity. Action is conveyed with minimal imagery and brushstrokes, just as brevity and clarity form the basis for all good writing.

Thus "Fuji From the Mountains of Izu" mirrors qualities of good written communication, while also reinforcing the global theme of the book.

Technical Writing
A Practical Approach

FIFTH EDITION

William Sanborn Pfeiffer
Southern Polytechnic State University

I.C.C. LIBRARY

Prentice
Hall

Upper Saddle River, New Jersey
Columbus, Ohio

Library of Congress Cataloging-in-Publication Data

Pfeiffer, William S.

   Technical writing: a practical approach / William S. Pfeiffer.—5th ed.

     p. cm.

   Includes index.

   ISBN 0–13–093891-2 (pbk.)

    1. English language—Technical English—Problems, exercises, etc. 2. English
language—Rhetoric—Problems, exercises, etc. 3. Technical writing—Problems, exercises,
etc. I. Title.

PE1475.P47 2003

808'.0666—dc21

2001058075

Editor in Chief: Stephen Helba
Executive Editor: Frank I. Mortimer, Jr.
Developmental Editor: Ohlinger Publishing Services
Media Development Editor: Michelle Churma
Production Editor: Louise N. Sette
Copy Editor: Mariellen Hanrahan
Design Coordinator: Diane Ernsberger
Cover Designer: Bryan Huber
Cover art: © Art Resource
Production Manager: Brian Fox
Marketing Manager: Tim Peyton

This book was set by Carlisle Communications, Ltd. It was printed and bound by R.R.
Donnelley & Sons Company. The cover was printed by The Lehigh Press, Inc.

Photo credits: "McDuff" color insert: Page 2: Photo courtesy of Ohio Department of Natural
Resources; Page 3: Photo courtesy of Schlumberger/Sedco Forex; Page 4: Photo courtesy of
Grant Medical Center, Columbus, Ohio; Page 5: Photo property of the Houston Oilers; Page 6:
© 1993 Terence E. Seidel; Page 7: Photo courtesy of Universitat of Hohenheim; Page 8: Photo
courtesy of Tony Stone Images.

Pearson Education Ltd., *London*
Pearson Education Australia Pty. Limited, *Sydney*
Pearson Education Singapore Pte. Ltd.
Pearson Education North Asia Ltd., *Hong Kong*
Pearson Education Canada, Ltd., *Toronto*
Pearson Educación de Mexico, S.A. de C.V.
Pearson Education—Japan, *Tokyo*
Pearson Education Malaysia Pte. Ltd.
Pearson Education, *Upper Saddle River, New Jersey*

**Copyright © 2003, 2000, 1997, 1994, 1991 by Pearson Education, Inc., Upper Saddle
River, New Jersey 07458.** All rights reserved. Printed in the United States of America. This
publication is protected by Copyright and permission should be obtained from the publisher
prior to any prohibited reproduction, storage in a retrieval system, or transmission in any
form or by any means, electronic, mechanical, photocopying, recording, or likewise. For information regarding permission(s), write to: Rights and Permissions Department.

10 9 8 7 6 5 4 3 2 1

ISBN 0-13-093891-2

Preface

It is a good thing, perhaps, to write for the amusement of the public, but it is a far higher and nobler thing to write for their instruction, their profit, their actual and tangible benefit.

Mark Twain, "Curing a Cold"

Perhaps it was in a moment of weakness that Mark Twain—an amusing writer if there ever was one—wrote this passage about the value of practical writing. Whatever his motivation, he clearly acknowledges the importance of writing that makes the world function, the kind of prose we call "technical writing" today.

Every student who plans to work in business and industry must master the art of technical writing. Your employer will expect as much. Just as importantly, skill in writing will give you personal satisfaction and will help you advance throughout your life. Indeed, in this new century, effective writing remains a major criterion for success in all professions.

To this end, this book stresses one simple principle: you learn to write best by doing as much writing as possible. What follows is a summary of the main features of this book. Some were brought forward from the previous edition and others are new features developed for this edition.

The fifth edition adds features that make it more readable and usable, without changing what has made the book work in all its editions—numbered guidelines, good models, practical assignments, and clear writing.

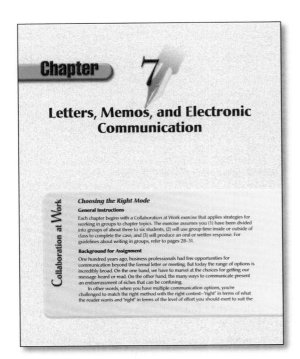

New Features to This Edition

The fifth edition of *Technical Writing: A Practical Approach* includes six new features:

● 1. "Collaboration at Work" Exercises

Every chapter begins with a "Collaboration at Work" exercise.

These exercises engage the student's interest in chapter content by getting teams to complete a simple project. Now all 15 chapters have "bookends" of collaborative exercises—the "Collaboration at Work" at the start and the "Communication Challenge" at the end.

● 2. International Assignments

Each chapter's set of exercises ends with an "international communication assignment."

The "international communication assignments" are marked by an icon in the margin. The assignment requires students to use information from the chapter to solve a problem related to communicating across cultures and nations.

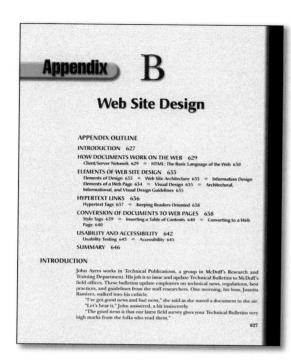

● *3. Appendix on Web Site Design*

Appendix B introduces students to issues of design, construction, use, and evaluation of web sites.

● *4. Expanded Email Coverage*

Chapter 7 includes a revised set of guidelines, an "ABC Format" for email documents, and an additional email model.

This expanded coverage of email is critically important as email continues to replace internal and external correspondence at an amazing rate. Writers need a clear understanding of how to use email properly.

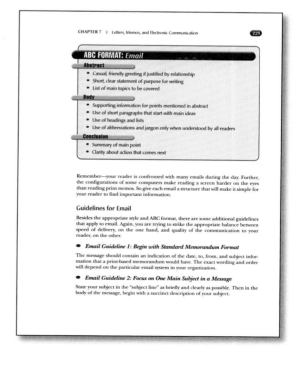

● 5. Revised Research Chapter

Chapter 13 has been thoroughly updated to respond to rapid changes in Internet research and databases.

This chapter includes a new research example topic—hybrid and battery-powered vehicles—that runs throughout the entire chapter. Also new to the chapter is a model of a short research report that uses the hybrid vehicle topic.

● 6. Expanded Coverage of Style and Usage

Chapter 15 adds information on tone, and the Handbook (Appendix C) includes 30 new entries on matters of correct usage.

Other changes in the fifth edition include (1) more assignments, (2) more information on electronic job searches in Chapter 14, (3) outlines of chapter headings at the beginning of each chapter, (4) additional web site support at www. prenhall.com/pfeiffer, and (5) the use of four-color layout throughout the book.

Core Features of This Book

Technical Writing: A Practical Approach is built on a core of five basic features.

● 1. Focus on Process and Product

This book has you practicing writing early (Chapter 1).

The text immerses you in the process of technical writing while teaching you practical formats for getting the job done. After initial chapters that discuss the planning, drafting, and editing process, the bulk of the book provides guidance for writing the most common types of documents. A final set of chapters and appendices covers special topics such as oral presentations, research, and style.

2. A Simple ABC Pattern for All Documents

The "ABC format"—Abstract, Body, and Conclusion —will guide your work in this course and throughout your career.

This underlying three-part structure provides you with a convenient handle for designing almost every technical document you encounter.

3. McDuff, Inc.—A Fictional Company

McDuff, Inc. creates a fictional company for the classroom.

Not all students have experience working in a professional or technical organization, so Mc-Duff supplies a realistic backdrop for many of the book's examples and assignments. By placing your reader or yourself in the role of a McDuff employee, you give concrete importance to the abstract principle of "writing for your reader."

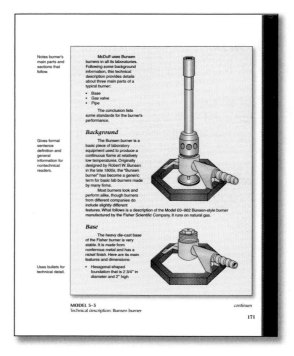

● 4. Numbered Guidelines

Many sets of short, numbered guidelines make this book easy to use while you complete class projects.

Each set of guidelines will take you through the process of finishing assignments, such as writing a proposal, doing research on the Internet, constructing a bar chart, and preparing an oral presentation.

● 5. Annotated Models

The text contains models grouped at the end of chapters on color-edged pages for easy reference.

Annotations in the margins are highlighted in color and show you exactly how the sample documents illustrate the guidelines set forth in the chapters.

Additional Features

Beyond features already described, the following six additional features define the book's mission and demonstrate its utility in the classroom.

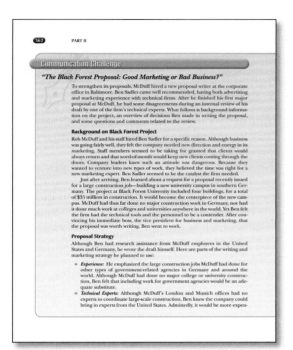

● 1. Communication Challenges

Every chapter includes a McDuff case study, with related questions, at the end of the chapter text.

Called a "Communication Challenge," each case describes a communication problem that relates to the material in its respective chapter. Your instructor may use these case studies as a springboard for class discussion or for collaborative assignments.

● 2. Writing Handbook

This book provides a well-indexed, alphabetized handbook on grammar, mechanics, and usage.

The writing handbook gives you quick access to rules for eliminating editing errors during the revision process.

● 3. Coverage of International Communication

Because globalism continues to transform the business world, this book includes suggestions for understanding other cultures and for writing in an international context.

• 4. Information on English as a Second Language

An ESL section is included in the Handbook.

A growing number of technical communication students are from countries or cultures where English is not the first language. The ESL section of the Handbook focuses on three main problem areas: articles, prepositions, and verb use. It also applies ESL analysis to an excerpt from a technical report.

• 5. Assignments on Ethics

To reinforce the ethical guidelines described in Chapter 2, each chapter includes an ethics assignment.

No one can escape the continuous stream of ethical decisions required of every professional almost every day—such as deciding what to include in your resume or choosing what tone to adopt in a proposal—which is why the text challenges you to address ethical issues in these assignments.

PART II

Whether you write an individual or a group report, follow this general procedure:

- Assemble a list of 5 or 10 problems that you have observed at your school. These problems might concern (1) the physical campus (as in poor design of parking lots or inadequate lab space), (2) the curriculum (as in the need to update certain courses), (3) extracurricular activities (as in the need for more cultural or athletic events), or (4) difficulties with campus support services (as in red tape during registration).
- Work with your instructor to choose the one topic for which you can find the most information and for which you can develop the context described here.
- Collect information in whatever ways seem useful—for example, site observations, surveys to students, follow-up phone calls, or interviews.
- Submit progress reports at intervals requested by your instructor. (Consult guidelines in the "Progress/Periodic Reports" section of chapter 8.)
- Consider your *role* to be the one that you, in fact, have—a student or a group of students at the school. Then select as your *reader(s)* the school officials who would actually be in charge of solving the problem you have identified. (You may or may not end up sending the report. Follow the advice of your instructor in this matter.) The *purpose* of this report will be to explain, in great detail, all aspects of the problem *and to* form conclusions as to its cause. If it seems appropriate, you may take one further step to suggest recommendations for a solution—*if* your research has taken you this far. In any case, detail and also tact are important criteria.

8. **McDuff-Based Formal Report.** For this assignment you will place yourself into a role of your choosing at McDuff, Inc. Use the following procedure, which may be modified by your instructor:

- Review the section at the beginning of this chapter that lists McDuff cases for formal reports to get a sense of when formal reports are used at companies like McDuff.
- Review the McDuff information in chapter 2, especially with regard to the kinds of jobs people hold at the company and the kinds of projects that are undertaken.
- Choose a specific job that you could assume at McDuff—based on your academic background, your work experience, or your career interests.
- Choose a specific project that (a) could conceivably be completed at McDuff by someone in the role you have chosen, (b) would result in a formal report directed either inside or outside the company, and (c) would be addressed to a complex audience at two or three of the levels indicated on the Planning Form at the end of the book.
- Be sure you have access to information that will be used in this simulated report—for example, from work experience, from a term paper or class project in another course, or from your interviews of individuals already in the field. (For this assignment, you may want to talk with a professional, such as a recent graduate in your major.)
- Prepare a copy of the Planning Form at the end of the book for your instructor's approval—*before* proceeding further with the project.
- Complete the formal report, following the guidelines in this chapter.

9. **Ethics Assignment.** Illustrations on cover pages of formal reports are one strategy for attracting the readers' attention to the document. Note the use of an illustration in Model 9–1, along with the rationale on page 303. Do you think Model 9–1 uses its graphics in an ethically sound way to engage the reader with the report? Why or why not? How do you determine whether a cover page illustration is an appropriate persuasive tool, on the one hand, or an inappropriate attempt to manipulate the reader, on the other? Give hypothetical examples.

• 6. Appendix on Online Documents

Appendix A addresses issues raised by the increasing prevalence of online documents in the workplace.

Many professionals, not just technical writers, are writing documents that will be read by their audience online, often never to be placed in hard copy. This new medium requires that we change the way we think about designing "pages" (screens), to help the reader gain quick access to all parts of the document.

Supplements

The book is supported by the following supplements.

- Instructor's Resource Manual that includes Sample Syllabi, Chapter Summaries, Chapter Tests with Answer Keys, Editing Revisions, Planning Form and Letterhead (blank for copying), Resources for New Teachers of Technical Writing, and over 100 Transparency Masters of the book's Writing Guidelines and some of the Models.

- Transparency Acetates of most of the book's Models and ABC Formats.
- PowerPoint Slides of all the book's Models, Guidelines, and ABC Formats.
- Companion Web Site.

Acknowledgments

This edition and previous editions benefited from contributions by many colleagues, of whom the following deserve special mention. Shawn Tonner—who revised Chapter 13 in the two previous editions—thoroughly updated the same chapter for this edition, with the help of her husband and my colleague, Mark Stevens. Hattie Schumaker added guidelines on color and upgraded the examples in Chapter 11. Mike Hughes, an expert on online documentation, wrote two of the appendices: "Online Technical Documents" and "Web Site Design." Ken Rainey helped with coverage of email in Chapter 7. And Jeff Orr contributed a section on English as a Second Language to the book's Handbook in Appendix C.

Many other friends and colleagues offered excellent suggestions used in this book. I want to thank Saul Carliner, George Ferguson, Alan Gabrielli, Bob Harbort, Dory Ingram, Becky Kelly, Chuck Keller, Jo Lundy, Minoru Moriguchi, Randy Nipp, Lisa A. Rossbacher; Betty Oliver Seabolt, Herb Smith, James Stephens, Mark Stevens, Lavern Smith, John Ulrich, Steven Vincent, and Tom Wiseman. In addition, these reviewers from other schools provided their assistance with the project: Theresa A. James, South Florida Community College; Andrew Jones, University of California, Davis; Richard N. Lewis, Jr., Sandhills Community College; and Susan S. Rehwaldt, Southern Illinois University.

Four companies allowed me to use written material gathered during my consulting work: Fugro-McClelland, Law Engineering and Environmental Services, McBride-Ratcliff and Associates, and Westinghouse Environmental and Geotechnical Services. Though this book's fictional firm, McDuff, Inc., does have features of the world I observed as a consultant, I want to emphasize that McDuff is truly an invention.

I also appreciate the help of the following students for allowing me to adapt their written work for use in this book: Michael Alban, Becky Austin, Corey Baird, Natalie Birnbaum, Cedric Bowden, Gregory Braxton, Ishmael Chigumira, Bill Darden, Jeffrey Daxon, Rob Duggan, William English, Joseph Fritz, Jon Guffey, Sam Harkness, Gary Harvey, Lee Harvey, Hammond Hill, Sudhir Kapoor, Steven Knapp, Wes Matthews, Kim Meyer, James Moore, Chris Owen, Scott Lewis, James Porter, James Roberts, Mort Rolleston, Chris Ruda, Barbara Serkedakis, Tom Skywark, Tom Smith, DaTonja Stanley, James Stephens, Chris Swift, and Jeff Woodward.

For all five editions, it has been my good fortune to have the same extraordinary developmental editor, my friend and colleague, Monica Ohlinger. Many of the ideas that have made this book successful for over a decade were hers. Also, I want to give special thanks to my Prentice Hall editors, Steve Helba and Frank Mortimer, for their continuing faith in my book. Other Prentice Hall people who contributed greatly to the project include Louise Sette, production editor, and Diane Ernsberger, design coordinator.

Finally, deepest thanks go to my family—Evelyn, Zachary, and Katie—for their love and support throughout this and every writing project I take on.

A Final Note to Students

I'll close this preface with the same comment made in the last edition: At the start of my classes, I sometimes ask students to describe their professional goals for the next 10 years. As you might expect, their comments suggest they hope to rise to important positions in the workplace and make genuine contributions to their professions. Such long-term thinking is crucial, keeping you on course in your life.

Yet, ultimately, the way you handle the small details of daily life most influences the real contribution you make in the long run. If you do good work, believe in what you do, and communicate well with others—both interpersonally and in writing—success will come your way. The author Robert Pirsig put it this way in his 1974 work, *Zen and the Art of Motorcycle Maintenance:*

> The place to improve the world is first in one's own heart and head and hands, and then work outward from there.

I believe—and this book tries to show—that clear, concise, and honest writing is one of the most powerful tools of your heart, head, *and* hands.

Contents

3   Organizing Information   74

4   Page Design   100

8   Informal Reports   258

9   Formal Reports   296

10 Proposals and Feasibility Studies   342

PART III   409

15 Style in Technical Writing   592

Appendix A: Online Technical Documents   619

Appendix B: Web Site Design   627

Part

One

Chapter 1

Process in Technical Writing

Collaboration at Work

Outline for a Consulting Report

General Instructions

Each chapter begins with a Collaboration at Work exercise that applies strategies for working in groups to chapter topics. The exercise assumes you (1) have been divided into groups of about three to six students, (2) will use group time inside or outside of class to complete the case, and (3) will produce an oral or written response. For guidelines about writing in groups, refer to pages 28–34.

Background for Assignment

Assume you and your group members compose one of several teams from a private consulting firm. The firm has been hired to help plan a hotel/conference center to be built on your campus. Although the center will cater to some private clients, most customers will be associated with your institution—for example, parents of students, candidates for teaching or administrative positions, and participants in academic conferences.

Obviously, a project of this sort requires careful planning. One step in the process is to assess the needs of various people and groups that will occupy the center.

CHAPTER OUTLINE

Following are listed just a few of the many groups or departments whose needs should be considered:

Accounting	Landscaping	Registration
Catering	Maintenance	Sales and marketing
Computing	Procurement	Security
Housekeeping	Recreation	Training

The topics range broadly because the facility will have multiple purposes for a diverse audience.

Group Assignment

The consulting firm—of which your group is a part—will issue a joint report that describes the needs of all groups who will work in the new hotel/conference center. Assume that your team's task is to produce just a portion of the *outline*—not the text—of the report. Your outline will address one or more of the needs reflected in the previous list, or other needs of your choosing that have not been listed. Your instructor will indicate whether you are to refer to the section in this chapter entitled "Writing an Outline."

*G*ood communication skills are essential in any career you choose. Jobs, promotions, raises, and professional prestige result from your ability to present both written and visual information effectively. With so much at stake, you need a simple road map to direct you toward writing excellence. *Technical Writing: A Practical Approach* is such a map.

The four chapters in Part 1 give you an overview of technical writing and prepare you to complete the assignments in this book:

Chapter 1: Defines technical writing and describes the process of writing and of integrating visual images into documents. As shown in Figure 1–1, the technical writing process has three main parts: planning, drafting, and revising. Careful completion of this *process* is the best guarantee of a successful *product*—the final document.

Chapter 2: Introduces you to the fictional company of McDuff, Inc., and focuses on ethical and international issues in corporate communication. McDuff provides the framework for most examples and assignments in this book.

Chapter 3: Discusses the organization of information for diverse readers, who are often very busy and unfamiliar with a project.

Chapter 4: Describes the elements of page design and techniques you can use to enhance the readability of documents.

So that you can begin writing early in the course, preliminaries are kept to a minimum. Now let's take a closer look at the nature of technical writing.

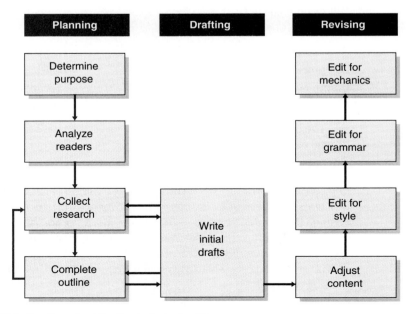

FIGURE 1–1 Flowchart for the technical writing process

GETTING FROM HERE TO THERE

You may have learned how to write short essays in previous writing courses. This book helps you transfer that basic knowledge to the kind of writing done on the job. Career writing is so practical, so well grounded in common sense, that it will seem to proceed smoothly from your previous work. This section highlights features of both traditional academic writing and job-related technical writing.

Features of Academic Writing

Writing you have done in school probably has had these characteristics:

- **Purpose:** Demonstrating what you know about the topic, in a way that justifies a high grade
- **Your knowledge of topic:** Less than the teacher who evaluates the writing
- **Audience:** Teacher who requests the assignment and who will read it from beginning to end
- **Criteria for evaluation:** Depth, logic, clarity, unity, and grammar
- **Statistical and graphic support:** Sometimes used to explain and persuade

Academic writing requires that you use words to display your learning to someone who knows more about the subject than you do. Because this person's job is to evaluate your work, you have what might be called a "captive audience." Here are two common situations in which academic writing would be appropriate, along with brief examples:

● *Example 1: Essays for English Class*

In a typical high school or college writing class, you write short essays based on your experience or readings. Instructors expect you to (1) state the topic, purpose, and main points in the introduction, (2) develop the supporting points in the body, and (3) wrap up the paper in the conclusion. Your reader wants a coherent, unified, grammatically sound, and interesting piece of writing. What follows is the first paragraph from one such essay. The student was asked to attend and then comment on a local cultural event.

> The recent exhibit at Atlanta's High Museum of Art, "Masterpieces of the American West," gives the viewer a small glimpse of the beautiful land and fascinating peoples that confronted the explorer-artists who ventured west in the 1800s. This paper will describe my response to three works—Catlin's *Mandan Dance,* Bierstadt's *Rocky Mountain Waterfall,* and Remington's *Turn Him Loose, Bill.* The first portrays the Indian culture; the second, Western landscape; and the third, the encroaching civilization from the East.

● *Example 2: Exams and Papers in General Studies Courses*

Courses like history and science often require that you write analytical papers and exams. As in essays for a writing class, these papers should start with a topic statement that outlines main points and then proceed with supporting paragraphs. For example, you may be asked to describe both the immediate and long-term reasons for a particular event. Here is the first paragraph of a cause-effect essay in environmental science. The writer must define and give causes for the "greenhouse effect."

> The term *greenhouse effect* refers to the global warming trend of the earth. It occurs when the sun's reflected heat is unable to escape the earth's atmosphere into space, because of a buildup of carbon monoxide and other gases in the atmosphere. This essay will describe these main reasons for the dramatic increase in the greenhouse effect: the burning of rain forests in the Amazon, accelerated emission of auto pollutants around the world, and increases in the use of coal-fired plants.

In both examples, the *purpose* is to demonstrate knowledge, and the *audience* is someone already familiar with the subject or approach. In this sense, academic writing shows your command of information to someone more knowledgeable about the subject than you are. The next section examines a different kind of writing—the kind you will be doing in your career and in this course. The purpose and audience differ considerably from those of academic writing.

Features of Technical Writing

The rules for writing shift somewhat when you begin your career. Employees unprepared for this change often flounder for years, never quite understanding the new rules. *Technical writing* is a generic term for all written communications done on the job—whether in business, industry, or other professions. It is particularly identified with jobs in technology, engineering, science, the health professions, and other fields with specialized vocabularies. The terms *technical writing, professional writing, business writing,* and *occupational writing* all mean essentially the same thing—writing done in your career. Here are the main characteristics of technical writing:

- **Purpose:** Getting something done within an organization (completing a project, persuading a customer, pleasing your boss, etc.)
- **Your knowledge of topic:** Usually greater than that of the reader
- **Audience:** Often several people, with differing technical backgrounds
- **Criteria for evaluation:** Clear and simple organization of ideas, in a format that meets the needs of busy readers
- **Statistical and graphic support:** Frequently used to explain existing conditions and to present alternative courses of action

Contrast these features with those of academic writing, listed earlier. In particular, note the following main differences:

1. Technical writing has a practical role on the job, whereas academic writing aims only to display your knowledge.
2. Technical writing is done by an informed writer conveying needed information both verbally and visually to an uninformed reader, whereas academic writing is done by a student as the learner for a teacher as the source of knowledge.
3. Technical writing often is read by many readers, whereas academic writing aims to satisfy only one person, the teacher.

Correspondence: In-House or External
- Memos to your boss and to your subordinates
- Routine letters to customers, vendors, etc.
- "Good news" letters to customers
- "Bad news" letters to customers
- Sales letters to potential customers
- Electronic mail (email) messages to coworkers or customers over a computer network

Short Reports: In-House or External
- Analysis of a problem
- Recommendation
- Equipment evaluation
- Progress report on project or routine periodic report
- Report on the results of laboratory or field work
- Description of the results of a company trip

Long Reports: In-House or External
- Complex problem analysis, recommendation, or equipment evaluation
- Project report on field or laboratory work
- Feasibility study

Other Examples
- Proposal to boss for new product line
- Proposal to boss for change in procedures
- Proposal to customer to sell a product, service, or idea
- Proposal to funding agency for support of research project
- Abstract or summary of technical article
- Technical article or presentation
- Operation manual or other manual
- Web site

FIGURE 1–2 Examples of technical writing

Finally, technical writing places greater emphasis on techniques of organization and format that help readers find important information as quickly as possible.

Figure 1–2 lists some typical on-the-job writing assignments. While not exhaustive, the list does include many of the writing projects you will encounter. Also, see Figure 1–3 for an example of a short technical document.

Besides projects that involve writing, your career will also bring you speaking responsibilities, such as formal speeches at conferences and informal presentations at meetings. The term *technical communication* is used to include the writing, speaking, and development of graphics required to communicate your ideas on the job.

Although technical communication plays a key role in the success of all technical professionals and managers, the amount of time you devote to it will depend on your job. A 1989 survey of technical managers gives some idea of the time involved. Conducted by the National Aeronautics and Space Administration (NASA), the survey canvassed managers in profit-making and nonprofit organizations in the field of aeronautics. As Figure 1–4 shows, 100% of the profit managers

MEMORANDUM

DATE: December 6, 2002
TO: Holly Newsome
FROM: Michael Allen *MA*
SUBJECT: Printer Recommendation

Introductory Summary

Recently you asked for my evaluation of the Hemphill LaserFast printer currently used in my department. Having analyzed the printer's features, print quality, and cost, I am quite satisfied with its performance.

Features

Among the LaserFast's features, I have found these five to be most useful:
1. Portrait and landscape print modes
2. Selectable character sizes (10/12/14 or 18 characters per inch)
3. 1.5 megabytes of print-buffer memory
4. Selectable paper sizes (letter, legal, half-letter)
5. Print speed of six pages per minute
In addition, the LaserFast printer is equipped with two built-in fonts and accepts several additional fonts from specially purchased cartridges. This combination of features makes the LaserFast a most versatile printer.

Print Quality

The Hemphill LaserFast printer produces laser-sharp clarity that rivals professional typeset quality. The print resolution is an amazing 300 x 300 dots per inch, among the highest attainable in current desktop printers. This memo was printed on LaserFast, and, as you can see, the quality speaks for itself.

Cost

Considering the features and quality, the LaserFast is an excellent printer for the money. At a retail price of $2500, it is one of the lowest-priced laser printers. In fact, the highest-quality daisy wheel and dot matrix printers cost just $500 less than the LaserFast, yet neither offers the superior features or quality of a laser printer.

Conclusion

On the basis of my observation, I strongly recommend that our firm continue to use and purchase the LaserFast printer. Please call me at ext. 204 if you want further information about this excellent machine.

FIGURE 1–3 Short report

and 98% of the nonprofit managers in the study consider technical communication a "somewhat important" or "very important" part of their jobs.

Even more telling is the amount of time the NASA survey respondents spend communicating. Figure 1–4 indicates that they use (1) over one-third of their work

TABLE 1. Importance of Technical Communications				
	Profit Managers		Nonprofit Managers	
How Important	No.	%	No.	%
Very	86	92.5	43	84.3
Somewhat	7	7.5	7	13.7
Not at all	0	0.0	1	2.0
Total	93	100.0	51	100.0

TABLE 2. Time Spent Communicating Technical Information to Others				
	Profit Managers		Nonprofit Managers	
Time Spent Per Week, Hours	No.	%	No.	%
5 or less	13	14.3	9	18.0
6 to 10	33	36.2	16	30.0
11 to 20	37	40.7	21	42.0
21 or more	8	8.8	5	10.0
Total	91	100.0	51	100.0
Mean	13.5		13.9	

TABLE 3. Time Spent Working with Technical Information Received from Others				
	Profit Managers		Nonprofit Managers	
Time Spent Per Week, Hours	No.	%	No.	%
5 or less	8	8.7	6	12.0
6 to 10	42	46.2	23	46.0
11 to 20	36	39.6	18	36.0
21 or more	5	5.5	3	6.0
Total	91	100.0	50	100.0
Mean	13.0		13.0	

FIGURE 1–4 Data from NASA aeronautics survey

Source: Thomas E. Pinelli et al., *Technical Communications in Aeronautics: Results of an Exploratory Study—An Analysis of Profit Managers' and Nonprofit Managers' Responses* (Washington, D.C.: National Aeronautics and Space Administration, NASA TM-101626, October 1989), 71. (Available from NTIS Springfield, Va.)

time conveying information *to* others and (2) another one-third working with technical information sent to them *by* others. Based on a 40-hour week, therefore, both groups spend roughly *two-thirds* of their working week on job duties associated with technical communication.

Now that you know the nature and importance of technical writing, the next section examines the first part of the planning stage: determining a document's purpose.

DISCOVERING YOUR PURPOSE

Kate Paulsen works as a training supervisor for the Boston office of McDuff, Inc., a firm described in more detail in chapter 2. The company is growing so quickly that hiring, training, and retraining employees have become major goals. Kate recently flew to Cleveland to attend a workshop sponsored by a major professional training organization. The workshop emphasized a new in-house procedure for surveying the training needs of a company's employees. After returning to Boston, Kate must write her manager a trip report that describes the survey technique. She ponders three different approaches to the report:

- **Giving an overview** of the survey procedure she studied during the three-day workshop—stressing a few key points so that her manager could decide whether to inquire further
- **Providing details** of exactly how the survey procedure could be applied to her firm—with enough specifics for her manager to see exactly how the survey could be used at McDuff
- **Proposing** that the procedure for conducting the needs survey be used at McDuff—in language that argues strongly for adoption

For Kate, the first step is to decide what she wants to accomplish. Likewise, every piece of *your* writing should have a specific reason for being. The purpose may be dictated by someone else or selected by you. In either case, it must be firmly understood *before* you start writing. Purpose statements guide every decision you make while you plan, draft, and revise.

Kate Paulsen's three choices indicate some of your options, but there are others. Your choice of purpose will fall somewhere within this continuum:

Neutral, objective statement ⟵⟶ Persuasive, subjective statement

For example, when reporting to your boss on the feasibility of adding a new wing to your office building, you should be quite objective. You must provide facts that can lead to an informed decision by someone else. If you are an outside contractor proposing to construct such a wing, however, your purpose is more persuasive. You will be trying to convince readers that your firm should receive the construction contract.

When preparing to write, therefore, you need to ask yourself two related questions about your purpose.

Question 1: Why Am I Writing This Document?

This question should be answered in just one or two sentences, even in complicated projects. Often the resulting purpose statement can be moved "as is" to the beginning of your outline and later to the first draft.

For example, Kate Paulsen finally decides on the following purpose statement, which becomes the first passage in her trip report: "This memo will highlight main features of the training needs survey introduced at the workshop I attended in Cleveland. I will focus on several possible applications you might want to consider for our office training." Note that Kate's purpose would rest about halfway across the persuasive continuum shown earlier. Though she will not be strongly advocating McDuff's use of the survey, she will be giving information that suggests the company might benefit by using it.

● *Question 2: What Response Do I Want from Readers?*

The first question about purpose leads inevitably to the second about results. Again, your response should be only one or two sentences long. Though brief, it should pinpoint exactly what you want to happen as a result of your document. Are you just giving data for the file? Will information you provide help others do their jobs? Will your document recommend a major change?

In Kate Paulsen's case, she decides on this results statement: "Though I'm not yet sure if this training survey is worth purchasing for McDuff, I want my boss to consider it." Unlike the purpose statement, the results statement may not go directly into your document. Kate's statement hints at a "hidden" agenda that may be implicit in her trip report but will not be explicitly stated. This statement, written for her own use, becomes an essential part of her planning. It is a concrete goal for her to keep in mind as she writes.

The answers to these two questions about purpose and results are included on the Planning Form your instructor may ask you to use for assignments. Figure 1–5 on pages 12 and 13 includes a copy of the form, along with instructions for using it. The last page of this book contains another copy you can duplicate for use with assignments.

Having established your purpose, you are now ready to consider the next part of the writing process: audience analysis.

ANALYZING YOUR READERS

One cardinal rule governs all on-the-job writing:

> WRITE FOR YOUR READER, NOT FOR YOURSELF.

This rule especially applies to science and technology because many readers may know little about your field. In fact, experts on writing agree that most technical writing assumes too much knowledge on the part of the reader. The key to avoiding this problem is to examine the main obstacles readers face and adopt a strategy for overcoming them.

This section (1) highlights problems that readers have understanding technical writing, (2) suggests techniques to prevent these problems, and (3) describes some

PLANNING FORM

NAME: _____ ASSIGNMENT: _____

I. Purpose: Answer each question in one or two sentences.

   A. Why are you writing this document? _____
   _____

   B. What response do you want from readers? _____
   _____

II. Reader Matrix: Fill in names and positions of people who may read the document.

	Decision-Makers	Advisers	Receivers
Managers			
Experts			
Operators			
General Readers			

III. Information on Individual Readers: Answer these questions about selected members of your audience. Attach additional sheets as is necessary.

   1. What is this reader's technical or educational background?
   _____
   _____

   2. What main question does this person need answered?
   _____
   _____

   3. What main action do you want this person to take?
   _____
   _____

   4. What features of this person's personality might affect his or her reading?
   _____
   _____

   5. What features does this person prefer in
      Format? _____
      Style? _____
      Organization? _____

IV. Outline: Attach an outline (topic) to use in drafting the **BODY** of this document.

FIGURE 1–5   Planning Form for all technical documents

Instructions for Completing "Planning Form"

The Planning Form is for your use in preparing assignments in your technical writing course. It focuses only on the planning stage of writing. Complete it before you begin your first draft.

1. Use the Planning Form to help plan your strategy for all writing assignments. Your instructor may or may not require that it be submitted with assignments.

2. Photocopy the form on the back page of this book or write the answers to questions on separate sheets of paper, whatever option your instructor prefers. (Your instructor may hand out enlarged, letter-size copies of the form, which are included in the Instructor's Resource Manual.)

3. Answer the two purpose questions in one or two sentences each. Be as specific as possible about the purpose of the document and the response you want—especially from the decision-makers.

4. Note that the reader matrix classifies each reader by two criteria: (a) technical levels (shown on the vertical axis) and (b) relationship to the decision-making process (shown on the horizontal axis). Some of the boxes will be filled with one or more names, while others may be blank. How you fill out the form depends on the complexity of your audience and, of course, on the directions of your instructor.

5. Refer to Chapter 2 for any McDuff positions and titles you may want to use in the reader matrix, if your paper is based on a simulated case from McDuff, Inc.

6. Note that questions in the "Information on Individual Readers" section can be filled out for one or more readers, depending on what your instructor requires.

7. Complete the topic outline after you have collected whatever information or research your document requires. The outline should be specific and should include two or three levels (see Figure 1–8 and Figure 1–9).

FIGURE 1–5 *continued*

main classifications of technical readers. At first, analyzing your audience might seem awkward and even unproductive. You are forced out of your own world to consider that of your reader, before you even put pen to paper. The payoff, however, will be a document that has clear direction and gives the audience what it wants.

Obstacles for Readers

As purchasing agent for McDuff, Inc., Charles Blair must recommend one automobile sedan for fleet purchase by the firm's sales force and executives. First, he will conduct some research—interviewing car firm representatives, reading car evaluations in consumer magazines, and inquiring about the needs of his firm's salespeople. Then he will submit a recommendation report to the selection committee consisting of the company president, the accounting manager, several salespeople, and the supervisor of company maintenance. As Charles will discover, readers of all backgrounds often have these four problems when reading any technical document:

1. Constant interruptions
2. Impatience finding information they need
3. A different technical background from the writer
4. Shared decision-making authority with others

If you think about these obstacles every time you write, you will be better able to understand and respond to your readers.

● *Obstacle 1: Readers Are Always Interrupted*

As a professional, how often will you have the chance to read a report or other document without interruption? Such times are rare. Your reading time will be interrupted by meetings and phone calls, so a report often gets read in several sittings. Aggravating this problem is the fact that readers may have forgotten details of the project.

● *Obstacle 2: Readers Are Impatient*

Many readers lose patience with vague or unorganized writing. They are thinking "What's the point?" or "So what?" as they plod through memos, letters, reports, and proposals. They want to know the significance of the document right away.

● *Obstacle 3: Readers Lack Your Technical Knowledge*

In college courses, the readers of your writing are professors who usually have knowledge of the subject on which you are writing. In your career, however, you will write to readers who lack the information and background you have. They expect a technically sophisticated response, but in language they can understand. If you write over their heads, you will not accomplish your purpose. Think of yourself as an educator; if readers do not learn from your reports, you have failed in your objective.

● *Obstacle 4: Most Documents Have More Than One Reader*

If you always wrote to only one person, technical writing would be much easier than it is. Each document could be tailored to the background, interests, and technical education of just that individual. However, this is not the case in the actual world of business and industry. Readers usually share decision-making authority with others, who may read all or just part of the text. Thus you must respond to the needs of many individuals—most of whom have a hectic schedule, are impatient, and have a technical background different from yours.

Ways to Understand Readers

Obstacles to communication can be frustrating. Yet there are techniques for overcoming them. First, you must try to find out exactly what information each reader needs. Think of the problem this way—would you give a speech without learning about the background of your audience? Writing depends just as much, if not more, on such analysis. Follow these four steps to determine your readers' needs:

● *Audience Analysis Step 1: Write Down What You Know about Your Reader*

To build a framework for analyzing your audience, you need to write down—not just casually think about—the answers to these questions for each reader:

1. What is this reader's technical or educational background?
2. What main question does this person need answered?
3. What main action do you want this person to take?
4. What features of this person's personality might affect his or her reading?
5. What features does this reader prefer in
    Format?
    Style?
    Organization?

The Planning Form in Figure 1–5 includes these five questions.

● *Audience Analysis Step 2: Talk with Colleagues Who Have Written to the Same Readers*

Often your best source of information about your readers is a colleague where you work. Ask around the office or check company files to discover who else may have written to the same audience. Useful information could be as close as the next office.

● *Audience Analysis Step 3: Find Out Who Makes Decisions*

Almost every document requires action of some kind. Identify decision-makers ahead of time so that you can design the document with them in mind. Know the needs of your *most important* reader.

● *Audience Analysis Step 4: Remember That All Readers Prefer Simplicity*

Even if you uncover little specific information about your readers, you can always rely on one basic fact: Readers of all technical backgrounds prefer concise, simple writing. The popular KISS principle (Keep It Short and Simple) is a worthy goal.

Technical Level	Decision-Making Level		
	Decision-Makers	Advisers	Receivers
Managers			
Experts			
Operators			
General Readers			

FIGURE 1–6   Reader matrix

Types of Readers

You have learned some typical problems readers face and some general solutions to these problems. To complete the audience-analysis stage, this section shows you how to classify readers by two main criteria: knowledge and influence. Specifically, you need to answer two questions about every potential reader:

1. How much does this reader already know about the subject?
2. What part will this reader play in making decisions?

Then use the answers to these questions to plan your document. Figure 1–6 (adapted from the Planning Form in Figure 1–5) provides a reader matrix by which you can quickly view the technical levels and decision-making roles of all your readers. For complex documents, your audience may include many of the 12 categories shown on the matrix. Also, you may have more than one person in each box—that is, there may be more than one reader with the same background and decision-making role.

Technical Levels.   On-the-job writing requires that you translate technical ideas into language that nontechnical people can understand. This task can be very complicated because you often have several readers, each with different levels of knowledge about the topic. If you are to "write for your reader, not for yourself," you must identify the technical background of *each* reader. Four categories will help you classify each reader's knowledge of the topic.

● *Reader Group 1: Managers*

Many technical professionals aspire to become managers. Once into management, they may be removed from hands-on technical details of their profession. Instead, they manage people, set budgets, and make decisions of all kinds. Thus you should assume that management readers will not be familiar with fine technical points, will have forgotten details of your project, or both. These managers often need:

- Background information
- Definitions of technical terms
- Lists and other format devices that highlight points
- Clear statements about what is supposed to happen next

In chapter 3, we will discuss an all-purpose "ABC format" for organization that responds to the needs of managers.

● *Reader Group 2: Experts*

Experts include anyone with a good understanding of your topic. They may be well educated—as with engineers and scientists—but that is not necessarily the case. In the example mentioned earlier, the maintenance supervisor with no college training could be considered an "expert" about selecting a new automobile for fleet purchase. That supervisor will understand any technical information about car models and features. Whatever their educational levels, most experts in your audience need:

- Thorough explanations of technical details
- Data placed in tables and figures
- References to outside sources used in writing the report
- Clearly labeled appendices for supporting information

● *Reader Group 3: Operators*

Because decision-makers are often managers or technical experts, these two groups tend to get most of the attention. However, many documents also have readers who are operators. They may be technicians in a field crew, workers on an assembly line, salespeople in a department store, or drivers for a trucking firm—anyone who puts the ideas in your document into practice. These readers expect:

- A clear table of contents for locating sections that relate to them
- Easy-to-read listings for procedures or instructions
- Definitions of technical terms
- A clear statement of exactly how the document affects their jobs

● *Reader Group 4: General Readers*

General readers, also called "laypersons," often have the least amount of information about your topic or field. For example, a report on the environmental impact of a toxic waste dump might be read by general readers who are home owners in the surrounding area. Most will have little technical understanding of toxic wastes and associated environmental hazards. These general readers often need:

- Definitions of technical terms
- Frequent use of graphics like charts and photographs
- A clear distinction between facts and opinions

Like managers, general readers need to be assured that (1) all implications of the document have been put down on paper and (2) important information has not been buried in overly technical language.

Decision-Making Levels.   Figure 1–6 shows that your readers, whatever their technical level, also can be classified by the degree to which they make decisions based on your document. Pay special attention to those most likely to use your report to create change. Use three levels to classify your audience during the planning process:

● *First-Level Audience: Decision-Makers*

The first-level audience must act on the information. If you are proposing a new fax machine for your office, first-level readers will decide whether to accept or reject the idea. If you are comparing two computer systems for storing records at a hospital, the first-level audience will decide which unit to purchase. If you are describing electrical work your firm completed in a new office building, it will decide whether the project has fulfilled agreed-upon guidelines.

In other words, decision-makers translate information into action. They are usually, but not always, managers within the organization. One exception occurs in highly technical companies, wherein decision-makers may be technical experts with advanced degrees in science or engineering. Another exception occurs when decision-making committees consist of a combined audience. For example, the deacons' committee of a church may be charged with the task of choosing a firm to build a new addition to the sanctuary.

● *Second-Level Audience: Advisers*

This second group could be called "influencers." Although they don't make decisions themselves, they read the document and give advice to those who will decide. Often the second-level audience comprises experts such as engineers and accountants, who are asked to comment on technical matters. After reading the summary, a decision-making manager may refer the rest of the document to advisers for their comments.

● *Third-Level Audience: Receivers*

Some readers do not take part in the decision-making process. They only receive information contained in the document. For example, a report recommending changes in the hiring of fast-food workers may go to the store managers after it has been approved, just so they can put the changes into effect. This third-level audience usually includes readers defined as "operators" in the previous section—that is, those who may be asked to follow guidelines or instructions contained in a report.

Using all this information about technical and decision-making levels, you can analyze each reader's (1) technical background with respect to your document and (2) potential for making decisions after reading what you present. Then you can move on to the research and outline stages of writing.

COLLECTING INFORMATION

Having established a clear sense of purpose and your readers' needs, you're ready to collect information for writing. Although you may want to use a scratch outline to

Primary	Secondary
1. **Interviews** 2. **Surveys** 3. **Laboratory Work** 4. **Field Work** 5. **Personal Observation**	1. **Bibliographies** (lists of possible sources—in print or on computer data bases) 2. **Periodical Indexes** (lists of journal and magazine articles, by subject) 3. **Newspaper Indexes** 4. **Books** 5. **Journals** 6. **Newspapers** 7. **Reference Books** (encyclopedias, dictionaries, directories, etc.) 8. **Government Reports** 9. **Company Reports**

FIGURE 1–7   Research sources

guide the research process, a detailed outline normally gets written after you have collected research necessary to support the document.

This section lays out a general strategy for research. Details about research are included in chapter 13 ("Technical Research").

● *Research Step 1: Decide What Kind of Information You Need*

There are two types of research—primary and secondary. Primary research is that which you collect on your own, whereas secondary information is generated by others and found in books, periodicals, or other sources. Figure 1–7 gives examples of both types. Use the kind of research that will be most helpful in supporting the goals of your project. Here are two examples:

- **Report context for using primary research:** A recommendation report to purchase new light tables for the drafting department is supported by your survey of the office drafters. All of them recommend the brand of table you have selected.
- **Report context for using secondary research:** Your report on light tables depends on data found in several written sources, such as an article in a mechanical engineering journal that contrasts features of three tables. On the basis of this article, you recommend a particular table.

● *Research Step 2: Devise a Research Strategy*

Before you start searching through libraries or surfing the Internet, you need a plan. In its simplest form, this plan may list the questions that you expect to answer in your quest for information. For example, a research strategy for a report on office chairs for word-processing operators might pose these questions:

- What kind of chair design do experts in the field of workplace environment recommend for word processors?
- Are there any data that connect the design of chairs with the efficiency of operators?

- Have any specific chair brands been recommended by experts?
- Is there information that suggests a connection between poor chair design and specific health problems?

● *Research Step 3: Record Notes Carefully*

See chapter 13 for the variety of resources available to you at a well-stocked library. Once you have located the information you need in these sources, you must be very careful incorporating it into your own document. As chapter 13 denotes, you must clearly distinguish direct quotations, paraphrasing, and summaries in your notes. Then, when you are ready to translate these notes into a first draft, you will know exactly how much borrowed information you have used and in what form.

● *Research Step 4: Acknowledge Your Sources*

The care that you took in step 3 must be accompanied by thorough acknowledgment of the specific sources. Chapter 13 demonstrates one citation system.

● *Research Step 5: Keep a Bibliography for Future Use*

Consider any research you do for a writing project as an investment in later efforts. Even after your research for a project is complete and you have submitted the report, keep active files on any subjects that relate to your work. Update these files every time you complete a research-related project, such as the two mentioned previously on chair design and light tables. If you or a colleague wants to examine the subject later, you will have developed your own data base from which to start.

WRITING AN OUTLINE

After determining purpose and audience and completing your research, you are ready to write an outline. Outlines are one method for planning a piece of writing, especially long documents. They do not have to be pretty; they just have to guide your writing of the draft. If you conscientiously use outlines now, you will find it easier to organize and write documents of all kinds throughout your career. Refer to the following steps in preparing functional outlines. Figures 1–8 and 1–9 show the outline process in action.

● *Outline Step 1: Record Your Random Ideas Quickly*

At first, ideas need *not* be placed in a pattern. Just jot down as many major and minor points as possible. For this exercise, try to use only one piece of paper, even if it is oversized. Putting points on one page helps prepare the way for the next step, in which you begin to make connections among points.

● *Outline Step 2: Show Relationships*

Next you need to connect related ideas. Using your brainstorming sheet, follow these three steps:

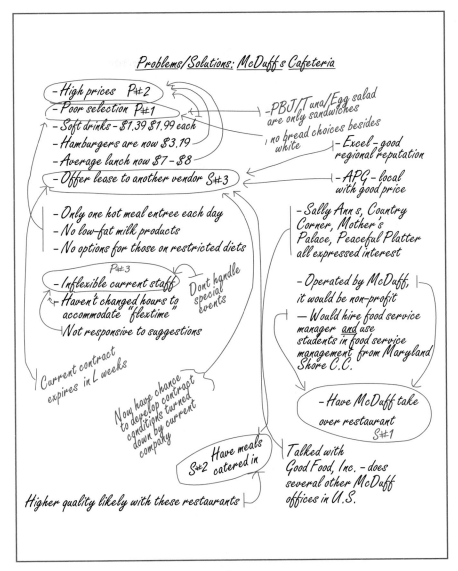

FIGURE 1–8 The outlining process: Early stage

1. Circle or otherwise mark the points that will become main sections.
2. Connect each main point with its supporting ideas, using lines or arrows.
3. Delete material that seems irrelevant to your purpose.

Figure 1–8 shows the results of applying steps 1 and 2 to a writing project at McDuff, Inc., the company used throughout this book. Diane Simmons, office services manager at the Baltimore branch, plans to recommend a change in food service. She uses the brainstorming technique to record her major and minor points.

PROBLEMS AND SOLUTIONS: CURRENT CAFETERIA IN BUILDING

I. Problem #1: Poor selection
   A. Only one hot meal entree each day
   B. Only three sandwiches—PBJ, egg salad, and tuna
   C. Only one bread—white
   D. No low-fat milk products (milk, yogurt, LF cheeses, etc.)
   E. No options for those with restricted diets
II. Problem #2: High prices
   A. Soft drinks from $1.39 to $1.99 each
   B. Hamburgers now $3.19
   C. Average lunch now $7–$8
III. Problem #3: Inflexible staff
   A. Unwilling to change hours to meet McDuff's flexible work schedule
   B. Have not acted on suggestions
   C. Not willing to cater special events in building
IV. Solution #1: End lease and make food service a McDuff department
   A. Hire food service manager
   B. Use students enrolled in food service management program at
       Maryland Shore Community College
   C. Operate as nonprofit operation—just cover expenses
V. Solution #2: Hire outside restaurant to cater meals in to building
   A. Higher quality likely
   B. Initial interest by four nearby restaurants
       1. Sally Ann's
       2. Country Corner
       3. Mother's Palace
       4. Peaceful Platter
VI. Solution #3: Continue leasing space but change companies
   A. Initial interest by three vendors
       1. Excel—good regional reputation for quality
       2. APG—close by and local, with best price
       3. Good Food, Inc.—used by two other McDuff offices with good
           results
   B. Current contract over in two months
   C. Chance to develop contract not acceptable to current company

FIGURE 1–9   The outlining process: Later stage

First, she circles the six main ideas. As it happens, these ideas include three main problems and three possible solutions, so she labels them P#1 through P#3 (problems) and S#1 through S#3 (solutions). Second, she draws arrows between each main point and its related minor points. In this case, there is no material to be deleted. Although the result is messy, it prepares her for the next step of writing the formal outline.

Like Diane Simmons, you will face one main question as you plan your outline: What pattern of organization best serves the material? Chapter 3 presents an "ABC format" that applies to overall structure. Each document should start with an **A**bstract (summary), move to the **B**ody (discussion), and end with a **C**onclusion. However, outlines usually cover only the *body* of a document. Here are some common patterns to consider in outlining the middle part, or body, of a report or proposal. The examples all relate to a decision by McDuff, Inc., to revamp its photocopying system:

- **Chronological:** The writer wants to describe the step-by-step procedure for completing a major photocopying project, from receiving copy at the copy center to sending it out the next day.
- **Parts of an object:** The writer wants to provide a part-by-part description of the current photocopying machine, along with a list of parts that have been replaced in three years of service calls.
- **Simple to complex or vice versa:** The writer wants to describe problems associated with the current photocopying procedure, working from minor to major problems or vice versa.
- **Specific to general (inductive):** The writer wants to begin by listing about 20 individual complaints about the present photocopying machine, later arriving at three general groupings into which all the individual complaints fall.
- **General to specific (deductive):** The writer wants to lead off with some generalizations about photocopying problems at McDuff, followed by a description of incidents related to poor copies, missed deadlines, and high service bills.

● *Outline Step 3: Draft a Final Outline*

Once related points are clustered, it is time to transform what you have done into a somewhat ordered outline. (See Figure 1–9, which continues with the report context used in Figure 1–8.) This step allows you to (1) refine the wording of your points and (2) organize them in preparation for writing the draft. Although you need not produce the traditional outline with Roman numerals, etc., some structure is definitely needed. Abide by these basic rules:

- **Depth:** Make sure every main point has enough subpoints so that it can be developed thoroughly in your draft.
- **Balance:** When you decide to subdivide a point, have at least two breakdowns (because *any* object that is divided will have at least two parts). This same rule applies to headings and subheadings in the final document. (In fact, a good outline will provide you with the wording for headings and subheadings. The outline even becomes the basis for a table of contents in formal documents.)
- **Parallel Form:** For the sake of consistency, phrase your points in either topic or sentence form. Sentences give you a head start on the draft, but they may lock you into wording that needs revision later. Most writers prefer the topic approach; topics take up less space on the page and are easier to revise as you proceed through the draft.

● *Outline Step 4: Consider Where to Use Graphics*

The time to consider using graphics in your document is at the planning stage of the writing process, not at the drafting or revising stage. In this way, graphic communication will combine with text to become an integral part of the document. Too often, however, graphics such as charts, pictures, and tables appear to be a mere afterthought—because they probably were. Instead, you should use the outline stage to plan a strategy for developing graphics that will complement your text.

In the final outline in Figure 1–9, for example, the writer might discover several opportunities for reinforcing textual information with visual language. Here are a few possibilities:

- Chart showing the increase in cafeteria prices over the last three years
- Table contrasting prices for a few lunch items at the current cafeteria and at the four restaurants mentioned in section V. B. of the outline
- Map showing the location of the four nearby restaurants
- Chart showing the relative costs for the contract with the vendors listed in section VI. A. of the outline

As a side benefit, the exercise of planning graphics at the outline stage may uncover weaknesses in your argument—that is, places where you need to develop further statistical support. Chapter 11 will cover the use of graphics in detail.

WRITING INITIAL DRAFTS

With your research and outline completed, you are ready to begin the draft. This stage in the writing process should go quickly if you have planned well. Yet many writers have trouble getting started. The problem is so widespread that it has its own name—"writer's block." If you suffer from it, you are in good company; some of the best and most productive writers often face the "block."

In business and industry, the worst result of writer's block is a tendency to delay the start of writing projects, especially proposals. These delays can lead to rushed final drafts and editing errors. Outlining and other planning steps are wasted if you fail to complete drafting on time. The suggestions that follow can help you to start writing and then to keep the words flowing.

● *Drafting Step 1: Schedule at Least a One-Hour Block of Drafting Time*

Most writers can keep the creative juices flowing for at least an hour *if* distractions are removed. Rather than writing for three or four hours with your door open and thus with constant interruptions, schedule an hour or two of uninterrupted writing time. Most other business can wait an hour, especially considering the importance of good writing to your success. Colleagues and staff members will adjust to your new strategy for drafting reports. They may even adopt it themselves.

● *Drafting Step 2: Do Not Stop to Edit*

Later, you will have time to revise your writing; that time is not now. Instead, force yourself to get ideas from the outline to paper or computer screen as quickly as possible. Most writers have trouble getting back into their writing pace once they have switched gears from drafting to revising.

● *Drafting Step 3: Begin with the Easiest Section*

In writing the body of the document, you need not move chronologically from beginning to end. Because the goal is to write the first draft quickly, you may want to start with the section that flows best for you. Later, you can piece together sections and adjust content.

● *Drafting Step 4: Write Summaries Last*

As already noted, the outline used for drafting covers just the body of the document. Only after you have drafted them should you write overview sections like summaries. You cannot summarize a report until you have actually completed it. Because most writers have trouble with the summary—a section that is geared mainly to decision-makers in the audience—they may get bogged down if they begin writing it prematurely.

REVISING DRAFTS

You may have heard the old saw, "There is no writing, only rewriting." In technical writing, as in other types of communication, careful revision breeds success. The term *revision* encompasses four tasks that transform early drafts into final copy:

1. Adjusting and reorganizing content
2. Editing for style
3. Editing for grammar
4. Editing for mechanics

Following are some broad-based suggestions for revising your technical prose. For more details, consult chapter 15, "Style in Technical Writing," or the Handbook at the end of the book. Also, chapter 4 examines the use of word processing during the revision process.

● *Revision Step 1: Adjust and Reorganize Content*

In this step, go back through your draft to (1) expand sections that deserve more attention, (2) shorten sections that deserve less, and (3) change the location of sentences, paragraphs, or entire sections. The use of word processing has made this step considerably easier than it used to be.

● *Revision Step 2: Edit for Style*

The term *style* refers to changes that make writing more engaging, more interesting, more readable. Such changes are usually matters of choice, not correctness. For example, you might want to:

- Shorten paragraphs
- Rearrange a paragraph to place the main point first
- Change passive voice sentences to active
- Shorten sentences

- Define technical terms
- Add headings, lists, or graphics

One stylistic error deserves special mention because of its frequency: long, convoluted sentences. As a rule, you should simplify a sentence if its meaning cannot easily be understood in one reading. Also, be wary of sentences that are so long you must take a breath before you complete them.

● *Revision Step 3: Edit for Grammar*

You probably know your main grammatical weaknesses. Perhaps comma placement or subject-verb agreement gives you problems. Or maybe you have trouble distinguishing couplets like imply/infer, effect/affect, or complementary/complimentary. In editing the document for grammar, focus on the particular errors that have given you problems in the past.

● *Revision Step 4: Edit for Mechanics*

Your last revision pass should be for mechanical errors such as misspelled words, misplaced pages, incorrect page numbers, missing illustrations, and errors in numbers (especially cost figures). Word processing software can help prevent some of these errors, such as most misspellings, but computer technology has not eliminated the need for at least one final proofing check.

This four-stage revision process will produce final drafts that reflect well on you, the writer. Here are two final suggestions that apply to all stages of the process:

1. Depend on another set of eyes besides your own. One strategy is to form a partnership with another colleague, whether in a technical writing class or on the job. In this arrangement, you both agree that you will carefully review each other's writing. This buddy system works better than simply asking favors of friends and colleagues. Choose a colleague in whom you have some confidence and from whom you can expect consistent editing quality. However, never make changes suggested by another person unless you fully understand the reason for doing so. After all, it is *your* writing.

2. Remember the importance of completing each step separately. Revising in stages yields the best results.

WRITING IN GROUPS

Writing can seem like a lonely act at times. Your own experience in school may reinforce the image of the solitary writer—with sweat on brow—toiling away on research, outlines, and drafts. In fact, this description does *not* typify much writing in the working world outside college. Writing in teams is the rule rather than the exception in many professions and organizations.

Group writing (also called collaborative writing) can be defined in this way:

> **Group writing:** the effort by two or more people to produce one document, with each member *sharing* in the writing process. The term assumes all members actually help with the drafting process, as opposed to (1) the writing of a document by one person, after all group members have met to discuss the project, or (2) group editing of something written by one person. The team must have clear goals and effective leadership to achieve results. Group work can be done in person, over the phone, or through electronic mail (email) on the computer.

This section further defines group writing, highlighting benefits of the strategy and noting some pitfalls to avoid. Then you are given guidelines for using group writing in your classes and on the job.

Benefits and Drawbacks of Group Writing

Most organizations rely on people working together throughout the writing process to produce documents. The success of writing projects depends on information and skills contributed by varied employees. For example, a proposal writing team may include technical specialists, marketing experts, graphic designers, word processors, and technical editors. The company depends on all these individuals working together to produce the final product, a first-rate proposal.

In group writing, however, the whole is greater than the sum of the parts. In other words, benefits go beyond the collective specialties and experience of individual group members. Participants create *new* knowledge as they plan, draft, and edit their work together. They become better contributors and faster learners simply by being a part of the social process of a team. Discussion with fellow participants moves them toward new ways of thinking and inspires them to contribute their best. This collaborative effort yields ideas, writing strategies, and editorial decisions that result from the mixing of many perspectives.

Of course, group writing does have drawbacks. Most notably, the group must make decisions without falling into time traps that slow down the process. There must be procedures for getting everyone's ideas on the table *and* for reaching decisions on time. A leader with good interpersonal skills will help the group reach its potential, whereas an indecisive or autocratic leader will be an obstacle to progress. Good leadership rests at the core of every effective writing team.

In addition to good leadership, shared decision-making is at the heart of every successful writing team. Group writing is not one writer simply getting information from many people before he or she writes a draft. Nor is it one person writing a draft for the editorial red pen of individuals at higher levels. These two models may have their place in some types of company writing, but they do not constitute group writing. Instead, participants in a group must work together during the planning, drafting, and revising stages of writing. Although the degree of collaboration may vary, all forms of group writing differ considerably from the model of one person writing with only occasional help from others.

Guidelines for Group Writing

This section offers six pointers for group writing, to be used in this course and throughout your career. The suggestions concern the writing process as well as interpersonal communication.

● *Group Guideline 1: Get to Know Your Group*

Most people are sensitive about strangers evaluating their writing. Before collaborating on a writing project, therefore, learn as much as you can about those with whom you will be working. Drop by their offices before your first meeting, or talk informally as a group before the writing process begins. In other words, first establish a personal relationship. This familiarity will help set the stage for the spirited dialogue, group criticism, and collaborative writing to follow.

● *Group Guideline 2: Set Clear Goals and Ground Rules*

Every writing group needs a common understanding of its objectives and procedures for doing business. Either before or during the first meeting, these questions should be answered:

1. What is the group's main objective?
2. Who will serve as team leader?
3. What exactly will be the leader's role in the group?
4. How will the group's activities be recorded?
5. How will responsibilities be distributed?
6. How will conflicts be resolved?
7. What will the schedule be?
8. What procedures will be followed for planning, drafting, and revising?

The guidelines that follow offer suggestions for answering the preceding questions.

● *Group Guideline 3: Use Brainstorming Techniques for Planning*

The term *brainstorming* means to pool ideas in a *nonjudgmental* fashion. In this early stage, participants should feel free to suggest ideas without criticism by colleagues in the group. This nonjudgmental approach does not come naturally to most people. Thus the leader may have to establish ground rules for brainstorming before the group proceeds.

Here is one sample approach to brainstorming:

Step 1: The group recorder takes down ideas as quickly as possible.
Step 2: Ideas are written on large pieces of paper affixed to walls around the meeting room so all participants can see how major ideas fit together.
Step 3: Members use ideas as springboards for suggesting others.
Step 4: The group takes some time to digest ideas generated during the first session, before meeting again.

Results of a brainstorming session might look much like a nonlinear outline produced during a solo writing project (see Figure 1–8). The goal of both is to generate as many ideas as possible; these ideas can be culled and organized later.

● *Group Guideline 4: Use Storyboarding Techniques for Drafting*

Storyboarding helps propel participants from the brainstorming stage toward completion of a first draft. It also makes visuals an integral part of the document. Originating in the screenwriting trade in Hollywood, a storyboard is a sheet of paper that contains (1) one draft-quality illustration and (2) a series of sentences about one topic. (See Figure 1–10.) As applied to technical writing, the technique involves six main steps:

Step 1: The group or its leader assembles a topic outline from ideas brought forth during the brainstorming session.

Step 2: All group members are given one or more topics to develop on storyboard forms.

Step 3: Each member works independently on the boards, creating an illustration and a series of subtopics for each main topic (see Figure 1–10).

Step 4: Members meet again to review all completed storyboards, modifying them where necessary and agreeing on key sentences.

Step 5: Individual members develop draft text and related graphics from their own storyboards.

Step 6: The group leader or the entire group assembles the draft from the various storyboards.

● *Group Guideline 5: Agree on a Thorough Revision Process*

As with drafting, all members usually help with revision. Team editing can be difficult, however, as members strive to reach consensus on matters of style. Here are some suggestions for keeping the editing process on track:

- Avoid making changes simply for the sake of individual preference.
- Search for areas of agreement among group members, rather than areas of disagreement.
- Make only those changes that can be supported by accepted rules of style, grammar, and usage.
- Ask the group's best all-round stylist to do a final edit.

This review will help produce a uniform document, no matter how many people work on the draft.

● *Group Guideline 6: Use Computers to Communicate*

When team members are at different locations, computer technology can be used to complete some or all of the project. Team members must have personal computers and the software to connect their machines to a network, allowing members to send and receive information on-line. This section describes three specific computer applications that can improve communication among members of a group-writing project: email, computer conference, and groupware. Chapter 4 ("Page Design") will discuss how the individual writer can use computers to plan, draft, and revise copy. Chapter 7 ("Letters, Memos, and Electronic Communication") will discuss stylistic features of electronic mail.

DOCUMENT TITLE: McDuff's Training Needs
STORYBOARD TOPIC: Results of employee survey
STORYBOARD WRITER: Susan Hernandez

1. In one sentence, summarize this section of the document.

The recent survey of employees showed a strong preference for nontechnical over technical training.

2. In sentence form, include the key points to be developed in this document section. Use same order that points will appear in document.

A. The greatest interest was in the area of sales and marketing training—engineers, in particular, feel deficient here.

B. Many employees also wanted further training in project management—with emphasis on scheduling, accounting practices, and basic management.

C. The third most called-for training area was communication skills—that is, report writing, grammar, and oral presentations.

D. Many employees want training in stress management to reduce or manage on-the-job pressures and make work more enjoyable.

E. The fifth area of interest was technical training in the respondents' own area of expertise.

3. Include an illustration that supports the text in this document section.

Percent of respondents expressing strong interest in various types of training

90% Sales & marketing training
70% Project management
55% Communication skills
45% Stress management
25% Technical training

Caption: Training Interests of McDuff Employees

(The 5 most popular training topics, according to company-wide employee survey.)

FIGURE 1–10 Completed storyboard

- **Electronic mail (email):** Individuals can send and receive messages from their office computers or from remote locations. Like written memos, email messages usually include the date, sender, receiver, and subject. Unlike written memos, the style is often quite informal. Messages are sent at a time convenient for the sender and saved until a time that readers check their mail.
- **Computer conference:** Members of a group can make their own comments and respond to comments of others on a specific topic or project. Computer conferences may be open to all interested users or open only to a particular group. For the purposes of group writing, the conference probably would be open only to members of the writing team. A leader may be chosen to monitor the contributions and keep the discussion focused. Contributions may be made over a long period, as opposed to a conventional face-to-face meeting wherein all team members are present at the same time. Accumulated comments in the conference can be organized or indexed by topic. The conference may be used to brainstorm and thus to generate ideas for a project, or it may be used for comments at a later stage of the writing project.
- **Groupware:** Team members using this software can work at the same time, or different times, on any part of a specific document. Groupware that permits contributions at the same time is called "synchronous"; that which permits contributions at different times is called "asynchronous." Because team members are at different locations, they may also be speaking on the phone at the same time they are writing or editing with synchronous groupware. Such sophisticated software gives writers a much greater capability than simply sending a document over a network for editing or comment. They can collaborate with team members on a document at the same time, almost as if they were in the same room. With several windows on the screen, they can view the document itself on one screen and make comments and changes on another screen.

Computers can be used to overcome many obstacles for writers and editors in different locations. Indeed, electronic communication can help to accomplish all the guidelines just noted. Specifically, (1) email can be used by group members to get to know each other, (2) email or a computer conference can be used to establish goals and ground rules, (3) synchronous, or real-time, groupware can help a team brainstorm about approaches to the project (and may, in fact, encourage more openness than a face-to-face brainstorming session), (4) computer conferences combined with groupware can approximate the storyboard process, and (5) either synchronous or asynchronous groupware can be used to approximate the editing process.

Granted, such techniques lack the body language used in face-to-face meetings. Yet when personal meetings are not possible, computerized communication can provide a substitute that allows writers in different locations to work together to meet their deadline.

Of course, computers can create problems during a group writing project, if you are not careful. When different parts of a document have been written and stored by different writers, your group must be vigilant during the final editing and proofreading stages. Before submitting the document, review it for consistency and correctness.

Communication Challenge

"A Group Project Gone Bad"

Professor Sam Crosby's technical writing class was assigned a group-writing project. Each of the five groups was to (1) learn about the documents produced by a particular office on their campus, Irwin State, (2) produce a written report on their findings, and (3) give a short presentation to highlight their findings to the class. Professor Crosby announced that he would give the same grade to all members of the same group. As part of his introduction to the assignment, he covered the guidelines for group writing included in this chapter. The groups had about 10 days from the time they were assigned the project, on Wednesday, until the time they had to hand in the paper and give a speech, the following Friday.

As is sometimes the case in group projects, one of the teams in Crosby's class encountered problems with the assignment. What follows are details about (1) Group 4's makeup as well as its field activities during the project, (2) the group's preparation for the oral and written report, (3) response to the grade by members of the group, and (4) questions for discussion.

Group 4 Fieldwork

Group 4—composed of Hester Pinn, Bob Simpson, Hector Lopez, and Miranda Arons—was assigned to the Public Relations (PR) Department. To complete its project, the group planned to (1) conduct some library research on PR writing, (2) interview the director of the PR staff, (3) write a short report on its research and interviews, and (4) deliver a panel presentation in which all four members would speak. At its first meeting in class on Wednesday, Group 4 selected a team leader (Miranda), set up a tentative schedule of activities, and agreed to meet the next day at noon at the library to begin its research.

The next day all members except Bob showed up on time. After waiting 10 minutes, the other three members went to the library computer terminals and began locating research sources. By the time Bob arrived a half hour later, the group had jotted down eight periodical sources. They divided up the list of sources and split up to find, read, and photocopy two articles each. Later they reconvened to discuss what they found. They agreed to write one-paragraph summaries of their articles and meet again the next day in the library.

When they met the next day, everyone arrived on time. However, Hester brought only one of the two summaries, noting that the other article was too difficult to read and summarize in the time she had. She promised to finish the second summary that night. Then all four group members proceeded to read the seven summaries and agree on a list of questions to ask the PR director on the content, format, and production of PR documents at Irwin State. Bob, however, wanted to make the report more interesting for the class and suggested they expand their questions to include several about career possibilities in public relations for Irwin State graduates. Although the other team members worried that his questions would be off the assigned topic, they agreed to include them in the interview.

Group 4 arranged to meet the PR director, Rachelle Cochran, on Monday at noon in the cafeteria. Both Bob and Hester arrived late, so Miranda and Hector conducted the first half of the interview on their own. Then Bob and Hester arrived, the interview was completed, and the PR director returned to her office.

Work on the Oral Presentation and Written Report

After Ms. Cochran left the cafeteria, Group 4 stayed in the cafeteria a few more minutes to discuss remaining tasks and to write a scratch outline for the report. Failing to get volunteers, Miranda said she would write the first draft of the report and present it to the others for editing at a meeting at the campus word processing center the next day, Tuesday. Both Hector and Hester said they had other obligations on Tuesday and Wednesday, so Miranda reluctantly agreed to delay the editing meeting until Thursday, the day before the project was due.

As for the panel presentation, the group decided that (1) Bob and Hester would give an overview of the library research conducted by the group, (2) Hector would discuss the results of the interview with Rachelle Cochran, and (3) Miranda would summarize the findings and conclusions of the project. Because the report was being edited the day before the deadline, they decided there would be no time to practice the presentation together. Instead, the other three members said they would call Miranda to summarize what they intended to cover. If she saw inconsistencies in the four parts, she could suggest changes. In short, they planned to "wing it" during their panel.

The members of Group 4 submitted their report and gave their presentation on time. Professor Crosby gave them a "C" on the written report, noting that there were some good ideas but that the editing needed work and the paper lacked a clear focus. He gave them a "C−" on the presentation, commenting that the panel seemed poorly coordinated.

Response to the Group Grade

Bob and Hester seemed satisfied with the grades and quickly left the classroom. Hector was disappointed. But because he had never participated in a successful group project, he had low expectations and shrugged his shoulders as he left the room. Miranda, however, was furious about the grade. She didn't know who to blame—the instructor, the other group members, or herself. Later that day she visited Professor Crosby and complained that having three weak members on her team doomed her to a mediocre project. Then she asked if she could complete an extra credit paper to replace her group project grade or at least to be added

Continued

in with her other grades. Professor Crosby turned her down. He explained that as group leader, she should have exerted more leadership. Furthermore, he noted that if there were major problems, she should have told him before the project was submitted.

Questions for Discussion

Consider the following questions for class discussion. Base your responses on details from the case study and on information from this chapter's guidelines on group work.

1. From beginning to end, what are the various ways that this group project broke down?
2. What, if anything, could have been done to ensure, or at least to encourage, better participation by all group members?
3. One member of the group, Bob Simpson, clearly did not participate to the same degree as the others, and several other group members were lacking at times. What could have been done? Did the group leader, Miranda, have any special responsibilities in this regard?
4. Professor Crosby assigned Group 4 members with a final grade of "C" on the report and "C−" on the panel presentation. Which of the responses to the grade by the group members were appropriate? In particular, discuss Miranda's strategy in her meeting with Professor Crosby.
5. Did Professor Crosby provide adequate direction in the assignment and give an appropriate response to Miranda in their meeting?
6. What are some specific techniques that can be used within a group for keeping all members contributing adequately to the completion of the task?
7. What can a professor do to ensure or encourage the most effective operation of groups in a writing project?

Team writing may play an important part in your career. If you use the preceding techniques, you and your team members will build on each other's strengths to produce top-quality writing.

CHAPTER SUMMARY

Technical writing refers to the many kinds of writing you will do in your career. In contrast to most academic writing, technical writing aims to get something done (not just to demonstrate knowledge), relays information from someone more knowledgeable about the topic (you) to someone less knowledgeable about it (the reader), and is read by people from mixed technical and decision-making levels.

For each writing project, you should complete a three-stage process of planning, drafting, and revising. Planning involves understanding your purpose, knowing the readers' needs, collecting information, and outlining major and minor points. In the drafting stage, you use the outline to write a first draft as quickly as possible—without stopping to make changes. Finally, the revision process requires that you adjust content and then edit for style, grammar, and mechanics.

Technical writing can be completed by you alone or by you as a member of a writing team. The latter approach is common in companies, for it exploits the strengths of the varied professionals within an organization. In group writing you work closely with your team during the planning, drafting, and revising processes.

Access the Technical Writing Web site by pointing your browser at www.prenhall.com/pfeiffer. Here, you will find additional resources to aid you in learning and applying the information from this chapter.

ASSIGNMENTS

Your instructor will indicate whether assignments 1 through 6 should serve as the basis for class discussion, for a written exercise, or for both. Assignments 7, 8, 9, 10, and 12 require a written response. Assignment 11 requires a brief oral presentation by groups.

1. **Features of Academic Writing.**

   Option A. Select an example of writing that you wrote for a high school or college course other than this one. Then prepare a brief analysis in which you explain (1) the purpose of the writing sample, (2) the audience for which it was intended, and (3) the ways in which it differs from technical writing, as defined in this chapter.

   Option B. As an alternative to using your own example, complete the assignment by using the following example. Assume that the passage was written as homework or as an in-class essay in an environmental science class in college.

There are many different responses that are possible in the event toxic waste contamination is suspected or discovered at a site. First, you can simply monitor the site by periodically taking soil and/or water samples to check for contamination. This approach doesn't solve the problem and may not prove politically acceptable when contamination is obvious to the community, but it does help determine the extent of the problem. A second approach—useful when contamination is likely or proved—is to contain the toxic waste by sealing off the site in some fashion, such as by building barriers between it and the surrounding area or by "capping" it in some manner (as in the case of a toxic waste pit). Basically, this alternative depends on the ability to isolate the toxic substances effectively. A third strategy, useful when the contamination is liquified (as with toxic groundwater), is to pump the water from under the ground or from surface ponds and then transport it to treatment systems.

A fourth method is appropriate when toxic substances need to be treated on-site, in which case they can be incinerated or they can be solidified at the site in some way. Then they can be placed in a landfill at the site. Fifth, waste can be hauled to another location where it can be incinerated or placed in some kind of secure landfill—when an off-site disposal approach is needed.

2. **Features of Technical Writing.**

   Option A. Locate an example of technical writing (such as by borrowing it from a family member or an acquaintance who works in a technical profession). Then prepare a brief analysis in which you explain (1) the purpose for which the piece was written, (2) the apparent readers and their needs, (3) the way in which the example differs from typical academic writing, and (4) the relative success with which the piece satisfies this chapter's guidelines.

   Option B. Using the following brief example of technical writing, prepare the analysis requested in Option A.

3. **Purpose and Audience.** The following examples deal with the same topic in four different ways. Using this chapter's guidelines on purpose and audience, determine the main reason for which each excerpt was written and the technical level of the intended readers.

   A. You can determine the magnitude of current flowing through a resistor by use of this process:
   - Connect the circuit (power supply, resistor, ammeter, voltmeter).
   - Set the resistor knob to a setting of "1."
   - Turn the voltage adjusting knob to the left until it stops rotating.
   - Switch the voltmeter to "on" and make sure it reads 0.00 volts.
   - Switch the power supply to "on."
   - Slowly increase the voltage on the voltmeter from 0 to 10 volts.
   - Take the reading from the ammeter to determine the amount of current flowing through the resistor.

   B. After careful evaluation of several testers, I strongly recommend that Langston Electronics Institute purchase 100 Mantra Multitesters for use in our laboratories in Buffalo, Albany, and Syracuse.

   C. Selected specifications for the Ames Multitester are as follows:
   - Rangers 43
   - DC Voltage 0–125–250mV 1.25–2.5–10–5–125–500–1000V
   - AC Voltage 0–5–25–125–250–500–1000V
   - DC Current 0–25–50μA–2.5–5–25–50–250–500mA–10amperes
   - Resistance . . . . .0–2K–20K–200K–20 Mega ohms
   - Decibels . . . .−20 to +62 in db 8 ranges
   - Accuracy . . . .±3% on DC measurements
      ±4% on AC measurements
      ±3% on scale length on resistance
   - Batteries . . . . .one type AA penlight cell
   - Fuse . . . . .0.75A at 250V

   Note that the accuracy rate for the Ames is within our requirements of ±6% and is considerably lower than the three other types of testers currently used by our staff.

   D. Having used the Ames Multitester in my own home laboratory for the last few months, I found it extremely reliable during every experiment. In addition, it is quite simple to operate and includes clear instructions. As a demonstration of this operational ease, my 10-year-old son was able to follow the instructions that came with the device to set up a functioning circuit.

4. **Interview.** Interview a friend, relative, or recent college graduate who works as a technical professional or manager. Gather specific information on these topics:
   - The percentage of the workweek spent on writing
   - Types of documents that are written and their purpose
   - Specific types of readers of these documents

```
DATE:        June 15, 2002
TO:          Pat Jones, Office Coordinator
FROM:        Sean Parker, Word Processing Operator
SUBJECT:     New Word Processing Software
```

Introductory Summary

As you requested, I have examined the WordWonder word processing software we are considering. On the basis of my observations, I recommend we secure one copy of WordWonder and test it in our office for two months. Then after comparing it to the other two packages we have tested, we can choose one of the three word processing packages to use throughout the office.

Features of WordWonder

As we agreed, my quick survey of WordWonder involved reading the user's manual, completing the orientation disk, and meeting with a salesperson from the company. Here are the five features of the package that seemed most relevant to our needs:

1. Formatting Flexibility: WordWonder includes diverse "style sheets" to meet our needs in producing reports, proposals, letters, memos, articles, and even brochures. By engaging just one command on the keyboard, the user can change style sheets—whereby the program will automatically place text in a specified format.

2. Mailers: For large mailings, we can take advantage of WordWonder's "Mail Out" feature that automatically places names from mailing lists on form letters.

3. Documentation: To accommodate our staff's research needs, WordWonder has the capacity to renumber and rearrange footnotes as text is being edited.

4. Page Review: This package's "PagePeek" feature permits the user to view an entire written page on the screen. Without having to print the document, he or she can then see how every page of text will actually look on the page.

5. Tables of Contents: WordWonder can create and insert page numbers on tables of contents, created from the headings and subheadings in the text.

Conclusion

Though I gave WordWonder only a brief look, my survey suggests that it may be a strong contender for use in our office. If you wish to move to the next step of starting a two-month office test, just let me know. Then I will make arrangements with the manufacturer for us to receive a complimentary trial copy.

5. **Contrasting Styles.** Find two articles on the same topic in a professional field that interests you. One article should be taken from a newspaper or magazine of general interest, such as one you would find on a newsstand. The other should be from a magazine or journal written mainly for professionals in the field you have chosen. Now contrast the two articles according to purpose, intended audience, and level of technicality.

6. **Contrasting Audiences.** Photocopy three articles from the same Sunday issue of a local or national newspaper. Choose each article from a different section of the paper—for

example, you could use the sections on automobiles, business, travel, personal computers, national political events, local events, arts, editorials, or employment. Describe the intended audience of each article. Then explain why you think the author has been successful, or unsuccessful, in reaching the particular audience for each article.

7. **Rewrite for Different Audience.** Locate an excerpt from a technical article or textbook, preferably on a topic that interests you because of your background or college major. Rewrite all or part of the selection so that it can be understood by readers who have no previous knowledge of the topic.

8. **Group Writing #1.** Your instructor will divide you into groups of three to five members for completing this assignment. Follow the guidelines for group writing in this chapter. (If your instructor decides that the storyboarding rule is impractical for this exercise, bypass that step.)

Your objective is to write a brief evaluation of the teaching effectiveness of either the room in which your class is held or some other room or building of your instructor's choice. In following the tasks listed in this chapter, the group must establish criteria for evaluation, apply these criteria, and report on the results.

Your brief report should have three parts: (1) a one-paragraph summary of the room's effectiveness, (2) a list of the criteria used for evaluation, and (3) details of how the room met or did not meet the criteria you established.

Most of us who fly in planes either take the safety of this particular mode of transportation for granted or resign ourselves to the fact that if a plane does manage to crash then we cannot do anything about it so we tend either to try hard to enjoy the flight while we're on it or to keep so busy during it such that our mind is preoccupied with other thoughts; however, when we do hear about a plane emergency that has occurred, we tend to wonder how we ourselves would respond in a similar situation or event. An example or instance for possible consideration might be the April 4, 1979, incident in which a flight traveling from New York to Minneapolis experienced a severe emergency situation that caused the plane to go into a nosedive, from its cruising altitude of almost 40,000 feet down to about a mile from the surface of the earth, while at the same time rolling completely over two times and diving at a speed so fast that passengers experienced G forces that forced them into the back of their seats without the ability to move their bodies. The captain, who was an experienced pilot, and his copilot and his flight engineer tried a number of different emergency techniques and responses during the over 40 seconds that they were careening toward the earth, but nothing seemed to work until at the last minute as a last resort they lowered the landing gear, which slowed the plane down to the point where the pilot was able to level off somewhat and prevent a fiery explosion into the earth, and after making a number of other emergency moves, the pilot was able to land at the Detroit airport without any serious injuries to the passengers or flight crew. It appeared after some investigating that the incident was caused by leaks in the plane's hydraulic system, which then caused various failures in the plane that affected its ability to fly correctly, and although these kinds of events do not happen very often or frequently, when they do occur, they make us wonder how we would ourselves respond if we were in the cockpit or if we were among the passengers or if we were among the flight attendants.

Besides preparing the written report, be prepared to discuss the relative effectiveness with which the group followed this chapter's guidelines for group writing. What problems were encountered? How did you overcome them? How would you do things differently next time?

9. **Group Writing #2.** For this assignment, use either the draft on page 38 or the draft of a paper provided by your instructor. Your instructor will divide the class into working groups. Using this chapter's guidelines for revision and for group work, revise the draft. Focus your efforts on (a) eliminating excess words, (b) shortening sentences, and (c) dividing the passage into several paragraphs.

10. **Computerized Communication #1.** If your campus computer facilities permit, set up an email system with members of a writing group to which you have been assigned by your instructor. Decide on a topic upon which you and your team members will comment. Each member should make four comments, or "postings." At least one posting should be an original comment, and at least one should be a response to another member's comment. Print the group's copy and submit it to your instructor. Depending on the instructions you have been given, this assignment may be independent or it may be related to a larger group-writing assignment.

11. **Computerized Communication #2.** Using the working groups your instructor has established, collect information on collaborative learning and then make a brief oral presentation on your findings to the entire class. Your sources may involve print media or computer sources such as the Internet.

12. **Ethics Assignment.** Write a short essay in which you describe one or more ethical problems that might arise from collaborative writing. First review the "Writing in Groups" section of this chapter, and then interview at least one person who has written collaboratively on the job. Each problem you cite should be supported either by a simulated case from the workplace or by a real example gained from your interview(s).

13. **International Communication Assignment.** This chapter offers guidelines on group writing because collaborative communication is essential for success in most careers. However, world cultures differ in the degree to which they use and require collaboration on the job. For this assignment, interview someone who is from a culture different from your own. Using information supplied by this informant, write a brief essay in which you (1) describe the importance of collaboration in the individual's home culture and workplace, (2) give specific examples of how and when collaborative strategies would be used, and (3) modify or expand this chapter's "Guidelines for Group Writing" to suit the culture you are describing, on the basis of suggestions provided by the person you interviewed.

McDuff, Inc.: Ethics and Globalism in the Workplace

Defining Culture

General Instructions

Each chapter begins with a Collaboration at Work exercise that applies strategies for working in groups to chapter topics. The exercise assumes you (1) have been divided into groups of about three to six students, (2) will use time inside or outside of class to complete the case, and (3) will produce an oral or written response. For guidelines about writing in groups, refer to pages 28–34.

Background for Assignment

Assume you and your group members work for a communication consulting firm hired by your school. Your task is to improve communications at your institution— both external communication (for example, to prospective students, prospective employees, and the community) and internal communication (for example, between and among current students, faculty, and administrators).

Before your group can begin to develop an action plan, you would like to describe the current "culture" of the school. (See the definition of "company culture" on page 43 of this chapter.) When applied to a college or university, the term *culture* might include some of the following features, among others:

CHAPTER OUTLINE

- History of the school
- Type of institution and variety of academic programs
- Typical background of students
- Academic structure
- Types of interaction among faculty and staff
- Enrollment patterns
- Extracurricular life on campus
- Relationship with community outside the school

Group Assignment

Your group will choose one or more features of the school's culture to describe. (Alternatively, your instructor may assign specific features to each group, such that the combined descriptions of all groups will present a composite of the school's culture.) The members of your group should (1) discuss a plan for developing the description, (2) collect information in ways that seem most appropriate (for example through interviews, from written documents, or simply from discussion among your group members), and (3) assemble information into a cohesive response. Your main goal is to produce an objective observation, not to argue a point.

*C*hapter 1 defines technical writing and outlines the writing process. This second chapter (1) introduces you to the fictional company, McDuff, Inc., used in examples and assignments throughout this book and (2) brings ethics and globalism into the context of effective communication. Then chapters 3 and 4 cover organization and page design, respectively. Together, this four-chapter package provides the foundation for your work in the rest of the book.

The use of McDuff is intended to yield two main benefits for you as a student:

- **Real-world context:** McDuff provides you with an extended case study in modern technical communication. By placing you in actual working roles, the text prepares you for writing and speaking tasks ahead in your career.
- **Continuity:** The use of McDuff material will lend continuity to class assignments and discussions throughout the term. Your frequent use of this international organization in assignments and class will emphasize the connections among all on-the-job assignments.

Thus McDuff gives you a window into an international organization similar to one where you may work soon. The rest of this chapter addresses six main topics related to McDuff—(1) corporate culture, (2) globalism, (3) background information on McDuff and its types of projects, (4) activities and positions at the corporate and branch offices, (5) typical writing tasks at McDuff, and (6) some ethical guidelines for writing and working on the job.

CORPORATE CULTURE IN THE NEW CENTURY

As a preface to information about McDuff, the first part of this section presents three features common to the culture of any organization that may employ you. Then the second part concentrates on one corporate trend of special note: the drive for quality.

Elements of a Company's Culture

For simplicity here, we'll use the term *company* or *firm* to refer to any organization where you may work. As noted in chapter 1, the writing you do in a company differs greatly from the writing you do in college. The stakes on the job are much higher than a grade on your college transcript. Writing will directly influence the following:

- Your performance evaluations
- Your professional reputation
- Your company's productivity and success in the marketplace

Given these high stakes, let's look at typical features of the organizations where you may spend your career.

Starting a job is both exciting and, sometimes, a bit intimidating. Although you look forward to practicing skills learned in college, you also wonder just how you will fare in new surroundings. Soon you discover that any organization you join has its own personality. This personality, or "culture," can be defined as follows:

> **Company culture:** term for the main features of life at a particular company. A company's culture is influenced by the firm's history, type of business, management style, values, attitude toward customers, and attitude toward its own employees. Taken together, all features of a particular company's culture create a definable quality of life within the working world of that company.

Before examining the culture at McDuff, let's look more closely at three features mentioned in the preceding definition: a firm's history, its type of business, and its management style.

⬤ *Feature 1: Company History*

A firm's origin often is central to its culture. On the one hand, the culture of a 100-year-old steel firm will depend on accumulated traditions to which most employees are accustomed. On the other hand, the culture of a recently established software firm may depend more on the entrepreneurial spirit of its founders. Thus the facts, and even the mythology, of a company's origin may be central to its culture, especially if the person starting the firm remained at the helm for a long time. WalMart, for instance, possesses a culture very much connected to the dreams, aspirations, and open management style of its founder, the late Sam Walton. And IBM, while facing many changes in the computer marketplace during its history, still reflects the strong research-and-development orientation of its early leaders.

⬤ *Feature 2: Type of Business*

Culture is greatly influenced by a company's type of business. Many computer software firms, for example, are known for their flexible, nontraditional, and sometimes chaotic culture. Such firms encourage constant change and innovation. The industry's well-known competitiveness probably inspires this cultural trait. Some of the large computer hardware firms, however, have a culture focused more on tradition, formality, and custom. Yet both hardware and software companies share the same cultural trait of a high level of customer service.

⬤ *Feature 3: Management Style*

A major component of a company's culture is its style of leadership. Some companies run according to a rigid hierarchy, with all decisions coming from the top. Other companies have fewer top-down pronouncements from upper management. Instead, they involve a wide range of employees in the decision-making process. As you might expect, most firms have a decision-making culture somewhere between these two extremes.

These three features give you some idea of what makes up the culture of any company. A company's culture influences who is hired and promoted at the firm, how decisions are made, and even how company documents are written and reviewed. Now let's examine one particular feature—the emphasis on quality—which is becoming a part of the life of many companies throughout the world, including McDuff.

The Search for Quality

"Total quality management (TQM)," "continuous improvement," and similar phrases have become buzzwords in today's workplace. They signal industry's effort to give customers better goods and services. The many spokespersons in this movement agree that the search for quality must permeate every part of an organization. It especially applies to employee productivity and customer service. This section highlights several features of this quality revolution, with special emphasis on relevance to written and oral communications.

● *Quality Feature 1: Putting the Customer First*

Most quality experts agree that any successful company must keep its finger on the pulse of customers, never taking them for granted. Putting the customer first starts with the design of products and services. It also is reflected in the efficiency and sincerity with which complaints are handled.

In the area of communication, this responsiveness to the customer can be seen in the way a writer strives to understand the reader's needs before a report is written. The reader may be a manager within the writer's company or a customer outside the firm. From the quality perspective, both internal and external readers alike are "customers" who deserve first-class treatment.

● *Quality Feature 2: Stressing Teamwork Over Internal Competition*

Many quality experts believe we spend too much time pitting employees against each other, as in the way we distribute merit raises and give commissions on sales. These experts suggest that companies should focus attention on team goals to promote the good of the entire company.

In communication, teamwork can translate into the need to increase the use of group writing (see chapter 1). Collaborative efforts help draw on specialties of writers, technical experts, editors, graphics specialists, and others to produce the best possible document. A culture that encourages group writing reduces the "lone ranger" mentality in communication. Employees work *together* for the good of the final written product.

● *Quality Feature 3: Giving People the Freedom to Do Their Jobs*

Too often, management gets in the way of employee productivity by drawing a box around each person's responsibilities. Quality-centered firms, instead, give everyone more power to suggest changes that will improve the overall business. If employees believe their opinions are important to the overall plan of the company, they will be motivated to work harder and smarter.

In communication, this quality feature might mean that a good word processor should feel comfortable recommending improvements in writing style, rather than simply typing words without regard for the quality of the product. It might also mean that new hires, right out of school, would be encouraged to apply writing skills they learned in college to improve the company's report format, rather than simply following whatever format is handed to them by supervisors. Quality-based firms give all employees the authority to suggest improvements in the way the company does business.

● *Quality Feature 4: Thinking Long Term, Not Short Term*

Most quality experts agree that focusing on short-term profits works against the long-term well-being of organizations. Instead, companies should cultivate permanent, trusting relationships with suppliers and customers. Although such relationships may not produce immediate profits or savings, they will pay off in the end by reducing cyclical ups and downs in profits. Another long-term strategy is to support company training of all types for *all* levels of employees. In particular, people can be cross-trained in other jobs to avoid burnout, to help them understand more about the firm's business, and to prepare them for career advancement.

In communication, investing in the long term can mean training employees to write and speak well. One such investment is to produce an up-to-date writing manual. This manual should specify guidelines for format and style and provide excellent models of all company documents. Most importantly, it should be followed by employees at all levels.

The concern for quality may change the culture of companies throughout the world, including the ones where you may work. Now let's explore another corporate trend today—the move toward globalism.

THE GLOBAL WORKPLACE

Very possibly, you will end up working for an organization that does some of its business beyond the borders of its home country. It may even have many international offices, as does McDuff. Such organizations face opportunities and challenges of diversity among employees, customers, or both. They seek out employees who are able to view issues from a perspective outside their own cultural bias, which we all have. This section examines work in the global workplace, with emphasis on suggestions for writing for readers in different cultures.

Communication has entered what might be seen as its newest frontier—intercultural and international communication. More than ever before, industries that depend on good communication have moved beyond their national borders into the global community. Some people criticize internationalism and the so-called shrinking of the planet. They worry about the possible fusion of cultures and loss of national identities and uniqueness. Some people welcome the change, and others believe that globalism, ironically, will drive people back into their own native cultures for relief from the invasion of globalism. This text avoids value judgments about the phenomenon of internationalism and instead offers practical suggestions for dealing with it.

Strategies for Understanding Cultures

In studying other cultures, we need to avoid extremes of focusing exclusively on either the differences or similarities among cultures. On the one hand, emphasizing differences can lead to inaccurate stereotypes; large generalizations about people can be misinformed and thus can impede, rather than help, communication. On the other hand, emphasizing similarities can tend to mask important differences by assuming we are all alike—one big global family. The truth is somewhere in between. All cultures have both common features and distinctive differences that need to be studied. Such study helps set the stage for establishing productive ties outside one's national borders, in fields such as technical communication.

Exactly how do we go about studying features of other cultures? Traditionally, there are two ways. One only touches the surface of cultural differences by offering simplistic do's and don'ts such as the following:

1. In Japan, always bow as you greet people.
2. In Mexico, be sure to exchange pleasantries with your client before you begin to discuss business.
3. In Germany, do not be a minute late for an appointment.
4. In China, always bring gifts that are nicely wrapped.

These and hundreds of other such suggestions may be useful in daily interactions, but they do not create cultural understanding and often present inaccurate stereotypes of the way people operate.

The other, more desirable, approach goes below the surface to the "deep structure" of culture. It requires that we understand not only *what* people do but also *why* they do it. Although learning another language certainly enhances one's ability to learn about another culture, linguistic fluency does not in and of itself produce cultural fluency. One must go beyond language to grasp one essential point:

> People in different cultures have different ways of thinking, different ways of acting, and different expectations in communication.

To be sure, there are a few basic ethical guidelines evident in most cultures with which you will do business. But other than these core values, differences abound that should be studied by employees of multinational firms. Then these differences must be reflected in communication with colleagues, vendors, and customers. At the close of their excellent text, Iris Varner and Linda Beamer list five recommendations to help organizations with intercultural business communication:

1. "Train employees in intercultural business communication skills and distribute this training . . . to . . . employees at all levels. . . .

2. Send more people to foreign subsidiaries and don't restrict travel to top executives. Foreign travel should not be a perk but [instead] should meet specific business goals. . . .

3. Train employees in intercultural communication skills early in their careers. At this point employees are less costly and more flexible. . . .

4. Carefully evaluate employees as they are hired. . . . Interpersonal skills, language ability, a sense of adventure, and an open attitude may be much more important than specific technical skills. . . .

5. Above all, encourage a climate of excitement and adventure."*

Perhaps you will work for a firm that takes Varner and Beamer's enlightened view toward interculturalism. To prepare you for that possibility, this section includes some questions to ask about those with whom you communicate outside your own culture. Consider these questions to be a starting point for your journey toward understanding communication in the global workplace.

Question 1—*Work:* What are their views about work and work rules?

Question 2—*Time:* What is their approach to time, especially with regard to starting and ending times for meetings, being on time for appointments, expected response time for action requests, hours of the regular workday, etc.?

Question 3—*Beliefs:* What are the dominant religious and philosophical belief systems in the culture, and how do they affect the workplace?

Question 4—*Gender:* What are their views of equality of men and women in the workplace, and how do these views affect their actions?

Question 5—*Personal Relationships:* What degree of value is placed on close personal relationships among people doing business with each other?

Question 6—*Teams:* What part does teamwork have in their business, and, accordingly, how is individual initiative viewed?

Question 7—*Communication Preferences:* What types of business communication are valued most—formal writing, informal writing, formal presentations, casual meetings, email, phone conversations?

Question 8—*Negotiating:* What are their expectations for the negotiation process, and, more specifically, how do they convey negative information?

Question 9—*Body Language:* What types of body language are most common in the culture, and how do they differ from your own?

Question 10—*Writing Options:* What writing conventions are most important to them, especially in prose style and the organization of information? How important is the design of the document in relationship to content and organization?

To be sure, asking these questions does not mean we bow to attitudes that conflict with our own ethical values, as in the equal treatment of women in the workplace. It only means that we first seek to comprehend cultures with which we are dealing before we operate within them. Intercultural knowledge translates into

---

*Iris Varner and Linda Beamer, *Intercultural Communication in the Global Workplace* (Chicago: Irvin, 1995), 306–308. The ten questions in this section are gleaned from information in two excellent sources for the student of international communication: the Varner/Beamer text and *International Business Communication*, by David A. Victor (New York: HarperCollins, 1992).

power in the international workplace. If we are aware of diversity, then we will be best prepared to act.

It might help to see how some of these issues were addressed by Sarah Logan, a marketing specialist who transferred to McDuff's Tokyo office three years ago. In her effort to find new clients for McDuff's services, she discovered much about the Japanese culture that helped her and her colleagues do business in Japan. For example, she learned that Japanese workers at all levels depend more on their identification with a group than on their individual identity. Thus Sarah's marketing prospects in Japan felt most comfortable discussing their work in a corporate department or team, rather than their individual interests or accomplishments—at least until a personal relationship was established.

Sarah learned that an essential goal of Japanese employees is what they call "Wa"—harmony among members of a group and, for that matter, between the firm and those doing business with it. Accordingly, her negotiations with the Japanese often took an indirect path. Personal relationships usually were established and social customs usually observed before any sign of business occurred. A notable exception, she discovered, occurred among the smaller, more entrepreneurial Japanese firms, where employees often displayed a "Western" predisposition toward getting right down to business.

She also discovered that Japanese business is dominated by men more than in her own culture and that there tends to be more separation of men and women in social contexts. Although this cultural feature occasionally frustrated her, she tried to focus on understanding behavior rather than judging it from her own perspective. Moreover, she knew Japan is making changes in the role of women. Indeed, her own considerable success in getting business for McDuff suggested that Japanese value ability and hard work most of all.

Like Sarah Logan, you should enter every intercultural experience with a mind open to learning about those with whom you will work. Adjust your communication strategies so that you have the best chance of succeeding in the international marketplace. Intercultural awareness does *not* require that you jettison your own ethics, customs, or standards. Instead, it provides you with a wonderful opportunity to learn about, empathize with, and show respect for the views of others.

Strategies for Communicating Internationally

This section includes two sets of suggestions for helping you communicate effectively among people from other cultures. The first set, taken from an article by Brent Rubin, describes personality traits that help us communicate better across cultures. The second includes a few simple "default" guidelines for writing and designing documents—if you don't already have information about the specific needs of the audience.

● *Guidelines for Cross-Cultural Fluency*

The communicator who wants to develop cross-cultural fluency should follow seven basic guidelines, according to Brent Rubin.

1. **Be flexible.**

   Example: Westerners dealing with Asian cultures should shift to being *less* concerned about getting negotiations over with quickly and *more* concerned with establishing long-term relationships. As more than one Japanese has said, "First we make a friend; then we make a contract."

2. **Be nonjudgmental about the business at hand.**

   Example: We need to listen carefully to those from another culture before asserting our own views about the project—and then we should do so with discretion. This feature of calm and balance is especially difficult for individuals to develop who come from a culture that values debate, lively exchanges of ideas, and opinionated statements.

3. **Be tolerant of ambiguity.**

   Example: We must be able to work comfortably and without anxiety when the rules are unclear and the path to consensus is not immediately in view. Culturally savvy communicators should respond to new situations without showing visible discomfort, even though their minds are working hard to assess the importance of an intercultural exchange.

4. **Be respectful.**

   Example: Our demeanor should reflect genuine, positive feelings for our counterparts that go beyond the requirements of simple business dealings. Many cultures do not separate working and social relationships in the manner that this division occurs in the U.S. and some other Western countries.

5. **Be nonjudgmental about personal beliefs.**

   Example: We must avoid seeing our own political, religious, or personal ideas as the only reality and instead be open to the fact there may be many paths to knowledge—not just our own.

6. **Be empathetic.**

   Example: We should display the ability to project ourselves into the lives of our counterparts at the negotiation table, trying to understand their needs and perspective.

7. **Be able to take turns.**

   Example: We must avoid either overwhelming the conversation with our views, on the one hand, or sitting back in reflective silence, on the other. The goal is an exchange of information and views.

As Brent Rubin puts it, we should aim to be international sojourners whose goals are to "facilitate the transfer of skills, increase the likelihood of meaningful social interaction, and heighten the probability of achieving a sense of personal and professional accomplishment and growth" (Brent D. Rubin, "Guidelines for Cross-Cultural Communication Effectiveness," pp. 36–46, in *Toward Internationalism,* 2nd ed., Louise Fiber Luce and Elise C. Smith, eds., Heinle and Heinle Publishers, 1987).

⬤ *Guidelines for Writing*

When writing documents for other cultures, remember that your work will not be read in the cultural context in which it was written. For that matter, you may lose control of the document altogether if it is translated into a language that you do not know. Here are some basic guidelines to reduce the risk of misunderstanding:

1. **Remove ambiguity.** It is best to write in simple language—with relatively simple syntax and short sentences—so that ideas cannot be misunderstood.
2. **Define technical terms.** All good technical writing includes well-defined terminology, but this feature is especially important in international writing. A glossary remains an effective tool for helping international readers.
3. **Use many headings and subheadings.** Readers outside your culture will be searching for visual markers that help them find information quickly. They need help wading through blocks of text.
4. **Use much white space.** Besides making the document easier to read, white space gives room for a translated document to "grow" without adding pages, for most translations produce an expansion of text.
5. **Include visuals.** Graphics are a universal language that allows readers entry into the meaning of your document, even if they have difficulty with the text.
6. **Avoid slang terms and idioms.** A nonnative speaker—or someone from outside the United States—may be unfamiliar with phrases you use everyday. The ever popular sports metaphors such as "ballpark estimate," "hitting a homerun," and "let's punt on this" present obvious obstacles for some readers. Use phrasing that requires little cultural context.

BACKGROUND AND TYPES OF PROJECTS

Today McDuff, Inc., is trying hard to develop a company culture based on a concern for quality and intercultural awareness. The firm's management believes such an effort is crucial to the success of the firm. This section takes a detailed look at McDuff, first with a brief overview of its history and then with a description of its major project types.

History of McDuff, Inc.

Just out of Georgia Tech in 1950, Rob McDuff spent several years as a civil engineer in the U.S. Army Corps of Engineers during the Korean War. Then in 1954, he started a small engineering consulting firm in Baltimore, Maryland, where he had grown up. This firm's specialty was doing consulting work for construction firms and real estate developers. Specifically, McDuff, Inc., tested soils and then recommended foundation designs for structures that were being proposed.

In the early days, Rob McDuff did much of the fieldwork himself. He also analyzed the data, wrote reports, and did the marketing for new business. Work progressed well, and his new company earned a reputation for high-quality service. Now Rob McDuff nostalgically looks back upon those days as some of the most satisfying of his career. He had seized the opportunity to fulfill a dream that many people still have today: starting a business and then using skill, hard work, and imagination to make it grow.

**McDuff Sequence of Tasks
Typical Construction Job:
1954-1958**

1. **Reviewing** whatever information was already on file about the construction site

2. **Visiting** the site to observe and record surface features such as rock formations or waste dumps

3. **Drilling** one or more borings (deep, cylindrical holes) into the earth, using special equipment to collect the dirt and rock samples from various depths

4. **Testing** the samples back in the office laboratory

5. **Analyzing** the laboratory data to come up with recommendations for foundation design

6. **Writing** a report that records project activities and specific recommendations for the client

7. **Observing** construction activities, like the pouring of concrete slabs, to be sure correct procedures are followed

---

FIGURE 2–1   McDuff sequence of tasks, typical construction job

From its founding in 1954 until about 1958, the company worked mostly for construction firms in the Baltimore area. Each job—whether a building, dam, or highway—involved tasks like those listed in Figure 2–1. The work was not glamorous. Yet it provided an important service. By the early 1960s, the firm enjoyed a first-rate reputation. It had offices in Baltimore and Boston and about 80 employees.

McDuff, Inc., kept growing steadily, with a large spurt in the mid-1960s and another in the 1980s. The first was tied to increased oil exploration in all parts of the world. Oil firms needed experts to test soils, especially in offshore areas. The results of these projects were used to position oil rigs at locations where they could withstand rough seas. The second growth period was tied to environmental work required by the federal government, state agencies, and private firms. McDuff became a major player in the waste-management business, consulting with clients about ways to store or clean up hazardous waste. The third growth period has moved the firm into diverse service industries, such as security systems, hotel management, and landscaping.

Today, after about fifty years of business, McDuff, Inc., has about 2,500 employees. There are nine offices in the United States and six overseas, as well as a corporate headquarters in Baltimore that is separate from the Baltimore branch office. McDuff performs a wide variety of work. What started as a technical consulting engineering firm has expanded into a firm that does both technical and nontechnical work for a variety of customers.

Projects

Every company must improve its products and services to stay in business. McDuff is no exception. If it had stayed just with soils testing work, the company would be stagnant today. Periodic slowdowns in the construction and oil industries would have taken their toll. Fortunately, the company diversified. These are its eight main project areas today:

1. Soils work on land: Still the company's bread-and-butter work, these projects involve (a) taking soil samples, (b) testing samples in the laboratory, and (c) making design recommendations for foundations and other parts of office buildings, dams, factories, subdivisions, reservoirs, and mass transit systems. When done well, this kind of work helps to prevent later problems, like cracks in building walls.

2. Soils work at sea: Now a smaller market than it was in the 1970s and early 1980s, this geological and engineering work used to be done exclusively for oil and gas companies. It helped them to place offshore platforms at safe locations or to select drilling locations with the best chance of hitting oil. Now, however, McDuff also is hired by countries and states who want to preserve the ecologically sensitive offshore environment. By collecting and analyzing data from its ship, *Dolphin,* McDuff helps clients decide whether an offshore area should be preserved or developed.

3. Construction monitoring: Besides designing parts of structures, McDuff also helps observe construction. Here are some services it offers during the construction process:

- Checking the quality of concrete being poured into structures like cooling towers for nuclear power plants
- Watching construction workers to make sure they follow proper procedures
- Testing the strength of concrete and other foundation materials, once they are put in place

4. Construction management: About 10 years ago, McDuff got into the business of actually supervising projects other than its own jobs. Large construction companies hire McDuff to orchestrate all parts of a project so that it is completed on time. The work involves these activities:

- Establishing a schedule
- Observing the work of subcontractors
- Regularly informing contractors about job progress

5. Environmental management: In the mid-1970s, Rob McDuff began to realize that garbage—all kinds of it—could mean big business for his firm. Suddenly the United States and other countries faced major problems caused by the volume of current wastes and by improper disposal of wastes since World War II. As McDuff's fastest growing market, environmental management work can involve one or more of these tasks:

- Testing surface soil and water for toxic wastes
- Drilling borings to see if surface pollution has filtered into the groundwater
- Designing a cleanup plan

- Supervising the cleanup
- Predicting the impact of proposed projects on the environment
- Analyzing the current environmental health of wetlands, beaches, national forests, lakes, and other areas

As Rob McDuff had hoped, managing wastes and determining the environmental impact of proposed projects proved to be excellent markets. Although there is growing competition, the company got into the business early enough to establish a good reputation for reliable, affordable work.

6. Equipment development: Here the firm departs from its traditional emphasis on services and instead produces products. The ED group, as it is known, designs and builds specialized equipment, both for McDuff's own project needs and for its clients. As the company's newest and most innovative group, it takes on a variety of projects. For example, it is building mechanisms as diverse as a prototype for a new device to test water pollution levels, on the one hand, and a new instrument gauge to install in tractors, on the other. Although the ED group is based in the Baltimore corporate office, it is mobile enough to go to other offices—even in other countries—to complete projects.

7. Training: McDuff entered the training business about five years ago, when it realized that there was a good market for technical training in skills represented by the firm. Recently the Training Department also started offering nontechnical training in areas such as report writing, since the company employs several writers who are excellent trainers.

8. Miscellaneous service industries: Once it achieved growth in fields clearly related to its original mission, McDuff began seeing opportunities for starting or buying out companies that provide services only indirectly related to civil engineering. The three most prominent examples are corporate and residential landscaping, security (both systems and staff), and hotel management. These businesses have grown rapidly and created a more diverse group of employees at McDuff. Yet the firm's management is careful about such acquisitions, for it does not want to enter markets it does not understand.

These eight project areas reflect McDuff's diversity. Though starting as a traditional engineering firm, McDuff has sought out new markets and become a scrappy competitor in many areas. Still company president at age 73, Rob McDuff likes to think that the entrepreneurial spirit thrives in this company he started out of his basement in the 1950s.

CORPORATE AND BRANCH OFFICES

Headquarters

McDuff, Inc., has fifteen branch offices and a corporate headquarters. Though not a large company by international standards, it has become well known within its own fields. The company operates as a kind of loose confederation. Each office enjoys a good measure of independence. Yet some corporate structure is required for these purposes:

1. To coordinate projects that involve employees from several offices
2. To prevent duplication of the same work at different offices
3. To ensure fairness, consistency, and quality in the handling of human resource issues throughout the firm (salaries, benefits, workload, and so forth)

The corporate office gives special attention to problems related to international communications. Among its non-U.S. clients and employees, it must respond to differences in cultures and ways of doing business. This effort can mean the difference between success or failure in negotiating deals, completing projects, hiring employees, and so forth.

The corporate office in Baltimore is housed in a building across the street from the Baltimore branch office. In Rob McDuff's mind, this separation is important. He likes to keep the mostly "overhead" functions of the corporate office distinct from the mostly profit-generating functions of the branches. Also, he believes the physical separation is symbolic to offices outside of Baltimore, which already suspect that the large Baltimore branch office receives special treatment from corporate headquarters. Figure 2–2 shows an organization chart for this office. What follows is a brief description of the responsibilities of each service group in the corporate office.

● *Service 1: Human Resources*

The Human Resources Department performs mostly personnel-related tasks. Its main work covers these four fields:

1. Employment: The office handles job advertisements, ensures that branch offices follow government guidelines that apply to hiring, gives legal advice on workers' compensation, and visits college campuses to recruit prospective graduates.

2. Benefits: Two staff members are responsible for handling company benefits. They send information to employees, check the accuracy of benefit deductions from paychecks, stay current about the newest benefits available for McDuff employees, and make recommendations to the corporate staff about benefit changes.

3. Safety: Given the company's interest in waste management, safety is crucial. The company hired a manager of safety in 1980 to complete these tasks:

- Educate employees about the importance of safe work practices
- Provide proper equipment and training
- Visit job sites to make safety checks
- Respond to questions by government agencies

4. Compensation: The compensation expert has three main duties:

- Monitor salaries at all offices to ensure some degree of uniformity within the job classifications
- Research salary guidelines in all professions represented in the firm, to make sure the company's salaries are competitive

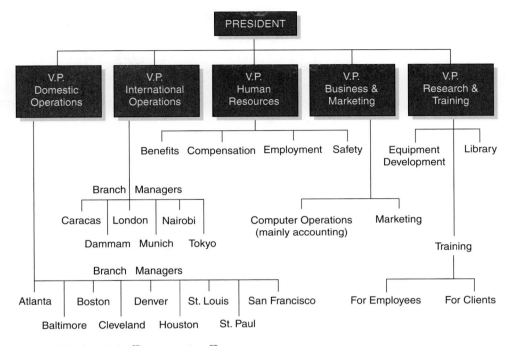

FIGURE 2–2 McDuff's corporate office

- Monitor branch offices to make sure performance evaluation interviews are done each year for each employee, before salary decisions are made

● Service 2: Project Coordination

With fifteen offices spread over a wide geographical area, the company sometimes falls into the trap of the left hand not knowing what the right hand is doing. Specifically, individual offices may not know what project resources exist in another office. If a large project requires experts from several offices, the corporate office can assemble the team. To make this system work, the office keeps an accurate record of current projects at each office.

● Service 3: Marketing

By working closely with clients, the engineers and scientists at every branch help to secure repeat work from current clients. Yet their technical responsibilities keep them from spending much time on marketing for *new* clients. The corporate office, however, has a marketing and proposal-writing staff that works extensively on seeking new business. Also, it helps branch offices write proposals requested by current clients.

● *Service 4: Computer Operations*

The corporate office houses the company's two mainframe computers. These units are used mainly for the firm's accounting databases. Each branch office has terminals tied into the mainframes, giving them direct access to corporate databases in Baltimore. All employees can reach each other through email. Of course, each branch also uses its own stand-alone computer systems for word processing and some other functions.

● *Service 5: Research*

The term *research* at McDuff covers the services of the Equipment Development (ED) Group and the library. The ED lab is housed in the corporate office; it is the only group in the building that could be considered profit-generating. However, this young department is a start-up operation that does not make much money yet. It also retains the important "overhead" function of designing and building tools and other mechanisms that McDuff employees use on their projects. As for the McDuff library, two full-time librarians maintain a modest corporate collection that includes these items:

- Copies of all company reports and proposals since 1954
- Over 100 technical, business, and general periodicals
- Many reference works in technology and the sciences

McDuff employees around the world can receive information the same day they request it—by either phone or fax. In addition, each branch office usually keeps a small reference and periodical collection for its own use.

● *Service 6: Training*

As noted earlier, McDuff now performs two types of training: in-house courses for its own employees, and external training for clients who need training in a number of technical and nontechnical subjects in which McDuff is proficient. The Training Department directs these efforts. Also, it helps McDuff employees find useful outside training or college courses.

Branches

Each McDuff branch is unique in its particular combination of technical and nontechnical positions. Yet all 15 branches include a common management structure, as shown in Figure 2–3. A branch manager, who reports to one of two corporate vice presidents, supervises a group of four or more department managers. These managers, in turn, supervise the technical and nontechnical employees at the branch.

Branch positions below the manager can be grouped into four categories:

1. Technical professional
2. Nontechnical professional

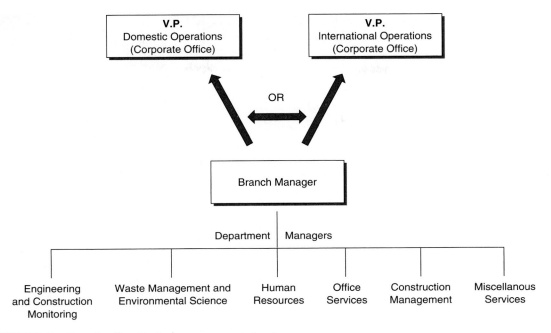

FIGURE 2–3   Branch office: Typical management structure

3. Technical staff
4. Nontechnical staff

Figures 2–4 through 2–7 list some of the positions in these four groupings. Although all these employees are under the supervision of their respective branch managers, some interact closely with employees at the corporate level. For example, each human resources manager reports to his or her respective branch manager *and* works closely with the corporate vice president of human resources.

Most McDuff positions designated "professional" (Figures 2–4 and 2–5) require bachelor's degrees or higher in an appropriate field. Positions designated "staff" require at least a high-school education and sometimes more (such as a certificate program, vocational-technical training, or two-year college degree). Although these categories apply just to our fictional firm of McDuff, some may resemble jobs that exist in a real organization where you now work or will work.

Now that you have viewed McDuff's structure, you need to be made aware of a continuing management problem at the company—one common to many firms with widely spread offices and a corporate office. Branch employees often question whether the corporate people really understand and respond to their needs. Indeed, it is easy to feel misunderstood and even neglected when you work hundreds or thousands of miles from the company's hub. Effective written and spoken

Position	Typical Education (Minimum)	Main Duties
1. Department Manager	• B.S. in engineering or science • M.S. in engineering or science or M.B.A.	• Oversees entire technical department in engineering or science
2. Project Manager	• B.S. in engineering or science or engineering technology	• Oversees entire projects in engineering, waste management, construction, etc.
3. Research Engineer (Equipment Development)	• B.S. in engineering or design • M.S. or Ph.D. in engineering or design	• Designs new tools, mechanisms, or other equipment at ED lab at corporate office
4. Field Engineer	• B.S. in engineering or engineering technology	• Completes site work for projects and then completes remaining work at office
5. Field Scientist	• B.S. in biology, chemistry, environmental science, etc.	• Completes site work for hazardous waste projects and then completes remaining work back at office
6. Landscape Architect	• B.A. or B.S. in landscape architecture	• Designs and oversees construction of landscape plans

FIGURE 2–4   Sample positions: Technical professionals

communication can go a long way toward bridging the gaps between McDuff's local and central offices.

WRITING AT MCDUFF

Good writing is crucial to McDuff's existence. First, one of its main products is a written report. After the company completes a technical project, the project report stands as a permanent statement about, and reflection of, the quality of McDuff's work. Second, many company projects result from written proposals. Third, most routine activity within the firm is preceded or followed by memos, reports, in-house proposals, and manuals. As an employee at McDuff, you would be writing to readers in these groups:

- Superiors at your own McDuff branch
- Subordinates at your McDuff branch
- Employees at other branches or at the corporate office
- Clients
- Subcontractors and vendors

Position	Typical Education (Minimum)	Main Duties
1. Office Services Manager	• B.S. or B.A. in business	• Oversees accounting, word processing, purchasing, physical plant, etc.
2. Human Resources Manager	• B.S./B.A. in liberal arts or in human resources	• Oversees benefits, safety employment, compensation
3. Technical Writer	• B.S./B.A. in technical communication or in liberal arts	• Helps write and edit reports, proposals, and other branch documents
4. Marketing Specialist	• B.S./B.A. in business or in liberal arts	• Writes to and visits potential clients • Helps with proposals
5. Training Specialist	• B.S./B.A. in education or in liberal arts	• Works with corporate office to plan in-house training and external training for clients
6. Security Manager	• B.S. in criminal justice	• Oversees design and implementation of a security plan
7. Hotel Manager	• B.S. in business or hotel management	• Manages hotel property

FIGURE 2–5   Sample positions: Nontechnical professionals

Position	Typical Education (Minimum)	Main Duties
1. Field/Lab Technician	• Vo-tech or associate's degree (in technical field)	• Recovers samples from site • Completes lab tests
2. Computer Operator	• Vo-tech or associate's degree (in technical field)	• Inputs data
3. Field Hand	• High school diploma	• Operates and maintains equipment • Orders and picks up supplies
4. Research Technician	• Vo-tech or associate's degree (in technical field)	• Assists research engineers in work at the Equipment Development Lab
5. Warehouse Supervisor	• Vo-tech or associate's degree (in technical field)	• Keeps track of, and maintains, equipment in a branch office

FIGURE 2–6   Sample positions: Technical staff

Position	Typical Education (Minimum)	Main Duties
1. Word Processing Operator	• High school diploma or associate's degree	• Produces drafts of all company documents
2. Secretary	• High school diploma or associate's degree	• Handles paperwork for professional workers • Does some typing • Has some client contact
3. Receptionist	• High school diploma	• Oversees all of switchboard operation
4. Library Assistant	• High school diploma	• Helps librarian with cataloging, ordering books, etc.
5. Training Assistant	• High school diploma	• Helps orchestrate training activities of all kinds

FIGURE 2–7   Sample positions: Nontechnical staff

As pointed out in chapter 1, you often write to a mixed group of readers, all with different needs and backgrounds. Likewise, at McDuff, readers of the same document could come from more than one of the groups just listed. For example, assume McDuff's corporate training manager needs to send a memo to 20 employees officially confirming their attendance at an upcoming training seminar at the corporate office. Coming to Baltimore from all domestic offices, these employees need a seminar schedule as well as information about the course. Copies of the memo would have to be sent to (1) the participants' managers, who need to be reminded that they will be minus an employee for three days; (2) the vice president for research and training, who likes to be made aware of any company-wide training; and (3) a training assistant, who needs to make room arrangements for the seminar.

This memo is not unique. Most documents at McDuff and other companies are read by persons from different levels. Listed next are more examples of McDuff writing directed to diverse readers. Some of these projects resemble the examples and assignments in later chapters.

Examples of Internal Writing

1. Memo about changes in benefits—from a manager of human resources at a branch to all employees at that branch
2. Memo about changes in procedures for removing asbestos from buildings—from a project manager to field engineers and technicians

3. Orientation booklet on McDuff—from the manager of employment to all new employees at the firm
4. Internal proposal for funds to develop a new piece of equipment—from a technician to the Equipment Development lab manager
5. Draft of a project report—written by a project manager for review by a department manager (before being submitted to the client)
6. Long report on future markets for McDuff—from the vice president of business and marketing to all 2,500 McDuff employees
7. Memo on new procedure for compensating domestic employees who work on overseas projects—from the corporate manager of compensation to all branch managers
8. Manual on new accounting procedures—from the corporate manager of computer operations to all branch managers
9. Article on an interesting environmental project at a national park—from a project manager to all employees who read the company's monthly newsletter
10. Trip report on a professional conference—from a biologist in the environmental science area to the corporate manager of training

Examples of External Writing

1. Sales letter—to potential client
2. McDuff brochure describing technical services—to potential client
3. Proposal—to potential client
4. Progress report—to client
5. Final project report—to client
6. Refresher letter—to previous client
7. Complaint letter—to supplier
8. Article on technical subject—for technical periodical
9. Training manual—for client
10. Affirmative-action report—for government

Writing and speaking tasks, like many responsibilities at McDuff, present ethical dilemmas. The last section of this chapter describes an ethical framework within which technical work—and technical writing, in particular—can be completed at organizations where you may work.

ETHICS ON THE JOB

This section outlines the ethical context in which all workers, like those at McDuff, do their jobs. The goal is (1) to present one main ethical principle and four related guidelines for the workplace and (2) to show how ethical guidelines can be applied to a specific activity at McDuff—writing. Then at the end of the chapter, and throughout this book, you will find assignments in which your own ethical decisions play an important role.

Ethical Guidelines for Work

Like your personal life, your professional life will hold many opportunities for demonstrating your views of what is right or wrong. There is no way to escape these ethical challenges. Most of them occur daily, without much fanfare, but cumulatively they compose our personal approach to morality. Thus our belief systems, or lack thereof, will be revealed by the manner in which we respond to this continuous barrage of ethical dilemmas.

Obviously, not everyone in the same company—let alone the same industry or profession—has the same ethical beliefs. Nor should they. After all, each person's understanding of right and wrong flows from individual experiences, upbringing, religious beliefs, and cultural values. Some "ethical relativists" even argue that ethics only makes sense as a descriptive study of what people *do* believe, not a prescriptive study of what they *should* believe. Yet there are some basic ethical guidelines that, in this author's view, should be part of the decision-making process in every organization. These guidelines apply to small employers, just as they apply to large multinational organizations. Although they may be displayed in different ways in different cultures, they should transcend national identity, cultural background, and family beliefs. In other words, these guidelines represent what, ideally, should be the *core* values for employees at international companies like McDuff.

The guidelines in this section flow from one main tenet that Peter Singer calls the principle of Equal Consideration of Interests (ECI):

> **ECI:** You should make judgments and act in ways that treat the interests and well-being of others as no less important than your own.*

Note that the ECI principle resembles similar principles espoused by religions and philosophies worldwide. That fact makes ECI especially useful as a bedrock principle for multinational organizations. Now let's examine four guidelines that flow from this principle.

● *Ethics Guideline 1: Be Honest*

First, you should relate information accurately and on time—to your colleagues, to customers, and to outside parties, such as government regulators. This guideline also means you should not mislead listeners or readers by leaving out important information that relates to a situation, product, or service. In other words, give those with whom you communicate the same information that you would want presented to you.

This guideline does *not* prescribe the manner or form in which information will be delivered. Indeed, issues such as format, organization, and presentation will

---

*This definition is a slightly condensed form of the one in Singer's *Practical Ethics* (Cambridge, England: Cambridge Univ. Press, 1979) 19, as included in Raymond S. Pfeiffer and Ralph P. Forsberg's *Ethics on the Job* (Belmont, CA: Wadsworth, 1993) 4. The four guidelines that follow are paraphrased from six rules included in *Ethics on the Job,* 12–17.

change from culture to culture. The need for accurate and timely information, however, will not change.

● *Ethics Guideline 2: Do No Harm*

The most healthy, productive, and enjoyable workplaces are those with a positive, constructive atmosphere. One way to achieve such a working environment is to avoid words or actions calculated to harm others. For example, avoid negative, rumor-laden conversations that hurt feelings, spread unsupported information, or waste time.

Of course, different cultures and countries differ in the degree to which personal, familiar chatting takes place at work. But this cultural difference does not change the fact that you should consider the impact words and deeds have on colleagues, clients, and competitors. Our ideal goal should be to make the working world a better place at the end of each day; however, a minimum goal should be to leave the world at least as good as we found it.

● *Ethics Guideline 3: Keep Your Commitments*

People expect that you will keep your word. Be careful about the commitments you make to superiors, subordinates, customers, and others. When you do make a legitimate commitment, follow through on it. "Legitimate" means a commitment wherein you do not violate other ethical guidelines, such as those of being honest and doing no harm.

Cultural guidelines differ on exactly what makes up a commitment, and you should be sensitive to such variations. In one country, a comment in a meeting may seal an agreement from the perspective of some participants. In another, multiple legal contracts are required. Make sure you know what constitutes a commitment with your audience, and then make sure you abide by it.

● *Ethics Guideline 4: Be Independent*

Make no mistake, teamwork will be crucial to the success of organizations for which you work. Group efforts, however, do not relieve you from the personal responsibility for making decisions that come with your job and then accepting the resulting blame or credit. We cannot simply "go along with" group decisions if we have ethical reservations. This phenomenon of excessive conformity within teams has its own name—"groupthink"—and it can be dangerous. Effective teamwork is more important than ever, but it is only as valid as the strength of individual contributors.

If you do business in non-Western countries, such as Japan, you will learn that some cultures emphasize teamwork to the point where individuals seem to be absorbed into the fabric of the group. If you are part of such teams, you may need to alter your style to adapt to a pattern of decision-making that is highly consensual. Yet such cultural adjustments don't change the importance of being assertive, when appropriate, and taking individual responsibility. Although you should be cooperative in teamwork, you cannot sacrifice your own values on the altar of group consensus. Be clear and direct, while still listening and adjusting to others' views.

Now let's examine the manner in which ethical considerations play a part in the *writing* responsibilities at companies such as McDuff, Inc.

Ethics in Writing

In your career you should develop and apply your own code of ethics, making certain it follows the four guidelines already noted. Writing—whether on paper, audiotape, videotape, or computer screen—presents a special ethical challenge for demonstrating your personal code of ethics. Along with speaking, there may be no more important way you display your beliefs during your career. The following section will (1) list some ethical questions related to specific documents and (2) provide responses based on the ethical guidelines noted earlier.

Being honest, doing no harm, keeping commitments, showing independence of thought and action—all four of these ethical guidelines apply to written communication. Here are some typical examples from the working world of McDuff, followed by some discussion of your legal obligations in writing.

● *Ethics Questions in McDuff Writing*

Each of the six pairs that follow presents an ethical dilemma regarding a specific document at McDuff, followed by an answer to each problem.

Lab report: Should you mention a small, possibly insignificant percentage of the data that was collected but that doesn't support your conclusions?

> **Answer:** Yes. Readers deserve to see all the data, even (and perhaps especially) any information that doesn't support your conclusion. They need a true picture of the lab study so that they can draw their own conclusions.

Trip report: Should you mention the fact that one client you visited expressed dissatisfaction with the service he received from your group?

> **Answer:** Yes. Assuming that your report was supposed to present an accurate reflection of your activities, your reader deserves to hear about *all* your client contacts—good news and bad news. You can counter any critical comments by indicating how your group plans to remedy the problem.

Proposal: Should you include cost information, even though cost is not a strong point in your proposal?

> **Answer:** Almost assuredly yes. Most clients expect complete and clear cost data in a proposal. It is best to be forthright about costs, even if they are not your selling point. Then you can highlight features that are exemplary about your firm so that the customer is encouraged to look beyond costs to matters of quality, qualifications, scheduling, experience, etc.

Feasibility study: Should you list all the criteria you used in comparing three products, even though one criterion could not be applied adequately in your study?

> **Answer:** Yes. It is unethical to adjust criteria after the fact to accommodate your inability to apply them consistently. Besides, information about a project dead end may be useful to the reader.

Technical article: Should you acknowledge ideas you derived from another article, even though you quoted no information from the piece?

> **Answer:** Yes. Your reliance on *all* borrowed ideas should be noted, whether the ideas are quoted, paraphrased, or summarized. The exception is common knowledge, which is general information that is found in many sources. Such common knowledge need not be footnoted.

Statement of qualifications (SOQ): Should you feel obligated to mention technical areas in which your firm does *not* have extensive experience?

> **Answer:** Probably not, as long as you believe the customer is not expecting such information in the statement of qualifications. Ethical guidelines do not require you to tell everything about your firm, especially in a marketing document like an SOQ. They only require that you provide the information that the client requests or expects.

Of course, many other types of technical writing require careful ethical evaluation. You might even consider performing an "ethical review" during the final process of drafting a document. Other parts of this book cover topics that apply to specific stages of such an ethical review, as well as to ethics in spoken communication. For ethics in argumentation, see chapter 5; for ethics in instructions, see chapter 6; for ethics in the use of graphics, see chapter 11; for ethics in the research process, see chapter 13; and for ethics in negotiation, see chapter 14.

● *Legal Issues in Writing*

Some countries, such as the United States, have a fairly well-developed legal context for writing. That means you have to pay great attention to detail as you apply ethical principles to the writing process. This section highlights some common guidelines:

- **Acknowledge Sources for Information Other Than Common Knowledge**
  As noted in the "technical article" example in the previous section, you are obligated to provide sources for any information other than common knowledge. "Common knowledge" is usually considered to be factual, nonjudgmental information that could be found in general sources about a subject. The sources for any other types of information beyond this definition must be cited in your document. Chapter 13 offers more detail about plagiarism and the format for citing sources.

- **Ask Written Permission Before Borrowing Extensive Text**
  Generally, it is best to seek written permission for borrowing more than a few hundred words from a source, especially if the purpose of your document is profit. This so-called fair use is, unfortunately, not clearly defined and subject to varying interpretations. It is best to (a) consult a reference librarian or other expert for an up-to-date interpretation of the application of fair use to your situation and (2) err on the side of conservatism by asking permission to use information, if you have any doubt.

- **Seek Written Permission Before Borrowing Graphics**
  This guideline applies to any nontextual element, whether it is borrowed directly from the original or adapted by you from the source. Even if the graphic

Communication Challenge

"McDuff's Moscow Buyout: Global Dilemmas"

McDuff has steadily grown to the point where it now has 15 branches, along with its corporate office in Baltimore, Maryland. Though conservative in its approach to growth, the McDuff corporate staff is still looking for ways to add international offices so that it can expand its business. A recent international opportunity is being considered by Rob McDuff and his staff. What follows is (1) background on the possible buyout of a Moscow firm, (2) issues that the buyout study has raised, and (3) questions and comments for discussion.

Background on the Moscow Buyout

With the breakup of the former Soviet Union in December 1991, the 15 republics that made up one of the largest and most powerful countries in the world became a series of separate nations. Actually, economic changes had begun several years earlier with the start of "glasnost" (the new openness to other cultures and ideas) and "perestroika" (the economic restructuring of the country). Out of the economic change that followed came many new companies run by a new class of entrepreneurs called "biznesmeny." One member of this group, Vlad Gorky, joined with some former government engineers and scientists to establish an engineering firm called, in translation, Moscow Technical Services (MTS). Since 1991 the firm has grown to 75 employees.

Interested in the potential for work in Russia and other former Soviet republics, Rob McDuff has met with Vlad Gorky several times in the United States and in Moscow. The stage has been set for McDuff, Inc., to buy out MTS soon. All government and regulatory requirements have been met in both the United States and Russia. The only remaining obstacle is approval by (1) a McDuff committee consisting of high-level corporate staff and representative employees from various levels and offices and (2) an MTS committee made up of the six principals of that firm. Both groups are meeting this week to discuss important issues related to the purchase. The thorniest of these issues are summarized below.

Problems with the Buyout

In many ways, the buyout appears to be a "win-win" for both firms. McDuff wants an avenue into the growing markets of Russia—especially in construction and environmental work. MTS, for its part, needs additional capital to expand its operation and wants the stability associated with being part of an international firm. Yet the buyout committees in both firms are struggling with the following issues, among others:

Language. When it started acquiring international offices, McDuff adopted a policy that all internal documents would be written in English, to reduce communication barriers within the company. Most external documents, like reports and proposals, are also in English, unless of course clients request otherwise. The policy of English-only internal documents bothers the MTS committee. It sees no reason why MTS cannot continue to write internal memos, reports, procedures, and other documents in Russian. Although most MTS employees have a fair reading and writing knowledge of English and all employees have some, there is the issue of pride at work. Moreover, the MTS buyout committee sees no strong rationale for the English-only policy. To them it smacks of the kind of arbitrary centralized control that they remember all too well.

Work Rules. Like many new entrepreneurial firms around the world, MTS has remained unstructured in its approach to work rules—office hours, lunch breaks, vacation time, job duties, policies and procedures, etc. The firm was started by a creative, innovative scientist who has attracted many like-minded colleagues. Work rules were not on the top of their list. McDuff, however, has a Human Resources Manual that is quite specific about issues such as work hours (whether regular or flextime), vacation time, office dress, required training, and safety. Because the manual has been adopted by all 15 branch offices, McDuff would want it to be used in the Moscow branch.

Perceptions of Undue Influence. McDuff knows that many of the MTS principals, especially those in the environmental field, were formerly a part of the communist technical bureaucracy. They still have strong friendships with present-day officials and do a considerable amount of work for the government. Indeed, MTS continues to hire technical specialists who leave the government for higher-paying jobs at the firm. This "revolving door" practice is considered "business as usual" by the firm, providing MTS with well-educated workers who know where the work is. Although no one at McDuff is uncomfortable with being friends with clients, some members of the McDuff buyout committee are worried about the perception, if not the reality, of undue influence at MTS caused by close associations with, and the "revolving door" hiring of, government officials. At the very least, they worry that the buyout may appear to be an effort by McDuff to gain Russian projects through unfair means. That perception may be held by Russian officials or it may be held by international competitors of McDuff.

Questions and Comments for Discussion

Answer the following questions from your own point of view. Before doing so, however, make sure you have carefully considered the perspective of both McDuff and MTS.

1. Is McDuff's English-only policy justified? Is there any compromise that would satisfy both groups in this dispute?

Continued

2. Elaborate on some of the general language problems multinational firms can face.
3. The use of English does not by itself break down communication barriers with colleagues and customers at global firms. That is, English is spoken around the world by people from many different cultures. Its use does not mean that people necessarily think, write, or speak by the same conventions. Examine this view. Putting aside obvious dialect and vocabulary differences, how can one's culture and national background affect the use of English in writing and speaking?
4. How can the work rules controversy be solved? How might the way this issue is resolved affect the success of the entire buyout? Indeed, give your opinion on the degree to which common work rules and practices are important at McDuff's domestic offices.
5. Give your views of the dilemma posed by the close association of MTS and the government. What are the real and perceived problems here? Which problems most concern McDuff and why? Is there a solution that would satisfy both sides?

is not copyrighted, such as in an annual report from a city or county, you should seek permission for its use.

● **Seek Legal Advice When You Cannot Resolve Complex Questions**
Some questions, such as the use of trademarks and copyright, fall far outside the expertise of most of us. In such cases it is best to consult an attorney who specializes in such law. Remember that the phrase "ignorance is bliss" has led many a writer into problems that could have been prevented by seeking advice when it was relatively cheap—at the beginning. Concerning U.S. copyrights in particular, you might first want to consult free information provided by the U.S. Copyright Office at its Web site (http://lcweb.loc.gov/copyright/).

In the final analysis, acting ethically on the job means thinking constantly about the way in which people will be influenced by what you do, say, and write. Peter Singer's ECI principle embodies this approach perfectly: "Make judgments and act in ways that treat the interests and well-being of others as no less important than your own." These "others" can include your colleagues, your customers, your employers, or the general public. Show them your very best self.

CHAPTER SUMMARY

This book uses the fictional firm of McDuff, Inc., to lend realism to your study of technical writing. The many McDuff examples and assignments give you a purpose, an audience, and an organizational context that simulate what you will face in your career.

Like other organizations where you might work, McDuff has developed its own personality, or "culture." A company's culture can be influenced by many features including its history, type of business, and management style. Two particular features that many organizations have in common today are (1) an interest in improving the quality of their services or products and (2) the need to operate in a global environment.

This chapter looks specifically at the culture of McDuff, Inc. Though started as an engineering consulting firm with a narrow focus, McDuff is now an international company with 2,500 employees, 15 branches, and a corporate office. The firm works in seven main project areas: soils engineering on land, soils engineering at sea, construction monitoring, construction management, waste management (environmental science), equipment development, and training. The color insert gives summary information about specific McDuff projects in all seven areas.

McDuff's corporate office is in Baltimore, Maryland. It helps the branch offices in the areas of human resources, project coordination, marketing, computer operations, research, and training. Each of the 15 offices is run by a branch manager. At each branch, employees are grouped into four categories: technical professionals, nontechnical professionals, technical staff, and nontechnical staff.

McDuff employees at all levels do a good deal of writing, both to superiors and subordinates within the organization and to clients and other outside readers. Documents often have multiple readers with different backgrounds, making writing even more challenging. McDuff employees who meet this challenge will have the best chance of doing valuable work for the company and succeeding in their careers.

Another major concern at McDuff—and at all organizations—is ethical behavior in the workplace. Companies and their employees should follow some basic ethical guidelines in all their work, including communication with colleagues and customers.

www.prenhall.com/pfeiffer **Access the Technical Writing Web site by pointing your browser at www.prenhall.com/pfeiffer. Here, you will find additional resources to aid you in learning and applying the information from this chapter.**

ASSIGNMENTS

1. **Company Profile.** Having read the information in this chapter about McDuff, conduct your own profile of a multinational company in your region. Collect information from some sources such as the following: corporate annual reports, newspaper or magazine articles, or personal contacts. Consider some or all of the following subtopics: company history, types of projects, corporate structure, common types of writing produced, and special features of the company (such as an international market or workforce). Your instructor will indicate whether your report should be presented orally or in writing.

Assignments 2, 3, 4, 5, and 6 can be completed either as individual exercises or as group projects, depending on the directions of your instructor. Prepare a response that can be delivered as an oral presentation for discussion in class.

Analyze the context of each case by considering what you learned in chapter 1 about the context of technical writing *and* what you learned in this chapter about McDuff. In particular, answer these five questions:

- What is the purpose of the document to be written?
- What result will you hope to achieve by writing it?
- Who will be your readers and what will they want from your document?

- What method of organization will be most useful?
- What tone and choice of language will be most effective?

2. **Analysis: Memo Changing Supplies Policy.** As the office services manager at the St. Louis office, you have a problem. In the last fiscal year, the office has used significantly more bond paper, computer paper, pens, mechanical pencils, eraser fluid, and file folders than in previous fiscal years. After going back through the year's projects, you can find no business-related reason why the office has bought $12,000 more of these items. Given that everyone has easy access to the supplies, you have concluded that some employees are taking them home. Putting the best face on it, you assume they may be "borrowing" supplies to complete company business they take home with them, then just keeping items at home. Putting the worst face on it, you wonder whether some employees are stealing from the company.

   After consulting with the branch manager and some other managers, you decide to restrict access to office supplies. Starting next month, these supplies must be signed out through secretaries in the various departments. First, you plan to meet with the secretaries to explain how to make the system work. Then on the following day, you will send a memo explaining the change to *all employees*.

   Would you change your approach in this memo if it were to be sent to a *specific* audience in the office? Why or why not? Answer this question with regard to the four employee groups shown in Figure 2–4 ("Technical Professionals"), Figure 2–5 ("Nontechnical Professionals"), Figure 2–6 ("Technical Staff"), and Figure 2–7 ("Nontechnical Staff").

3. **Analysis: Letter Requesting Testimonials.** As a writer in the corporate marketing department, you spend a good deal of your time preparing materials to be used in sales letters, brochures, and company proposals. Yesterday you were assigned the task of asking 20 customers if they will write "testimonial letters" about their satisfaction with McDuff's work. In all cases, these clients have used McDuff for many projects and have informally expressed satisfaction with the work. Now you are going to ask them to express their satisfaction in the form of a letter, which McDuff could use as a testimonial to secure other business.

   Your strategy is to write a "personalized" form letter to the 20 clients, and then follow it up with phone calls.

4. **Analysis: Memo on Inventory Control.** For five years, you have supervised the supply warehouse at the Houston office. Your main job is to maintain equipment and see that it is returned after jobs are completed. When checking out equipment, each project manager is supposed to fill out part of a project equipment form that lists all equipment used on the job and the date of checkout. Upon returning the equipment, the project manager should complete the form by listing the date of return and any damage, no matter how small, that needs to be repaired before the equipment is used again. This equipment ranges from front-end loaders and pickup trucks to simple tools like hammers, wrenches, and power drills.

   Lately you have noticed that many forms you receive are incomplete. In particular, project managers are failing to record fully any equipment damage that occurred on the job. For example, if someone fails to report that a truck's alignment is out, the truck will not be in acceptable shape for the next project for which it could be used.

   Your oral comments to project managers have not done much good. Apparently, the project managers do not take the warehouse problem seriously, so you believe it is time to put your concerns in writing. The goal is to inform all technical professionals who manage projects that from now on the form must be correctly filled out. You have no authority, as such, over the managers; however, you know that their boss would be very concerned about this problem if you chose to bring it to his or her attention.

At this point, you have decided to ask nicely one more time—this time in writing. You want your memo to emphasize issues of safety and profitability, as well as the need to follow a procedure that has helped you to maintain a first-rate warehouse.

5. **Analysis: Memo Report on Flextime.** As branch manager of the Atlanta office, you have always tried to give employees as much flexibility as possible in their jobs—as long as the jobs got done. Recently you have had many requests to adopt flextime. In this arrangement, the office would end its standard 8:00 A.M. to 4:30 P.M. workday (with a half-hour lunch break). Instead, each employee would fit her or his eight-hour day within the following framework: 7:00 A.M. to 8:30 A.M. arrival, a half hour or full hour for lunch, and 3:30 P.M. to 5:30 P.M. departure.

   Two conditions would prevail if flextime were adopted. First, each employee's supervisor would have to agree on the hours chosen, since the supervisor would need to make sure that departmental responsibilities were covered. Second, each employee would "lock in" a specific flextime schedule until another was negotiated with the supervisor. In other words, an employee's hours would not change from day to day.

   Before you spend any more time considering this change, you want to get the views of employees. You decide to write a short memo report that (1) explains the changes being considered and the conditions (see previous paragraph); (2) solicits their views in writing, by a certain date; and (3) asks what particular work hours they would prefer, if given the choice. Also, you want your short report to indicate that later there may be department meetings and finally a general office meeting on the subject, depending on the degree of interest expressed by employees in their memos to you.

6. **Ethics Assignment.** For this assignment your instructor will place you into a group, with the goal of presenting an oral or written report.

   Option A: The Society for Technical Communicators (STC) is the main U.S. professional association for technical communicators. Its ethical guidelines that follow are intended both for those who are permanent employees of organizations and also for communicators who work as consultants and contractors. Evaluate the quality, usefulness, and appropriateness of these guidelines by answering the following questions:

   a. What do the guidelines suggest about the role of technical communicators in the workplace?

   b. How would you adjust the depth, breadth, or balance of the items presented, if at all?

   c. How does the document satisfy, or fail to satisfy, the ethical guidelines discussed in this chapter, such as the ECI principle?

   d. Are all guidelines and terms clear to the reader?

   e. How might the role of the U.S. technical communication professional, as described in the guidelines, differ from the role of technical communicators in several other cultures outside the United States?

   Option B: Your group is to investigate the ethical climate in one or more organizations that are in the same type of business. You may decide to (a) collect company codes of ethics, (b) do research on ethical guidelines issued by professional associations to which the organizations belong, (c) interview employees about ethical decisions they face on the job, and/or (d) read any available information on ethics related to the companies or profession.

   Option C: For this option your group will select (or be assigned) one of the seven projects sheets in the color insert. Perform a "brainstorming" session in your group by which you arrive at numerous potential ethical dilemmas related to your project. For example, you may want to consider some of these concerns: (a) decisions to be made by and about employees on the job, (b) technical questions related to the project, (c) interaction with clients, and (d) communication with any parties or agencies that are not directly connected with the project but that may be influenced by it.

STC Ethical Guidelines for Technical Communicators

Introduction

As technical communicators, we observe the following ethical guidelines in our professional activities. Their purpose is to help us maintain ethical practices.

Legality

We observe the laws and regulations governing our professional activities in the workplace. We meet the terms and obligations of projects we undertake. We ensure that all terms of our contractual agreements are consistent with the STC Ethical Guidelines.

Honesty

We seek to promote the public good in our activities. To the best of our ability, we provide truthful and accurate communications. We dedicate ourselves to conciseness, clarity, coherence, and creativity, striving to address the needs of those who use our products. We alert our clients and employers when we believe material is ambiguous. Before using another person's work, we obtain permission. In cases where individuals are credited, we attribute authorship only to those who have made an original, substantive contribution. We do not perform work outside our job scope during hours compensated by clients or employers, except with their permission; nor do we use their facilities, equipment, or supplies for personal gain. When we advertise our services, we do so truthfully.

Confidentiality

Respecting the confidentiality of our clients, employers, and professional organizations, we disclose business-sensitive information only with their consent or when legally required. We acquire releases from clients and employers before including their business-sensitive information in our portfolios or before using such material for a different client or employer or for demonstration purposes.

Quality

With the goal of producing high quality work, we negotiate realistic, candid agreements on the schedule, budget, and deliverables with clients and employers in the initial project planning stage. When working on the project, we fulfill our negotiated roles in a timely, responsible manner and meet the stated expectations.

Fairness

We respect cultural variety and other aspects or diversity in our clients, employers, development teams, and audiences. We serve the business interests of our clients and employers, as long as such loyalty does not require us to violate the public good. We avoid conflicts of interest in the fulfillment of our professional responsibilities and activities. If we are aware of a conflict of interest, we disclose it to those concerned and obtain their approval before proceeding.

Professionalism

We seek candid evaluations of our performance from clients and employers. We also provide candid evaluations of communication products and services. We advance the technical communication profession through our integrity, standards, and performance.

(Approved by the STC Board of Directors, April 1995)

THIS COLOR INSERT PROVIDES A GLIMPSE into the world of McDuff, Inc. In addition to the map below, the insert contains information about seven specific projects completed by McDuff.

Each page includes an illustration and project overview. Assignments throughout the book ask you to make use of information from the project sheets.

Worldwide Locations of McDuff, Inc., Offices

■ U.S. Locations

1. Corporate headquarters—Baltimore, Maryland
2. Baltimore, Maryland
3. Boston, Massachusetts
4. Atlanta, Georgia
5. Houston, Texas
6. Cleveland, Ohio
7. St. Paul, Minnesota
8. St. Louis, Missouri
9. Denver, Colorado
10. San Francisco, California

■ Non-U.S. Locations

11. Caracas, Venezuela
12. London, England
13. Munich, Germany
14. Nairobi, Kenya
15. Dammam, Saudi Arabia
16. Tokyo, Japan

M McDuff, Inc.

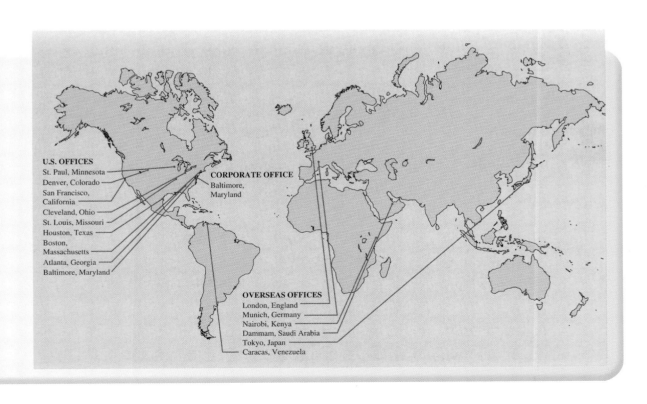

U.S. OFFICES
St. Paul, Minnesota
Denver, Colorado
San Francisco, California
Cleveland, Ohio
St. Louis, Missouri
Houston, Texas
Boston, Massachusetts
Atlanta, Georgia
Baltimore, Maryland

CORPORATE OFFICE
Baltimore, Maryland

OVERSEAS OFFICES
London, England
Munich, Germany
Nairobi, Kenya
Dammam, Saudi Arabia
Tokyo, Japan
Caracas, Venezuela

PROJECT: Sentry Dam

CLIENT: Sanborne County Water Authority

■ **Brief Project Description**

McDuff worked from January through April of 1998 on field and laboratory work preceding construction of Sentry Dam. After submitting its geological and engineering report, the McDuff team worked with the water district on final dam design. It also gave some help during construction. The dam was completed in July of 1999. Because Sentry is a high-hazard dam—meaning that its failure would cause loss of life—safety was crucial.

■ **Main Technical Tasks**

- Drilled 20 soil borings at the site to sample soil and rock
- Drilled 15 test wells to find the depth to groundwater and to check on water seepage in the dam area
- Lab tested the soil, shale, and other samples from borings to evaluate the strength of material on which the dam would rest
- Designed an overflow spillway that would be anchored in strong bedrock, not weak shale
- Monitored water seepage in the foundation and dam during construction
- Certified the dam's safety after construction

■ **Main Findings or Benefits**

- Completed field and lab work on schedule and at budget
- Designed innovative concrete dam spillway that bypassed weak shale and connected with strong bedrock

■ **Brief Project Description**

During the spring of 1999, McDuff used its drillship *Dolphin* to examine the ocean floor over a 10-mile stretch off the Cameroon coast. McDuff collected data on site and then tested and analyzed samples at its labs. After sending the report on the study, McDuff met with the client concerning conclusions and recommendations for further offshore use of the coastline.

■ **Main Technical Tasks**

- Kept *Dolphin* on-site for two months to map the seafloor, to drill borings, and to observe ocean habitats
- Used sonar to develop a profile of the surface of the ocean floor and its near-surface geology (return time of sound waves helped gauge the depth to the floor and to sediments below the floor)
- Drilled successfully for samples from *Dolphin's* drilling platform, often in difficult weather
- Analyzed samples from the borings to estimate geological age and stability of the ocean floor
- Viewed ocean life and geology firsthand at some locations, using a small submersible craft with a one-person crew

■ **Main Findings or Benefits**

- Concluded that most of the zone was too environmentally and geologically sensitive to be used for offshore drilling of oil and gas
- Found two locations where a pipeline might be safely placed, with minimum damage to ocean life and minimum risk of geological disturbance (such as earthquake or ocean avalanche)

PROJECT: Monitored Construction of General Hospital
CLIENT: Floor County, Florida

■ Brief Project Description

McDuff was hired to observe construction of General Hospital to make sure that all construction work was done according to agreed-on standards and legal specifications. McDuff had one or more employees on site continuously during the entire construction process.

■ Main Technical Tasks

- Checked quality of masonry, steel, and other materials used in the building's foundation and main structure
- Ascertained that heating, plumbing, and other systems were installed according to legal codes and according to contract
- Served as liaison between medical staff and construction personnel so that interior construction was done correctly
- Completed final "sign-off" for entire facility before it opened

■ Main Findings or Benefits

- Guaranteed client that all materials and procedures used in construction were up to contract standards
- Spent over 50 extra hours (without charge) consulting with doctors, nurses, and other technical staff about interior construction and placement of major pieces of equipment
- Billed client at 10% below proposed fee because of other work recently contracted with the county for the same calendar year

PROJECT: *Managed Construction of Nevada Gold Dome*
CLIENT: **City of Rondo, Nevada**

■ Brief Project Description

Rondo, Nevada, recently received permission to start a new football team, the Nevada Gamblers. The city is building a domed stadium, called the Gold Dome, for the new franchise. Because of concerns about scheduling the various construction firms and subcontractors involved in the project, the client hired McDuff to manage the entire project—from groundbreaking to occupancy.

■ Main Technical Tasks

- Established construction schedule that would allow for maximum overlapping work by different subcontractors
- Held daily coordination meetings with a group of principals that represented every contractor on site that day
- Handled schedule delays by dealing immediately with contractors and subcontractors
- Used McDuff's patented "TimeTrack" scheduling software to stay aware of the entire flow of work at site

■ Main Findings or Benefits

- Had entire project completed two weeks before the deadline
- Saved client $100,000 by eliminating proposed overtime work that was not needed once scheduling was fine-tuned
- Arranged schedule so that team owners and staff members could complete a walkthrough of the facility at key points during the construction process

PROJECT: **Examined Big Bluff Salt Marsh**

CLIENT: **State of Georgia**

■ Brief Project Description

Georgia hired McDuff to explore the environmental quality of the Big Bluff Salt Marsh, located on Paradise Island off the Atlantic coast. The island is owned by the state. Developers have approached the state about buying island land for building condominiums and other tourist-related structures. McDuff's fieldwork, lab work, and research resulted in a report about the level of development that the marsh can tolerate.

■ Main Technical Tasks

- Took inventory of wildlife and grasses throughout the marsh
- Consulted with wetlands experts around the U.S. about notable features of Big Bluff Salt Marsh
- Tested soil and water for current levels of pollution
- Researched two other Atlantic coast salt marshes where development has occurred to determine compatibility of development and marshes

■ Main Findings or Benefits

- Concluded that Big Bluff Salt Marsh serves as a nursery and feeding ground for fish caught commercially along the coast
- Learned from Environmental Protection Agency (EPA) that any major development of Big Bluff Salt Marsh could be a violation of federal wetlands policy and thus could be challenged by the EPA
- Recommended that the state sell land next to the marsh only if it would be used for low-impact activities—such as day trips by visitors—and not for construction of homes and businesses

PROJECT:
Designed and Installed Control Panel for Nuclear Plant

CLIENT:
Russian Government

■ Brief Project Description

Since the breakup of the old Soviet Union, some individual republics have sought Western assistance in updating their nuclear plants. McDuff's safety experts and mechanical engineers were hired to design a new control panel and to retrofit it into an existing plant. The panel was designed, manufactured, installed, and tested by McDuff—with the help of several subcontractors.

■ Main Technical Tasks

- Spent one week at site observing operators using old panel
- Hired ergonometric and nuclear power experts to help evaluate old panel design and to suggest features of new design
- Designed and manufactured panel
- Installed panel at Russian plant and observed one full week of testing, when panel was used at plant under simulated conditions
- Remained on site for three days after full power was resumed so we could continue training operators on use of new panel

■ Main Findings or Benefits

- Designed panel that international experts considered to be as safe as any currently in use
- Stayed on schedule, keeping the plant out of use only two weeks

PROJECT: **Designed and Taught Seminar in Technical Writing**
CLIENT: **Government of Germany**

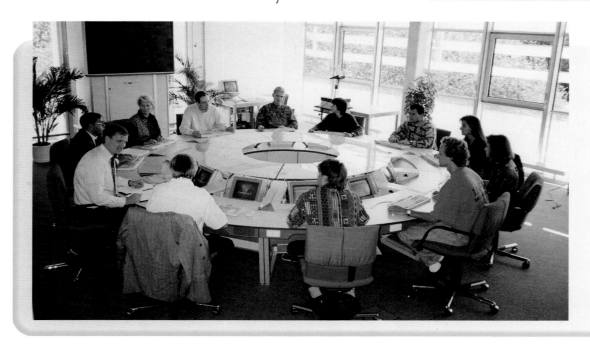

■ Brief Project Description

The government of Germany has greatly increased the number of technical experts in departments in the city of Bonn. McDuff's Munich office was selected to design and teach an in-house technical writing seminar for 20 mid-level government employees in Bonn. They work in agriculture, health, and engineering. Though in different fields, the individuals write the same types of reports.

■ Main Technical Tasks

- Met with members of the seminar and their managers to determine needs of the group
- Examined sample reports from all participants
- Studied the government's style guidelines
- Designed and taught a three-day seminar, using a manual of guidelines and samples tailored for the group
- Evaluated actual on-the-job reports written by participants after the seminar

■ Main Findings or Benefits

- Received "very good" to "excellent" ratings from all 20 participants on the written critiques completed on the last day of the course
- Wrote a final report to client that documented improvement shown in participants' reports written after the seminar, as compared with those written before the seminar

7. **International Communication Assignment.** Refer to the 10 questions in "The Global Workplace" section of this chapter. Using them as the basis for your investigation, conduct your own research project on the cultural features of employees of a specific country. Consider using some or all of the following sources: campus library, travel agencies, consulate offices, international students office on your campus, or individuals who have worked in or visited the country. Your instructor will indicate whether your report should be presented orally or in writing.

Chapter 3

Organizing Information

Organizing the Catalog

General Instructions

Each chapter begins with a Collaboration at Work exercise that applies strategies for working in groups to chapter topics. The exercise assumes you (1) have been divided into groups of about three to six students, (2) will use group time inside or outside of class to complete the case, and (3) will produce an oral or written response. For guidelines about writing in groups, refer to pages 28–34.

Background for Assignment

Here the term *organization* means the arrangement of information—such as purpose statements, supporting details, conclusions, and recommendations. As explained in this chapter, you should aim to choose patterns of organization that fit the context. In particular, they should respond to the needs of the specific readers.

Your college or university catalog is an example of a document that includes varied information, varied readers, and, in many cases, varied patterns of organization. Standard topics covered in catalogs often include the following:

- Accreditation organization, status, and guidelines
- Mission of the institution

- Admissions procedures
- Academic departments
- Degree programs
- Course descriptions
- Financial aid
- Extracurricular activities
- Academic regulations

To add to the complexity, different sections of the catalog may have been written by different writers. However, usually one or two people are responsible for coordinating and editing the entire document.

Group Assignment

Your group either will choose—or will be assigned—one or more sections of your institution's catalog. Your task is to (1) describe the manner in which information is organized, (2) speculate about the rationale the writer had for the pattern(s) selected, and (3) develop an opinion as to whether the patterns meet the needs of the catalog's main readers.

*T*om Kent asks the department secretary to hold his calls. Closing his door, he reaches for the report draft written by one of his staff members and sits down to read it. As a McDuff manager for 10 years, he has reviewed and signed off on every major report written by members of his department. Of all the problems that plague the drafts he reads, poor organization bothers him the most.

This problem is especially annoying at the beginning of a document and the beginning of individual sections. Sometimes he has no idea where the writer is going. His people don't seem to understand that they are supposed to be "telling a story," even in a technical report. Grammar and style errors are annoying to him, but organization problems are much more troublesome. They require extensive rewriting and time-consuming meetings with the report writer. Reaching for his red pen, Tom hopes for the best as he begins to read yet another report.

You, too, will face internal reviewers like Tom Kent when you write on the job. To help you avoid organization problems, this chapter offers strategies for organizing information as you plan, draft, and revise your writing. It builds on the discussion of the three stages of writing covered in chapter 1. Then the next chapter will complete your introduction to technical writing by showing you how to use effective page design to keep readers' attention.

IMPORTANCE OF ORGANIZATION

In a survey of engineering professionals, respondents named "organizing information" the most important topic for any undergraduate technical writing course [Pinelli, T. E., M. Glassman, R. O. Barclay, and W. E. Oliu. 1989. *Technical communications in aeronautics: Results of an exploratory study—an analysis of profit managers' and nonprofit managers' responses.* (NASA TM–101626, p. 28.) Washington, DC: National Aeronautics and Space Administration]. This research is backed up by the experience of many communication consultants—including the author of this textbook, who for 18 years has helped companies improve their employees' writing. Overwhelmingly, these firms have cited poor organization as the main writing problem among both new and experienced employees. That concern underlies all the suggestions in this chapter.

As you learned in chapter 1, your documents will be read by varied readers with diverse technical backgrounds. Chapter 2 displays this technical range within McDuff and refers to an even broader technical spectrum among McDuff's clients. Given this reader diversity, this chapter aims to answer one essential question: How can you best organize information to satisfy so many different people?

Figure 3–1 shows you three possible options for organizing information for a mixed technical audience, but only one is recommended in this book. Some writers, usually those with technical backgrounds themselves, choose Option A. They direct their writing to the *most* technical people. Other writers choose Option B. They respond to the dilemma of a mixed technical audience by finding the lowest common denominator—that is, they write to the level of the *least* technical person. Option A and B each satisfies one segment of readers at the expense of the others.

Option A
Organize information
for technical readers

Option B
Organize information
for *less technical* readers

Option C (preferred)
Organize information
for *all* readers

FIGURE 3–1   Options for organizing information

Option C is preferred in technical writing for mixed readers. It encourages you to organize documents so that **all** readers—both technical and nontechnical—get what they need. The rest of this chapter provides strategies for developing this option. It describes general principles of organization and guidelines for organizing entire documents, individual document sections, and paragraphs.

THREE PRINCIPLES OF ORGANIZATION

Good organization starts with careful analysis of your audience. Most readers are impatient and skip around as they read. Think about how you examine a weekly newsmagazine or an airline magazine. You are likely to take a quick look at articles of special interest to you; then you might read them more thoroughly, if there is time. That approach also resembles how *your* audience treats technical reports and other work-related documents. If important points are buried in long paragraphs or sections, busy readers may miss them. Three principles respond realistically to the needs of your readers:

● *Principle 1: Write Different Parts for Different Readers*

The longer the document, the less likely it is that anyone will read it from beginning to end. As shown in Figure 3–2, they use a "speed-read" approach that includes these steps:

Step 1: Quick scan. Readers scan easy-to-read sections like executive summaries, introductory summaries, introductions, tables of contents, conclusions, and recommendations. They pay special attention to beginning and ending sections, especially in documents longer than a page or two, and to illustrations.

Step 2: Focused search. Readers go directly to parts of the document body that will give them what they need at the moment. To find information quickly, they search for format devices like subheadings, listings, and white space in margins to guide their reading. (See chapter 4 for a discussion of page design.)

Step 3: Short follow-ups. Readers return to the document, when time permits, to read or re-read important sections.

STEP 1: Quick Scan

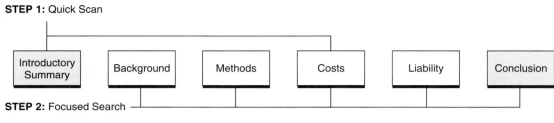

STEP 3: Short Follow-Ups
Can involve any section, especially the Introductory Summary

FIGURE 3–2   Sample speed-read approach to short proposal

Your job is to write in a way that responds to this nonlinear, episodic reading process of your audience. Most importantly, you should direct each section to those in the audience most likely to read that particular section. Shift the level of technicality as you move from section to section *within* the document, to meet the needs of each section's specific readers. On the one hand, managers and general readers favor less technical language and depend most heavily on overviews at the beginning of documents. On the other hand, experts and operators expect more technical jargon and pay more attention than others to the body sections of documents.

Of course, you walk a thin line in designing different parts of the document for different readers. Although technical language and other stylistic features may change from section to section, your document must hang together as one piece of work. Common threads of organization, theme, and tone must keep it from appearing fragmented or pieced together.

This approach breaks the rules of nontechnical writing, which strives for dogged consistency throughout the same document. However, technical writing marches to the beat of a different drummer.

● *Principle 2: Emphasize Beginnings and Endings*

Suspense fiction relies on the interest and patience of readers to piece together important information. The writer usually drops hints throughout the narrative before finally revealing who did what to whom. Technical writing operates differently. Busy readers expect to find information in predictable locations without having to search for it. Their first-choice locations for important information are as follows:

- The beginning of the entire document
- The beginnings of report sections
- The beginnings of paragraphs

The reader interest curve in Figure 3–3 reflects this focus on beginnings. But the curve also shows that the readers' second choice for reading is the ends of documents, sections, and paragraphs. That is, most readers tend to remember best the first and last things they read. The ending is a slightly less desirable location than the beginning because it is less accessible, especially in long sections or documents. Of

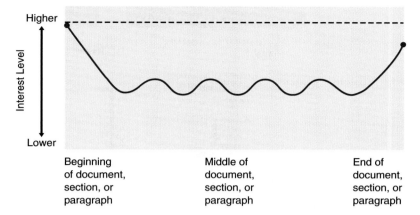

FIGURE 3–3   Reader interest curve

course, some readers inevitably will read the last part of a document first, for they may have the habit of fanning pages when first seeing a document. Their thumb first locks on the last section of the report. Thus, although there is no guarantee that the first document section will be read first, you can be fairly sure that *either* the beginning *or* the ending will get first attention.

Emphasizing beginnings and endings responds to the reading habits and psychological needs of readers. At the beginning, they want to know where you're heading. They need a simple "road map" for the rest of the passage. In fact, if you don't provide something important at the beginnings of paragraphs, sections, and documents, readers will start guessing the main point themselves. It is in your best interest to direct the reader to what *you* consider most important in what they are about to read, rather than to encourage them to guess at the importance of the passage. At the ending, readers expect some sort of wrap-up or transition; your writing shouldn't simply drop off. The following paragraph begins and ends with such information:

> *The proposed word-processing software has two other features that will help our writers: a dictionary and a thesaurus.* When the built-in dictionary is engaged, it compares each word in a document with the same word in the system's dictionary. Differences are then highlighted so that they can be corrected by the operator of the system. The thesaurus also can help our writers by offering alternative word selections. When just the right word is escaping the writer, he or she can trigger the system to provide a list of related words or synonyms. *Both the dictionary and thesaurus are very quick and thus far superior to their counterparts in book format.*

The first sentence gives readers an immediate impression of the two topics to be covered in the paragraph. The paragraph body explores details of both topics. Then the last sentence flows smoothly from the paragraph body by reinforcing the main point about features of the dictionary and thesaurus.

Why does this top-down pattern, which seems so logical from the reader's perspective, frequently get ignored in technical writing? The answer arises from the difference between the way you complete your research or fieldwork and the way busy

FIGURE 3–4   Technical *work* vs. technical *writing*

readers expect results of your work to be conveyed in a report. Figure 3–4 illustrates this difference. Having moved logically from data to conclusions and recommendations in technical work, many writers assume they should take this same approach in their report. They reason that the reader wants and needs all the supporting details before being confronted with conclusions and recommendations that result from these data.

Such reasoning is wrong. Readers want the results placed first, followed by details that support your main points. Of course, you must be careful not to give *detailed* conclusions and recommendations at the beginning; most readers want and expect only a brief summary. This overview will provide a framework within which readers can place the details presented later. In other words, readers of technical documents want the "whodunit" answer at the beginning. Recall the motto in chapter 1: *Write for your reader, not for yourself.* Now you can see that this rule governs the manner in which you organize information in everything you write.

● *Principle 3: Repeat Key Points*

You have learned that different people focus on different sections of a document. Sometimes no one carefully reads the *entire* report. For example, managers may have time to read only the summary, whereas technical experts may skip the leadoff sections and go directly to "meaty" technical sections with supporting information. These varied reading patterns require a *redundant* approach to organization—you must repeat important information in different sections for different readers.

For example, assume you are a McDuff employee in Denver and are writing a report to the University of Colorado on choosing sites for several athletic fields. Having examined five alternatives, your report recommends one site for final consideration. Your 25-page report compares and contrasts all five alternatives according to criteria of land cost, nearness to other athletic locations, and relative difficulty of grading the site and building the required facilities. Given this context, where will your recommendation appear in the report? Here are five likely spots:

1. Executive summary
2. Cost section in the body
3. Location section in the body
4. Grading/construction section of the body
5. Concluding section

Our assumption, you recall, is that few readers move straight through a report. Because they often skip to the section most interesting to them, you need to make main sections somewhat self-contained. In the University of Colorado report, that would mean placing the main recommendation at the beginning, at the end, and at one or more points within each main section. In this way, readers of all sections would encounter your main point.

What about the occasional readers who read all the way through your report, word for word? Will they be put off by the restatement of main points? No, they won't. Your strategic repetition of a major finding, conclusion, or recommendation gives helpful reinforcement to readers always searching for an answer to the "So what?" question as they read. Fiction and nonfiction may be alike in this respect—writers of both genres are "telling a story." The theme of this story must periodically reappear to keep readers on track.

Now we're ready to be more specific about how the three general principles of organization apply to documents, document sections, and paragraphs.

ABC FORMAT FOR DOCUMENTS

You have learned the three principles of organization: (1) write different parts of the document for different readers, (2) emphasize beginnings and endings, and (3) repeat key points. Now let's move from principles to practice. Here we will develop an all-purpose pattern of organization for writing entire documents. (The next major section covers document sections and paragraphs.)

Technical documents should assume a three-part structure that consists of a beginning, a middle, and an end. This book labels this structure the "ABC Format" (for **A**bstract, **B**ody, and **C**onclusion). Visually, think of this pattern as a three-part diamond structure, as shown in Figure 3–5:

- **Abstract:** A brief beginning component is represented by the narrow top of the diamond, which leads into the body.
- **Body:** The longer middle component is represented by the broad, expansive portion of the diamond figure.
- **Conclusion:** A brief ending component is represented by the narrow bottom of the diamond, which leads away from the body.

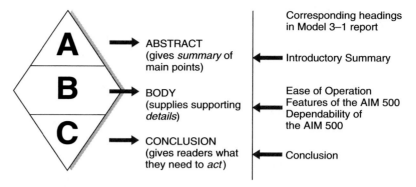

FIGURE 3–5   ABC format for all documents

Model 3–1 (pp. 96–97) includes a memo report that conforms to this structure. The following sections discuss the three ABC components in detail.

Document Abstract: The "Big Picture" for Decision-Makers

Every document should begin with an overview. As used in this text, *abstract* is defined as follows:

> **Abstract:** brief summary of a document's main points. Although its makeup varies with the type and length of the document, an abstract usually includes (1) a clear purpose statement for the document, (2) the most important points for decision-makers, and (3) a list or description of the main sections that follow the abstract. As a capsule version of the entire document, the abstract should answer readers' typical mental questions, such as the following: "How does this document concern me? What's the bottom line? So what?"

Abstract information is provided under different headings, depending on the document's length and degree of formality. Some common headings are "Summary," "Executive Summary," "Introductory Summary," "Overview," and "Introduction." The abstract may vary in length from a short paragraph to a page or so. Its purpose, however, is always the same: to provide decision-makers with highlights of the document.

For example, assume you are an engineer who has evaluated environmental hazards for the potential purchaser of a shopping mall site. Here is how the summary might read:

> As you requested, we have examined the possibility of environmental contamination at the site being considered for the new Klinesburg Mall. Our field exploration revealed two locations with deposits of household trash, which can be easily cleaned up. Another spot

has a more serious deposit problem of 10 barrels of industrial waste. However, our inspection of the containers and soil tests revealed no leaks.

Given these limited observations and tests, we conclude that the site poses no major environmental risks and recommend development of the mall. The rest of this report details our field activities, test analyses, conclusions, and recommendations.

You have provided the reader with a purpose for the report, an overview of important information for decision-makers, and a reference to the four sections that follow. In so doing, you have answered the following questions, among others, for the readers:

- What are the major risks at the site?
- Are these risks great enough to warrant not buying the land?
- What major sections does the rest of the report contain?

This general abstract, or overview, is mainly for decision-makers. Highlights must be brief, yet free of any possible misunderstanding. On some occasions, you may need to state that further clarification is included in the text, even though that point may seem obvious. For example, if your report concerns matters of safety, the overview may not be detailed enough to prevent or eliminate risks. In this case, state this point clearly so that the reader will not misunderstand or exaggerate the purpose of the abstract.

Later chapters in this book contain guidelines for writing the following specific types of abstracts:

- Introductory summaries for short reports (chapter 8)
- Executive summaries for formal reports (chapter 9)
- Introductory summaries for short proposals (chapter 10)
- Executive summaries for formal proposals (chapter 10)
- Abstracts of technical articles (chapter 13)

Document Body: Details for All Readers

The longest part of any document is the body. As used in this book, the body is defined as follows:

> **Body:** the middle section(s) of the document providing supporting information to readers, especially those with a technical background. Unlike the abstract and conclusion, the body component allows you to write expansively about items such as (1) the background of the project, (2) field, lab, office, or any other work upon which the document is based, and (3) details of any conclusions, recommendations, or proposals that might be highlighted at the beginning or end of the document. The body answers this main reader question: "What support is there for points put forth in the abstract at the beginning of the document?"

Managers may read much of the body, especially if they have a technical background and if the document is short. Yet the more likely readers are technical specialists who

(1) verify technical information for the decision-makers or (2) use your document to do their jobs. In writing the body, use the following guidelines:

- **Separate fact from opinion.** Never leave the reader confused about where opinions begin and end. Body sections usually move from facts to opinions that are based on facts. To make the distinction clear, preface opinions with phrases such as "We believe that," "I feel that," "It is our opinion that," and the like. Such wording gives a clear signal to readers that you are presenting judgments, conclusions, and other nonfactual statements. Also, you can reinforce the facts by including data in graphics.
- **Adopt a format that reveals much structure.** Use frequent headings and subheadings to help busy readers locate important information immediately. (Chapter 4 covers these and other elements of page design.)
- **Use graphics whenever possible.** Use graphics to draw attention to important points. Today more than ever, readers expect visual reinforcement of your text, particularly in more persuasive documents like proposals. (Chapter 11 deals with graphical elements in technical documents.)

By following these guidelines, which apply to any document, you will make detailed body sections as readable as possible. They keep ideas from becoming buried in text and show readers what to do with the information they find.

Document Conclusion: Wrap-Up Leading to Next Step

Your conclusion deserves special attention, for readers often recall first what they have read last. We define the conclusion component as follows:

> **Conclusion:** the final section(s) of the document bringing readers—especially decision-makers—back to one or more central points already mentioned in the body. Occasionally, it may include one or more points not previously mentioned. In any case, the conclusion provides closure to the document and often leads to the next step in the writer's relationship with the reader.

The conclusion component may have any one of several headings, depending on the type and length of the document. Possibilities include "Conclusion," "Closing," "Closing Remarks," and "Conclusions and Recommendations." The chapters in Part 2 of this text describe the options for short and long documents of many kinds. In general, however, a conclusion component answers these sorts of questions:

- What major points have you made?
- What problem have you tried to solve?
- What should the reader do next?
- What will you do next?
- What single idea do you want to leave with the reader?

Because readers focus on beginnings and endings of documents, you want to exploit the opportunity to drive home your message—just as you did in the abstract. Format can greatly affect the impact you make on decision-makers. Although specific formats vary, most conclusions take one of these two forms:

- **Listings:** This format is especially useful when pulling together points mentioned throughout the document. Whereas the abstract often gives readers the big picture in narrative format, the conclusion may instead depend on listings of findings, conclusions, and/or recommendations. (Chapter 4 gives suggestions on using bulleted and numbered listings.)
- **Summary paragraph(s):** When a listing is not appropriate, you may want to write a concluding paragraph or two. Here you can leave readers with an important piece of information and make clear the next step to be taken.

Whichever alternative you choose, your goal is to return to the main concerns of the most important readers—decision-makers. Both the abstract and conclusion, in slightly different ways, should respond to the needs of this primary audience.

TIPS FOR ORGANIZING SECTIONS AND PARAGRAPHS

First and foremost, the ABC format pertains to the organization of entire documents. Yet the same "beginning-middle-end" strategy applies to the next smaller units of discourse—document sections and paragraphs. In fact, you can view the entire document as a series of interlocking units, with each responding to reader expectations as viewed on the reader interest curve in Figure 3–3.

Document Sections

As mentioned earlier, readers often move from the document abstract to the specific body sections they need to solve their problem or answer their immediate question. Just as they need abstracts and conclusions in the whole document, they need "mini-abstracts" and brief wrap-ups at the start and finish of each major section.

To see how a section abstract works, we must first understand the dilemma of readers. Refer to Model 3–2 on page 98, which contains one section from a long report. Some readers may read it from beginning to end, but others might not have the time or interest to do so in one sitting. Instead, they would look to a section beginning for an abstract, and then move around within that section at will. Thus the beginning must provide them with a map of what's ahead. Here are the two items that should be part of every section abstract:

1. **Interest grabber:** a sentence or more that captures the attention of the reader. Your grabber may be one sentence or an entire paragraph, depending on the overall length of the document.

2. **Lead-in:** a list, in sentence or bullet format, that indicates main topics to follow in the section. If the section contains subheadings, your lead-in may include the same wording as the subheadings and be in the same order.

The first part of the section gives readers everything they need to read on. First, you get their attention with a grabber. Then you give them an outline of the main points to follow so that they can move to the part of the section that interests them most. As in Model 3–2 on page 98, the section abstract immediately precedes the first subheading when subheads are used.

Sections also should end with some sort of closing thought, rather than just dropping off after the last supporting point has been stated. For example, you can (1) briefly restate the importance of the information in the section or (2) provide a transition to the section that follows. Model 3–2 takes the latter approach by suggesting the main topic for the next section. Whereas the section lead-in provides a map to help readers navigate through the section, the closing gives a sense of an ending so that readers are ready to move on.

Paragraphs

Paragraphs represent the basic building blocks of any document. Organizing them is not much different in technical writing than it is in nontechnical prose. Most paragraphs contain these elements:

1. **Topic sentence:** This sentence states the main idea to be developed in the paragraph. Usually it appears first. Do not delay or bury the main point, for busy readers may read only the beginnings of paragraphs. If you fail to put the main point there, they may miss it entirely.
2. **Development of main idea:** Sentences that follow the topic sentence develop the main idea with examples, narrative, explanation, or other details. Give the reader concrete supporting details, not generalizations.
3. **Transitional elements:** Structural transitions help the paragraph flow smoothly. Use transitions in the form of repeated nouns and pronouns, contrasting conjunctions, and introductory phrases.
4. **Closing sentence:** Most paragraphs, like sections and documents, need closure. Use the last sentence for a concluding point about the topic or a transitional point that links the paragraph with the one following it.

Model 3–3 on page 99 shows two paragraphs that follow this pattern of organization. The paragraphs are from a McDuff recommendation report. McDuff was hired to suggest ways for a hospital to modernize its physical plant. Each paragraph is a self-contained unit addressing a specific topic, while being linked to surrounding paragraphs (not shown) by theme and transitional elements.

This suggested format applies to many, but not to all, paragraphs included in technical documents. In one common exception, you may choose to delay statement of a topic sentence until you engage the reader's attention with the first few sentences. In other cases, the paragraph may be short and serve only as an attention

Communication Challenge

"Telecommuting: The Last Frontier?"

Calling themselves the "Commute Group," five managers at McDuff's Boston office have been meeting to discuss telecommuting (that is, permitting some or all employees to do part of their work at home). The branch manager, Janet Remington, has expressed interest in the group's work and suggested that group members write a report proposing a pilot project at the branch. The report would be read by Janet and by members of the McDuff corporate staff in Baltimore—especially Kerry Camp, vice president for domestic operations, Janet's boss. Any change in branch work schedules must be approved by corporate headquarters.

The Commute Group now must decide (1) what to include in its report to Janet Remington and (2) how to *organize* information for maximum impact. What follows are some details on the audience for the report, the group's reasons for favoring telecommuting, some problems discussed by the group, and questions that remain about the organization of the report. Although the group has made progress in discussing telecommuting, it has been unable to decide on a structure for its report.

Report Audience

The group has spent much time discussing what points would be most persuasive with the primary audience, Janet Remington and Kerry Camp. Janet has been open to new ideas since being chosen for the manager job a year ago. She meets often with all departments in the office and shows a genuine interest in creating a more comfortable workplace. For example, she recently accepted recommendations by department managers to purchase office chairs and desks that will allow employees to work with less physical strain.

Yet Janet knows that her boss, Kerry Camp, will evaluate her largely on the financial performance of the branch, not on the comfort level of her staff. Indeed, Kerry is a 60-year-old engineer, with a master's degree in finance, who keeps a close eye on the "bottom line" of each branch. He is interested in exploring new work practices only if they may improve employee productivity. More than likely, he will be the final decision-maker about the pilot project.

Rationale for Pilot Project

The Commute Group spent much time discussing two topics: branch jobs that would be best suited to telecommuting and specific arguments in support of a telecommute policy.

Group members agreed that employees who do much independent work, especially on the computer, would be the best candidates for a pilot project. In particular, members of the technical and scientific staff often spend half their

Continued

days at personal computers, either performing technical calculations or drafting sections of reports and proposals.

Next the group discussed reasons for adopting a telecommute pilot project. The group first met to discuss the issue after a series of horrible rush hours over the holiday season in December. Bad weather forced most of the 125 branch employees either to miss some workdays during the period or to arrive up to two hours late several days. Most employees already have a one-way commute of at least one hour, since there is little affordable housing close to the office location in downtown Boston. Thus the heavy holiday traffic prompted the discussion about telecommuting.

In its deliberations, the Commute Group focused mostly on the kind of work that could be done by employees at home. What follows are some of the points discussed by the group, in random order.

- Telecommuting will save time either by eliminating commuting (on days the employee works exclusively at home) or by reducing commuting time (on days when the employee comes to the office for part of the day and thus avoids one or both rush-hour periods that day).
- Employees can write and edit reports and proposals at home for several hours at a time, without the usual office interruptions of meetings, phone calls, drop-in visitors, etc. Some experts claim that writers are most productive during the drafting stage if they have uninterrupted blocks of writing time.
- Morale will improve as long as there is a clear rationale for adopting the policy and selecting participants for the pilot project. Employees chosen should be those who work well independently, whose jobs can be handled through telecommuting, and who have already made significant contributions to their departments.
- If McDuff adopts a telecommuting policy after the pilot project, the firm may attract an additional pool of excellent employees.
- Telecommuting will improve some employees' productivity by allowing them to work when they are recovering from an illness at home or when family members are ill—in other words, times when the employee would normally be on sick leave.
- The company would benefit from the increase in computer literacy among both the telecommuters and those who work with them back at the office. The firm would begin to take advantage of the considerable investment it already has made in computer technology—personal computers, networking, software for electronic mail, Internet connections, modems, etc. In particular, email would become a way of life. Up until now, many employees have been reluctant to replace time-consuming meetings, phone calls, and memos with email.
- If telecommuting were to become a regular way of doing business, it might reduce the amount of work space needed at the office and thus reduce overhead. For example, several employees could share the same office work space if much of their work time were spent at home.

- Even some noncomputer tasks, like phone calls to clients, could be done best in the quiet environment of the home, as opposed to the hectic environment of the office where noise and interruptions are a part of doing business.
- If a telecommuting policy were adopted, McDuff would gain public support by showing that it is part of the solution to the central problems of traffic congestion and air pollution. Even some potential clients might be attracted by the firm's progressive policies.

Possible Problems with Telecommuting

The Commute Group also addressed problems that might arise with the pilot project and with telecommuting in general. Group members were unsure how or if the problems should be woven into the fabric of the report. Here are some concerns that were discussed:

- The right employees must be selected for the pilot project. Whereas some employees might improve their productivity at home, others might find it difficult to stay on task, either because of their own work habits or because of their home environment. Some kind of appropriate screening device would be in order.
- The branch must determine how to evaluate the success of the pilot project, perhaps by some combination of (a) self-evaluation by the employee, (b) performance evaluations by the employees' supervisors, (c) productivity assessment by the corporate office, or (d) opinions gathered by surveying employees who are not part of the pilot project but who interact regularly with the employees who are telecommuting.
- Good communication is central to the project. Employees must be involved in selecting participants, planning the study, conducting the project on a day-to-day basis, and evaluating its success.

Organization of the Report

The Commute Group has agreed on the audience for the report, the likely qualifications for participation in the pilot study, advantages of telecommuting, and some possible problems with the study and with telecommuting in general. But the group has *not* resolved two main questions: (1) what part of the information assembled should be included in the report and (2) what order this information should assume. In other words, the group must wrestle with matters of *organization*. Indeed, disagreements about these two issues created a stumbling block in the group's work.

Assume that you have been called in by the Commute Group to help get it back on track. Answer the following questions, remembering that you are not to be concerned with specific report sections or headings described later in this book. Instead, this exercise concerns only the generic ABC (**A**bstract/**B**ody/**C**onclusion) structure explained in this chapter.

Continued

1. Briefly, how would you describe an overall ABC structure that would work in this report? Your answer should take into account the intended purpose and audience of the report.
2. More specifically, what points would you suggest be included in the abstract component? What issues need *not* be addressed? Why?
3. What points would you suggest be included in the body? Why? In what order? Explain the rationale for the order you suggest. If you have excluded some information discussed by the committee, explain why.
4. Given what you've read so far, what one main purpose should be served by the conclusion component of the group's report? To accomplish this purpose, what information should be included? Why?
5. What issues, if any, remain to be discussed by the Commute Group before it writes its recommendation report?

grabber or a transitional device between several longer paragraphs. Yet for most paragraphs in technical writing, the beginning-middle-end model described here will serve you well. Remember these other points as well as you organize paragraphs in technical writing:

- **Length:** Keep the typical length of paragraphs at around 6 to 10 lines. Many readers won't read long blocks of text, no matter how well organized the information may be. If you see that your topic requires more than 10 lines for its development, split the topic, and develop it in two or more paragraphs.
- **Listings:** Use short listings of three or four items to break up long paragraphs. Readers lose patience when they realize information could have been more clearly presented in listings. The next chapter offers detailed suggestions on using lists.
- **Use of Numbers:** Paragraphs are the worst format for presenting technical data of any kind, especially numbers that describe costs. Readers may ignore or miss data packed into paragraphs. Usually tables or figures would be a more clear and appropriate format. Also, be aware that some readers may think that cost data couched in paragraph form represent an attempt to hide important information.

This chapter mostly concerns the ordering of ideas within paragraphs, sections, and whole documents. Good organization helps make your writing successful. Organization alone, however, will not win the day. Readers also expect a visually appealing document. The next chapter describes technical devices for creating the best possible design of your pages.

CHAPTER SUMMARY

Good technical writing calls on special skills, especially in organization. Writers should follow three guidelines for organizing information: (1) write different parts of the document for different readers, (2) place important information at the beginnings and endings, and (3) repeat key points throughout the document.

This chapter recommends the "ABC format" for organizing technical documents. This format includes an **A**bstract (summary), a **B**ody (supporting details), and a **C**onclusion (wrap-up and transition to next step). The abstract section is particularly important because most readers give special attention to the start of a document.

Individual sections and paragraphs also require attention to organization. Sections need overviews and closing passages so that busy readers can find information quickly. Most paragraphs should contain a topic sentence, supporting details, transitional words and phrases, and a closing sentence that leads into the next paragraph.

Access the Technical Writing Web site by pointing your browser at www.prenhall.com/pfeiffer. Here, you will find additional resources to aid you in learning and applying the information from this chapter.

ASSIGNMENTS

1. **Overall Organization.** Find an example of technical writing directed to more than one reader. Prepare a written or an oral report (your instructor's choice) that explains how well the excerpt follows this chapter's guidelines for organization.

2. **Evaluating an Abstract.** Read the following abstract and evaluate the degree to which it follows the guidelines in this chapter.

> As one of the buyers for Randall Auto Parts, I constantly search for new products that I feel can increase our sales. I recently attended an electronics convention to see what new products were available. One product that caught my eye was the new Blaupunkt BMA5350B amplifier. I recommend that we adopt this amplifier into our line of car audio products.
>
> This proposal supports my recommendation and includes the following sections:
> 1. Features of the Blaupunkt BMA5350B
> 2. Customer Benefits
> 3. Cost
> 4. Conclusions

3. **Section Organization.** As a graphics specialist at McDuff, you have written a recommendation report on ways to upgrade the graphics capabilities of the firm. One section of the report describes a new desktop publishing system, which you believe will make McDuff proposals and reports much more professional looking. Your report section describes technical features of the system, the free training that comes with purchase, and the cost.

   Write a lead-in paragraph for this section of your report. If necessary, invent additional information for writing the paragraph.

4. **Paragraph Organization: Analysis.** Select a paragraph from each of four different articles taken from periodicals in your campus library. Choose one from a nationally known

newspaper (like *The New York Times*), one from a popular magazine (like *Time* or *Sports Illustrated*), one from a business magazine (like *Forbes* or *Business Week*), and one from a technical journal (like *IEEE Transactions on Professional Communication*).

Explain in writing how each of the paragraphs does or does not follow the top-down pattern of organization discussed in this chapter. If a paragraph does not follow the top-down pattern, indicate whether you believe the writer made the right or wrong decision in organizing the paragraph. In other words, was there a legitimate reason to depart from the ABC pattern? If so, what was the reason? If not, how would you revise the paragraph to make it fit the ABC model?

5. **Paragraph Organization: Writing.** With the following list of related information, write a paragraph that follows the organizational guidelines in this chapter. Use all the information, change any of the wording when necessary, and add appropriate transitions. Assume that the paragraph is part of an internal McDuff document suggesting ways to improve work schedules.

- Four-day weeks may lower job stress—employees have long weekends with families and may avoid worst part of rush hour.
- A four-day, 10-hour-a-day workweek may not work for some service firms, where projects and clients need five days of attention.
- Standard five-day, eight-hour-a-day workweeks increase on-the-job stress, especially considering commuter time and family obligations.
- McDuff is considering a pilot program for one office, whereby the office would depart from the standard 40-hour workweek.
- There are also other strategies McDuff is considering to improve work schedules of employees.
- The 40-hour workweek came into being when many more families had one parent at home while the other worked.
- Some firms have gone completely to a four-day week (with 10-hour days).
- McDuff's pilot program would be for one year, after which it would be evaluated.

6. **Writing an Abstract: Individual Work.** The short report, on the following page, lacks an abstract that states the purpose and provides the main conclusion or recommendation from the body of the report. Write a brief abstract for this report.

7. **Writing an Abstract: Conventional Group Work.** For this assignment your instructor will divide the class into groups. Using one of the seven project sheets included in the color insert, your group will write a generic abstract for a report of the completed project. Follow the guidelines in this chapter.

8. **Writing an Abstract: Group Work Using Computer Communication.** As in assignment 7, for this assignment you will (a) work in groups established by your instructor, (b) write a generic abstract for a report on one of the seven project sheets included in the color insert, and (c) follow the abstract guidelines included in this chapter. In addition, you are to conduct at least part of your team business by email. The degree to which your team uses email will depend on the technical resources of team members and the campus. At a minimum, you should plan for each member to send a message to every other member concerning, for example, the drafting or editing process. At a maximum, and if computer resources permit, you may develop on-screen "windows" whereby you conduct a conversation with each fellow member in one window and make changes in text in another window. The point of this assignment, in other words, is that team members can use email in a substantive way to communicate with each other in the completion of group projects.

DATE: June 13, 2002
TO: Ed Simpson
FROM: Jeff Radner
SUBJECT: Creation of an Operator Preventive Maintenance Program

The Problem
The lack of operator involvement in the equipment maintenance program has caused the reliability of equipment to decline. Here are a few examples:

◆ A tractor was operated without adequate oil in the crankcase, resulting in a $15,000 repair bill after the engine locked.
◆ Operators have received fines from police officers because safety lights were not operating. The bulbs were burned out and had not been replaced. Brake lights and turn-signal malfunctions have been cited as having caused rear-end collisions.
◆ A small grass fire erupted at a construction site. When the operator of the vehicle nearest to the fire attempted to extinguish the blaze, he discovered that the fire extinguisher had already been discharged.

When the operator fails to report deficiencies to the mechanics, dangerous consequences may result.

The Solution
The goal of any maintenance program is to maintain the company equipment so that the daily tasks can be performed safely and on schedule. Since the operator is using the equipment on a regular basis, he or she is in the position to spot potential problems before they become serious. For a successful maintenance program, the following recommendations should be implemented:

◆ Hold a mandatory four-hour equipment maintenance training class conducted by mechanics in the motor pool. This training would consist of a hands-on approach to preventive maintenance checks and services at the operator level.
◆ Require operators to perform certain checks on a vehicle before checking it out of the motor pool. A vehicle checklist would be turned in to maintenance personnel.

The attached checklist would require 5 to 10 minutes to complete.

Conclusion
I believe the cost of maintaining the vehicle fleet at McDuff will be reduced when potential problems are detected and corrected before they become serious. Operator training and the vehicle pretrip inspection checklist will ensure that preventable accidents are avoided. I will call you this week to answer any questions you may have about this proposal.

9. **Ethics Assignment.** Read this chapter's "Communication Challenge" section entitled "Telecommuting: The Last Frontier?" (You may also want to conduct some library or Internet research on telecommuting *or* discuss the concept with someone who works full time.) Then write a short essay or report that examines (1) any personal ethical dilemmas that may arise for McDuff employees if the Boston office adopts a telecommuting policy and (2) possible solutions to the ethical problems you discuss.

McDuff, Inc.
Fleet Maintenance Division
Vehicle Checklist
Pretrip Inspection

Inspected by: _____   Date: _____

Vehicle #: _____   Odometer: _____

Fluid Levels, Full/Low   Comments

_____ Engine Oil   _____

_____ Transmission Fluid   _____

_____ Brake Fluid   _____

_____ Power Steering   _____

_____ Radiator Level   _____

Before Cranking Vehicle

_____ Tire Condition   _____

_____ Battery Terminals   _____

_____ Fan Belts   _____

_____ Bumper and Hitch   _____

_____ Trailer Plug-in   _____

_____ Safety Chains   _____

After Cranking Vehicle

_____ Parking Brakes   _____

_____ Lights   _____

_____ All Gauges   _____

_____ Seat Belts   _____

_____ Mirrors/Windows/Wipers   _____

_____ Clutch   _____

_____ Fire Ext. Mounted and Charged   _____

_____ Two-Way Radio Working   _____

Additional Comments:_____

_____

10. **International Communication Assignment.** In the ABC pattern, the beginning of a document ("A" or Abstract) includes a clear purpose statement, summary of main points for decision-makers, and brief description or listing of information to follow in the document. Although this pattern works in most situations—especially in the United States and many Western countries—there may be contexts and cultures in which it is not the best choice. For example, the front-end location of important points may appear abrupt and even offensive in some cultures. Using a country or culture outside the United States, describe a context in which you would depart from the strict ABC pattern. Be specific about purpose, readers, and preferred organizational pattern. Also, give the source upon which you base your conclusions—family experience, business experience, interviews, books, Internet research, etc.

MEMORANDUM

DATE: September 5, 2002
TO: Danielle Firestein
FROM: Barbara Ralston *BR*
SUBJECT: Recommendation for AIM 500 Fax

INTRODUCTORY SUMMARY

The purpose of this report is to present the results of the study you requested on the AIM 500 facsimile (fax) machine. I recommend purchase of additional AIM 500 machines, when needed, because they deliver fast, dependable service and include features we need most. This report includes the following sections: Ease of Operation, Features of the AIM 500, Dependability of the AIM 500, and Conclusion.

EASE OF OPERATION

The AIM 500 is so easy to operate that a novice can learn to transmit a document to another location in about two minutes. Here's the basic procedure:

1. Press the button marked TEL on the face of the fax machine. You then hear a dial tone.
2. Press the telephone number of the person receiving the fax on the number pad on the face of the machine.
3. Lay the document facedown on the tray at the back of the machine.

At this point, just wait for the document to be transmitted—about 18 seconds per page to transmit. The fax machine will even signal the user with a beep and a message on its LCD display when the document has been transmitted. Other more advanced operations are equally simple to use and require little training. Provided with the machine are two different charts that illustrate the machine's main functions.

The size of the AIM 500 makes it easy to set up almost anywhere in an office. The dimensions are 13 inches in width, 15 inches in length, and 9.5 inches in height. The narrow width, in particular, allows the machine to fit on most desks, file cabinets, or shelves.

FEATURES OF THE AIM 500

The AIM 500 has many features that will be beneficial to our employees. In the two years of use in our department, the following features were found to be most helpful:

Automatic redial
Last number redial memory
LCD display

MODEL 3–1
ABC format in whole document

Preset dialing
Group dialing
Use as a phone

Automatic Redial. Often when sending a fax, the sender finds the receiving line busy. The redial feature will automatically redial the busy number at 30-second intervals until the busy line is reached, saving the sender considerable time.

Last Number Redial Memory. Occasionally there may be interference on the telephone line or some other technical problem with the transmissions. The last number memory feature allows the user to press one button to automatically trigger the machine to retry the number.

LCD Display. This display feature clearly shows pertinent information, such as error messages that tell a user exactly why a transmission was not completed.

Preset Dialing. The AIM 500 can store 16 preset numbers that can be engaged with one-touch dialing. This feature makes the unit as fast and efficient as a sophisticated telephone.

Group Dialing. Upon selecting two or more of the preset telephone numbers, the user can transmit a document to all of the preset numbers at once.

Use as a Phone. The AIM 500 can also be used as a telephone, providing the user with more flexibility and convenience.

DEPENDABILITY OF THE AIM 500

Over the entire two years our department has used this machine, there have been no complaints. We always receive clear copies from the machine, and we never hear complaints about the documents we send out. This record is all the more impressive in light of the fact that we average 32 outgoing and 15 incoming transmissions a day. Obviously, we depend heavily on this machine.

So far, the only required maintenance has been to change the paper and dust the cover.

CONCLUSION

The success our department has enjoyed with the AIM 500 compels me to recommend it highly for additional future purchases. The ease of operation, many exceptional features, and record of dependability are all good reasons to buy additional units. If you have further questions about the AIM 500, please contact me at extension 3646.

MODEL 3–1
continued

ADDITIONAL FEATURES OF MAGCAD

This report has presented two main advantages of the MagCad Drawing System: ease of correction and multiple use of drawings. However, there are two other features that make this system a wise purchase for McDuff's Boston office: the selective print feature and the cost.

Selective Printing

When printing a MagCad drawing, you can "turn off" specific objects that are in the drawing with a series of keystrokes. The excluded items will not appear in the printout of the drawing. That is, the printed drawing will reflect exactly what you have temporarily left on the screen, after the deletions. Yet the drawing that remains in the memory of the machine is complete and ready to be reconstructed for another printout.

The selective print feature is especially useful on jobs where different groups have different needs. For example, in a drawing of a construction project intended only for the builder, one drawing may contain only land contours and the building structures. If the same drawing is going to the paving company, we may need to include only land contours and parking lots. In each case, we will have used the selective print feature to tailor the drawing to the specific needs of each reader.

This feature improves our service to the client. In the past, we either had to complete several different drawings or we had to clutter one drawing with details sufficient for the needs of all clients.

Cost of MagCad

When we started this inquiry, we set a project cost limit of $12,000. The MagCad system stays well within this budget, even considering the five stations that we need to purchase.

The main cost savings occurs because we have to buy only one copy of the MagCad program. For additional work stations, we need pay only a $400 licensing fee per station. The complete costs quoted by the MagCad representative are listed below:

1. MagCad Version 5	$5,000
2. Licenses for five additional systems	1,500
3. Plotter	2,000
4. Installation	1,000
TOTAL	$9,500

With the $2,500 difference between the budgeted amount and the projected cost of the system, we could purchase additional work stations or other peripheral equipment. The next section suggests some add-ons we might want to purchase later, once we see how the MagCad can improve our responsiveness to client needs.

MODEL 3–2
ABC format in document section

Conversion to a partial solar heating and cooling system would upgrade the hospital building considerably. In fact, the use of modern solar equipment could decrease your utility bills by up to 50%, using the formula explained in Appendix B. As you may know, state-of-the-art solar systems are much more efficient than earlier models. In addition, equipment now being installed around the country is much more pleasing to the eye than was the equipment of ten years ago. The overall effect will be to enhance the appearance of the building, as well as to save on utility costs.

We also believe that changes in landscaping would be a useful improvement to the hospital's physical plant. Specifically, planting shade trees in front of the windows on the eastern side of the complex would block sun and wind. The result would be a decrease in utility costs and enhancement of the appearance of the building. Of course, shade trees will have to grow for about five years before they begin to affect utility bills. Once they have reached adequate height, however, they will be a permanent change with low maintenance. In addition, your employees, visitors, and patients alike will notice the way that trees cut down on glare from the building walls and add "green space" to the hospital grounds.

MODEL 3–3
ABC format in paragraphs

Chapter 4

Page Design

Collaboration at Work

Design of the Campus Paper

General Instructions

Each chapter begins with a Collaboration at Work exercise that applies strategies for working in groups to chapter topics. The exercise assumes you (1) have been divided into groups of about three to six students, (2) will use group time inside or outside of class to complete the case, and (3) will produce an oral or written response. For guidelines about writing in groups, refer to pages 28–34.

Background for Assignment

Chapter 3 used the term *organization* to refer to the structure of ideas in your writing. This chapter introduces the term *page design* to refer to the arrangement of all visual elements—including text—on the page. As you will learn, effective page design incorporates features such as white space, headings, lists, font size and type, and color. Choices in page design can greatly influence the interest readers show in a document.

This exercise concerns page design in a document with which you may be familiar—the campus print or online newspaper. Typical features of a college or university paper include the following:

- Campus news and events

- Local, regional, or local news
- Updates on academic programs
- Updates on student organizations and activities
- Editorials
- Letters to the editor
- Advertisements

How these features are presented through effective (or ineffective) page design helps determine whether members of the campus community read the paper and take it seriously as journalism.

Group Assignment

The purpose of this assignment is to evaluate the page design of your campus newsletter. (If your campus has no paper, use a similar document available on or near campus.) Your group has two options for this assignment. Option 1 is to review a current issue of your campus newspaper, note elements of its page design, and determine whether these elements serve a useful purpose. Option 2 is to compare and contrast the effectiveness of the page design of two different campus newspapers or two different issues of the same paper.

*T*he four chapters in Part 1 cover basics you need to know before moving to the applications in Parts 2 and 3. The first chapter describes the technical writing process, with emphasis on writing for your reader. The second introduces McDuff, Inc., the fictional company used throughout the book, and the third deals with organizing information. This chapter covers page design, another basic building block in technical writing. Here's an operating definition:

> **Page design:** a term that refers to formatting options used to create clear, readable, and visually interesting documents. Some of these options include judicious use of white space, headings, lists, in-text emphasis, fonts, and color. Many firms combine these elements into style sheets that allow employees to produce documents in uniform, consistent formats.

This chapter starts with brief sections on the benefits and potential drawbacks of using computers to design pages. The rest of the chapter presents guidelines and examples for page design. As you read the material, remember that you can use this chapter during all three stages of the writing process: planning, drafting, and revising.

COMPUTERS IN THE WRITING PROCESS

Using a computer can speed up all three stages of writing: planning, drafting, and revising. Two main advantages are that (1) you can change text and graphics continuously and easily and (2) you have many typographical choices. Such choices give you the freedom to communicate your message in the most convincing way. Here are suggestions for integrating computers into all three writing stages.

Using Computers to Plan

As noted in chapter 1, the planning stage of writing requires that you determine your purpose, consider your readers' needs, collect information, and construct an outline. Computers are especially useful during the last two steps, collecting information and outlining.

Word processing can speed up the collection of information on your topic. The research procedure, detailed in chapter 13, may require that you assemble information from many sources. If you are taking notes next to your computer, you can enter them directly into your system. Each one can be placed on a separate "page" and then indexed and organized by topic, in preparation for writing the first draft. Of course, such computerized note taking is practical only if you are reading sources next to a terminal. Some software programs are more helpful than others in making such note-taking functions easy to use.

You can also use computers to place ideas in outline form. Words and phrases in your outline can be added and moved quickly—without the constant scratching out and rewriting necessary with handwritten outlines. Clustering ideas and arranging points in logical order become as simple as touching the keyboard. Some software programs provide the hierarchal outline format on the screen, just awaiting your insertion of wording. For extensive outlines, you can set up a separate file for each main topic and, again, use the keyboard to shift back and forth among files as you build the underlying structure of your document.

In summary, the ability to write and edit on screen can help you collect information and construct outlines. Yet many writers prefer to complete research notes and outlines by hand. The old-fashioned approach still offers the advantage of writing nonlinear (i.e., messy) outlines. Outlined points and research notes can be stretched all over paper of any size or even pinned to the walls, giving you a "big picture" effect that small, ever-changing screens may not. Choose the best method for you, but at least be aware of what options computers offer.

Using Computers to Draft

Though helpful during the planning process, word processing shows its main strength as you write early drafts. Whether you touch-type or plunk away with two fingers, the word processor will keep you writing faster. For example, the common "word-wrap" feature automatically moves you from line to line without need of carriage returns. Most importantly, you add and delete text very quickly. You simply "scroll" forward or backward on the screen to the point where you want to make changes. Here are some other advantages of writing early drafts on the computer:

- Systems with multiple windows may keep your outline on the screen while you write the related draft section.
- When you cannot think of a word or phrase, you can type in an easily recognizable symbol and then return later to insert the word(s).
- When several alternative passages are possible, you can type in all of them, with brackets or other associated symbols. Then during the revision stage, you can make your final choice.

As noted in the chapter 1 discussion on collaborative writing, the computer helps you combine your work with that of others in team-writing efforts. Individual writers in the group can produce draft material at their separate terminals, which are part of one network. At that point, individual files can be "pulled up" on the screens of other group members for revision and commentary. Later, separate files can be merged into one draft for revision. Thus both groups and individuals benefit from the computer at the drafting stage of writing.

Using Computers to Revise

Drafting requires that you get ideas onto paper (or screen) as quickly as possible, to avoid writer's block. When you shift to the revising stage, the aim is to select options and

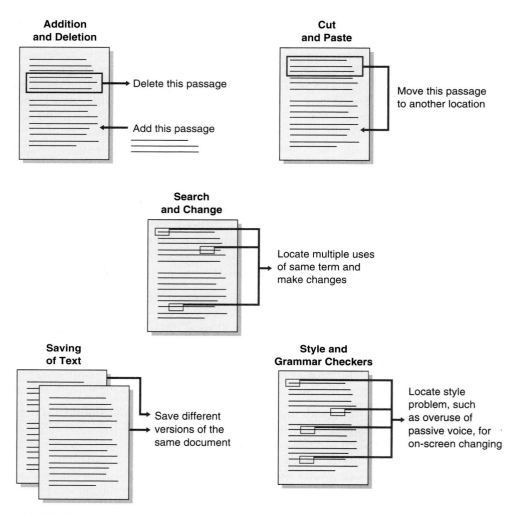

FIGURE 4–1 Sample changes during revising

change text in whatever way best meets the readers' needs. As shown in Figure 4–1, computer systems offer these sorts of tools for revision:

- **Addition and deletion:** You can add or delete text by placing the cursor—an icon that shows where you are in the text—at the appropriate point. This technique eliminates the time-consuming process of scribbling changes into margins by hand, deleting words, and then retyping drafts.
- **Cut-and-paste:** This feature allows you to move sentences, paragraphs, graphics, or entire sections within one document, or among several. No longer must you literally cut and paste sections of typed or handwritten copy. Cut-and-paste allows you easily to revise text to produce the most logical arrangement of material.

- **Search and change:** Used correctly, this feature can save considerable time during revision. It helps you locate every instance of a particular word or phrase throughout the document. For example, assume that you want to check your draft to make certain that you used *data* consistently as a plural noun, not a singular noun. Simply enter the word *data* on the keyboard so that your system will scroll to each section of the text where the word is used. Then you can make necessary changes.

- **Saving of text:** Often you may be uncertain what version of a paragraph or section is most appropriate for the final draft. In this case you can save multiple versions in a file for later reference. A note or an asterisk on your screen will trigger your memory that another version of the text is available. Later you can select the best option and delete the others.

- **Style and grammar checkers:** Now there are editing programs that use the "search and change" function to locate awkward or incorrect passages in your writing. Once you see the highlighted passage, you can decide whether to change it. This software is sophisticated enough to identify errors or constructions like these: split infinitives ("to randomly select"), passive voice ("the recommendations were made"), clichés ("skyrocketing costs"), sexist pronouns ("an engineer should complete his reports on time"), and overuse of "to be" verbs (such as "is," "are," "was," and "were"). Another useful feature can analyze the "readability" of your style by, for example, determining the average number of words per sentence.

  Interactive software also helps you locate grammar, proofreading, and spelling errors. Examples include double words ("and and"), incorrect capitalization ("THey"), wrong spacing, incorrect punctuation, and subject-verb agreement errors. Another feature is a thesaurus. It can suggest alternative word choices and even give the etymology of words. Some programs allow you to add particular grammar or style rules that you want applied to your writing.

Computers allow changes as quickly as you can touch the keyboard. Besides speed, there is the advantage of convenience. Copy can be corrected until the last moment before printing. If errors are found, another draft can be printed. Ironically, it is the very ease of editing during word processing that can lead to misuse and inefficiency of computers during the writing process. The next section notes several problems and ways to avoid them.

Problems to Avoid

The following actual case typifies problems that can arise with computers during the writing process. At one engineering firm in the Southwest a few years ago, engineers usually produced their own report drafts in longhand or by cutting and pasting typed copy from previous reports. This hard copy was delivered to a central word processing department, where operators produced first drafts for later editing. In theory, the operators were to see each report only two or three times—once during the initial inputting of information and another time or two when changes were incorporated into the final draft. In fact, however, writers constantly made changes that required multiple printing of the reports—often five or six times instead of two or three. As a result, productivity decreased, and word processing expenses increased. The very ease of word processing had led to inefficiency in the writing process.

The company solved its productivity problem by forming a review committee of both users and operators, which resulted in the adoption of some new procedures. For example, the group agreed to limit the number of revision cycles, requiring writers to give more attention to each draft. In the same way, you can save time and take full advantage of computer technology in writing by following these guidelines:

- **Work in stages.** Note the suggestions in chapter 1 for completing stylistic, grammatical, and mechanical editing in stages—so that errors are caught at the appropriate time.
- **Limit number of printings.** Print as often as you need to print, but don't be wasteful. In other words, make sure you have edited enough to justify another printing. One hidden cost of writing today is the proliferation of word processing drafts. Resources are squandered, and time is added to the writing process.
- **Use the buddy system.** Ask a colleague to help with the editing process so that another pair of eyes sees your work. Your closeness to the project may keep you from spotting errors, no matter how much technology is at your command.
- **Know the limitations of style programs.** Remember that software programs to correct spelling and other mechanical errors will not eliminate all errors. You still need to watch for problems with words like *effect/affect, complement/compliment,* and *principal/principle,* as well as misspellings of personal names and other proper nouns. You might want to use the "search and change" feature of your system to locate trouble spots.

If used with common sense, computers can be a powerful tool as you plan, draft, and revise documents. You will spend more time preparing the content of writing and less time on mechanics.

However, there is one more hazard brought about by this new technology. Computers allow far more typographical variations than most readers want or need. For example, systems give you the option of using one or all of the following techniques for typographical emphasis: underlining, bold, full caps, italics, and increased type size. Too much variety in one document can confuse rather than clarify a passage. The next two sections provide guidelines for effective page design.

GUIDELINES FOR PAGE DESIGN

As one expert says, often you will write for readers who are "in a hurry, frustrated, and bored, and who would prefer to get the information needed from text *in any other way but reading*" (my emphasis) [Schriver, K.A. Fourth Quarter, 1989. Document design from 1980 to 1989: Challenges that remain. *Technical Communication,* p. 319]. Your challenge is to respond to this prejudice by making pages interesting to the eye. A document may offer your reader great opportunities but *never* get read. Why? Because it doesn't look *readable*.

Good organization, as pointed out in the last chapter, can fight readers' indifference by giving information when and where they want it. But to get and keep readers interested, you must use effective page design—on each page of your document. Each page needs the right combination of visual elements to match the needs of your readers and the purpose of the document.

Many firms know the benefits of "visual imaging" and develop company style sheets for all frequently used documents including letters, memos, various types of reports, and proposals. Once developed, they are assembled into a style manual and distributed for general use. To make universal use easier they are often loaded into the company's text editing software for easy access.

Using style sheets saves time and reinforces the firm's corporate image. Their use also makes it possible for members of writing teams to work independently while adhering to corporate formatting guidelines. Use of style sheets also ensures that readers see a clear relationship among ideas because headers and subheaders are used consistently.

If the company you work for does not use a style manual, you will want to use these elements of page design to develop your own style sheets: (1) paper, (2) white space, (3) headings, (4) lists, (5) in-text emphasis, (6) fonts, and (7) color.

Paper

Your selection of paper may be as simple as using the stock purchased by the company for use on your printer, or as sophisticated as finding a paper that is the correct weight, finish, or design for the document you are producing. In most instances, your documents will be printed one-sided on stationery that bears your company's logo or letterhead. In all cases, the paper should meet the needs of your printer.

However, if your document is very long—for instance a lengthy report, a policy manual, or a procedure manual—you may want to print the document on both sides of the paper. Your decision to print a two-sided document will permit readers to find the information they need more easily. Also, by reducing the document's thickness, you will make it appear less intimidating. If you use both sides, the paper you select should be thick enough and opaque enough to prevent images from bleeding through and distorting the message on the opposite side.

Occasionally, you may want to use a paper that is preprinted with a design that meets your special needs: an announcement, a sales campaign, or perhaps even a special report. Your document design may need to be completely different from the style sheets developed by your company.

White Space

The term *white space* simply means the open places on the page with no text or graphics—literally, the *white* space. Experts have learned that readers are attracted to text because of the white space that surrounds it, as with a newspaper ad that includes a few lines of copy in the middle of a white page. Readers connect white space with important information.

FIGURE 4–2   Use of white space: margins   **FIGURE 4–3**   Use of white space: columns

In technical writing you should use white space in a way that (1) attracts attention, (2) guides the eye to important information on the page, (3) relieves the boredom of reading text, and (4) helps readers organize information. Here are some opportunities for using white space effectively:

1. Margins: Most readers appreciate generous use of white space around the edges of text. Marginal space tends to "frame" your document, so the text doesn't appear to push the boundaries of the page. Good practice is to use $1''$ to $1\frac{1}{2}''$ margins, with more space on the bottom margin. When the document is bound on the left, as in Figure 4–2, also place more space in the left margin than at the bottom—to account for space lost in the binding process.

2. Columns: Long lines can be an obstacle to keeping the readers' attention. Eyes get weary of overly long lines, so some writers add double columns to their design options. This "book look," as shown in Figure 4–3, uses white space between columns to break up text and thus reduce line length.

3. Hanging Indents: Instead of using columns to increase white space, some writers place headers and subheads at the left-hand margin, and indent the text block an additional inch or so, as shown in Figure 4–4. Headers and subheads force the readers' eyes and attention to the text block.

4. Line space: When choosing single, double, or $1\frac{1}{2}$ line spacing, consider the document's length and degree of formality. Letters, memos, short reports, and other documents read in one sitting are usually single spaced, with one line space between paragraphs. Longer documents, especially if they are formal, are usually $1\frac{1}{2}$-space or double-spaced, usually without, but sometimes with, extra spacing between paragraphs. (See Figure 4–5.)

FIGURE 4–4   Use of white space: hanging indents

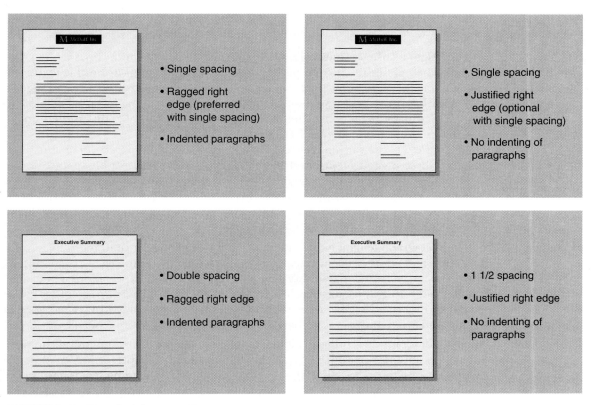

FIGURE 4–5   Use of white space: lines

Poor Format: One long
paragraph on page

Better Format: Several
paragraphs on page

FIGURE 4–6   Use of white space: paragraphs

5. Right-justified versus ragged edge: To justify or not to justify lines is often the question. In right-justified copy, all lines are the same length—as on this textbook page. In ragged-edge copy, lines are variable length. Some readers prefer ragged-edge copy because it adds variation to the page, making reading less predictable for the eye. Yet many readers like the professional appearance of right-justified lines, especially in formal documents. Both views have merit.

As a rule, (1) use ragged edge on densely packed, single-spaced documents and (2) use either ragged or justified margins on $1\frac{1}{2}$-space and double-spaced documents, depending on reader preference or company style (see Figure 4–5). However, only use justified margins if your word processing software maintains uniform spacing between words and between letters within words. It is unnerving to read justified text with inconsistent spacing at these points.

6. Paragraph length: New paragraphs give readers a chance to regroup, as one topic ends and another begins. These shifts also have a visual impact. The amount of white space produced by paragraph lengths can shape reader expectations. For example, two long paragraphs suggest a heavier reading burden than do three or four paragraphs of differing lengths. Most readers skip long paragraphs, so vary paragraph lengths and avoid putting more than 10 lines in any one paragraph (see Figure 4–6).

7. Paragraph indenting: Another design decision involves indenting the first lines of paragraphs. As with ragged-edge copy, most readers prefer indented paragraphs because the extra white space creates visual variety. Reading text is hard work for the eye. You should take advantage of any opportunity to keep the reader's attention (see Figure 4–5).

8. In-text graphics: Any illustration within the text needs special attention. Whereas chapter 11 provides a complete discussion of graphics, the section that follows discusses their placement for visual appeal. Here are some pointers:

- Make sure there is ample white space between any in-text graphic and the text. If the figure is too large to permit adequate margins, reduce its size.
- When you have the choice, place in-text graphics near the top of the page. That position gives them the most attention.
- When a graphic doesn't fit well on a page with text, place it on its own page where there will be adequate space. Normally it appears on the page following the first reference to it.
- Pay special attention to page balance when graphics will be included on multi-column pages, two-page spreads, or both.
- Draw rough sketches of the layout for the entire document so that you can use white space consistently and persuasively from start to finish.

9. Pagination: You may place page numbers at either the top or the bottom of the page and at either the center or outside edge of the page. Your readers must be able to find page numbers easily; at the same time, they should not detract from the appearance of the document.

In very long documents, you may want to page each section with the section number, and number pages within the section beginning with "1." For instance, the first page of section two would appear as "2–1." This style makes it easier to compile sections of material developed by different writers and to make revisions without having to renumber an entire lengthy document. It is especially effective when the document will be revised periodically, for it saves cost and printing time.

10. Heading space and lines: White space helps the reader connect related information immediately. Always have slightly more space above a heading than below it. That extra space visually connects the heading with the material into which it leads. In a double-spaced document, for example, you would add a third line of space between the heading and the text that came before it.

In addition, some writers add a horizontal line across the page above headings, to emphasize the visual break. The next section will cover other aspects of headings.

In summary, well-used white space can add to the persuasive power of your text. As with any design element, however, it can be overused and abused. Make sure there is a *reason* for every decision you make with regard to white space on your pages.

Headings

Headings are brief labels used to introduce each new section or subsection of text. They serve as (1) a signpost for the reader who wants to know the content, (2) a "grabber" to entice readers to read documents, and (3) a visual "oasis" of white space where the reader gets relief from text.

As a general rule, every page of any document over one page should have at least one heading. Readers need these markers to find their way through your writing. Models throughout this text show how headings can be used in short and long documents.

Of course, heading formats differ greatly from company to company and even from writer to writer. With all the typographical possibilities of word processing, there is incredible variety in typeface, type size, and the use of bold, underlining, and capitals. Here are some general guidelines that apply:

1. Use your outline to create headings and subheadings. A well-organized outline lists major and minor topics. With little or no change in wording, they can be converted to headings and subheadings within the document. As with outlines, you need to follow basic principles of organization.

- First, if you have one subheading, you must have at least one more at that same level—anything that is divided has at least two parts.
- Second, the number of subheadings should be one indication of the relative length or importance of the section. Be consistent with your approach to headings throughout the document.

2. Use substantive wording. Headings give readers an overview of what content will follow. They entice readers into your document; they can determine whether readers—especially those who are hurried and impatient—will read or skip over the text. Strive to use concrete rather than abstract nouns, even if the heading must be a bit longer. Note the improvements in the following revised headings:

Original:	"Background"
Revised:	"How the Simmons Road Project Got Started" or "Background on Simmons Road Project"
Original:	"Discussion"
Revised:	"Procedure for Measuring Toxicity" or "How to Measure Toxicity"
Original:	"Costs"
Revised:	"Production Costs of the FastCopy 800" or "Producing the FastCopy 800: How Much?"

3. Maintain parallel form in wording. Headings of equal value and degree should have the same grammatical form, as shown in the following:

A) *Headings with Parallel Form*
Scope of Services
Schedule for Fieldwork
Conditions of Contract
B) *Headings That Lack Parallel Form*
Scope of Services
How Will Fieldwork Be Scheduled?
Establish Contract Conditions

You don't have to be a grammar expert to see that the three headings in "B" are in different forms. The first is a noun phrase, the second is a question, and the third is an action phrase beginning with a verb. Because such inconsistencies distract the reader, you should make headings in each section uniform in wording.

4. Establish clear hierarchy in headings. Whatever typographical techniques you choose for headings, your readers must be able to distinguish one heading level from another. Visual features should be increasingly more striking as you move up the ranking of levels. Figure 4–7 shows several heading options that reflect such distinctions. Here are specific guidelines for using such typographical distinctions:

- *Use larger type size for higher-level headings.* You want readers to grasp quickly the relative importance of heading levels as they read your document. Type size

1. Four Levels, Formal Report

LEVEL 1 HEADING

In this example, four level headings are used. Level 1 is set in 18 point Arial Black, a sans serif type font, is in all caps, and is centered. The body is set in 12 point Caslon BdBT. The contrast between the sans serif and serif type fonts ensures there is a clear distinction between headings and body text.

LEVEL 2 HEADING

The Level 2 heading uses the same type font, Arial Black. To clearly identify this level as being subordinate to Level 1, however, it is set in 14 point and is positioned at the left-hand margin. No change is needed or used in the body text.

Level 3 Heading

The Level 3 heading, once again, uses the same type font, Arial Black. Once again the level is clearly identified as being subordinate to the preceding level. It is set in 12 point, uses initial capital and lowercase letters, and is positioned at the left-hand margin. No change is needed or used in the body text.

Level 4 Heading: The Level 4 heading uses the same style font, Arial Black, and is clearly identified as being subordinate to the preceding level. Although it uses initial capitals and lowercase characters and is set in 12 point, it is indented, followed by a colon, and the text begins on the same line as the heading.

FIGURE 4–7   Some heading options

2. Three levels,
Formal Report:
Option A

OTHER HEADING OPTIONS
Level 1 Heading

In this example, four level headings are used. Level 1 is set in 18 point Arial Black, a sans serif type font, is in all caps, and is centered. The body is set in 12 point Caslon BdBT. The contrast between the sans serif and serif type fonts ensures there is a clear distinction between headings and body text.

Level 2 Heading

The Level 2 heading uses the same type font, Arial Black. To clearly identify this level as being subordinate to Level 1, however, it is set in 14 point and is positioned at the left-hand margin. No change is needed or used in the body text.

Level 3 Heading. The Level 3 heading, once again, uses the same type font, Arial Black. Once again the level is clearly identified as being subordinate to the preceding level. It is set in 12 point, uses initial capital and lowercase letters, and is positioned at the left-hand margin. No change is needed or used in the body text.

3. Three levels,
Formal Report:
Option B

Level 1 Heading

In this example, four level headings are used. Level 1 is set in 18 point Arial Black, a sans serif type font, is in all caps, and is centered. The body is set in 12 point Caslon BdBT. The contrast between the sans serif and serif type fonts ensures there is a clear distinction between headings and body text.

LEVEL 2 HEADING

The Level 2 heading uses the same type font, Arial Black. To clearly identify this level as being subordinate to Level 1, however, it is set in 14 point and is positioned at the left-hand margin. No change is needed or used in the body text.

Level 3 Heading. The Level 3 heading, once again, uses the same type font, Arial Black. Once again the level is clearly identified as being subordinate to the preceding level. It is set in 12 point, uses initial capital and lowercase letters, and is positioned at the left-hand margin. No change is needed or used in the body text.

FIGURE 4–7 *continued*

4. Three levels, informal (letter or memo) report

Level 1 Heading

In this example, four level headings are used. Level 1 is set in 18 point Arial Black, a sans serif type font, is in all caps, and is centered. The body is set in 12 point Caslon BdBT. The contrast between the sans serif and serif type fonts ensures there is a clear distinction between headings and body text.

Level 2 Heading

The Level 2 heading uses the same type font, Arial Black. To clearly identify this level as being subordinate to Level 1, however, it is set in 14 point and is positioned at the left-hand margin. No change is needed or used in the body text.

Level 3 Heading. The Level 3 heading, once again, uses the same type font, Arial Black. Once again the level is clearly identified as being subordinate to the preceding level. It is set in 12 point, uses initial capital and lowercase letters, and is positioned at the left-hand margin. No change is needed or used in the body text.

5. Two levels, informal (letter or memo) report: Option A

LEVEL 1 HEADING

In this example, four level headings are used. Level 1 is set in 18 point Arial Black, a sans serif type font, is in all caps, and is centered. The body is set in 12 point Caslon BdBT. The contrast between the sans serif and serif type fonts ensures there is a clear distinction between headings and body text.

Level 2 Heading

The Level 2 heading uses the same type font, Arial Black. To clearly identify this level as being subordinate to Level 1, however, it is set in 14 point and is positioned at the left-hand margin. No change is needed or used in the body text.

6. Two levels, informal (letter or memo) report: Option B

Level 1 Heading

In this example, four level headings are used. Level 1 is set in 18 point Arial Black, a sans serif type font, is in all caps, and is centered. The body is set in 12 point Caslon BdBT. The contrast between the sans serif and serif type fonts ensures there is a clear distinction between headings and body text.

Level 2 Heading. The Level 2 heading uses the same type font, Arial Black. To clearly identify this level as being subordinate to Level 1, however, it is set in 14 point and is positioned at the left-hand margin. No change is needed or used in the body text.

7. One level, informal (letter or memo) report: Option A

LEVEL 1 HEADING

In this example, four level headings are used. Level 1 is set in 18 point Arial Black, a sans serif type font, is in all caps, and is centered. The body is set in 12 point Caslon BdBT. The contrast between the sans serif and serif type fonts ensures there is a clear distinction between headings and body text.

8. One level, informal (letter or memo) report: Option B

Level 1 Heading

In this example, four level headings are used. Level 1 is set in 18 point Arial Black, a sans serif type font, is in all caps, and is centered. The body is set in 12 point Caslon BdBT. The contrast between the sans serif and serif type fonts ensures there is a clear distinction between headings and body text.

FIGURE 4–7 *continued*

fixes this relative importance in their minds so that they can easily find their way through your material both the first time and upon rereading it. The examples in Figure 4–7 reflect two and sometimes three different type sizes in some of the examples. The incremental upgrading of type size helps readers determine the relative importance of the information.

- *Use heading position to show ranking.* In formal documents, your high-level headings can be centered. The next two or three levels of headings are at or off the left margin, as shown in Figure 4–7. Be sure these lower-level headings also use other typographical techniques, such as size, to help the reader distinguish levels.

- *Use typographical techniques to accomplish your purpose.* Besides type size and position, previously mentioned, you can vary heading type with features such as these:

Uppercase and lowercase

Bold type

Underlining

Changes in type font

With this embarrassment of riches, writers must be careful not to overdo it and create "busy" pages of print. Use only those features that will look good on the page and provide an easy-to-grasp hierarchy of levels for the reader.

- *Consider using decimal headings for long documents.* Decimal headings include a hierarchy of numbers for every heading and subheading listed in the table of contents. Many an argument has been waged over their use. People who like them say that they help readers find their way through documents and refer to subsections in later discussions. People who dislike them say that they are cumbersome and give the appearance of "bureaucratic" writing.

Unless decimal headings are expected by your reader, use them only with formal documents that are fairly long. Include at least three heading levels. Following is the normal progression of numbering in decimal headings for a three-level document:

```
1.0  xxxxxxxxxxxxx
     1.1  xxxxxxxxxx
          1.1.1  xxxxxxxxxx
          1.1.2  xxxxxxxxxx
     1.2  xxxxxxxxxx
          1.2.1  xxxxxxxxxx
          1.2.2  xxxxxxxxxx
2.0  xxxxxxxxxxxxx
     2.1  xxxxxxxxxx
          2.1.1  xxxxxxxxxx
          2.1.2  xxxxxxxxxx
     2.2  xxxxxxxxxx
3.0  xxxxxxxxxxxxx
```

Lists

Technical writing benefits from the use of lists. Readers welcome your efforts to cluster items into lists for easy reading. In fact, almost any group of three or more related points can be made into a bulleted or numbered listing. Here are some points to consider as you apply this important feature of page design:

1. **Typical uses:** Lists emphasize important points and provide a welcome change in format. Because they attract more attention than text surrounding them, they are usually reserved for these uses:

    Examples

    Reasons for a decision

    Conclusions

    Recommendations

    Steps in a process

    Cautions or warnings about a product

    Limitations or restrictions on conclusions

2. **Number of items:** The best lists are those that subscribe to the rule of short-term memory. That is, people can retain no more than five to nine items in their short-term memory. A listing over nine items may confuse rather than clarify an issue. Consider placing 10 or more items in two or three groupings, or grouped lists, as you would in an outline. This format gives the reader a way to grasp information being presented.

3. **Use of bullets and numbers:** The most common visual clues for listings are numbers and bullets (enlarged dots or squares like those used in the following listing). Here are a few pointers for choosing one or the other:

    ● *Bullets:* Best in lists of five or fewer items, unless there is a special reason for using numbers.
    ● *Numbers:* Best in lists of over five items *or* when needed to indicate an ordering of steps, procedures, or ranked alternatives. Remember that your readers sometimes will *infer* sequence or ranking in a numbered list.

4. **Format on page:** Every listing should be easy to read and pleasing to the eye. These specific guidelines cover practices preferred by most readers:

    ● *Indent the listing.* Although there is no standard list format, readers prefer lists that are indented farther than the standard left margin. Five spaces is adequate.
    ● *Hang your numbers and bullets.* Visual appeal is enhanced by placing numbers or bullets to the left of the margin used for the list, as done with the items here.
    ● *Use line spaces for easier reading.* When one or more listed items contain over a line of text, an extra line space between listed items can enhance readability.
    ● *Keep items as short as possible.* Depending on purpose and substance, lists can consist of words, phrases, or sentences—like the list you are reading. Whichever format you choose, pare down the wording as much as possible to retain the impact of the list format.

5. **Parallelism and lead-ins:** Make the listing easy to read by keeping all points grammatically parallel and by including a smooth transition from the lead-in to the listing itself. (The term *lead-in* refers to the sentence or fragment preceding the listing.) "Parallel" means that each point in the list is in the same grammatical form, whether a complete sentence, verb phrase, or noun phrase. If you change form in the midst of a listing, you take the chance of upsetting the flow of information.

EXAMPLE:
"To complete this project, we plan to do the following:

- Survey the site
- Take samples from the three boring locations
- Test selected samples in our lab
- Report on the results of the study"

The listed items are in verb form (note introductory words "survey," "take," "test," "report"). An alternative would be to put them in noun form, with a slightly different lead-in.

EXAMPLE:
"To complete the project, we will perform the following activities:

- Surveying the site
- Taking samples from the three boring locations
- Testing selected samples in our lab
- Reporting on the results of the study"

6. **Punctuation and capitalization:** Although there are acceptable variations on the punctuation of lists, preferred usage includes a colon before a listing, no punctuation after any of the items, and capitalization of the first letter of the first word of each item. Refer to the alphabetized Handbook under "Punctuation: Lists" for alternative ways to punctuate lists.

7. **Overuse:** With listings it *is* possible to have too much of a good thing. Too many lists on one page can create a distracting, fragmented effect. One rule of thumb is to use no more than one or two lists per page. Too many lists on the same page force the reader to decide which one deserves attention first.

In-Text Emphasis

Sometimes you want to emphasize an important word or phrase within a sentence. Computers give you these options: underlining, boldface, italics, and caps. The least effective are FULL CAPS and *italics,* for both are difficult to read within a paragraph and distracting to the eye. The most effective highlighting techniques are <u>underlining</u> or **boldface;** they add emphasis without distracting the reader. This writer's preference is <u>underlining,</u> which draws attention to the word without the overstatedness of bold copy (which can also be hard to read).

Whatever typographical techniques you select, use them sparingly. They can create a "busy" page that leaves the reader confused about what to read. Excessive in-text emphasis also detracts from the impact of headings and subheadings, which should be receiving significant attention.

FONTS AND COLOR

Besides page format, you have something else in your word processing bag of tricks: changes in the size and type of font you use in the text itself, and the use of color.

Size of Type

Traditionally, type size has been measured in "points" (72 points to an inch). When you go to your font-selection menu on your computer screen, the sizes may be listed as such: 9, 10, 12, 14, 18, and 24. Other features of your software will allow you to expand type even farther, for specialty uses.

New Century Schoolbook
9 point
10 point
12 point
14 point
18 point
24 point

Times Roman
9 point
10 point
12 point
14 point
18 point
24 point

Helvetica
9 point
10 point
12 point
14 point
18 point
24 point

Despite these many options, most technical writing is printed off the desktop in 10- or 12-point type. When you are choosing type size, however, be aware that the actual size of letters varies among font types. Some 12-point type appears larger than other 12-point type. Differences stem from the fact that your selection of a font affects (1) the thickness of the letters, (2) the size of lowercase letters, and (3) the length and style of the parts of letters that extend above and below the line. The following examples show the differences in three common fonts. Note that the typeface used in setting the text of this book is 10-point Meridien.

Before selecting your type size, run samples on your printer so that you are certain of how your copy will appear in final form.

Font Types

Your choice of fonts may be either prescribed by your employer or determined by you on the basis of (1) the purpose of the document, (2) the image you want to convey, and (3) your knowledge of the audience. Think of this step as selecting just the right piece of art for your home or office.

Font types are classified into two main groups:

- Serif fonts: Characters have "tails" at the ends of the letterlines.
- Sans serif fonts: Characters do not have tails. (See Figure 4–8.)

If you are able to choose your font, the obvious advice is to use the one that you know is preferred by your readers. A phone call or a look at documents generated by your reader may help you. If you have no reader-specific guidelines, here are three general rules to follow:

1. Use serif fonts for regular text in your documents. The tails on letters make letters and entire words more visually interesting to the reader's eye. In this sense, they serve the same purpose as ragged-edge copy—helping your reader move smoothly through the document.

2. Consider using another typeface—sans serif—for headings. Headings benefit from a clean look that emphasizes the white space around letters. Sans serif type helps attract attention to these elements of organization within your text.

3. Avoid too many font variations in the same document. There is a fine line between interesting font variations and busy and distracting text. But there *is* a line. Your rule of thumb might be to use no more than two fonts per document, one for text and another for headings and subheadings.

FIGURE 4–8   Font types

"The St. Paul Format Guide: Trouble in the River City"

Frustrated by inconsistency in report formats, Elaine Johnson, a department manager at McDuff's St. Paul office, decided to take action. After collecting examples of office reports with diverse page designs, she met with her friend and branch manager, Randall DiSalvo, to complain. "Enough is enough, Randall," Elaine said as the meeting started. "The technical staff produces all kinds of formats, the word processing staff doesn't know what designs are approved, and the clients get a fragmented image of the firm. Let's decide on one page design for reports and stay with it."

After an hour's talk, Elaine and Randall agreed that the office needed a style sheet to describe the required format for each document type written at the St. Paul branch. Busy with many other tasks, Randall told Elaine that he didn't have time to supervise the project. So he gave Elaine the authority to design what he wanted to be called the "Format Guide." However, first she had to meet with all department managers and a few other employees about the project. Also, he asked that her first version of the guide be a modest one that covered only brief letter reports. Later the guide could be expanded.

What follows are details about (1) Elaine's process of gathering information, (2) some actual guidelines she decided to include in the Format Guide, and (3) some problems that arose with the project.

Soliciting Opinions from Around the Office

The same morning she met with Randall, Elaine met with all five of her fellow department managers in the St. Paul office. Gathering in a meeting room overlooking the Mississippi River, they agreed immediately that the format problem needed to be solved.

The managers then concurred with the branch manager's idea about starting small—i.e., covering only short letter reports now but later adding formal reports, proposals, manuals, letters, and memos, along with suggestions on style and grammar. Knowing that Elaine was one of the best writers and editors in the office, the managers said they were comfortable with her writing the manual herself. She could draw from whatever information she gathered from around the office and whatever guidelines she collected from her research on the subject. When the draft was complete, she would run it by them for their comments. Then it would go to Randall for final approval before distribution to all branch employees.

Elaine could hardly believe it. In one morning, five department managers and the branch manager had reached consensus about the nature of the format problem and its solution. Buoyed by her success, Elaine was almost able to look

Continued

beyond the fact that she had been given the job of writing the manual. Always the one to get a job done quickly, however, she moved to the following steps in the next few days:

- Notepad in hand, she interviewed all seven word processors and secretaries about their preferences in page design.
- She emailed all members of the professional staff, asking them to respond in writing in three days if they had specific preferences about the "look" they wanted to achieve in their letter reports.
- She called a local chapter of the Society for Technical Communication (STC), asking for references on page design.
- She located the four sources she received from STC and read them cover to cover.

By the end of the following week, Elaine was ready to begin writing the first draft of what would become the Format Guide for McDuff's St. Paul office.

Elaine's Format for Short Reports

Elaine quickly developed a clear idea of what features should be part of McDuff's short reports. To be sure, some of what she heard from department managers and other employees was at odds with her own views. For example, her preference for ragged-right margins in letter reports differed from that of many colleagues. When there were differences of opinion, she made decisions on the basis of her research and the level of response from those employees she surveyed. The following information summarizes some of the guidelines included in her draft:

1. Font choice should be 12-point New Century Schoolbook.
2. There should be 1.2″ margins on the sides, a $\frac{1}{2}$″ margin on the top, and a 1″ margin on the bottom (except for the first page where letterhead requires the use of a top margin of 2″ and a bottom margin of 1″).
3. Paragraphs should be block style without indenting of the first line.
4. Text should have ragged-right margins, not justified-right margins.
5. Text should be single spaced, with double spacing between paragraphs.
6. Arrangement of date, inside address, and report title should be the same as that used in Model 8–6 in this text.
7. Every page after the first page should include a header. Placed in the top right corner (see margin guideline above), the header should include three single-spaced items: the company name (McDuff, Inc.—St. Paul), the project number (e.g., McDuff Project #134), and the date (e.g., July 29, 2002).
8. The heading system should follow the pattern shown in example 4 in Figure 4–7 of this chapter.

9. Bulleted and numbered lists should be indented $\frac{1}{2}''$ from the left margin, with double spacing before and after the list and between items in the list.

Once her draft was approved by the branch manager, Elaine had 95 copies printed and distributed to all employees of the St. Paul office. In a cover memo with the Format Guide, she indicated that (1) the guide should be considered the new model for all letter reports leaving the office but that (2) it was a pilot draft that the office would review after six months. Further, she noted that the guide would be expanded later to include other documents.

Problems and Questions

Having finished her project by her deadline, Elaine was pleased. There would be clear guidelines for the staff, and the office would reap the rewards of a more efficient process of producing letter reports. As you look back on Elaine's activities and the guidelines she developed, consider the following questions and comments for discussion:

1. Elaine did a good job of seeking opinions of branch employees before she began her draft. Would it have been useful to consult with McDuff customers while the guide was being developed? Why or why not?
2. Should Elaine have "tested" the usefulness of the Format Guide before it was issued to all employees or was her "pilot draft" approach adequate? If you think further testing was needed, what specifically would you have suggested?
3. Elaine chose to issue the final Format Guide through the office mail, with a cover memo. Was this strategy ideal? If so, why? If not, in what other way might she have introduced the manual?
4. Using Elaine's nine format guidelines, edit the letter report included in Model 8–6 in this textbook. Do you think the revision is in a better format than the original? Why or why not?
5. Elaine's manual provides a fairly rigid set of guidelines, as shown by the guidelines excerpt included in this case. Do you think a company should require employees to follow such a narrowly prescribed format? Why or why not? Give the advantages and disadvantages of each point of view.
6. One engineer called Elaine to complain that the new guidelines did not allow him to use decimal numbered headings and subheadings in his short report. He said that he preferred such headings, and he suspected that his clients did as well. If you were in Elaine's position, how would you respond to this complaint?
7. If you were designing a McDuff Format Guide for short reports, what are some of the guidelines you would include, considering your own personal preferences?

Because font selection is an important tool for developing graphics as well as page format, fonts are covered in more detail in chapter 11.

Color

Use of color, like fonts, is a graphic design tool that should reflect the tone, mood, and image of your documents. When used effectively, it focuses your reader's attention on important details. When used indiscriminately—inserted into a document just to show that color can be used—it is distracting.

You'll want to limit your use of color in routine documents for two reasons: (1) compared to black and white, printing color is very expensive and should be used only when it is cost effective, and (2) compared to black and white, printing color documents is very slow and must be scheduled with longer lead times. Use of color is discussed in detail in chapter 11.

CHAPTER SUMMARY

This chapter shows you how to apply principles of page design to your assignments in this class and your on-the-job writing. The term *page design* refers to the array of formatting options you can use to improve the visual effect of your document.

The last several decades have seen an incredible change in the way documents are produced. Today individual writers working at their personal computers write, edit, design, and print sophisticated documents. Used judiciously, computers are an effective tool during the planning, drafting, and revising stages.

Effective page design requires that you use specific elements such as white space, headings, listings, and in-text emphasis. White space draws attention to adjacent items. Headings quickly lead the reader to important points and subpoints. Listings emphasize related groups of points. And conservative use of in-text emphasis, like underlining and bold print, can draw attention to items within sentences and paragraphs.

Another strategy for page design is to change the size and type of fonts in your documents. Like other strategies, this one must be used with care so that your document does not become too "busy." Page design remains a technique for highlighting content, not a substitute for careful organization and editing.

www.prenhall.com/pfeiffer

Access the Technical Writing Web site by pointing your browser at www.prenhall.com/pfeiffer. Here, you will find additional resources to aid you in learning and applying the information from this chapter.

ASSIGNMENTS

1. **Group Evaluation of Page Design.** Working in small groups, analyze the effectiveness of the page design of the document in Model 4–1 on pages 128–129. Your instructor will indicate whether you should prepare a written or an oral report of your findings. Give specific support for your praise or criticism. (Assignment 4 uses this same memorandum for a writing exercise.)

2. **Individual Evaluation of Page Design.**

 Option A. Visit your library and locate an example of technical writing, such as a government document or a company's annual report. Use the guidelines in this chapter to analyze the document's page design. Your instructor will indicate whether your report should be oral or written.

 Option B. Use the guidelines in this chapter to evaluate the page design of one of the color inserts in chapter 2 of this textbook. What works well? What could be improved? Be specific in your comments.

3. **Individual Practice in Page Design.** As a manager at McDuff, you have just finished a major report to a client. It gives recommendations for transporting a variety of hazardous materials by sea, land, and air. The body of your report contains a section that defines the term *stowage plan* and describes its use. Given your mixed technical and nontechnical audience, this basic information is much needed. What follows is the *text* of that section. Revise the passage by applying any of this chapter's principles of page design that seem appropriate—such as adding headings, graphics, lists, and white space. If you wish, you also can make changes in organization and style. *Optional:* If your class has access to email, transmit your version to another student to receive his or her response.

In the chemical shipping industry, a stowage plan is a kind of blueprint for a vessel. It lists all stowage tanks and provides information about tank volume, tank coating, stowed product, weight of product, loading port, and discharging port. A stowage plan is made out for each vessel on each voyage and records all chemicals loaded. The following information concerns cargo considerations (chemical properties and tank features) and some specific uses of the stowage plan in industry.

The three main cargo considerations in planning stowage are temperature, compatibility, and safety. Chemicals have physical properties that distinguish them from one another. To maintain the natural state of chemicals and to prevent alteration of their physical properties, a controlled environment becomes necessary. Some chemicals, for example, require firm temperature controls to maintain their physical characteristics and degree of viscosity (thickness) and to prevent contamination of the chemicals by any moisture in the tanks. In addition, some chemicals, like acids, react violently with each other and should not be stowed in adjoining, or even neighboring, tanks. In shipping, this relationship is known as chemical compatibility.

The controlled environment and compatibility of chemicals have resulted in safety regulations for the handling and transporting of these chemicals. These regulations originate with the federal government, which bases them on research done by the private manufacturers. Location and size of tanks also determine the placement of cargo. A ship's tanks are arranged with all smaller tanks around the periphery of the tank grouping and all larger tanks in the center. These tanks, made of heavy steel and coated with zinc or epoxy, are highly resistant to most chemicals, thereby reducing the chance of cargo contamination. Each tank has a maximum cargo capacity, and the amounts of each chemical are matched with the tanks. Often chemicals to be discharged at the same port are staggered in the stowage plan layout so that after they are discharged the ship maintains its equilibrium.

The stowage plan is finalized after considering the cargo and tank characteristics. In its final form, the plan is used as a reference document with all information relevant to the loading/discharging voyage recorded. If an accident occurs involving a ship, or when questions arise involving discharging operations, this document serves as a visual reference and brings about quick decisions.

4. **Group Practice in Page Design.** Working in small groups, prepare a redesigned version of the memorandum in Model 4–1 (pp. 128–129). If your class is being held in a computer lab, present your group's version on-screen. If you are not using a lab, present your version on an overhead transparency.

5. **Group Practice in Page Design: Using Computer Communication.** This assignment is feasible only if you and your classmates have access to software that will allow you to post messages to team members, edit on screen, and send edited copy back and forth. Your task is to add appropriate page design features to either (a) the "stowage plan" excerpt in assignment 3 or (b) any other piece of straight text permitted for use by your instructor. Choose a team leader who will collect and collate the individual edits. Choose another group member to type or scan the excerpt into the computer and then "mail" the passage to other group members. Then each person should add the features desired and mail the edited document to the team leader, who will collate the revisions and mail the new version to team members for a final edit. Throughout this process, participants may conduct email "conversations" about the draft and resolve differences, if possible, before sending drafts to the leader. The group may need one or two short meetings in person, but most business should be conducted via the computer. The goal is to arrive at one final version for your group.

6. **Individual or Group Practice in Organization and Page Design.** The list that follows mostly includes exact wording or paraphrased excerpts from the National Center for Environmental Health Publication No. 01-0164—March 2001. (Some information has been slightly altered to accommodate this assignment, and much information in the publication has been left out.) Assume that the points are to be included as a *section* of a report you are producing. (NOTE: You are *not* being asked to produce a complete technical report, a subject covered in chapters 8 and 9.) First, arrange the information in an order that generally follows the ABC format described in chapter 3, eliminating any possible redundant information or any items that do not seem to fit the report section you are producing. Second, make adjustments in wording or style you consider appropriate. Third, add appropriate elements of page design that have been covered in this chapter.

    A. This report will be followed by yearly updates. Future reports will attempt to answer the following questions: (1) Are exposure levels increasing or decreasing over time? (2) Are public health efforts to reduce exposure working? and (3) Do certain groups of people have higher levels of exposure than others?

    B. Cotinine is a metabolite of nicotine that tracks exposure to environmental tobacco smoke (ETS) among nonsmokers—higher levels reflect more exposure to ETS.

    C. An environmental chemical is a chemical compound or chemical element in air, water, soil, dust, food, or other environmental media.

    D. Biomonitoring is the assessment of human exposure to environmental chemicals by measuring the chemicals (or their breakdown products) in human specimens, such as blood or urine.

    E. Because the sample size was relatively small and because the sampling was only conducted in 12 locations, the data cannot be considered conclusive. Additional studies should be conducted.

    F. It should be noted that just because people have an environmental chemical in their blood or urine does not mean that the chemical causes disease. Research studies, separate from this report, are required to determine which levels of the chemical may cause disease and which levels are of negligible health concern.

    G. The reduction in cotinine levels reflected in the attached table indicates a dramatic reduction in exposure of the general population to environmental tobacco smoke since 1988–1991.

H. Special populations of children at high risk for lead exposure (for example, those living in homes containing lead-based paint or lead-contaminated dust) remain a major public health concern.

I. The report provides new data on blood mercury levels among children aged 1-5 and among women of childbearing age (16-49).

J. This report needs to be updated each year. The results from all samplings of the 27 chemicals at the 12 locations are included on the attached table. Most of the data will not be considered conclusive—and therefore will not be commented on until we have results for additional years, which will give more support for comments on trends.

K. This report measures the exposure (through biomonitoring) of a sample population to 27 environmental chemicals, which include 13 metals (antimony, barium, beryllium, cadmium, cesium, cobalt, lead, mercury, molybdenum, platinum, thallium, tungsten, and uranium), 6 organophosphate pesticide metabolites, 7 phthalate metabolites, and cotinine.

L. The results showed that levels of some metabolites in the sample population were considerably higher than other metabolites, indicating a need for further study.

M. Compared with an adult, the fetus and child are usually more vulnerable to the effects of metals, such as mercury. One goal of collecting such data is to better estimate health risks for the fetus, children, and women of childbearing age from potential exposures to mercury.

N. One noteworthy conclusion is that blood lead levels continue to decline among U.S. children when considered as a group—highlighting the success of public health efforts to decrease the exposure of children to lead.

O. Since more than half of American youth are still exposed to ETS, it remains a major public health concern.

P. Plans are to expand the list of measured chemicals from 27 to about 100.

7. Ethics Assignment. Page design greatly influences the way people read documents—no matter what message is being delivered. Even a product, a service, or an idea that could be considered harmful to the individual or public good can be made to seem more acceptable by a well-designed piece of writing. Find a well-designed document that promotes what is, in your opinion, a harmful product, service, or idea. Explain why you think the writers made the design decisions they did. Although your example may include graphics, focus mainly on page design covered in this chapter.

8. International Communication Assignment. Collect one or more samples of business or technical writing that originate in—or are designed for—cultures outside the United States. (Use either print examples or examples found on the Internet.) Comment on features of page design in the samples. If applicable, indicate how such features differ from those evident in business and technical writing designed for an audience within the United States.

MEMORANDUM

DATE: August 19, 2002
TO: Randall Demorest, Dean
FROM: Kenneth Payne, Professor and Head *KP*
SUBJECT: BSTW Advisory Board

What? Lunch meetings between Advisory Board members and me
Why?   To get more Board support for the BSTW degree program
Who?   Each individual member at a separate luncheon
When? Fall 2002
How?   Allocation of $360 to pay for the lunches

Rationale

When we seek support for the college, we have to (1) make people feel that they will get something in return and (2) make them feel comfortable about us and our organization. As businesses have demonstrated, one way we can accomplish these goals is by taking potential donors to lunch.

As you and I have discussed, the B.S. in Technical Writing degree program (BSTW) needs to strengthen ties to its Advisory Board. We must ask Board participants to provide tangible support for the program *and* give them meaningful involvement in the work we are doing.

Method

The immediate need is to involve members of the Advisory Board in the coming year's program. I want to do this in two ways:

1. Plan carefully for a fall Board meeting
2. Discuss with each of them individually what we want to accomplish this year

Cost

To do the second item mentioned, I request an allocation of $360 so that I can take each member to lunch for an extended one-on-one discussion. I plan to discuss the needs of our program and each member's capabilities to support it.

MODEL 4–1
Page design in memorandum

Specifics

Each member of the Board will be asked individually to consider the following ways to contribute:

1. Continuing support for the internship program
2. Participation in the research project we began a year ago
3. Cooperative work experiences for BSTW faculty, possibly during the summer of 2003. Financial support for the following items:
   • The college's membership as a sponsoring organization in the Society for Technical Communication
   • Contributions——financial or otherwise——to library holdings in technical writing
   • Usability testing laboratory
   • A workshop series bringing to the campus some outstanding technical communicators (for example, Edward Tufte, expert in graphics; JoAnn Hackos, expert in quality management; and William Horton, expert in online documentation)

Benefits

What are Board members going to get from this?

Long range: A better BSTW program, which will produce better technical writers for them to hire

Immediately: Meaningful involvement in the program

Specifically: Training opportunities for their personnel through the workshops mentioned

My tentative plan for those workshops is to provide a one-day seminar for our students and a second seminar for employees of Advisory Board members. (We will allow them a number of participants based on how much they contribute to the workshops.)

Response Needed

Please let me know as soon as possible if money is available for the lunches. I hope to begin scheduling meetings within a week.

MODEL 4–1
continued

Part

Two

Chapter 5

Patterns of Organization

Revising the Core

General Instructions

Each chapter begins with a Collaboration at Work exercise that applies strategies for working in groups to chapter topics. The exercise assumes you (1) have been divided into groups of about three to six students, (2) will use group time inside or outside of class to complete the case, and (3) will produce an oral or written response. For guidelines about writing in groups, refer to pages 28–34.

Background for Assignment

Like other colleges and universities, your institution may require students to complete a core curriculum of required subjects. Some cores are virtually identical for students in all majors; others vary by major. Following is one example of a core curriculum (required for institutions in the University System of Georgia):

1. **Essential Skills** (9 semester hr): Includes two freshman composition courses and college algebra.
2. **Institutional Options** (4–5 semester hr): Includes courses of the institution's choosing, such as public speaking and interdisciplinary classes.

CHAPTER OUTLINE

Continued

3. **Humanities/Fine Arts** (10–11 semester hr): Includes courses such as literature surveys, art appreciation, music appreciation, and foreign language.

4. **Science, Mathematics, and Technology** (10–11 semester hr): Includes laboratory and non-lab classes in fields such as physics, chemistry, biology, and calculus.

5. **Social Sciences** (12 semester hr): Includes courses such as American history, world history, political science, and religion.

6. **Courses Related to Student's Program of Study** (18 semester hr): Includes lower-level classes related to the student's specific major. For example, an engineering major may be required to take extra math, whereas a technical communication major may be required to take introductory technical writing.

The core curriculum at your institution probably shares the same philosophy as that behind the Georgia core—if not some of the same courses. Goals include, among others, developing intellectual curiosity and providing a basic background in communication, critical thinking, culture, and scientific reasoning.

Group Assignment

Present an argument for keeping or changing the core curriculum at your institution. If your argument includes suggestions for deleting some courses and adding others, give sound reasons for your recommendations.

*P*art I gave you some basic background on technical writing and on McDuff, Inc., the example company used throughout this book. Building on that foundation, Part II focuses on specific types of technical writing. This chapter and the next one cover seven common patterns for organizing information. Then chapters 7–10 discuss the following formats for entire documents: letters and memos, short reports, long reports, and proposals and feasibility studies.

If you have already taken a basic composition course, you will see similarities between patterns studied in that course and those described in this chapter. Indeed, technical writing uses the same building blocks as all other good writing. What follows are separate sections on five of these main patterns: argument, definition, description, classification/division, and comparison/contrast. They are roughly arranged by the frequency with which they are used in technical writing, starting with the most common pattern. Each section contains these four parts:

1. Introduction about the pattern
2. Short case studies from McDuff
3. Writing guidelines
4. Extended example at the end of the chapter, with marginal comments

Then the assignments challenge you to use the chapter's five patterns in the context of some short letter or memo reports.

ARGUMENT

Good argument forms the basis for *all* technical writing. Some people have the mistaken impression that only recommendation reports and proposals argue their case to the reader, and that all other writing should be objective rather than argumentative. The fact is, every time you commit words to paper you are "arguing" your point. This text uses the following broad-based definition for argument:

> **Argument:** the strategies you use in presenting evidence to support your point *and* to support your professional credibility, while still keeping the reader's goodwill.

Thus even the most uncontroversial document, like a trip report, involves argument in the sense defined here. That is, a trip report would present information to support the fact that you accomplished certain objectives on a business trip. It also would show the reader, usually your boss, that you worked hard to accomplish your objectives. In fact, you hope that every document you write becomes a written argument for your own conscientiousness as a professional.

The strongest form of argument—called "persuasion"—tries to convince your reader to adopt a certain point of view or pursue a certain line of action (see chapter 10, "Proposals and Feasibility Studies"). In other words, persuasion seeks obvious *changes* in opinions or actions, whereas argument only presents evidence or logic to support a point of view. To help you use argument correctly, the next section provides you with some short case studies from McDuff, guidelines for applying argument, and an annotated example.

Short Cases from McDuff: Argument

Under the definition just given, many in-house and external documents could be considered argument. Here are three McDuff examples:

⬤ *Argument Case 1: Fire Control in the Everglades*

As a fire-control expert in McDuff's Atlanta office, you just completed a project for the state of Florida. You spent two weeks in the Everglades examining the likelihood for major fires this season. Last season, major fires and westerly winds caused soot and smoke to drift to Florida's tourist area. Your investigation suggests that fires are likely and, in fact, necessary this season. Regular, contained swamp fires help keep down underbrush and thus reduce the chance for much bigger fires. Limited fires will not damage wildlife or major trees, whereas major fires will. Now you must write a report to Florida authorities arguing that controlled fires in the Everglades are desirable, despite the general feeling that all fires should be prevented.

⬤ *Argument Case 2: Back Pain in the Boston Office*

The engineers, programmers, scientists, and other office workers in McDuff's Boston branch spend a lot of time in their chairs. Although all the office furniture is new and expensive, you and your Boston branch colleagues have experienced regular back pain since the new chairs arrived. Unfortunately, the furniture was ordered through the corporate office, so your complaint cannot be handled in a routine, informal way in your own office. As the branch manager, you have mentioned the problem to your boss, the vice president for domestic operations. Predictably, he asked you to "put it in writing." Although this memo report must explain the problem, you do not want to convey a complaining attitude. Instead, you must thoroughly and objectively document the problems associated with the arrival of the new chairs.

⬤ *Argument Case 3: Quality Control in the Labs*

As McDuff's quality-control manager, you operate out of the corporate office in Baltimore. Though friendly with all the branch managers, you have the sensitive job of making sure that the quality of McDuff's products and services remains high. During a recent road trip to every office, you noticed problems at the labs. In particular, testing procedures were not being followed exactly, equipment was sometimes not properly cleaned, and samples from the field were not always labeled clearly. Now you must write a memo report to all lab supervisors noting these lapses.

All three cases involve argument as defined in this chapter. That is, you must collect evidence to support your point, while still keeping the goodwill of your readers.

Guidelines for Writing Argument

Argumentative writing has a long tradition, from the rhetoric of ancient times to the political debate of today. This section describes six guidelines about argument that apply to your on-the-job writing.

● *Argument Guideline 1: Use Evidence Correctly*

Whether writing a proposal to a customer or a memo to your boss, you often move from specific evidence toward a general conclusion supported by the evidence. Called "inductive reasoning," this approach to argument requires that you follow some accepted guidelines. Here are three, along with brief examples:

- **Use points the reader can grasp.** For example, as the fire-control expert in Case 1, you would need to educate readers about how limited fires can benefit swamp ecosystems. Your report must use evidence that could be understood by a mixed audience of park technicians and bureaucrats.

- **Use points that are a representative sample.** For example, as the quality-control expert in Case 3, you would want to have collected evidence from *all* labs mentioned in the memorandum, before sending the memo calling for corrective action. As the writer in Case 2, you would want to have collected information from enough colleagues to support your point about the office furniture.

- **Have enough points to justify your conclusion.** Again from Case 3, you would want to have made *many* observations, not two or three, to justify the memo's conclusions about poor lab procedures.

● *Argument Guideline 2: Choose the Most Convincing Order for Points*

Chapter 3 points out that readers pay most attention to beginnings and endings, not the middle. Thus you should include your stronger points at the beginning and ending of an argumentative sequence, with weaker points in the middle. It follows that the strongest point usually appears in one of these two locations:

- **Last**—in arguments that are fairly short
- **First**—in arguments that are fairly long

Readers finish short passages in one sitting. You can expect them to stay attentive as you lead up to your strongest point. For long passages, however, the audience may take several sittings to read through your argument. For this reason, place the most important evidence at the beginning of longer arguments, where readers will see it before they get distracted.

For Case 2, assume you plan to write a fairly short memo about the 45 new office chairs, as a follow-up to your phone call to your manager at the corporate office. Your main points are that (1) employees have had medical problems leading to frequent sick leave, (2) productivity has suffered among those who are at work, (3) a local expert on health in the workplace has confirmed design flaws in the chairs, and (4) you have contacted a representative of the chair company, who has agreed to trade your chairs for a better model, for a net cost to McDuff of only $2000.

You decide that your fourth point is the strongest because it suggests a feasible solution. The next strongest is the first point because it deals with employee health. Given the brevity of your memo, therefore, you decide to place the medical point first, and the suggested solution last. In between will occur the point about reduced productivity and the expert's views. This arrangement holds the best chance for gaining your reader's support.

● *Argument Guideline 3: Be Logical*

Some arguments rest on a logical structure called a syllogism. When it occurs formally (which is rare), a syllogism may look something like this:

Major Premise: All chairs with straight backs are unacceptable for prolonged office work.

Minor Premise: The chairs at the Boston office have straight backs.

Conclusion: The chairs at the Boston office are unacceptable for prolonged office work.

In fact, syllogisms usually appear in an implied, less-structured fashion than the sample shown here. For example, recall that your McDuff chair report would include a testimonial from an outside expert in office furniture design. That person could have confirmed the major premise, leading to your statement of the minor premise and the conclusion. Indeed, this example shows that a syllogism is effective—that is, logical—*only* if the premises (1) can be verified as true and (2) will be accepted by the reader.

● *Argument Guideline 4: Use Only Appropriate Authorities*

In the technical world, experts often supply evidence that will win the day in reports, proposals, and other documents. Using Case 2 again, your argument would rely on opinions by the expert in workplace health. This testimonial, however, would work only if you could verify that (1) there is a legitimate body of research on the health effects of chair design and (2) the expert you have chosen has credentials that readers will accept.

You will encounter a similar rule in chapter 13, which covers proper documentation in research writing. Such papers also require that you cite sources that will be accepted as valid in the field about which you are writing.

● *Argument Guideline 5: Avoid Argumentative Fallacies*

Guidelines 1–4 mention a few ways you can err in arguing your point. Listed next are some other fallacies that can damage your argument. Accompanying each fallacy is an example from Case 2:

- **Ad hominem** (Latin: "To the man"): arguing against a person rather than discussing the issue. (Example: suggesting that the chairs should be replaced because the purchasing agent at the Baltimore office, who ordered the chairs, is surly and incompetent.)
- **Circular reasoning:** failing to give any reason for why something is or is not true, other than stating that it is. (Example: proposing that the straight-backed chairs should be replaced simply because they are straight-backed chairs.)
- **Either/or fallacy:** stating that only two alternatives exist—yours and another one that is much worse—when in fact there are other options. (Example: claiming that if the office does not purchase new chairs next week, 15 employees will quit.)
- **False analogy:** suggesting that one thing should be true because of its similarity to something else, when in fact the two items are not enough alike to justify the analogy. (Example: suggesting that the chairs in the Boston office should be replaced because the Atlanta office has much more expensive chairs.)
- **Hasty generalization:** forming a generalization without adequate supporting evidence. (Example: proposing that all Jones office furniture will be unacceptable, simply because the current chairs carry the Jones trademark.)
- **Non sequitur** (Latin: "It does not follow"): making a statement that does not follow logically from what came before it. (Example: claiming that if the office chairs are not replaced soon, productivity will decrease so much that the future of the office will be in jeopardy—even though there is no evidence that the chairs would have this sort of dramatic effect on business.)
- **Post hoc ergo propter hoc** (Latin: "After this, therefore because of this"): claiming a cause-effect relationship between two events simply because one occurred before the other. (Example: associating the departure of an excellent secretary with the purchase of the poorly designed chairs a week earlier—even though no evidence supports the connection.)

● *Argument Guideline 6: Refute Opposing Arguments*

Often you will be faced with arguments that oppose your views. Perhaps your proposition will cost a great deal of money, take considerable time to complete, or risk alienating employees with a sudden change in procedure. Effective argument requires that you strive to do the following:

- identify the strongest opposing points
- counter these points with effective refutation

In the process, be certain to maintain an even, unemotional tone that does not irritate those with opposing views. Your goal is to convince with facts and analysis. For example, you might choose to counter a point about the high short-term cost of your proposal with a response that emphasizes the long-term benefits to your organization.

Example of Argument

At the end of this chapter there is a complete memo report resulting from Case 2 discussed previously (see Model 5–1, pp. 167–168). As the branch manager in McDuff's Boston office, you are writing to your supervisor in the Baltimore corporate office. From your brief phone conversation, you expect that he supports the recommendation in your memo. Yet you know that he will show the memo to a few other corporate colleagues, including the procurement officer who ordered the chairs. Thus you cannot assume support. You must present good arguments, in a tactful manner.

DEFINITION

During your career, you will use technical terms known only to those in your profession. As a civil engineer, for example, you would know that a "triaxial compression test" helps determine the strength of soil samples. As a computer professional, you would know the meaning of RAM (random access memory), ROM (read only memory), and LAN (local area network). When writing to readers unfamiliar with these fields, however, you would need to define technical terms.

Good definitions can support findings, conclusions, and recommendations throughout your document. They also keep readers interested. Conversely, the most organized, well-written report will fall on deaf ears if it includes terms that readers do not grasp. "Define your terms!" is the frustrated exclamation of many a reader. For your reader's sake, then, you need to be asking questions like these about definitions:

- How often do you use them?
- Where should they be placed?
- What format should they take?
- How much information is enough and how much is too much?

To answer these questions, the following sections give guidelines for definitions and supply an annotated example. First, here are some typical contexts for definitions within McDuff, Inc.

Short Cases from McDuff: Definitions

These cases may help you visualize the use of definitions in your own on-the-job writing.

● *Definition Case 1: New Power Plant*

As a McDuff technician helping to build a power plant, you often use the term *turbine* in speech and writing. Obviously your coworkers and clients understand the term. Now, however, you are using it in a plant brochure to be sent to consumers.

For this general audience, you decide to define "turbine" as follows: "a machine that receives energy from a moving fluid and then uses this energy to make a shaft rotate." You accompany the definition with an illustration so that the unfamiliar audience can visualize how the turbine works.

● *Definition Case 2: Health Bulletin to Employees*

As a health and benefits specialist in McDuff's corporate office, you have watched the company's health insurance costs increase dramatically over the last five years. Much of this cost has not been passed on to the employee, resulting in a drain on profits. You have convinced management to support a major campaign to encourage healthy habits by employees. The campaign might include these free benefits: in-house quit-smoking seminars, stress-reduction sessions, and annual physicals.

Your first project, however, is a health memorandum to all employees. In it you must (1) explain the nature of the cholesterol problem and (2) invite employees to free cholesterol screening tests sponsored by the company. Your memo must provide clear definitions for terms like "good" cholesterol and "bad" cholesterol.

● *Definition Case 3: Forestry Report*

As a forestry and agriculture expert with McDuff's Denver office, you have coordinated a major study for the state of Idaho. Your job has been to recommend ways that a major forested region can still be used for timber with little or no damage to the region's ecological balance.

Although the report will go first to technical experts in Idaho's Department of Natural Resources, you have been told that it must be understood by other potential readers like politicians, bureaucrats, and citizen groups. Thus you have decided to include a glossary that defines terms such as these:

- **Silvaculture** (cultivation of forests to supply a renewable crop of timber)
- **Shelterwood cutting** (removal of only mature trees in several cuttings over a few decades)
- **Seed-tree cutting** (removal in one cutting of all but a few trees, which are left to regenerate the forest)
- **Clearcutting** (removal of all trees in an area in one cutting)*

These three cases all require well-placed definitions for their readers. When in doubt, insert definitions! Readers can always skip over ones they do not need.

Guidelines for Writing Definitions

Once you know definitions are needed, you must decide on their format and location. Again, consider your readers. How much information do they need? Where is this information best placed within the document? To answer these and other questions, here are five working guidelines for writing good definitions.

---

*Adapted from *Environmental Science: An Introduction*, 2/e by G. Tyler Miller, Jr. © 1988 by Wadsworth, Inc. Reprinted by permission of the publisher.

● *Definition Guideline 1: Keep It Simple*

Occasionally the sole purpose of a report is to define a term. Most times, however, a definition just clarifies a term in a document with a larger purpose. Your definitions should be as simple and unobtrusive as possible. Always present the simplest possible definition, with only that level of detail needed by the reader.

For example, in writing to a client on your land survey of her farm, you might briefly define a transit as "the instrument used by land surveyors to measure horizontal and vertical angles." The report's main purpose is to present property lines and total acreage, not to give a lesson in surveying, so this sentence definition is adequate. Choose from these three main formats (listed from least to most complex) in deciding the form and length of definitions:

- **Informal definition:** a word or brief phrase, often in parentheses, that gives only a synonym or other minimal information about the term
- **Formal definition:** a full sentence that distinguishes the term from other similar terms and that includes these three parts: the term itself, a class to which the term belongs, and distinguishing features of the term
- **Expanded definition:** a lengthy explanation that begins with a formal definition and is developed into several paragraphs or more

Guidelines 2–5 show you when to use these three options and where to put them in your document.

● *Definition Guideline 2: Use Informal Definitions for Simple Terms Most Readers Understand*

Informal definitions appear right after the terms being defined, often as one-word synonyms in parentheses. They give just enough information to keep the reader moving quickly. As such, they are best used with simple terms that can be adequately defined without much detail.

Here is a situation in which an informal definition would apply. McDuff has been hired to examine a possible shopping-mall site. The buyers, a group of physicians, want a list of previous owners and an opinion about the suitability of the site. As legal assistant at McDuff, you must assemble a list of owners in your part of the group-written report. You want your report to agree with court records, so you decide to include real-estate jargon such as "grantor" and "grantee." For your nontechnical readers, you include parenthetical definitions like these:

> All **grantors** (persons from whom the property was obtained) and **grantees** (persons who purchased the property) are listed on the following chart, by year of ownership.

This same McDuff report has a section describing creosote pollution found at the site. The chemist writing the contamination section also uses an informal definition for the readers' benefit:

> At the southwest corner of the mall site, we found 16 barrels of **creosote** (a coal tar derivative) buried under about three feet of sand.

The readers do not need a fancy chemical explanation of creosote. They only need enough information to keep them from getting lost in the terminology. Informal definitions perform this task nicely.

● Definition Guideline 3: Use Formal Definitions for More Complex Terms

A formal definition appears in the form of a sentence that lists (1) the **term** to be defined, (2) the **class** to which it belongs, and (3) the **features** that distinguish the term from others in the same class. Use it when your reader needs more background than an informal definition provides. Formal definitions define in two stages:

- First, they place the term into a *class* (group) of similar items.
- Second, they list *features* (characteristics) of the term that separate it from all others in that same class.

In the list of sample definitions that follows, note that some terms are tangible (like "pumper") and others are intangible (like "arrest"). Yet all can be defined by first choosing a class and then selecting features that distinguish the term from others in the same class.

Term	Class	Features
An arrest is	restraint of persons	that deprives them of freedom of movement and binds them to the will and control of the arresting officer.
A financial statement is	a historical report about a business	and is prepared by an accountant to provide information useful in making economic decisions, particularly for owners and creditors.
A triaxial compression test is	a soils lab test	that determines the amount of force needed to cause a shear failure in a soil sample.
A pumper	is a fire-fighting apparatus	used to provide adequate pressure to propel streams of water toward a fire.

This list demonstrates three important points about formal definitions. First, the definition itself must not contain terms that are confusing to your readers. The definition of "triaxial compression test," for example, assumes readers will understand the term *shear failure* that is used to describe features. If this assumption were incorrect, then the term would need to be defined. Second, formal definitions may be so long that they create a major distraction in the text. (See Guideline 5 for alternative locations.) Third, the class must be narrow enough so that you will not have to list too many distinguishing features.

● *Definition Guideline 4: Use Expanded Definitions
for Supporting Information*

Sometimes a parenthetical phrase or formal sentence definition is not enough. If readers need more information, use an expanded definition with this three-part structure:

- **An overview at the beginning**—including a formal sentence definition and a
  description of the ways you will expand the definition
- **Supporting information in the middle**—using headings and lists as helpful format devices for the reader
- **Brief closing remarks at the end**—reminding the reader of the definition's relevance to the whole document

Here are seven ways to expand a definition, along with brief examples:

1. **Background and/or history of term**—expand the definition of "triaxial compression test" by giving a dictionary definition of "triaxial" and a brief history of
   the origin of the test
2. **Applications**—expand the definition of "financial statement" to include a
   description of the use of such a statement by a company about to purchase con-
   trolling interest in another
3. **List of parts**—expand the definition of "pumper" by listing the parts of the
   device, such as the compressor, the hose compartment, and the water tank
4. **Graphics**—expand the description of the triaxial compression test with an illustration showing the laboratory test apparatus
5. **Comparison/contrast**—expand the definition of a term like "management by
   objectives" (a technique for motivating and assessing the performance of employees) by pointing out similarities and differences between it and other management techniques
6. **Basic principle**—expand the definition of "ohm" (a unit of electrical resistance
   equal to that of a conductor in which a current of one ampere is produced by a
   potential of one volt across its terminals) by explaining the principle of Ohm's
   Law (that for any circuit the electric current is directly proportional to the voltage and inversely proportional to the resistance)
7. **Illustration**—expand the definition of CAD/CAM (Computer-Aided Design/
   Computer-Aided Manufacturing—computerized techniques to automate the
   design and manufacture of products) by giving examples of how CAD/CAM is
   changing methods of manufacturing many items, from blue jeans to airplanes

Obviously, long definitions might seem unwieldy within the text of a report, or
even within a footnote. For this reason, they often appear in appendices, as noted in
the next guideline. Readers who want additional information can seek them out,
whereas other readers will not be distracted by digressions in the text.

● *Definition Guideline 5: Choose the Right Location for Your Definition*

Short definitions are likely to be in the main text; long ones are often relegated to
footnotes or appendices. However, length is not the main consideration. Think first

about the *importance* of the definition to your reader. If you know that decision-makers reading your report will need the definition, then place it in the text—even if it is fairly lengthy. If the definition only provides supplementary information, then it can go elsewhere. You have these five choices for locating a definition:

1. **In the same sentence as the term,** as with an informal, parenthetical definition
2. **In a separate sentence,** as with a formal sentence definition occurring right after a term is mentioned
3. **In a footnote,** as with a formal or expanded definition listed at the bottom of the page on which the term is first mentioned
4. **In a glossary at the beginning or end of the document,** along with all other terms needing definition in that document
5. **In an appendix at the end of the document,** as with an expanded definition that would otherwise clutter the text of the document

Example of Expanded Definition

Expanded definitions are especially useful in reports from technical experts to non-technical readers. McDuff's report writers, for example, often must explain environmental, structural, or geological problems to concerned citizens or nontechnical decision-makers. Figure 5–1 gives a definition that might appear in a report to a county government about the placement of a landfill.

Starts with formal sentence definition—including term, class, and features.

Indicates way definition will be developed (description of acute and chronic types).

Gives examples of first type.

Gives examples of second type.

A **toxic substance** is a chemical that is harmful to people or other living organisms. The effects from exposure to a toxic substance may be acute or chronic. Acute effects are those that appear shortly after exposure, usually to a large concentration or dose over a short time. Examples are skin burns or rashes, eye irritation, chest pains, kidney damage, headache, convulsions, and death.

Effects that are delayed and usually long-lasting are called "chronic" effects. They may not appear for months or years after exposure and usually last for years. Examples are cancer, lung and heart disease, birth defects, genetic defects, and nerve and behavioral disorders. Chronic effects often occur as a result of prolonged exposure to fairly low concentrations or doses of a toxin. However, they may occur as the delayed effects of short-term exposure to high doses.

FIGURE 5–1 Definitions in a report
Excerpted from *Living in the Environment: An Introduction to Environmental Science,* 6/e by G. Tyler Miller, Jr. © 1990 by Wadsworth, Inc. Reprinted by permission of the publisher.

DESCRIPTION

Technical descriptions, like expanded definitions, require that you pay special attention to *details*. In fact, you can consider a description to be a special type of definition that focuses on parts, functions, or other features. It emphasizes *physical* details. Descriptions can appear in any part of a document, from the introduction to the appendix. As a rule, however, detailed descriptions tend to be placed in the technical sections.

To help you write accurate descriptions, this section describes case studies, offers guidelines for writing descriptions, and introduces two annotated examples of equipment descriptions.

Short Cases from McDuff: Descriptions

Descriptions often appear as supporting information in the document body or in appendices. Here are three situations from McDuff in which a detailed description would add to the effectiveness of the complete document.

● *Description Case 1: Possible Harbors*

McDuff's San Francisco office has been hired to recommend possible locations for a new swimming and surfing park in northern California. Written to a county commission (five laypersons who will make the decision), your report gives three possible locations and your reasons for selecting them. Then it refers to appendices that give brief physical descriptions of the sites.

Specifically, your appendices to the report describe (1) surface features, (2) current structures, (3) types of soils gathered from the surface, (4) water quality, and (5) aesthetic features, such as quality of the ocean views. Language is kept nontechnical, considering the lay background of the commissioners.

● *Description Case 2: Sonar Testing Equipment*

A potential McDuff client, Rebecca Stern, calls you in your capacity as a geologist at McDuff's Baltimore office. She wants information about the kind of sonar equipment McDuff uses to map geological features on the seafloor.

This client has a strong technical background, so you write a letter with a detailed technical description of the McDuff system. The body of the letter describes the locations and functions of (1) the seismic source (a device, towed behind a boat, that sends the sound waves) and (2) the receiver (a unit, also towed behind the boat, that receives the signals).

● *Description Case 3: Asbestos Site*

McDuff's Cleveland office was hired to examine asbestos contamination in a large high school built in 1949. As a member of the investigating team, you found asbestos throughout the basement in old pipe coverings. Your final report to the school board provides conclusions about the level of contamination and recommendations

for removal. An appendix gives a detailed technical description of the entire basement, including a map with a layout of the plumbing system.

Guidelines for Writing Descriptions

Now that you know how descriptions fit into entire documents, here are some simple guidelines for writing accurate, detailed descriptions. Follow them carefully as
you prepare assignments in this class and on the job.

● *Description Guideline 1: Remember Your Readers' Needs*

The level of detail in a technical description depends on the purpose a description
serves. Give readers precisely what they need—but no more. In the harbor description in Case 1, the commissioners do not want a detailed description of soil samples
taken from borings. That level of detail will be reserved for a few sites selected later
for further study. Instead, they want only surface descriptions. Always know just how
much detail will get the job done.

● *Description Guideline 2: Be Accurate and Objective*

More than anything else, readers expect accuracy in descriptions. Pay close attention to details. (As noted previously, the *degree* of detail in a description depends on
the *purpose* of the document.) In the building-floor description in Case 3, for example, you would want to describe every possible location of asbestos in the school basement. Because the description will become the basis for a cost proposal to remove
the material, accuracy is crucial.

Along with accuracy should come objectivity. This term is more difficult to pin
down, however. Some writers assume that an objective description leaves out all
opinion. This is not the case. Instead, an objective description may very well include
opinions that have these features:

- They are based on your professional background.
- They can be justified by the time you have had to complete the description.
- They can be supported by details from the site or object being described.

For example, your description of the basement pipes mentioned in Case 3 might
include a statement like this: "Because there is asbestos wrapping on the exposed
pipes above the boiler, my experience suggests that asbestos wrapping probably also
exists around the pipes above the ceiling—in areas that we were not able to view."
This opinion does not reduce the objectivity of your description; it is simply a logical conclusion based on your experience.

● *Description Guideline 3: Choose an Overall Organization Plan*

Like other patterns discussed in this chapter, technical descriptions usually make up
only parts of documents. Nevertheless, they must have an organization plan that permits them to be read as self-contained, stand-alone sections. Indeed, a description
may be excerpted later for separate use.

Following are three common ways to describe physical objects and events. In all three cases, a description should move from general to specific. That is, you begin with a view of the entire object or event. Then in the rest of the description, you focus on specifics. Headings may be used, depending on the format of the larger document.

1. **Description of the parts:** For many physical objects, like the basement floor and coastal scene in the previous cases, you will simply organize the description by moving from part to part.
2. **Description of the functions:** Often the most appropriate overall plan relies on how things work, not on how they look. In the sonar example, the reader was more interested in the way that the sender and receiver worked together to provide a map of the seafloor. This function-oriented description would include only a brief description of the parts.
3. **Description of the sequence:** If your description involves events, as in a police officer's description of an accident investigation, you can organize ideas around the major actions that occurred, in their correct sequence. As with any list, it is best to place a series of many activities into just a few groups. Four groups of 5 events each is much easier for readers to comprehend than a single list of 20 events.

● *Description Guideline 4: Use "Helpers" Like Graphics and Analogies*

The words of a technical description need to come alive. Because your readers may be unfamiliar with the item, you must search for ways to connect with their experience and with their senses. Two effective tools are graphics and analogies.

Graphics respond to the desire of most readers to see pictures along with words. As readers move through your part-by-part or functional breakdown of a mechanism, they can refer to your graphic aid for assistance. The illustration helps you too, of course, in that you need not be as detailed in describing locations and dimensions of parts when you know the reader has easy access to a visual. Note how the diagrams in Models 5–2 and 5–3 on pages 169–173 give meaning to the technical details in the verbal descriptions.

Analogies, like illustrations, give readers a convenient handle for understanding your description. Put simply, an analogy allows you to describe something unknown or uncommon in terms of something that is known or more common. A brief analogy can sometimes save you hundreds of words of technical description. This paragraph description contains three analogies:

McDuff, Inc., is equipped to help clean up oil spills with its patented product, SeaClean. This highly absorbent chemical is spread over the entire spill by means of a helicopter, which makes passes over the spill much like a lawn mower would cover the complete surface area of a lawn. When the chemical contacts the oil, it acts like sawdust coming in contact with oil on a garage floor. That is, the oil is immediately absorbed into the chemical and physically transformed into a product that is easily collected. Then our nearby ship can collect the product, using a machine that operates much like a vacuum cleaner. This machine sucks the SeaClean (now full of oil) off the surface of the water and into a sealed container in the ship's hold.

First analogy

Second analogy

Third analogy

● *Description Guideline 5: Give Your Description the "Visualizing Test"*

After completing a description, test its effectiveness by reading it to someone unfamiliar with the material—someone with about the same level of knowledge as your intended reader. If this person can draw a rough sketch of the object or events while listening to your description, then you have done a good job. If not, ask your listener for suggestions to improve the description. If you are too close to the subject yourself, sometimes an outside point of view will help refine your technical description.

Examples of Description

This section introduces two descriptions from McDuff documents, one written at a U.S. office and the other produced overseas. The descriptions themselves are at the end of the chapter. The first example includes a description of physical parts and a brief operating procedure. (For a thorough discussion of process descriptions and instructions, see chapter 6.) The second includes only a description of physical parts. Both show the importance of graphics in technical descriptions.

● *Description 1: Blueprint Machine*

Donna Millsly, human resources coordinator at McDuff's St. Paul office, has been asked to assemble an orientation guide for new secretaries, office assistants, and other members of the office staff. The manual will contain descriptions and locations of the most common pieces of equipment in the office, sometimes with brief instructions for their use.

One section of her manual includes a series of short equipment descriptions organized by general purpose of the equipment. Model 5–2 on pages 169–170 includes a description of the office blueprint machine, which gets used by a wide variety of employees. Because the blueprint machine has an operating procedure that is not self-evident, the description includes a brief procedure that shows the manner in which the machine should be used.

● *Description 2: Bunsen Burner*

Fahdi Ahmad, the director of procurement at McDuff's office in Saudi Arabia, is changing his buying procedures. He has decided to purchase some basic lab supplies from companies in nearby developing nations, rather than from firms in industrialized countries. For one thing, he thinks this move may save some money. For another, he believes it will help the company get more projects from these nations, for McDuff will become known as a firm that pumps back some of its profits into the local economies.

As a first step in this process, Fahdi is providing potential suppliers with descriptions of some basic lab equipment used at McDuff, such as pH meters, laboratory scales, glass beakers, and burners. Model 5–3 on pages 171–173 presents a moderately detailed description of one such piece of equipment—a Bunsen burner. Fahdi selected a burner model that is being successfully used at many McDuff offices in the United States. The Model 5–3 description can serve as a starting point for suppliers.

However, McDuff and the suppliers realize that burners will have slightly different features, depending on the manufacturer.

CLASSIFICATION/DIVISION

In technical writing, you often perform these related tasks: (1) grouping lists of items into categories, a process called "classification," or (2) separating an individual item into its parts, a process called "division." In practice, the patterns usually work together. So that you can see classification and division in action, this section starts with some case studies from McDuff. Then it presents basic guidelines and an example for classification, followed by basic guidelines and an example for division.

Short Cases from McDuff: Classification/Division

The first case shows how you might use both classification and division in the same context. The other cases show only one pattern at work.

● *Classification/Division Case 1: Civil-Engineering Projects*

A major client needs some detailed information before deciding between McDuff and a competitor for a big project. Specifically, the firm wants a detailed description of the civil-engineering projects completed by the Caracas office over the last 10 years.

- First, you *divide* the office's civil-engineering capabilities into these five groupings, based on type of project: (1) city planning, (2) construction, (3) transportation, (4) sanitation, and (5) hydraulics. (In each case, you describe the type of work done. For example, hydraulics projects concern the use of water.)
- Second, you collect descriptions of the 214 projects completed.
- Third, you *classify* the 214 projects into the five groupings mentioned. Thus the combined process of division and classification gives you a way to organize details for the client.

● *Classification/Division Case 2: Birds Near a Sub Base*

The federal government wants to expand a submarine base off the U.S. southeastern coast. Plans went along fine until last year, when a major environmental group raised questions about the effect the expansion would have on bird nesting areas.

The concerned organization, BirdWatch, hired McDuff to determine the environmental impact of increased base traffic on the bird populations of a nearby national seashore. As project director, you determine that 61 species of birds on the island could be affected by the expansion. For the purposes of your report, you classify the 61 species by physical features into the following groupings: (1) loons; (2) grebes; (3) gulls and terns; (4) cranes, rails, and coots; and (5) ducks, geese, and swans. These classifications, often used in ornithological work, will help readers understand the significance of your environmental impact statement.

● *Classification/Division Case 3: Employee Grievances*

Rob McDuff wants to change the procedure by which employees can register grievances. He prefers that a committee be appointed to hear all grievances. Before he makes his final decision, however, he wants you to submit a summary of the types of grievances filed since the company was founded in 1954. Your painstaking research uncovers 116 separate grievances for which there is paperwork. For easier reading, your report to the company president classifies the 116 incidents into four main categories, by type: (1) firing or discharge from the firm; (2) use of seniority in layoffs, promotions, or transfers; (3) yearly performance evaluations; and (4) overtime pay and the issue of required overtime at some offices.

Guidelines for Classification

Classification helps you (and of course your reader) make sense out of diverse but related items. The process of outlining a writing assignment requires the use of classification. Outlining (as described in chapter 1) forces you to apply both classification and division to organize diverse information into manageable "chunks." The guidelines presented here provide a three-step procedure for classifying any group of related items.

● *Classification Guideline 1: Find a Common Basis*

Classification requires that you establish your groupings on one main basis. This basis can relate to size, function, purpose, or any other factor that serves to produce logical groupings. For example, here are the bases for groupings established in the three cases cited:

Case 1 basis: the *type of work* being done in the civil-engineering field at McDuff (such as sanitation or city planning)

Case 2 basis: the *physical features* of the marsh or ocean birds found at the site (note that the classifications in this case are not original but rather are ones commonly used in ornithology)

Case 3 basis: the *purpose for the grievance* (that is, the grievances can be grouped under four main reasons why the employees made formal complaints to the company)

● *Classification Guideline 2: Limit the Number of Groups*

In chapter 4, you learned that readers prefer groupings of no more than nine items—the fewer the better. This principle of organization, as well as the appropriateness of the basis, should guide your use of classification. Strive to select a basis that will result in a limited number of groupings.

In Case 1, another basis for classifying the 214 civil-engineering projects would be by country in which the project was located, rather than type of work. Assume that after applying this project-site basis, you end up with 18 different classifications (that is, countries) for the 214 projects. Given this unwieldy number of groupings, however, you should either (1) avoid using this basis or (2) reduce the number of classifications by grouping countries together, perhaps by continent. In other words, the number of classifications affects the degree to which this pattern of organization succeeds with the reader.

● *Classification Guideline 3: Carefully Classify Each Item*

The final step is to place each item in its appropriate classification. If you have chosen classifications carefully, this step is no problem. For example, assume that as McDuff's finance expert, you want to identify sources for funding new projects. Therefore, you need to collect the names of all commercial banks started in Pennsylvania last year. A reference librarian gives you the names of 42 banks. To impose some order, you decide to classify them by charter. The resulting groups are as follows:

1. National banks (those chartered in all states)
2. Non-Federal Reserve state banks (exclusively Pennsylvania banks that are not members of the Federal Reserve System)
3. Federal Reserve state banks (exclusively Pennsylvania banks that are members of the Federal Reserve System)

With three such classifications that do not overlap, you can put each of the 42 banks in one of the three classifications.

Example of Classification

Now apply the three-part classification strategy to a detailed problem at McDuff. You are on a committee to write an outline summary of McDuff's safety practices. Since much of the company's work involves hazardous materials, you want to place special emphasis on techniques used to protect workers from toxic chemicals, polluted air, asbestos fibers, and so forth. This outline will be used as an attachment to sales letters and proposals sent to potential clients, such as the federal government and various state governments. Good organization is crucial so that the reader can locate information quickly. At its first meeting, your committee comes up with this list of topics to be included in the outline:

1. Headgear provided by company
2. Complete physicals done yearly
3. Partial physicals done at six-month intervals
4. Blood monitoring throughout work at hazardous sites
5. State-of-the-art face masks and oxygen equipment
6. Twenty-hour safety program during employment orientation
7. Certified safety official on-site during hazardous projects
8. Pre-employment drug testing
9. Random drug testing of all workers in dangerous jobs
10. State-of-the-art hand and body protection equipment
11. Biweekly training sessions on safety-related topics
12. Pre-employment complete physical
13. Monthly in-house newsletter with safety column
14. Incentive awards for employees with safety suggestions
15. Company membership in national safety organizations

Next the committee (1) selects an appropriate basis (the *time* at which these safety precautions should take place), (2) establishes three main groupings, (3) classifies each of the 15 items into one of the groups, and (4) assembles the following outline for use in McDuff sales literature. (Numbers in parentheses refer to the topics just listed.)

I. Procedures before employment
    A. Twenty-hour safety program during orientation (6)
    B. Pre-employment drug testing (8)
    C. Pre-employment complete physical (12)
II. Procedures during a project
    A. Headgear provided by company (1)
    B. Blood monitoring throughout work at hazardous sites (4)
    C. State-of-the-art face masks and oxygen equipment (5)
    D. Certified safety official on-site during hazardous projects (7)
    E. State-of-the-art hand and body protection equipment (10)
III. Procedures that take place periodically during one's employment at McDuff
    A. Complete physicals done yearly (2)
    B. Partial physicals done at six-month intervals (3)
    C. Random drug testing of all workers in dangerous jobs (9)
    D. Biweekly training sessions on safety-related topics (11)
    E. Monthly in-house newsletter with safety column (13)
    F. Incentive awards for employees with safety suggestions (14)
    G. Company membership in national safety organizations (15)

Guidelines for Division

Division begins with an entire item that must be *broken down* or *partitioned* into its parts, whereas classification begins with a series of items that must be *grouped* into related categories. Division is especially useful when you need to explain a complicated piece of equipment to an audience unfamiliar with it. Follow these three guidelines for applying this pattern:

● *Division Guideline 1: Choose the Right Basis for Dividing*

Like classification, division means you must find a logical reason for establishing groups or parts. Assume, for example, that you are planning to teach a McDuff training seminar in project management. In dividing this four-day training seminar into appropriate segments, it seems clear to you that each day should cover one of these crucial parts of managing projects: meeting budgets, scheduling staff, completing written reports, and seeking follow-up work from the client. In this case, the principle of division seems easy to employ.

Yet other cases present you with choices. If, for example, you were planning a training seminar on report writing, you could divide it in these three ways, among others: (1) by *purpose of the report* (for example, progress, trip, recommendation), (2) by *parts of the writing process* (for example, brainstorming, outlining, drafting, revising), or (3) by *report format* (for example, letter report, informal report, formal report). Here you would have to choose the basis most appropriate for your purpose and audience.

● *Division Guideline 2: Subdivide Parts When Necessary*

As with classification, the division pattern of organization can suffer from the "laundry list" syndrome. Specifically, any particular level of groupings should have from three to seven partitions—the number that most readers find they can absorb. When you write more than that number, consider reorganizing information or subdividing it.

Assume that you want to partition a short manual on writing formal proposals. Your first effort to divide the topic results in nine segments: cover page, letter of transmittal, table of contents, executive summary, introduction, discussion, conclusions, recommendations, appendices. You would prefer fewer groupings, so you then establish three main divisions: front matter, discussion, and back matter, with breakdowns of each. In other words, your effort to partition was guided by every reader's preference for a limited number of groupings.

● *Division Guideline 3: Describe Each Part with Care*

This last step may seem obvious. Make sure to give equal treatment to each part of the item or process you have partitioned. Readers expect this sort of parallelism, just as they prefer the limited number of parts mentioned in Guideline 2.

Example of Division

You are a McDuff manager helping to prepare a client's report. The main task is to describe major types of offshore oil rigs. You decide to partition the subject on the basis of *environmental application,* which gives you five main types. In the following list, notice that each of the five is described in a parallel fashion: the rig's purpose, type of structure, and design with respect to wave strength.

Five Main Types of Offshore Oil Rigs

1. **Platform rig:** Generally for drilling at water depths of less than 1000 ft/ Permanent structure supported by steel and concrete legs driven into ocean floor/ Designed to withstand waves of about 50 ft
2. **Submersible rig:** Generally for drilling at water depths of less than 100 ft/ Temporary structure supported by large tanks that, when filled, go to the ocean floor and thus form the foundation for the columns extending to the deck structure above/ Designed to withstand waves of about 30 ft
3. **Semisubmersible rig:** Generally for drilling in extreme water depths of 1000 ft or more/ Temporary structure with large water-filled pontoons that keep it suspended just below the water surface, with anchors or cables extending to the ocean bottom/ Designed to withstand extreme wave heights of 90 ft or more
4. **Drillship:** Generally for drilling at water depths between 100 and 1000 ft/ Drilling operations take place through opening in the middle of the ship/ Designed to operate in relatively low wave heights of less than 30 ft
5. **Jack-up rig:** Generally for drilling at water depths between 25 and 500 ft/ Temporary structure with platformlike legs that can be jacked up and down to rest on the ocean surface, much like the jacking system used to elevate a car when changing a tire/Designed to withstand hurricane-force winds that produce waves of over 50 ft

COMPARISON/CONTRAST

Many writing projects obligate you to show similarities or differences between ideas or objects. (For our purposes, the word *comparison* emphasizes similarities, whereas the word *contrast* emphasizes differences.) In the real world of career writing, this technique especially applies to situations wherein readers are making buying decisions. To help you write effective comparisons, this section presents several cases from McDuff, puts forth some simple guidelines, and introduces an annotated example that you will find at the end of the chapter.

Short Cases from McDuff: Comparison/Contrast

One of the most common patterns at McDuff, comparison/contrast is used in many in-house and external reports. Examples of each are noted here.

● *Comparison/Contrast Case 1: Jamaica Hotels*

As an architect for McDuff, you have had the good fortune of traveling to Jamaica three times on business in the last year. Each time you stayed at a different hotel on Montego Bay. Now your boss, Byron Scarsdale, is going there for an extended business trip. Before making his travel arrangements, he asks you for a memorandum that compares and contrasts main features of the three hotels. Presumably, he will use the information to decide where he will stay.

● *Comparison/Contrast Case 2: Security Systems*

One of your jobs as McDuff's corporate purchasing agent is to advise managers. The manager of the London branch has asked that you send him a summary of similarities and differences among five high-quality infrared detectors. He wants to install several in a new computer lab, to detect movement of any after-hours intruders and then to send a signal to local police.

● *Comparison/Contrast Case 3: Poisons*

Your job in McDuff's toxicology laboratory brings you in contact with dozens of poisons. One client has asked you to compare and contrast the relative dangers of five poisonous chemicals found in the well water at the site where the client firm wants to build a warehouse. Your report will be used in making the decision to build or not to build.

Guidelines for Comparison/Contrast

The three cases just cited show you real contexts in which you will use the comparison/contrast pattern of organization. Now here are guidelines for writing effective comparisons and contrasts.

● *Comparison/Contrast Guideline 1: Remember Your Purpose*

When using comparison/contrast in on-the-job writing, your purpose usually falls into one of these two categories:

1. **Objective:** essentially an unbiased presentation of features wherein you have no real "axe to grind"
2. **Persuasive:** an approach wherein you compare features in such a way as to recommend a preference

You must constantly remember your main purpose. Either you will provide raw data that someone else will use to make decisions *or* you will urge someone toward your preference. In either case, the comparison must show fairness in dealing with all alternatives. Only in this way can you establish credibility in the eyes of the reader.

● *Comparison/Contrast Guideline 2: Establish Clear Criteria and Use Them Consistently*

In any technical comparison, you must set clear standards of comparison and then apply them uniformly. Otherwise, your reader will not understand the evaluation or accept your recommendation (if there is one).

For example, assume you are a McDuff field supervisor who must recommend the purchase of a bulldozer for construction sites. You have been asked to recommend just *one* of these models: Cannon-D, Foley-G, or Koso-L. After background reading and field tests, you decide upon three main criteria for your comparison: (1) pushing capacity, (2) purchase details, and (3) dependability. These three factors, in your view, are most relevant to McDuff's needs. Having made this decision about criteria, you then must discuss all three criteria with regard to *each* of the three bulldozers. Only in this way can readers get the data needed for an informed decision.

● *Comparison/Contrast Guideline 3: Choose the "Whole-by-Whole" Approach for Short Comparison/Contrasts*

This strategy requires that you discuss one item in full, then another item in full, and so on. Using the bulldozer example, you might first discuss all features of the Cannon, then all features of the Foley, and finally all features of the Koso. This strategy works best if individual descriptions are quite short so readers can remember points made about the Cannon bulldozer as they proceed to read sections on the Foley and then the Koso machines.

Keep these two points in mind if you select the whole-by-whole approach:

- Discuss subpoints in the same order—that is, if you start the Cannon description with information about dependability, begin the Foley and Koso discussions in the same way.
- If you are making a recommendation, move from least important to most important, or vice versa, depending on what approach will be most effective with your

reader. Busy readers usually prefer that you start with the recommended item, followed by the others in descending order of importance.

● *Comparison/Contrast Guideline 4: Choose the "Part-by-Part" Approach for Long Comparison/Contrasts*

Longer comparison/contrasts require readers to remember much information. Thus readers usually prefer that you organize the comparison around major criteria, *not* around the whole items. Using the bulldozer example, your text would follow this outline:

   I. Pushing Capacity
     A. Cannon
     B. Foley
     C. Koso
  II. Purchase Details
     A. Cannon
     B. Foley
     C. Koso
 III. Dependability
     A. Cannon
     B. Foley
     C. Koso

Note that the bulldozers are discussed in the same order within each major section. As with the whole-by-whole approach, the order of the items can go either from most important to least important or vice versa—depending on what strategy you believe would be most effective with your audience.

● *Comparison/Contrast Guideline 5: Use Illustrations*

Comparison/contrasts of all kinds benefit from accompanying graphics. In particular, tables are an effective way to present comparative data. For example, your report on the bulldozers might present some pushing-capacity data you found in company brochures.

Example of Comparison/Contrast

Using the bulldozer example, Model 5–4 on pages 174–175 presents a sample of the part-by-part pattern. Remember that such a comparison would be only part of a final report—in this case, one that recommends the Cannon-D. The complete report would include an introductory summary at the beginning and a list of conclusions and recommendations at the end (see chapter 8). This example could be written in either a whole-by-whole or part-by-part manner. The latter is used here to emphasize the importance of the three criteria for comparison.

Communication Challenge

"A Dome Gone South: Argument in Action"

When McDuff finished its project of managing construction of the Nevada Gold Dome (see Project #4 on color insert in chapter 2), all was well. The company more than met its completion deadline and saved its client, the stadium owners, about $100,000. Now, two years later (and a year after the comprehensive warranty expired), a number of problems have surfaced. The following sections (1) present the four main complaints McDuff has heard, (2) analyze the problem from the points of view of both the dome owners and McDuff, and (3) list questions and comments for you to consider in evaluating this "breakdown." The focus will be opportunities for using patterns of organization discussed in this chapter.

Problems with the Dome

Over the last few months, the Gold Dome owners have alerted McDuff to four problems they believe may have resulted from construction or material problems:

1. **Artificial turf:** The turf has pulled up at several spots around the edge of the field. Although dome maintenance staff tried a quick fix, the turf has pulled back up again and again.
2. **Ramp cracks:** Several one-inch cracks have developed in concrete ramps that lead from the team locker rooms to the field. Just last week two football players tripped on the cracks as they entered the field in front of TV cameras covering the start of a game. The embarrassment, as well as the possible danger to high-priced players, has the dome management fuming.
3. **Roof leak:** Leaks developed in the dome roof, caused by the ponding of rainwater on a sunken roof section near one of the spectator decks.
4. **Equipment damage:** Water entered one of the dome's control rooms, damaging three TV monitors and causing carpeting to mildew.

What brought the problems to a head was an incident before last Sunday's game. While team co-owners, Ted Burner and Jane Honda, were getting settled into their luxury box near the upper deck, a roof panel broke loose near the area of the leak and crashed onto their balcony, spilling wine and cheese on Ted and Jane. Needless to say, that event generated a meeting this week between McDuff and dome management.

McDuff and the Dome Management: Different Analyses

With all its training in Total Quality Management, McDuff believes that satisfying the customer is a crucial corporate objective. Besides being the ethical high road, this goal also generally leads to the company getting repeat business from the

Continued

same client or good recommendations to be used with new clients. In either case, good service helps business.

Yet the firm has not reached its current level of success by admitting error every time a client expresses dissatisfaction with a job. Sometimes other parties are at fault. Indeed, at the end of the meeting between McDuff and the dome management, McDuff representatives had clear disagreements about the possible causes of the four problems. Here are the main differences that arose from preliminary analyses by both parties.

- **Artificial turf**

  Dome: The dome managers believe that artificial turf was installed improperly, without adequate drying time allowed for the adhesive.

  McDuff: McDuff expects that the problem was caused wholly or in part by the dome's home team, the Nevada Gamblers. Two years ago the team began practicing on the surface before the end of the two-week curing period recommended by the installers.

- **Ramp cracks**

  Dome: Dome managers remember McDuff, Inc.'s, comments on a construction progress report, indicating that several trucks arrived late when concrete was being poured for the ramps. Thus there were longer than usual delays between truck pours for the ramps.

  McDuff: Despite the truck delays, McDuff satisfied the minimum professional standards for timeliness during the concrete pour. Unrelated to this timing problem, McDuff had recommended additional reinforcing bars for the ramp concrete during construction. However, dome management had only approved the minimum amount required by applicable standards.

- **Roof leak**

  Dome: Dome managers believe that a structural flaw in the roof support system caused water to pool on one small section of the roof, instead of rolling off into the drainage canals.

  McDuff: After observing the roof area, McDuff suspects that a major hailstorm two months ago damaged the roof panels, which caused ponding of rainwater and led to the subsequent leak.

- **Equipment damage**

  Dome: The route of the water suggests that it may have come through wall spacings from the area of the roof leak.

  McDuff: Another possibility is that the damage resulted from a custodial worker failing to turn off a faucet located above the control room in a maintenance area. When the open faucet was discovered, the maintenance room was cleaned up without anyone observing seepage into other rooms or wall space, but such seepage seems just as likely as seepage from the region of the roof leak.

Although the dome's comprehensive warranty has ended, McDuff and key contractors remain partially liable for five years from construction for problems that can be linked to defects in construction or materials. Even if this warranty were not in force, however, McDuff would want to go above and beyond the call to satisfy the client—if there were some reason for doubt or if corrections were not too expensive. At this point, McDuff has not decided how to proceed.

Questions and Comments for Response

In responding to the following questions and comments, use information previously supplied as well as any other rationale you think applies.

1. **Argument:** Cyndy Perlman, a member of McDuff's dome project team, believes that McDuff should correct all four problems. What arguments, from strongest to weakest, should she use to convince her skeptical boss of this path of action?

2. **Argument:** Mason Hunter, another member of the McDuff project team, believes that McDuff should bear *no* financial responsibility for correcting any of the four problems. What arguments, from strongest to weakest, should he use to convince his skeptical boss of this path of action?

3. **Argument:** In the case presentation, find one example each of what could be considered *non sequitur* and *post hoc ergo propter hoc* argumentative fallacies.

4. **Definition:** Do some research to provide formal sentence definitions for some or all of the following terms mentioned in the Gold Dome case study:

Warranty	Total Quality Management
Dome stadium	Progress report
Artificial turf	Reinforcing bars
Concrete	Hailstorm
TV monitor	

5. **Description:** Provide an outline of what you would include in a technical description of a domed athletic stadium.

6. **Classification/Division:** Separate into groups established by your instructor. Then brainstorm to come up with a list of at least 30 additional construction- or materials-related problems that could have occurred with a stadium like the Gold Dome. Finally, perform a classification exercise like the one presented in this chapter regarding McDuff's safety practices.

7. **Comparison/Contrast:** In the "McDuff and the Dome Management" section of this case study, a comparison/contrast is presented in a part-by-part format. Convert it to a whole-by-whole format. Which of the two organizational strategies do you consider most effective in this case?

CHAPTER SUMMARY

This chapter examines five common patterns of organization that make up reports, proposals, and correspondence. By studying these patterns, you will be better prepared to write the documents covered later in this book.

Argument is used when the writer needs to support points with evidence. (Persuasion, the strongest form of argument, occurs when the writer seeks to change the reader's opinions or actions.) To write effective arguments, follow these guidelines:

1. Use evidence correctly.
2. Choose the most convincing order for points.
3. Be logical.
4. Use only appropriate authorities.
5. Avoid argumentative fallacies.
6. Refute opposing arguments.

Definitions occur in technical writing in one of three forms: informal (in parentheses), formal (in sentence form with term, class, and features), and expanded (in a paragraph or more). These main guidelines apply:

1. Keep it simple.
2. Use informal definitions for simple terms most readers understand.
3. Use formal definitions for more complex terms.
4. Use expanded definitions for supporting information.
5. Choose the right location for your definition.

Description, like definition, depends on detail and accuracy for its effect. Careful descriptions usually include a lengthy itemizing of the parts of a mechanism or the functions of a term. Follow these basic guidelines for producing effective descriptions:

1. Remember your readers' needs.
2. Be accurate and objective.
3. Choose an overall organization plan.
4. Use "helpers" like graphics and analogies.
5. Give your description the "visualizing test."

Classification/division patterns help you organize groups of related items (classification) and break down an item into its parts (division). The guidelines for classification are as follows:

1. Find a common basis (for grouping diverse items).
2. Limit the number of groups.
3. Carefully classify each item.

Similarly, the guidelines for division are as follows:

1. Choose the right basis for dividing.
2. Subdivide parts when necessary.
3. Describe each part with care.

Comparison/contrast gives you an organized way to highlight similarities and differences in related items—whether you are simply presenting data or are attempting to argue a point. These main writing rules apply:

1. Remember your purpose.
2. Establish clear criteria and use them consistently.
3. Choose the whole-by-whole approach for short comparison/contrasts.
4. Choose the part-by-part approach for long comparison/contrasts.
5. Use illustrations.

Access the Technical Writing Web site by pointing your browser at www.prenhall.com/pfeiffer. Here, you will find additional resources to aid you in learning and applying the information from this chapter.

ASSIGNMENTS

Part 1: Short Assignments

The following short assignments can be completed either orally or in writing. Unless a group project is specifically indicated, an assignment can be either a group or an individual effort. Your instructor will give you specific directions.

1. **All Patterns of Organization—Recognition Exercise.** For this group assignment, your instructor will provide each group with a different packet of "junk mail" (catalogs, sales letters, promotions, etc.) and perhaps other documents such as memos or product information sheets. Your group will search for and evaluate examples of various patterns of organization in the documents. Then it will report its findings to the whole class.

2. **Argument.** Analyze the argumentative effectiveness, or lack thereof, in the following passage:

After a good deal of thought, I have decided not to accept the committee's recommendation to allow employees up to two weeks of unpaid vacation leave (in addition to whatever paid vacation leave the employee receives). My reasons are as follows:

a. The proposal obviously would cause the company to lose many customers, for we definitely would not have the staff to cover our daily operations in the office and in the field.

b. Just as the policy of flextime hours has caused some companies to fail to respond adequately to phone calls early in the morning, the leave plan would keep the office uncovered during days when an excessive number of employees were taking leave.

c. The salary of employees using unpaid vacation leave would decrease, and they would be less able to pay their bills. In effect, then, the leave time policy would hurt their families.

d. Unpaid vacation leave has been supported by some members of the state legislature, and we all know how little that organization knows about how to operate business. In fact, I heard the other day that our local state senator, who supports the idea of unpaid vacation leave, bounced 17 checks during the statehouse banking scandal several years ago. That will tell you something about his knowledge of business activities.

e. One of our competitors, Jonquil Engineering, adopted an unpaid vacation leave policy two years ago, and I have just learned that the firm's stock dropped 10% recently.

f. Unpaid leave has been supported by the Americans for Family, an organization whose president is Arlin Thomas. And we all know about his antics in the media. In the last few years, he has supported any cause that has come his way.

3. **Argument: McDuff Project.** For this assignment, use points from Project #7 (on the color insert in chapter 2) to complete one of the options below. Each option will also require that you create additional information that could relate to McDuff's technical writing seminar in Germany.

   Option A: **Creating an Effective Argument.** Assume you were one of the individuals attending the seminar. Create an argument you would present to your boss to justify offering another seminar for some members of your department.

   Option B: **Creating Argumentative Fallacies.** Assume you attended the seminar and are presenting an evaluation of it to your supervisor. Create an example of each of the argumentative fallacies included in this chapter. Your seven answers can be in support of the seminar, against it, or both.

4. **Definition.** Using the guidelines in this chapter, discuss the relative effectiveness of the following short definitions. Speculate on the likely audience the definitions are addressing. [Excerpted from Turner, R. S., R. B. Cook, H. Van Miegroet, D. W. Johnson, J. W. Elwood, O. P. Bricker, S. E. Lindberg, and G.M. Hornberger. September 1990. Watershed and lake processes affecting surface water acid-base chemistry (NAPAP Report 10). In *Acidic Deposition: State of Science and Technology*. National Acid Precipitation Assessment Program, 722 Jackson Place, NW, Washington, D.C.]

   a. Watershed—the geographic area from which surface water drains into a particular lake or point along a stream.

   b. Acid mine drainage—runoff having high concentrations of metals and sulfate and high levels of acidity resulting from the oxidation of sulfide minerals that have been exposed to air and water by mining activities.

   c. Steady-state model—a model in which the variables under investigation are assumed to reach equilibrium and are independent of time.

   d. Acidic lake or stream—a lake or stream in which the acid neutralizing capacity is less than or equal to zero.

   e. Biomass—the total quantity of organic matter in units of weight or mass.

   f. Detritus—dead and decaying organic matter originating from plants and animals.

   g. Hydrology—the science that deals with the waters of the earth—their occurrence, circulation, and distribution; their chemical and physical properties; and their relationship to living things.

   h. Plankton—plant or animal species that spend part or all of their lives carried passively by water currents.

   i. Mineral weathering—dissolution of rocks and minerals by chemical and physical processes.

5. **Definition: McDuff Projects.** This assignment can be completed as an individual or a group project. Select one of the projects in the color insert in chapter 2. Using the outline of information on the project as a starting point, do the following:

   • Conduct some research on the technical field reflected in the project.
   • Select some terms related to the field.
   • Write *either* short formal definitions *or* expanded definitions of the terms.

In assigning this project, your instructor will indicate (a) how many terms you should select and (b) whether you should write formal or expanded definitions.

6. **Description.** Write a description of a piece of equipment or furniture located in your classroom or brought to class by your instructor—for example, a classroom chair, an overhead projector, a screen, a three-hole punch, a mechanical pencil, or a computer floppy disk. Write the description for a reader totally unfamiliar with the item.

7. **Description: McDuff Project.** For this project put yourself in the position of someone who worked on the General Hospital project (Project #3 in color insert in chapter 2). Write a description of a piece of equipment that either (a) could have been used in the construction of the hospital or (b) could be housed in any modern medical facility such as General Hospital. This assignment may require a visit to a local hospital, construction company, or library with resources in the fields of construction or medical care.

8. **Classification/Division.** Divide into groups of 8 or 10 as selected by your instructor. Then, as a group, take an inventory of the career plans of each member of your group. Finally, using career plans as your basis, classify the group participants into a limited number of groupings. Be sure to adjust the size and focus of categories so that you have at least 2 students in each grouping. Present the group results to the entire class.

9. **Classification/Division: McDuff Projects.** For this assignment you will use the 33 "main technical tasks" listed on the seven projects on the color insert in chapter 2. Perform a classification exercise by finding a common basis, selecting an appropriate number of groups, and placing each of the 33 technical tasks into one of the groups.

10. **Comparison/Contrast.** Divide into groups of three to five students, as determined by your instructor. Select two members who will give the other group members information about their (a) academic career, (b) extracurricular activities, and (c) work experience. Using the information gathered during this exercise, develop an outline of either a whole-by-whole or part-by-part comparison/contrast.

11. **Comparison/Contrast: McDuff Projects.** Select two or three of the McDuff projects from the color insert in chapter 2. Perform a part-by-part comparison/contrast of the two or three projects. Use some or all of the information on the projects. If necessary, create some of your own information to supply adequate detail for your response.

Part 2: Longer Assignments

These assignments test your ability to use or to analyze patterns of organization covered in this chapter. Assignments 12–17 are placed in the context of short reports within McDuff, Inc. Use the following guidelines for these assignments:

- Write each exercise in the form of a *memo report* (if it is directed within McDuff, Inc.) or a *letter report* (if it is directed to an outside reader).
- Follow organization and design guidelines given in chapters 3 and 4, especially with regard to the ABC format (**A**bstract/**B**ody/**C**onclusion) and the use of headings. Chapter 8 gives thorough format guidelines for short reports, but such detail is not necessary to complete the assignments here.
- For each assignment, fill out a copy of the Planning Form at the end of the book. Invent any audience analysis information not included in the following so that you have a "real" person to whom you are writing.

Assignments 18 and 19 require essay responses.

12. **Argument for New Wastewater Specialist.** You are a wastewater specialist and manager of a four-person crew with McDuff's Houston office. Recently you have become concerned

about the amount of overtime worked by you and your group. Here are your main concerns, from most to least important. First, you are worried that excessive overtime might lead to errors by exhausted workers. The four employees on your crew often work at dangerous sites with poisonous chemicals and polluted water. Mistakes could lead to worker exposure to chemicals or errors in recording data or collecting samples. Second, three of your four field-workers have complained about their 50–55-hour weeks and the excessive time spent out of town on projects. Although they like the overtime pay, they would prefer to average just an extra five hours a week. You are concerned that they may quit to work at a competing firm. Third, you are convinced that there is enough work to support a fifth field employee, and still have three or four hours of overtime per employee per week. (Of course, you realize that there is always the risk that a future work slowdown would mean laying off an additional worker.)

Given these concerns, you want to write a short report to the branch manager, Elmore Lindley, describing the problems. You also want to suggest that McDuff hire another field-worker, or at least give the matter some study. Present this information to Lindley, using the argumentative strategies described in this chapter.

13. **Technical Definitions in Your Field.** Select a technical area in which you have taken course work or in which you have technical experience. Now assume that you are employed as an outside consulting expert, acting as a resource in your particular area to a McDuff manager not familiar with your specialty. For example, a food-science expert might provide information related to the dietary needs of oil workers working on an offshore rig for three months; a business or management expert might report on a new management technique; an electronics expert might explain the operation of some new piece of equipment that McDuff is considering buying; a computer programmer might explain some new piece of hardware that could provide supporting services to McDuff; and a legal expert might define "sexism in the workplace" for the benefit of McDuff's human resources professionals.

For the purpose of this report, develop a context in which you would have to define terms for an uninformed reader. Incorporate *one expanded definition* and *at least one sentence definition* into your report.

14. **Description of Equipment in Your Field.** Select a common piece of laboratory, office, or field equipment with which you are familiar. Now assume that you must write a short report to your McDuff supervisor, who wants it to contain a thorough physical description of the equipment. Later he or she plans to incorporate your description into a training manual for those who need to know how to use, and perform minor repairs on, the equipment. For the body of your description, choose either a part-by-part physical description or a thorough description of functions.

15. **Description of Position in Your Field.** Interview a friend or colleague about the specific job that person holds. Make certain it is a job that you yourself have *not* had. On the basis of data collected in the interview, write a thorough description of the person's position—including major responsibilities, reporting relationships, educational preparation, experience required, etc.

Now place this description in the context of a letter report to the manager of human resources at McDuff. Assume she has hired you, a technical consultant to McDuff, to submit a letter report that contains the description. She is preparing to advertise such an opening at McDuff but needs your report to write the job description and the advertisement. Because she has little firsthand knowledge of the position about which you are writing, you should avoid technical jargon.

16. **Classification/Division in Technical Courses.** For this assignment, you will need a catalog from your college or university or that of another school. First, select a number of

courses that have not yet been classified except perhaps by academic department. Using course descriptions in the catalog or any other information you can find, *classify* the courses using an appropriate basis (for example, the course level, topic, purpose in the department, prerequisites, etc.). Also *partition* one of the courses by using information gathered from a course syllabus or an interview with someone familiar with the course.

    The context for this exercise is a memo report written by you, a technical training specialist at McDuff's St. Louis office. Assume that the courses are taught at St. Louis Tech, a nearby college. A group of McDuff's St. Louis employees has expressed interest in further education in the area you have investigated. Your classification of the courses and your in-depth partition of one particular course will help these employees decide whether to consider enrolling.

17. **Comparison/Contrast for the Purchasing Agent.** For this memo report, select a category in the list following or another category approved by your instructor. Then choose three specific types or brand-name products that fit within the category and for which you can find data. Write either a whole-by-whole or a part-by-part comparison/contrast. Place this comparison/contrast pattern within the context of a short report from you, as a McDuff employee, to the company purchasing agent. You may or may not include a preference for one item or the other. However, understand that the purpose of your report is to give the reader the information needed to order one of the items in your comparison/contrast.

    Categories
    1. Briefcases
    2. Calculators
    3. Compact discs
    4. Credit cards
    5. Electric pencil sharpeners
    6. Electric sanders or saws
    7. Lawn mowers
    8. Mortgages
    9. Personal computers
    10. Photocopiers
    11. Pickup trucks
    12. Professional journals in a technical field
    13. Refrigerators
    14. Software programs
    15. Retirement plans
    16. Surveying methods
    17. Telephone systems or companies
    18. Types of savings accounts
    19. Vacuum cleaners
    20. VCRs

18. **Ethics Assignment.** Reread the chapter's "Communication Challenge" entitled "A Dome Gone South: Argument in Action." The questions at the end of the section ask you to consider arguments for and against McDuff taking responsibility for correcting the problem. Now focus exclusively on *ethical points* by writing an essay in which you explore issues of right and wrong that are evident in the Dome case. Specifically, what do you think is the firm's ethical responsibility in the situation? Support your views by (1) invoking the ECI Principle from chapter 2 (see p. 62), (2) using analogies to real-world case studies, and/or (3) creating hypothetical situations analogous to the Dome example.

19. **International Communication Assignment.** Elements of persuasion and argumentation differ from culture to culture. What might be considered appropriate in an argument targeted for one country could be considered inappropriate or even offensive in an argument aimed for another. Examine the strategies of persuasion of one particular country or culture other than your own. Then write an essay that shows how these strategies are similar to, or different from, the strategies employed in your own culture. Your inquiry may apply to written communication (like letters or reports), to oral communication (like speeches or negotiations), or to both. Possible sources of information include books on global cultures, Internet sources on international communication, and international friends and colleagues.

MEMORANDUM

DATE: February 25, 2002
TO: Kerry F. Camp, Vice President of Domestic Operations
FROM: Your Name, Boston Branch Manager
SUBJECT: Problems with Office Chairs

Approaches problem tactfully.

I enjoyed talking to you yesterday and look forward to seeing you at the Environmental Science Convention in New York. In the meantime, I am writing to ask your help with the chair problem I mentioned in our conversation. This memo will give you some background on our difficulties with the chairs. Also, I have presented a solution for you to consider.

SPECIFIC PROBLEMS WITH OFFICE CHAIRS

Mentions points of agreement— before moving toward argumentative issues.

About six months ago, we received new office furniture ordered by the corporate office. In many ways, it has been a considerable improvement over the 20-year-old furniture it replaced. The desks, cabinets, and credenzas have been especially well received. Besides being quite practical, they give a much more professional appearance to our office.

As I mentioned in our conversation, the chairs (Model 223) that accompany the new desks have not worked out as well as the other furniture. Here are the two main problems.

Increased Employee Illness

Presents evidence clearly— including statistic.

Starting about a month after the chairs arrived, I received complaints from ten employees about back pains they claimed were related to the new chairs. Taking a wait-and-see approach, I asked managers to have chairs adjusted and to give the furniture more break-in time. In the second month, however, I heard from a dozen additional employees about the chairs and noticed a 25 percent increase in sick days. Upon further inquiry, I learned that most managers attributed the increased sick time to back problems related to the office chairs.

Cites reliable authority.

Decreased Productivity

The increased sick leave presented a serious enough concern. Adding to the problem, however, was roughly a 10 percent decrease in productivity last month among employees who were not sick. I measured this decrease by three criteria:

States supporting statistic.

Gives basis for statistical information.

- Number of pages produced by the word-processing staff
- Number of billable hours logged by the professional staff
- Number of client calls made by the marketing staff

In my seven years managing this office, I have never witnessed so large a drop with no clear cause.

MODEL 5–1
Memo report—example using argument

continues

CONFIRMATION OF PROBLEM

Concerned about the health problems and lower productivity, I sought a recognized consultant in the area of workplace health and office design. Dr. Stacy Y. Stephens, professor of ergonomics at East Boston College, was referred to me when I called the local office of the Occupational Safety and Health Administration.

Dr. Stephens visited our office on January 17. She spent two days (1) interviewing and observing users of the new chairs, (2) examining the chairs in detail, and (3) meeting with me to offer her conclusions. At her suggestion, I called the supplier, Jones Office Furniture, to ask about the availability of other Jones chairs with different features. According to Dr. Stephens, the Jones Model 623 is much better suited to normal office use. Specifically, the Model 623 has these advantages:

- Excellent lower-lumbar support, compared to the inadequate support in the straight-backed Model 223
- Twelve inches of vertical adjustment, as opposed to the six inches available in Model 223
- Ball casters for easier rolling, as opposed to the Model 223 casters that move with difficulty on our office carpeting
- Adjustable back spring that provides limited movement backward, as opposed to the rigid Model 223

Considering Dr. Stephens' suggestions, I decided to seek more information from Jones Office Furniture Company.

PROPOSED SOLUTION

Last week, I met twice with Mr. Dan McCartney, the Boston sales representative for Jones Office Furniture. In response to our problem, he has offered this proposal. Jones will trade our 45 Model 223 chairs for 45 new Model 623s, for only an extra $2000 from us. Under this arrangement, he would be (1) crediting us for the full purchase of our four-month old Model 223s and (2) discounting the new chairs about $3000. I believe Jones is making this offer because it truly wants to retain our goodwill—and our future business.

CONCLUSION

As noted, this problem affects our employees' well-being and our office productivity. Would you please present my proposed solution to Ben Garner, vice president of business and marketing, so that he can approve the $2000 funding? We could have the new chairs within a week after approval.

Thanks for your help, Kerry. With relatively small investment, we should be able to solve our medical and productivity problems.

Cites authority and gives her credentials.

Mentions this expert's opinion *last*, because it is the strongest evidence in this short report.

Gives *specific* points to support expert's views.

Sets up clear contrast between two chair types, to support chair exchange.

Describes alternative clearly—giving necessary details about cost, etc.

Ends with request for action.

Closes with tone that encourages agreement on *mutual* problem.

MODEL 5–1
continued

TECHNICAL DESCRIPTION: BLUEPRINT MACHINE

A blueprint machine is a piece of office equipment used to make photographic repro-
ductions of architectural plans, technical drawings, and other types of figures. The blueprint
shows as white lines on blue paper. This description covers the physical parts of the machine
and a brief operation procedure.

PHYSICAL DESCRIPTION

The blueprint machine is contained in a metal cabinet measuring 36" wide, 10" tall, and
18" deep. Typically, the machine is placed on a cabinet containing blueprint supplies. The
most evident features of the machine are the controls and the paper path.

Blueprint Machine

Controls. The controls for the machine are located on the top right-hand side of the case, as
shown in the illustration below.

There are two knobs and one power switch. The topmost knob controls the speed of the paper
feed. The lower knob changes the direction of the paper. The power switch turns the machine
on and off.

MODEL 5–2
Detailed description

continues

(margin notes:)

Starts with a sentence defini-tion and brief overview of two sections that follow.

Begins section with dimensions of machine, before noting two main parts.

Uses three per-spectives in illustration, for clarity.

Describes three controls in same order that they are viewed—top to bottom.

Integrates pro-
cedure with il-
lustration.

Paper Path. The top of the machine has four horizontal slots. These slots are the openings to the paper path, as shown in the illustration below.

The first slot, beginning at the front of the machine, is the feed slot for the original and the unexposed blueprint paper. The second slot is the discharge for the original and the exposed print paper. The third slot is the feed for developing the exposed blueprint paper. The fourth slot is the discharge for the finished blueprint copy.

OPERATION OF THE BLUEPRINT MACHINE
Operating the blueprint machine involves setting the controls and making copies.

Includes brief
instructions that
help expand
upon description
of mechanism.

Setting the Controls:
1. Turn on the machine.
2. Set the paper path to forward.
3. Set the speed for the type of paper used.

Making Copies:
1. Place the original, face side up, on top of the blueprint paper, yellow side up.
2. Feed the two pieces of paper into the original feed slot.
3. Separate the original from the copy as they emerge from the discharge slot.
4. Feed the exposed blueprint paper into the developer slot.
5. Collect the copy as it emerges from the discharge slot.
6. Turn power off when desired copies have been made.

CONCLUSION
Blueprint machines are found in all McDuff offices and are used frequently. If a wide variety of employees understands how these machines function, our office will continue to run smoothly.

MODEL 5–2
continued

Notes burner's
main parts and
sections that
follow.

McDuff uses Bunsen
burners in all its laboratories.
Following some background
information, this technical
description provides details
about three main parts of a
typical burner:

- Base
- Gas valve
- Pipe

The conclusion lists
some standards for the burner's
performance.

Background

Gives formal
sentence
definition and
general
information for
nontechnical
readers.

The Bunsen burner is a
basic piece of laboratory
equipment used to produce a
continuous flame at relatively
low temperatures. Originally
designed by Robert W. Bunsen
in the late 1800s, the "Bunsen
burner" has become a generic
term for basic lab burners made
by many firms.

Most burners look and
perform alike, though burners
from different companies do
include slightly different
features. What follows is a description of the Model 03–962 Bunsen-style burner
manufactured by the Fisher Scientific Company. It runs on natural gas.

Base

The heavy die-cast base
of the Fisher burner is very
stable. It is made from
nonferrous metal and has a
nickel finish. Here are its main
features and dimensions:

Uses bullets for
technical detail.

- Hexagonal-shaped
  foundation that is 2 3/4" in
  diameter and 2" high

MODEL 5–3
Technical description: Bunsen burner

continues

- A 3/4"-diameter threaded cylinder at the top of the base
- A 1/2"-diameter hexagonal brass nut at the top of the cylinder, with a small hole in the center from which gas is emitted into the pipe

Jutting out from the side of the cylinder, parallel to the surface on which the base of the burner rests, is a tapered gas inlet 2 1/2" long. The inlet has a grooved surface that holds the gas tube securely to the burner.

Gas Valve

A valve is threaded vertically up into the bottom center of the base of the burner. It allows the user to adjust the volume of gas that flows from the gas inlet up through the base cylinder.

The valve looks like a car axle with only one wheel attached. The 2"-long stem, or "axle," rests on a round 3/4"-diameter base, or "wheel." Actually, this base is about the diameter and thickness of a U.S. nickel coin. It has serrated edges so that it can be twisted with ease. The 1/8"-diameter stem is threaded and screws vertically into the base of the burner.

When twisted clockwise, the valve closes and decreases gas flow. When twisted counterclockwise, it opens and increases gas flow.

Pipe

A 4 1/2"-long pipe extends straight up from the top of the base of the burner. Except for its flared ends, the pipe is 1/2" in diameter. The combining of gas and air at the bottom of the pipe produces a flame that emerges at the top.

The pipe threads on to the cylinder at the top of the burner base. The bottom of the pipe flares out to form an end piece with eight 3/16" holes drilled around its circumference. The holes let in air that mixes with gas entering from the base. When the pipe is turned clockwise, the amount of air entering the holes is reduced and the temperature of the flame is lowered. When the pipe is turned counterclockwise, the amount of air entering the holes is increased and the temperature of the flame is raised.

The flared top end of the pipe looks much like a socket piece for a socket wrench. Called a "flame retainer" by the manufacturer, it helps keep the flame from going out. Viewed vertically from above the burner, the top of the retainer is shaped like a wagon wheel with short spokes. The "spokes" are actually eight ports that open to the pipe hole but close off before they reach the circumference of the pipe. Spaced evenly on the circumference, these ports help control the flame.

MODEL 5–3
continued

…es analogies
…ch as
…eel, axle, and
…kel coin.

…scribes
…ovement to help
…ader understand
…rpose of part.

…cludes common
…sual terms like
…ared."

…eveals purpose
…parts through
…escription.

…mploys two more
…nalogies to help
…ss technical
…aders.

Performance Specifications

The Fisher version of the Bunsen burner offers the following performance standards:

Uses bullets for technical detail.

- Produces flame that is adjustable from 3/4″ to 12″
- Has heat output of over 1465w (5000 BTU/hr.)
- Consumes 5 cu ft hr. of natural gas
- Meets U.S., NIST, USTM, and ASTM standards

MODEL 5–3
continued

PART-BY-PART PATTERN COMPARISON OF BULLDOZERS

This is a comparison of the Cannon-D, Foley-G, and Koso-L bulldozers. The major criteria considered are pushing capacity, purchase details, and dependability.

Pushing Capacity

Both the Cannon-D and Foley-G have an excellent pushing capacity that is more than adequate for our U.S. construction projects. The Cannon has a pushing capacity of 1300 tons per hour (TPH) over 500 ft, whereas the Foley has a pushing capacity of 1100 TPH over 500 ft. Both figures are mentioned in brochures from their respective companies. Furthermore, they are confirmed by recent tests reported in the trade journal, *Bulldozer Unlimited.*

The Koso-L, however, lags behind its two competitors, being able to push only 1000 TPH over 500 ft. Furthermore, this figure was found only in the company's sales brochures, with no verification available in the trade journals I researched.

Purchase Details

The main elements of the purchase—price, warranty, and cost of extended warranty—vary considerably among the three machines. The basic purchase prices are as follows:

1. $250,000 for the Cannon-D
2. $300,000 for the Foley-G
3. $310,000 for the Koso-L

Concerning warranties, Cannon provides a complete parts-and-service warranty for 90 days, with an additional parts warranty on the drivetrain for another 12 months. Foley offers the same initial parts-and-service warranty, but it has a longer parts warranty on the drivetrain—18 months. Koso has a complete parts-and-service warranty for six months, with no additional parts warranty on the drivetrain.

Beyond these warranties that come with the machines, all three companies offer the same option of an additional two-year warranty that covers parts in the drivetrain. This additional warranty must be acquired at time of purchase and costs $4,000 for the Cannon, $4,000 for the Foley, and $7,000 for the Koso.

Dependability

Bulldozer downtime can cost the firm a good deal of money, either in project delays or in the added expense of renting another bulldozer, so dependability is an important criterion in comparing the three models. Although the trade journals I consulted contained no model-by-model comparison of dependability of the three bulldozers, I did find occasional references to features that affect reliability. In

---

Leads off by noting three dozers and main criteria.

Starts each section with Cannon (the preferred machine).

Follows descending order of importance (Cannon, then Foley, then Koso).

Lists three subpoints to be discussed in section.

Again, follows C-F-K order.

Uses separate paragraph for each main subpoint.

Mentions sources of information—to increase credibility.

MODEL 5–4
Part-by–part comparison/contrast: three bulldozers

addition, I sought out anecdotal evidence from nearby construction firms that have used the machines. Here is what I discovered.

Cannon-D. This machine is an established commercial bulldozer with a good reputation for reliability in the field. The "D" model is the latest version of a machine that began with the "A" model in 1954, so the company has had years to refine its technology. Interestingly, several years ago Cannon decided not to add sealed tracks to its list of features, even though this sealed approach is one of the newest track innovations in the industry. Even so, the Cannon-D generally will operate for 10,000 hours without major maintenance in the tracks. It remains to be seen if the decision about sealed tracks was the correct one.

Foley-G. While the Foley also has a good reputation for reliability in the field, the firm began making bulldozers only in 1970—16 years after Cannon started. Foley has the general reputation of being quicker than Cannon to introduce new technology into its bulldozer line. For example, two years ago it started using sealed tracks that lubricate themselves automatically. The track assembly should not need major maintenance for 15,000 hours, a 5000-hour improvement over Cannon. Of course, the newness of this advance suggests that there hasn't yet been enough time to judge the long-term effectiveness of sealed tracks.

Koso-L. This machine is definitely the "new guy on the block." Unlike the American-made Cannon and Foley, the Koso is produced in Korea by a company that got into business rather recently—1986. The Koso-L uses fairly traditional technology, not having incorporated sealed tracks into its design, for example. Despite the recent entry into the market, Koso has already built a strong reputation for reliability on projects in other countries. However, there are not yet enough data about performance on construction projects in the United States.

Uses subheadings because of large amount of information to relate.

Introduces Foley's main feature (technical innovation).

But reinforces Cannon preference by noting that sealed tracks are not fully tested.

Keeps option open for buying Koso later—but makes it clear that more data are needed.

MODEL 5–4
continued

Chapter 6

Process Descriptions and Instructions

Collaboration at Work

A Simple Test for Instructions

General Instructions

Each chapter begins with a Collaboration at Work exercise that applies strategies for working in groups to chapter topics. The exercise assumes you (1) have been divided into groups of about three to six students, (2) will use group time inside or outside of class to complete the case, and (3) will produce an oral or written response. For guidelines about writing in groups, refer to pages 28–34.

Background for Assignment

Writing instructions presents a challenge. The main problem is this: Although writers may have a good understanding of the procedure for which they are designing instructions, they have trouble adopting the perspective of a reader unfamiliar with the procedure. One way to test the effectiveness of instructions is to conduct your own "usability test." The following exercise will determine the clarity of instructions written by your group by asking another group to follow the instructions successfully.

Group Assignment

In this exercise your group will prepare a list of instructions for drawing a simple figure or object. The purpose is to write the list so clearly and completely that a classmate could draw the figure or object without knowing its identity. Following are instructions for completing the assignment:

1. Work with your group to choose a simple figure or object that requires only a relatively short set of instructions to draw.
   NOTE: Use a maximum of 15 steps.
2. Devise a list of instructions that your group believes cannot be misunderstood.
3. Test the instructions within your own group.
4. Exchange instructions with another group.
5. Attempt to draw the object for which the other group has written instructions.
   NOTE: Perform this test without knowing the identity of the object.
6. Talk with the other group about problems and suggestions related to the instructions.
7. Discuss general problems and suggestions with the entire class.

*L*ike other McDuff offices, McDuff's Denver office installed an electronic-mail system in the last few years. It permits employees to send and receive messages on their computer terminals. The idea seems to have pleased everyone. Messages are conveyed faster, and less paper is generated in the office. As office services manager, you met with a representative of the company that would install the system, shortly before the installation date. Your boss, Leonard Schwartz, expected a memo from you summarizing the installation process (see Model 6–1 on page 200). You also wrote a memo to all employees, giving them instructions about how to read electronic-mail messages (see Model 6–2, pp. 201–202).

This brief McDuff case study demonstrates two types of technical writing you will often face—process descriptions and instructions. Both are important patterns, but instructions will play the greater role in your career. As noted in one essay, "The field of technical writing has expanded greatly in the last ten years, and the new work is in writing *instructions*. . . . Many graduates of technical programs now find themselves writing *instructions* more often than they write descriptions and reports"* (italics mine). For this reason, instructions receive the most emphasis in this chapter.

Instructions and process descriptions share an important common bond. Both must accurately describe a series of steps leading toward a specific result. Yet they differ in purpose, audience, and format. This chapter (1) explores these similarities and differences, with specific reference to McDuff applications, (2) gives specific guidelines for developing both types, and (3) provides models to use in your own writing.

PROCESS DESCRIPTIONS VERSUS INSTRUCTIONS

In the McDuff example just given, the memo to your boss (Model 6–1) explained the process by which the vendor installed electronic mail. The other memo (Model 6–2) gave users the directions needed to read on-line mail. In other words, you write a process description to help readers *understand* what has been, is being, or will be done, whereas you write instructions to show readers how to perform the *process* themselves.

Process descriptions are appropriate when the reader needs to be informed about the action but does not need to perform it. If you suspect a reader may in fact be a "user" (that is, someone who uses your document to perform the process), always write instructions. Figure 6–1 provides a list of contrasting features of process descriptions and instructions; the two subsections that follow give these features some realism by briefly describing some McDuff contexts.

Process Descriptions at McDuff

Process descriptions provide information for interested readers who do not need instructional details. At times, describing a process may be the sole purpose of your doc-

---

*Reprinted by permission of the Modern Language Association of America from Janice C. Redish and David A. Schell, "Writing and Testing Usability Instructions," in *Technical Writing: Theory and Practice*, ed. Bertie E. Fearing and W. Keats Sparrow (New York: Modern Language Association of America, 1989), 63.

PROCESS DESCRIPTIONS

Purpose: Explain a sequence of steps in such a way that the reader understands a process

Format: Use paragraph descriptions, listed steps, or some combination of the two

Style: Use "objective" point of view ("2. The operator started the engine..."), as opposed to "command" point of view ("2. Start the engine...")

INSTRUCTIONS

Purpose: Describe a sequence of steps in such a way that the reader can *perform* the sequence of steps

Format: Employ numbered or bulleted lists, organized into subgroups of easily understandable units of information

Style: Use "command" point of view ("3. Plug the phone jack into the recorder unit"), as opposed to "objective" point of view ("3. The phone jack was plugged into the recorder unit")

FIGURE 6–1   Process descriptions versus instructions

ument, as in Model 6–1. More often, however, you use process description only as a pattern of organization within a document with a larger purpose. The following McDuff examples (1) show the supporting purpose of process descriptions and (2) reinforce the difference between process descriptions and instructions.

- **Accounting:** As an accountant at McDuff's corporate office, you have just finished auditing the firm's books. Now you must write a report to the vice president for business and marketing on the state of the firm's finances. Along with your findings, the vice president wants an overview of the procedure you followed to arrive at your conclusions.

- **Maintenance:** As a maintenance supervisor at McDuff's Saudi Arabian branch, you traveled to a construction site to repair a machine that tests the strength of concrete. The procedure requires billing the client for additional charges. Along with your bill, you send an attachment that summarizes the procedure you followed.

- **Laboratory work:** As lab supervisor at the St. Louis office, you spent all day Saturday in the lab assembling a new gas chromatograph needed to analyze gases. To justify the overtime hours, you write a memo to your manager describing the assembly process.

- **Welding inspection:** As an NDT (nondestructive testing) expert at McDuff's San Francisco office, you were hired by a California state agency to x-ray all welds at a bridge damaged by an earthquake. The text of your report gives test results; the appendices describe the procedures you followed.

- **Marketing:** As McDuff's marketing manager, you have devised a new procedure for tracking contacts with prospective clients (from first sales call to getting the job). You must write a memo to McDuff's two vice presidents for operations, briefly describing the process. Their approval is needed before the new marketing technique can be introduced at the 15 branch offices.

In each case, you are writing for a reader who wants to know what has happened or will happen, but who does *not* need to perform the process.

Instructions at McDuff

Think of instructions this way: They must provide users with a road map to *do* the procedure, not just understand it. That is, someone must complete a task on the basis of words and pictures you provide. Clearly, instructions present you, the writer, with a much greater challenge *and* risk. The reader must be able to replicate the procedure without error and, most importantly, with full knowledge of any dangers. The McDuff situations that follow reflect this challenge. Note that they parallel the case studies presented for process descriptions.

- **Accounting:** As McDuff's lead accountant for the last 20 years, you have always been responsible for auditing the firm's books. Because you developed the procedure yourself over many years, there is no comprehensive set of instructions for completing it. Now you want to record the steps so that other company accountants besides you can perform them.
- **Maintenance:** As a maintenance supervisor at McDuff's branch in Saudi Arabia, you must repair a piece of equipment for testing concrete. You have never disassembled this particular machine, and there are no manufacturer's instructions available. Therefore, the job takes you three full days. To help other employees perform this task in the future, you write a set of detailed instructions for making the repair.
- **Laboratory work:** As lab supervisor for McDuff's St. Louis office, you have assembled one of the two new gas chromatographs just purchased by the company. You are supposed to send the other unit to the Tokyo branch, where it will be put together by Japanese technicians. Unfortunately, the manufacturer's instructions are poorly written, so you plan to rewrite them for the English-speaking technicians at the Tokyo office.
- **Welding inspection:** As McDuff's NDT (nondestructive testing) expert at the San Francisco office, you have seen a large increase in NDT projects. Given California's aging bridges and constant earthquake activity, you persuaded your branch manager to hire several NDT technicians. Now you must write a training manual that will instruct these new employees on methods for inspecting bridge welds.
- **Marketing:** As McDuff's marketing manager, you have suggested a new approach for tracking sales leads. Having had your proposal approved by the corporate staff, you now need to explain the marketing procedure to technical professionals at all 15 offices. Your written set of instructions must be understood by technical experts in many fields, who have little if any marketing experience.

In each case, your instructions must describe steps so thoroughly that the reader will be able to replicate the process, *without* having to speak in person with the writer of the instructions. The next two sections give rules for preparing both process descriptions and sets of instructions.

GUIDELINES FOR PROCESS DESCRIPTIONS

You have already learned that process descriptions are aimed at persons who need to understand the process, not perform it. Process descriptions often have these purposes:

- Describing an experiment
- Explaining how a machine works
- Recording steps in developing a new product
- Describing what happened during a field test

In each case, follow these guidelines for creating first-rate process descriptions:

● *Process Guideline 1: Know Your Purpose and Audience*

Your intended purpose and expected audience influence every detail of your description. Here are some preliminary questions to answer before writing:

- Are you supposed to give just an overview or are details needed?
- Do readers understand the technical subject or are they laypersons?
- Do readers have mixed technical backgrounds?
- Does the process description supply supporting information (perhaps in an appendix) or is it the main part of the document?

Process descriptions are most challenging when directed to a mixed audience. In this case, write for the lowest common denominator—that is, for your least technical readers. It is better to write below the level of your most technical readers than to write above the level of your nontechnical readers.

For example, the process description in Model 6–3 on page 203 is directed to a mixed audience of city officials—some technical staff and some nontechnical political officials. It is contained in an appendix to a long McDuff report that recommends immediate cleanup of a toxic-waste dump. Note that the writer either uses nontechnical language or defines any technical terms used.

● *Process Guideline 2: Follow the ABC Format*

In chapter 3 you learned about the ABC format (**A**bstract/**B**ody/**C**onclusion), which applies to all documents. The abstract gives a summary, the body supplies details, and the conclusion provides a wrap-up or leads to the next step in the communication process. Whether a process description forms all or part of a document, it usually subscribes to the following version of the three-part ABC plan:

- The **Abstract** component includes three background items:

1. Purpose statement
2. Overview or list of the main steps that follow
3. List of equipment or materials used in the process

Steps of process
are embedded in
paragraph.

A. PARAGRAPH OPTION

The home owner should select rough-grade 2 x 4s for building the wooden form for
the patio. The form is just a box, with an open top and with the ground for the
bottom, into which concrete will be poured. First the four sides are nailed together,
and then the form is leveled with a standard carpenter's level. Finally, 2 x 4 stakes
are driven into the ground about every 2 or 3 ft on the outside of the form, to keep it
in place during the pouring of the concrete.

B. LIST OPTION

After brief lead-in,
steps of process
are placed in list
format.

Building a wooden form for a home concrete patio can be accomplished with some
rough-grade 2 x 4s. This form is just a box with an open top and the ground for the
bottom. Building involves three basic steps:

1. Nailing 2 x 4s into the intended shape of the patio
2. Leveling the box-shaped form with a standard carpenter's level
3. Nailing stakes (made from 2 x 4 lumber) every 2 or 3 ft at the outside edge of
   the form, to keep it in place during the pouring of the concrete

FIGURE 6–2 Two options for process description: (A) paragraph option; (B) list option

Model 6–3 includes all three, with a separate heading for equipment. First, the
purpose statement places the description in the context of the entire document.
Then the list of main steps gives readers a framework for interpreting details that follow. Finally, the list of equipment or materials provides a central reference point as
readers work through all the steps.

- The **Body** component of the process description moves logically through the steps
  of the process. By definition, all process descriptions follow a chronological, or
  step-by-step, pattern of organization. These steps can be conveyed in two ways:

1. **Paragraphs:** This approach weaves steps of the process into the fabric of typical
   paragraphs, with appropriate transitions between sentences. Use paragraphs
   when your readers would prefer a smooth explanation of the entire process,
   rather than emphasis on individual steps.
2. **List of steps:** This approach includes a list of steps, usually with numbers or bullets. Much like instructions, a listing emphasizes the individual parts of the
   process. Readers prefer it when they will need to refer to specific steps later on.

   Both paragraph and list formats have their places in process descriptions.
In fact, most descriptions can be written in either format. See Figure 6–2 for a
McDuff example showing both a paragraph and list description for the same
process of laying a concrete patio. As a public-service gesture, McDuff, Inc., has
written a pamphlet that briefly describes simple home improvements. It is
intended for owners of small homes who complete renovations with little or no
help from contractors. If home owners are interested in one of the projects, they
can write for detailed instructions to an address listed in the pamphlet.

   Building a concrete patio is one project covered; the process description
contains a subsection about constructing the wooden form into which concrete
is poured.

- The **Conclusion** component of a process description keeps the process from ending abruptly with the last step. Here you should help the reader put the steps together into a coherent whole. When the process description is part of a larger document, you can show how the process fits into a larger context (see Model 6–3).

● *Process Guideline 3: Use an Objective Point of View*

Process descriptions explain a process rather than direct how it is to be done. Thus they are written from an objective point of view—*not* from the personal "you" or "command" point of view common to instructions. Note the difference in these examples:

Process: The concrete is poured into the two-by-four frame.
or
The technician pours the concrete into the two-by-four frame.
Instructions: Pour the concrete into the two-by-four frame.

The process excerpts *explain* the steps, whereas the instructions excerpt *gives a command* for completing the instructions.

● *Process Guideline 4: Choose the Right Amount of Detail*

Only a thorough audience analysis will tell you how much detail to include. Model 6–3 (p. 203), for example, could have contained much more technical detail about the substeps for testing air quality at the site. The writer, however, decided that the city officials would not need more scientific and technical detail.

In supplying specifics, be sure to subdivide information for easy reading. In paragraph format, headings and subheadings can be used to make the process easier to grasp. In list format, an outline arrangement of points and subpoints may be appropriate. When such detail is necessary, remember this general rule of thumb: *Place related steps into groups of from three to seven points.* Readers find it easier to remember several groupings with subpoints, as opposed to one long list. Following are two rough outlines for a process description. The second is preferred in that it groups the many steps into three easily grasped categories.

Employment Interview Process
1. Interviewer reviews job description
2. Interviewer analyzes candidate's application
3. Candidate and interviewer engage in "small talk"
4. Interviewer asks open-ended questions related to candidate's resume and completed application form
5. Interviewer expands topic to include matters of personal interest and the candidate's long-term career plans
6. Interviewer provides candidate with information about the position (salary, benefits, location, etc.)
7. Candidate is encouraged to ask questions about the position
8. Interviewer asks candidate about her or his general interest, at this point, in the position
9. Interviewer informs candidate about next step in hiring process

Employment Interview Process

- **Preinterview Phase**

1. Interviewer reviews job description
2. Interviewer analyzes candidate's application

- **Interview**

3. Candidate and interviewer engage in "small talk"
4. Interviewer asks open-ended questions related to candidate's resume and completed application form
5. Interviewer expands topic to include matters of personal interest and the candidate's long-term career plans
6. Interviewer provides candidate with information about the position (salary, benefits, location, etc.)
7. Candidate is encouraged to ask questions about the position

- **Closure**

8. Interviewer asks candidate about his or her general interest, at this point, in the position
9. Interviewer informs candidate about next step in hiring process

- ### *Process Guideline 5: Use Flowcharts for Complex Processes*

Some process descriptions contain steps that are occurring at the same time. In this case, you may want to supplement a paragraph or list description with a flowchart. Such charts use boxes, circles, and other geometric shapes to show progression and relationships among various steps.

Model 6–4 on page 204, for example, shows a flowchart and an accompanying process description at McDuff. Both denote services that McDuff's London branch provides for oil companies in the North Sea. The chart helps to demonstrate that the geophysical study (mapping by sonar equipment) and the engineering study (securing and testing of seafloor samples) take place at the same time. Such simultaneous steps are difficult to show in a list of sequential steps.

GUIDELINES FOR INSTRUCTIONS

Rules change considerably when moving from process descriptions to instructions. Although both patterns are organized by time, the similarity stops there. Instructions walk readers through the process so that they can *do* it, not just understand it. It is one thing to explain the process by which a word processing program works; it is quite another to write a set of instructions for using that word processing program. This section explores the challenge of writing instructions by giving you some basic writing and design guidelines.

These guidelines for instructions also apply to complete operating *manuals,* a document type that many technical professionals will help to write during their careers.

Those manuals include the instructions themselves, as well as related information such as (1) features, (2) physical parts, and (3) troubleshooting tips. In other words, manuals are complete documents, whereas instructions can be part of a larger piece.

● *Instructions Guideline 1: Select the Correct Technical Level*

This guideline is just another way of saying you need to know *exactly* who will be reading your instructions. Are your readers technicians, engineers, managers, general users, or some combination of these groups? Once you answer this question, select language that every reader can understand. If, for example, the instructions include technical terms or names of objects that may not be understood, use the techniques of definition and description discussed in the previous chapter.

● *Instructions Guideline 2: Provide Introductory Information*

Like process descriptions, instructions follow the ABC format (**A**bstract/**B**ody/**C**onclusion) described in chapter 3. The introductory (or abstract) information should include (1) a purpose statement, (2) a summary of the main steps, and (3) a list or an illustration giving the equipment or materials needed (or a reference to an attachment with this information). These three items set the scene for the procedure itself.

Besides these three "musts," you should consider whether some additional items might help set the scene for your user:

- Pointers that will help with installation
- Definitions of terms
- Theory of how something works
- Notes, cautions, warnings, or dangers that apply to all steps

● *Instructions Guideline 3: Use Numbered Lists in the Body*

A simple format is crucial to the body of the instructions—that is, the steps themselves. Most users constantly go back and forth between these steps and the project to which they apply. Thus you should avoid paragraph format and instead use a simple numbering system. Model 6–5 on pages 205–206 shows a "before and after" example. The original version is written in paragraphs that are difficult to follow; the revised version includes nine separate, numbered steps.

● *Instructions Guideline 4: Group Steps under Task Headings*

Readers prefer that you group together related steps under headings, rather than present an uninterrupted "laundry list" of steps. Model 6–6 on pages 207–209 shows how this technique has been used in a fairly long set of instructions for operating an answering machine. Given the number of steps in this case, the writer has used a separate numbering system within each grouping.

Groupings provide two main benefits. First, they divide fragmented information into manageable "chunks" that readers find easier to read. Second, they give readers a sense of accomplishment as they complete each task, on the way to finishing the whole activity.

● *Instructions Guideline 5: Place One Action in a Step*

A common error is to "bury" several actions in a single step. This approach can confuse and irritate readers. Instead, break up complex steps into discrete units, as shown here:

- **Original:**

  Step 3: Fill in your name and address on the coupon, send it to the manufacturer within two weeks, return to the retail merchant when your letter of approval arrives from the manufacturer, and pick up your free toaster oven.

- **Revision:**

  Step 3: Fill in your name and address on the coupon.

  Step 4: Send the coupon to the manufacturer within two weeks.

  Step 5: Show your retail merchant the letter of approval after it arrives from the manufacturer.

  Step 6: Pick up your free toaster oven.

● *Instructions Guideline 6: Lead Off Each Action Step with a Verb*

Instructions should include the "command" form of a verb at the start of each step. This style best conveys a sense of action to your readers. Models 6–5 and 6–6 on pages 205–209 consistently use command verbs for all steps throughout the procedures.

● *Instructions Guideline 7: Remove Extra Information from the Step*

Sometimes you may want to follow the command sentence with an explanatory sentence or two. In this case, distinguish such helpful information from actions by giving it a label, such as "Note" or "Result" (for example, see Model 6–2, pp. 201–202).

● *Instructions Guideline 8: Use Bullets or Letters for Emphasis*

Sometimes you may need to highlight information, especially within a particular step. Avoid using numbers for this purpose, since you are already using them to signify steps. Bullets work best if there are just a few items; letters are best if there are many, especially if they are in a sequence. The revised version in Model 6–5 shows the appropriate use of letters, and Model 6–6 shows the use of bullets.

In particular, consider using bullets at any point at which users have an *option* as to how they will respond. The following example uses bullets in this way; it also eliminates the problem of too many actions being embedded in one step.

Part of Procedure for Firing Clay in a Kiln

(*Note:* A pyrometric "cone" is a piece of test clay used in a kiln, an oven for baking pottery. The melting of the small cone helps the operator determine that the clay piece has completed the firing process.)

- **Original:**

  Step 6: Check the cone frequently as the kiln reaches its maximum temperature of 1850 degrees. If the cone retains its shape, continue firing the clay and checking the cone frequently. When the cone begins to bend, turn off the kiln. Then let the kiln cool overnight before opening it and removing the pottery.

- **Revision:**

    Step 6: Check the cone frequently as the kiln reaches its maximum temperature of 1850 degrees.

    Step 7: Has the cone started to bend?

- If *no,* continue firing the piece of pottery and checking the cone frequently to see if it has bent.
- If *yes,* turn off the kiln.

    Step 8: Let the kiln cool overnight after turning it off.

    Step 9: Open the kiln and remove the pottery.

- ## *Instructions Guideline 9: Emphasize Cautions, Warnings, and Dangers*

Instructions often require drawing attention to risks in using products and equipment. Your most important obligation is to highlight such information. Unfortunately, professional associations and individual companies may differ in the way they use and define terms associated with risk. You must be certain to use language and/or graphics your reader will understand. If you have no specific guidelines, however, the definitions below can serve as "red flags" to the reader. The level of risk increases as you move from 1 to 3:

1. **Caution:** possibility of damage to equipment or materials
2. **Warning:** possibility of injury to people
3. **Danger:** probability of injury or death to people

If you are not certain that these distinctions will be understood by your readers, define the terms *caution, warning,* and *danger* in a prominent place before you begin your instructions.

As for placement of the actual cautions, warnings, or danger messages, here are your options:

- **Option 1:** *In a separate section right before the instructions begin.* This approach is most appropriate when you have a list of general warnings that apply to much of the procedure *or* when one special warning should be heeded throughout the instructions—for example, "DANGER: Keep main breaker on 'off' during entire installation procedure." Figure 6–3 shows such a warning at the start of instructions to install a security keypad.

- **Option 2:** *In the text of the instructions.* This approach works best if the caution, warning, or danger message applies to the step that immediately follows it. Thus users are warned about a problem *before* they read the step to which it applies. For an example, see the following:

    CAUTION: Use 220-grade sandpaper, to avoid scratching the surface of furniture.

- **Option 3:** *Repeatedly throughout the instructions.* This strategy is preferable with instructions that pose risk to the user over and over again. For example, steps 4, 9, 12A, and 22—appearing on several different pages—may *all* include the hazard

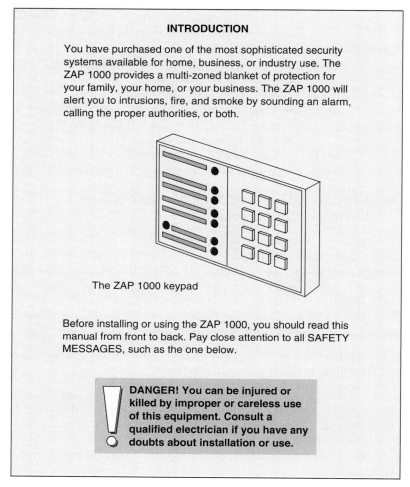

INTRODUCTION

You have purchased one of the most sophisticated security systems available for home, business, or industry use. The ZAP 1000 provides a multi-zoned blanket of protection for your family, your home, or your business. The ZAP 1000 will alert you to intrusions, fire, and smoke by sounding an alarm, calling the proper authorities, or both.

The ZAP 1000 keypad

Before installing or using the ZAP 1000, you should read this manual from front to back. Pay close attention to all SAFETY MESSAGES, such as the one below.

DANGER! You can be injured or killed by improper or careless use of this equipment. Consult a qualified electrician if you have any doubts about installation or use.

FIGURE 6–3 Example of "danger" message

of fatal electrical shock. Your "Danger" notice should appear in each step, as well as in the introduction to the document.

Give information about potential risks *before* the operator has the chance to make the mistake. Also, the caution, warning, or danger message can be made visually prominent by using one of the following techniques:

- Underlining: <u>Warning</u>
- Bold: **Warning**
- Full Caps: WARNING
- Italics: *Warning*

- Oversized Print: Warning
- Boxing: Warning
- Color: Warning
- Combined Methods: *Warning*
                                WARNING
                                WARNING

Color graphics are another effective indicator of risk. You have probably seen examples such as a red flame in a box for fire, a jagged line in a triangle for electrical shock, or an actual drawing of a risky behavior with an "X" through it. Although there are as yet no common standards for such graphics, consider using them to reinforce the message in your text about cautions, warnings, and dangers.

● *Instructions Guideline 10: Keep a Simple Style*

Perhaps more than any other type of technical writing, instructions must be easy to read. Readers expect a no-nonsense approach to writing that gives them required information without fanfare. Here are some techniques to use:

- Keep sentences short, with an average length of under 10 words.
- Use informal definitions (parenthetical, like this one) to define any terms not understood by all readers.
- Never use a long word when a short one will do.
- Be specific and avoid words with multiple interpretations (*frequently, seldom, occasionally,* etc.).

● *Instructions Guideline 11: Use Graphics*

Illustrations are essential for instructions that involve equipment. Place an illustration next to every major step when (1) the instructions or equipment is quite complicated or (2) the audience may contain poor readers or people who are in a hurry. Such word-picture associations create a page design that is easy to follow.

In other cases, just one or two diagrams may suffice for the entire set of instructions. The one reference illustration in Model 6–6 (pp. 207–209) helps the user of a message recorder locate parts mentioned throughout the instructions.

Another useful graphic in instructions is the table. Sometimes within a step you need to show correspondence between related data. For example, the instructions that follow would benefit from a list.

- **Original:**
    Step 3: Use pyrometric cones to determine when a kiln has reached the proper temperature to fire pottery. Common cone ratings are as follows: a Cone 018 corresponds to 1200°F; a Cone 07 corresponds to 1814°F; a Cone 06 corresponds to 1859°F; and a Cone 04 corresponds to 1940°F.

Communication Challenge

"McDuff's Home of Hope: The Good, the Bad, and the Ugly?"

Recently McDuff's Atlanta office decided to change its approach to charitable giving at the branch. Instead of supporting various regional charities, employees could participate in a local project of their own—converting an abandoned building into a homeless shelter called "Home of Hope." The idea seemed to be a creative way to make a personal contribution to the community. What follows is a description of the stages of the project and some questions and comments for discussion.

Project Planning

The process of making "Home of Hope" a reality began at McDuff-Atlanta's annual employee meeting last year. The human resources manager suggested that the office try a new approach to annual giving, and the office supervisors agreed to investigate. Eventually the office decided to purchase an abandoned brick building on an acre lot in downtown Atlanta, in an area where homeless people often congregated.

McDuff conducted a preliminary study of the land and building, calculating that the project would cost about $100,000. Management developed a formula whereby the company would pay a 20% mortgage down payment from its savings and carry the monthly mortgage note. Then over a one-year period, the employees—through their annual financial contributions and personal labor—would renovate the house and add landscaping. The managers developed a suggested sliding scale for what money employees should contribute, based on their salaries. Managers and supervisors also were asked to meet individually with each employee to encourage contributions.

Once McDuff bought the land, the firm began benefiting from excellent publicity on local radio and in the papers. The media championed this effort by an Atlanta employer.

Site Problems

It appeared that nothing could go wrong—but something did. Ironically, considering that McDuff itself does environmental work, the firm found an environmental problem with the land that had not been detected before purchase. Apparently, part of the site had been used as a dumping ground for old car batteries and for chemicals from a nearby dry cleaners. Both the batteries and a large portion of soil would have to be removed, adding $10,000 to the cost of the project.

Just as bad, there were environmental surprises in the building itself. The company found some asbestos and lead paint that had not been detected before purchase. Removal would cost about $5,000. The increase in the total project cost irritated many employees, some of whom had been skeptical about the project from the start.

Employee Involvement

What did seem to go well were the weekend work groups that the company set up for the coming year. A group of 5 to 10 employees worked a half day on each Saturday, meaning that most employees would end up working three or four Saturdays during the entire year-long project. Employees were strongly encouraged to participate, and about 85% of them signed up for the groups.

As noted previously, through meetings with managers and other means, employees were encouraged to contribute the amount suggested on the sliding scale. About 75% agreed to the amount suggested, 10% pledged more, 10% pledged less, and 5% pledged nothing. Pledges were drawn from paychecks over the one-year period.

Community Involvement

Once the lot was purchased and the renovation designed, McDuff worked with groups in the surrounding community, making sure that local people were informed about the project. One home-owners group from this working-class neighborhood raised questions about the project attracting even more homeless people to the area. The group worried that the possibility of increasing crime would lower the value of their homes. McDuff decided that an open community meeting was in order.

At the meeting at a local school, McDuff produced speakers who suggested that the home would actually help decrease crime by giving shelter, meals, and activities to people who otherwise would be vagrants. Although the answers seemed to satisfy many, McDuff officials were on the defensive and wished they had done more networking with local residents.

Final Preparations

Once the home and yard were finished, McDuff hired two permanent staff members and set up a group of volunteers from the community. Retired people were especially active as volunteers. The company also asked for, and received, an ongoing commitment of $15,000 a year from the city to pay half the salary of the Home of Hope director.

With these details handled, the home took in its first 25 residents several months ago. McDuff arranged for media coverage of the opening celebration, inviting a diverse group of company leaders. Of course, the company also made sure the event was covered in the McDuff corporate newsletter and by *EnviroNews*, a national news magazine in engineering and science.

Continued

Questions and Comments for Discussion

1. The McDuff corporate managers have expressed interest in the charity "model" developed by the Atlanta office. Specifically, they want the Atlanta human resources director to write a **process description** for the Home of Hope project. The description will be reviewed by all McDuff branch managers. What major points should be included in this process description? How will it differ from the way information is presented in the case just described?

2. Assume McDuff's corporate office has actually adopted a community-based charity option, such as that reflected by the Home of Hope project. Now it wants to provide project **instructions** for other urban offices that may want to build shelters. What major points will be emphasized in the instructions and in what order? What particular problems did the Atlanta office encounter, and how can the instructions be written to help other offices avoid such problems? In other words, how will the *ideal* set of instructions differ from the *actual* process that was performed?

3. Answer these questions first with regard to McDuff employees and second with regard to the community surrounding Home of Hope. What tactical mistakes, if any, were made by McDuff management in the process of promoting, communicating, and running this project? How could the problems have been avoided?

4. Are there any ethical problems revealed in the process described in this case? Specifically, how do you feel about the manner by which employees are encouraged to contribute to such causes?

5. Several large charity groups were disturbed that McDuff dropped them and instead involved employees in the Home of Hope. Give what you think would be the charities' point of view about the process described in this case.

- **Revision:**
  - **Step 3:** Use pyrometric cones to determine when a kiln has reached the proper temperature for firing pottery. Common cone ratings are as follows:

Cone 018	1200°F
Cone 07	1814°F
Cone 06	1859°F
Cone 04	1940°F

- ### *Instructions Guideline 12: Test Your Instructions*

Professional writers often test their instructions on potential users before completing the final draft. The most sophisticated technique for such testing involves a "usability laboratory." Here test subjects are asked to use the instructions or manual to perform the process, often while speaking their observations and frustrations (if any). The writers or lab personnel unobtrusively observe the process from behind a one-way mirror. Later they may review audiotaped or videotaped observations of the test subjects or they may interview these persons. This complex process helps writers to anticipate and then eliminate problems that users will confront when they follow written instructions.

Of course, you probably will not have access to a usability laboratory to test your instructions. However, you can adapt the following user-based approach to testing assignments in this class and projects in your career. Specifically, follow these four steps:

1. Team up with another class member (or a colleague on the job). This person should be unfamiliar with the process and should approximate the technical level of your intended audience.

2. Give this person a draft of your instructions and provide any equipment or materials necessary to complete the process. Of course, for the purposes of a class assignment, this approach would work only for a simple process with little equipment or few materials.

3. Observe your colleague following the instructions you have provided. You should record both your observations and any responses this person makes while moving through the steps.

4. Revise your instructions to solve problems your user encountered during the test.

CHAPTER SUMMARY

Both process descriptions and instructions share the same organization principle: time. That is, both relate a step-by-step description of events. Process descriptions address an audience that wants to be informed but does not need to perform the process itself. Instructions are geared specifically for persons who need to complete the procedure themselves.

In writing good process descriptions, follow these basic guidelines:

1. Know your purpose and audience.
2. Follow the ABC format.
3. Use an objective point of view.
4. Choose the right amount of detail.
5. Use flowcharts for complex processes.

For instructions, follow these twelve rules:

1. Select the correct technical level.
2. Provide introductory information.
3. Use numbered lists in the body.
4. Group similar steps under heads.
5. Place one action in a step.
6. Lead off each action step with a verb.
7. Remove extra information from the step.

8. Use bullets or letters for emphasis.
9. Emphasize cautions, warnings, and dangers.
10. Keep a simple style.
11. Use graphics.
12. Test your instructions.

 Access the Technical Writing Web site by pointing your browser at www.prenhall.com/pfeiffer. Here, you will find additional resources to aid you in learning and applying the information from this chapter.

ASSIGNMENTS

Part 1: Short Assignments

Assignments 1–5 can be completed either as individual exercises *or* as group projects, depending on the instructions you are given in class.

1. **Writing a Process Description—School-Related.** Your college or university has decided to evaluate the process by which students are advised about, and registered for, classes. As part of this evaluation, the registrar has asked a select group of students—you among them—to describe the actual process each of you went through individually during the last advising/registration cycle. These case studies collected from individual students, the customers, will be transmitted directly to a college-wide committee studying registration and advising problems.

   Your job is to give a detailed account of the process. Remain as objective as possible, without giving opinions. If you had problems during the process, the facts you relate will speak for themselves. Simply describe the process you personally experienced. Then let the committee members judge for themselves whether the steps you describe should or should not be part of the process.

2. **Writing a Process Description—McDuff Context.** Choose one of the projects from the color insert in chapter 2. For this assignment use (a) points listed in the "Main Technical Tasks" section of the project you have chosen or (b) related information you wish to supply from your own experience, reading, or imagination, or (c) both. Your assignment is to write a process description dealing with one or more of the bulleted points.

3. **Writing Instructions—McDuff Context.** Choose one of the projects from the color insert in chapter 2. For this assignment, conduct some research on either (a) one task or several related tasks in the "Main Technical Tasks" section or (b) a task that conceivably could be related to the project but is not specifically listed. Then write a set of numbered instructions for the task(s). Following are some sample tasks, with project references:

   - Project #1: Conducting a lab test to evaluate a soil sample
   - Project #2: Estimating the age of a geologic sample
   - Project #3: Inspecting all or part of a new building
   - Project #4: Running an effective meeting with subcontractors
   - Project #5: Testing soil or water for signs of pollution
   - Project #6: Evaluating the ergonometric features of a control panel
   - Project #7: Evaluating the effectiveness of a technical report

   If you prefer to write about less specific tasks, select a general topic related to a project. For example, here are four general topics derived from the specific ones just listed:

conducting a lab test, running an effective meeting, evaluating the ergonometric features of any product, and evaluating the effectiveness of any document.

4. **Writing Instructions—McDuff Context.** As an employee at the corporate office of McDuff, you just received the job of writing a set of instructions for completing performance appraisal reviews (PARs). The instructions will be included in a memo that goes to all supervisors at all branches of the firm, along with related forms. To help you get started on the instructions, you have been given a narrative description of the process (see the following). Your task is to convert this narrative into a simple set of instructions to go into the memorandum to supervisors.

PARs are conducted annually for each employee, during the anniversary month in which the employee was originally hired. Several days before the month in which the PARs are to be conducted, the corporate office will send each supervisor a list of employees in that supervisor's group who should receive PARs. The main part of the PAR process is an interview between the supervisor and the employee receiving the PAR. Before this interview takes place, however, the supervisor should give the employee a copy of the "McDuff PAR Discussion Guide," which offers suggestions for the topics and tone of a PAR interview. The supervisor completes a "PAR Report Form" after each interview and then sends a copy to corporate and to the employee, with the original staying in the personnel files of that respective supervisor's branch. If for any reason a PAR interview and report form are not completed in the required month, the supervisor must send a memo of explanation to the corporate Human Resources Department, with a copy to the supervisor's branch manager.

5. **Writing Instructions—School-Related.** In either outline or final written form, provide a set of instructions for completing assignments in this class. Consider your audience to be another student who has been ill and missed much of the term. You have agreed to provide her with an overview that will help her to plan and then write any papers she has missed.

Your instructions may include (1) highlights of the writing process from chapter 1 and (2) other assignment guidelines provided by your instructor in the syllabus or in class. Remember to present a generic procedure for all assignments in the class, not specific instructions for a particular assignment.

Part 2: Longer Assignments

These assignments test your ability to write and evaluate the two patterns covered in this chapter—process descriptions and instructions. Specifically, follow these guidelines:

- Write each exercise in the form of a letter report or memo report, as specified.
- Follow organization and design guidelines given in chapters 3 and 4, especially concerning the ABC format (**A**bstract/**B**ody/**C**onclusion) and the use of headings. Chapter 8 gives rules for short reports, but such detail is not necessary to complete the assignments here.
- Fill out a Planning Form (at the end of the book) for each assignment.

6. **Evaluating a Process Description.** Using a textbook in a technical subject area, find a description of a process. For example, a physics text might describe the process of waves developing and then breaking at a beach, an anatomy text might describe the process

of blood circulating, or a criminal justice text might describe the process of a criminal investigation.

Keeping in mind the author's purpose and audience, evaluate the effectiveness of the process description as presented in the textbook. Submit your evaluation in the form of a memo report to your instructor in this writing course, along with a copy of the textbook description.

For the purposes of this assignment, assume that your writing instructor has been asked by the publisher of the text you have chosen to review the book as an example of good or bad technical writing. Thus your instructor would incorporate comments from your memo report into his or her comprehensive evaluation.

7. **Writing a Process Description—School-Related.** Conduct a brief research project in your campus library. Specifically, use company directories, annual reports, or other library sources to find information about a company or other organization that could hire students from your college.

In a memo report to your instructor, (1) describe the process you followed in conducting the search and (2) provide an outline or paragraph summary of the information you found concerning the company or organization. Assume that your report will become part of a volume your college is assembling for juniors and seniors who are beginning their job search. These students will benefit both from information about the specific organization you chose and from a description of the process that you followed in getting the information—since they may want to conduct research on other companies.

8. **Writing a Process Description—McDuff Context.** As a project manager for McDuff's Atlanta office, you just found out that your office has been selected as one of the firms to help renovate Kiddieworld, a large amusement park in the Southeast. Before Kiddieworld officials sign the contract, however, they want you to report on the process McDuff uses to report and investigate accidents (since the project will involve some hazardous work). You found the following policy in your office manual, but you know it is not something you would want to send to a client. Take this stilted paragraph and convert it to a process description for your clients, in the form of a letter report. Remember: The readers will not be performing the process; they only want to understand it.

Accident reporting and investigation are an important phase of operating McDuff, Inc. The main purpose of an accident investigation and report is to gain an objective insight into facts surrounding the accident in order to improve future accident control measures and activities and to activate the protection provided by our insurance policies. It is therefore imperative that all losses, no matter how minor, be reported as soon as possible and preferably within 48 hours to the proper personnel. Specifically, all accidents must be reported orally to the immediate supervisor. For minor accidents that do not involve major loss of equipment or hospitalization, that supervisor has the responsibility of filling out a McDuff accident report form and then sending the form to the safety personnel at the appropriate branch office, who later sends it to the safety manager at the corporate office. For serious accidents that involve major loss of equipment or hospitalization of any individuals involved, the supervisor must call or telex the safety personnel at the appropriate branch office, who then should call or telex the safety manager at the corporate office. (A list of pertinent telephone numbers should be kept at every job site.) These oral reports will be followed up with a written report.

9. **Evaluation of Instructions.** Find a set of operating or assembly instructions for a VCR, microwave oven, CD player, computer, timing light, or other electronic device. Evaluate all or part of the document according to the criteria for instructions in this chapter.

   Write a memo report on your findings and send it, along with a copy of the instructions, to Natalie Bern. As a technical writer at the company that produced the electronic device, Natalie wrote the set of instructions. In your position as Natalie's supervisor, you are responsible for evaluating her work. Use your memo report either to compliment her on the instructions or to suggest modifications.

10. **User Test of Instructions.** Find a relatively simple set of instructions. Then ask another person to follow the instructions from beginning to end. Observe the person's activity, keeping notes on any problems she or he encounters.

    Use your notes to summarize the effectiveness of the instructions. Present your summary as a memo report to Natalie Bern, using the same situational context as described in assignment 7. That is, as Natalie's boss, you are to give her your evaluation of her efforts to produce the set of instructions.

11. **Writing Simple Instructions.** Choose a simple office procedure of 20 or fewer steps (for example, changing a printer ribbon, filling a mechanical pencil, adding dry ink to a copy machine, or adding paper to a laser printer). Then write a simple set of instructions for this process, in the form of a memo report. Your readers are assistants at the many offices of a large national firm. Consider them to be new employees who have no background or experience in office work and no education beyond high school. You are responsible for their training.

12. **Writing Complex Instructions, with Graphics—Group Project.** Complete this assignment as a group project (see the guidelines for group work in chapter 1). Choose a process connected with college life or courses—for example, completing a lab experiment, doing a field test, designing a model, writing a research paper, getting a parking sticker, paying fees, registering for classes, etc.

    Using memo report format, write a set of instructions for students who have never performed this task. Follow all the guidelines in this chapter. Include at least one illustration (along with warnings or cautions, if appropriate). If possible, conduct a user test before completing the final draft.

13. **Writing Instructions—McDuff Context.** McDuff does a good deal of environmental work around the country—cleaning up toxic-waste sites, building energy-efficient structures, removing asbestos from old buildings, and investigating construction sites to determine the most environmentally sound approach to design and construction. For business reasons—and also because of its sense of civic duty—the company encourages citizens to get directly involved in environmental action.

    As public relations manager for McDuff, you have just received an interesting assignment from the president, Rob McDuff. He wants you to prepare a set of instructions that will go out to citizen and school groups in the Baltimore-Washington area. In the form of a memo report, this document should give readers specific directions for recycling one or more types of waste. Your instructions should be directed toward a broad audience, of course. Moreover, they should give the kinds of details that allow someone to act without having to get more information.

    To get information for this report, you might consider (1) calling individuals in the waste-management department of your local government, (2) reading relevant articles from recent periodicals, or (3) checking an environmental science textbook at your college.

14. **Writing Instructions—Group Project with McDuff Context.** McDuff, Inc.'s, increasing international work has generated interest among the corporate staff in gaining ISO 9000

certification. (Based in Geneva, Switzerland, the International Organization for Standardization [ISO] helps organizations around the world develop standards in quality.) Your group will conduct some research on this topic of growing interest. Write a set of instructions for a company, like McDuff, that wishes to gain such certification. You may either (a) provide a generalized overview for completing the entire process or (b) focus on one limited, specific part of the process, such as the process for gaining certification for a particular product or service.

15. **Evaluating Document with Embedded Instructions.** A McDuff lab supervisor, Kerubo Awala, has very little time to train a new group of lab trainees who have little if any industrial lab experience. Although she has given the new recruits a detailed description of a Bunsen burner (see Model 5–3 in chapter 5), she also wants them to have a short, easy-to-read document that focuses on use and safety. For that purpose, she quickly assembled and distributed the document on the opposite page. Evaluate its effectiveness for her intended purpose and audience. How could it be improved?

16. **Ethics Assignment.** Examine a set of instructions for a household or recreational device that—either in assembly or use—poses serious risk of injury or death. Evaluate the degree to which the manufacturer has fulfilled its ethical responsibility to inform the user of such risk. You may want to consider the following questions, among others:
   a. Are risks adequately presented in text and/or graphic form?
   b. Are risk notices appropriately placed in the document?
   c. Is the document designed such that a user reading quickly could locate cautions, warnings, or dangers easily?

   If you have highlighted any ethical problems, also suggest solutions to these problems.

17. **International Communication Assignment.** Sets of instructions may reflect "cultural bias" of a particular culture or country. Such bias may be acceptable if the audience for the instructions shares the same background. However, cultural bias presents a problem (1) when the audience represents diverse cultures and backgrounds or (2) when the instructions need to be translated into another language by someone not familiar with cultural cues in the instructions. Following are just a few categories of information that can present cultural bias and possibly cause confusion:

   ● Date formats
   ● Time zones
   ● Types of monetary currency
   ● Units of measurement
   ● Address and telephone formats
   ● Historical events
   ● Geographical references
   ● Popular culture references
   ● Acronyms
   ● Legal information
   ● Common objects in the home or office

   [adapted from list on pp. 129–130 of Nancy L. Hoft's *International Technical Communication: How to Export Information About High Technology* (New York: John Wiley, 1995).]

   Choose a set of instructions that reflects several types of cultural bias, such as those included on the previous list. Point out the examples of bias and explain why they might present problems to readers outside a particular culture.

BUNSEN BURNER

Purpose

The Bunsen burner is used in experiments that require you to heat solutions in a glass container (flask).

Warning

The Bunsen burner produces an open flame that can burn you, burn other students, or set fire to flammable materials.

- Use it only when your lab instructor tells you.
- Apply the flame to glass or metal containers only.

Controls

This diagram shows where to connect the gas line to the Bunsen burner and where to light and control the flame.

Light **flame**

Adjust **flame temperature**

cooler   hotter

high

low

Adjust **flame size**

Connect **gas line**

MEMORANDUM

DATE: May 29, 2002
TO: Leonard Schwartz
FROM: Your Name
SUBJECT: New Electronic-Mail System

States purpose
clearly.

Yesterday I met with Jane Ansel, the installation manager at BHG Electronics, about our new electronic-mail system. Ms. Ansel explained the process by which the system will be installed. As you requested, this memo summarizes what I learned about that process.

BHG technicians will be at our offices on June 18 to complete these five tasks:

Describes five
main tasks,
using parallel
grammatical form.

1. Removing old cable from the building conduits
2. Laying cable to link remaining unconnected terminals with the central processing unit in the main frame
3. Installing software in the system that will give each terminal the capacity to operate the electronic-mail system
4. Testing each terminal to make sure that the system can operate from that location
5. Instructing selected managers on the use of the system

Confirms the
follow-up activities
they have already
discussed.

As you and I have agreed, next week I will send a memo to all office employees who will have access to electronic mail. That memo will mention the installation date and summarize the procedures for reading mail. Shortly thereafter, I will send another memo instructing them about sending electronic mail.

Gives reader
opportunity to
respond.

Let me know, Leonard, if you have further suggestions about how I can help make our transition to electronic mail as smooth as possible.

MODEL 6–1
McDuff process description: electronic mail

MEMORANDUM

DATE: June 5, 2002
TO: All Employees with Computer Terminals
FROM: Your Name
SUBJECT: Basic Instructions for Reading Electronic Mail

Gives clear purpose.

    Last month, you attended a brief seminar on the features of the new electronic-mail system. We have just learned that the system will be installed on June 18. This memo provides some basic instructions for reading mail sent to you on this system. Soon you will receive another set of instructions for sending electronic mail.

NOTE: In these instructions, the messages or prompts on your terminal screen appear in *italics*. Any key you push or response you type is shown in **bold** print.

Indicates what instructions do and *do not* cover.

1. Turn on the computer terminal.
   NOTE: The on/off button is on the right front corner of the unit.

2. Type in your terminal's number when the system requests it.
   Example: *terminal number:* **23**

Provides information to help reader understand instructions that follow.

3. Respond with your initials when the system asks for "login."
   Example: *login:* **wsp**

4. Give your password at the next system prompt.
   Example: *password:* **Tex**
   RESULT: After the system has verified your password, it will respond with one of two messages: either *no mail* or *yes, you have mail.*

Limits each step to *one* action.

5. Respond to the message in one of these two ways:

Separates action from results.

If…	Then…
■ Screen reads "no mail"	■ Press "e" for "exit" ■ Begin another task on the terminal
■ Screen reads "yes, you have mail"	■ Press "r" for "read" ■ Continue with these instructions

Explains options clearly.

6. Read the first screen of your message.

7. Press the **"return"** key to discover whether there are additional screens with that message.

MODEL 6–2
McDuff instructions: electronic mail

continues

Memo to: Employees with Computer Terminals.   Page 2

8. Do you wish to return to other screens in that message?
   • If yes, press the "backspace."
   • If no, continue with these instructions.

9. Enter one of the abbreviations from the following list when you are finished with a message:

   s = save message
   d = delete message
   p = print message

10. Press "**m**" (for "**move**") to move on to next message.

11. Press "**e**" (for "**exit**") when the system indicates no more messages.

12. Turn off the machine or begin another activity with the terminal.

   As noted earlier, the system will be installed on June 18. Feel free to call me if you have any questions about the installation or the instructions for reading messages.

Uses list to show choices.

Restates important date and shows reader how to get more information.

MODEL 6–2
continued

ABC format
begins with
abstract—with
purpose
statement and
summary of
appendix in this
paragraph.

Abstract ends
with list of
equipment used
in process that
follows.

Body section of
this process
uses paragraph
format and is
aimed at
nontechnical
audience.

Listing is used to
highlight
locations for
sampling.

Conclusion part
of ABC format
puts this
process in larger
context.

APPENDIX A: ON-SITE MONITORING

The purpose of monitoring the air is to determine the level of protective equipment needed for each day's work. This appendix gives an overview of the process for monitoring on-site air quality each day. Besides describing the main parts of the process, it notes other relevant information to be recorded and the manner in which data will be logged.

EQUIPMENT

This process requires the following equipment:

- Organic vapor analyzers (OVAs)
- Combustible-gas instruments
- Personal sampling devices

PROCESS

The project manager at the site is responsible for supervising the technician who performs the air-quality tests. At the start of every day, a technician uses an OVA to check the quality of air at selected locations around the site. Throughout the workday (at times specified by the project manager), the technician monitors the air with combustible-gas instruments and personal sampling devices. This monitoring takes place at these locations:

1. Around the perimeter of the site
2. Downwind of the site (to determine the extent of migration of vapors and gases)
3. Generally throughout the site
4. At active work locations within the site

Then at the end of every workday, the technician uses the OVA to monitor the site for organic vapors and gases.

CONCLUSION

Besides the air-quality data, the following information is collected by the technician at each sampling time: percent relative humidity, wind direction and speed, temperature, and atmospheric pressure. The project manager keeps records of air quality and weather conditions in dated entries in a bound log.

MODEL 6–3
Process description

Steps 1 and 2 are
shown in top
center portion of
flowchart.

Steps 3 and 4 are
shown in left and
right portions of
flowchart,
respectively.

Step 5 is shown in
bottom center
portion of
flowchart.

Flowchart shows
relationship among
steps occuring at
the same time.

COMBINED SITE INVESTIGATION

In helping to select the site for an offshore oil platform, McDuff recommends a combined site investigation. This approach achieves the best results by integrating sophisticated geophysical work with traditional engineering activities.

As the accompanying flowchart shows, a combined site investigation consists of these main steps:

1. Planning the program, with McDuff's scientists and engineers and the client's representatives
2. Reviewing existing data
3. Completing a high-resolution geophysical survey of the site, followed by a preliminary analysis of the data
4. Collecting, testing, and analyzing soil samples
5. Combining geophysical and engineering information into one final report for the client

The report from this combined study will show how geological conditions at the site may affect the planned offshore oil platform.

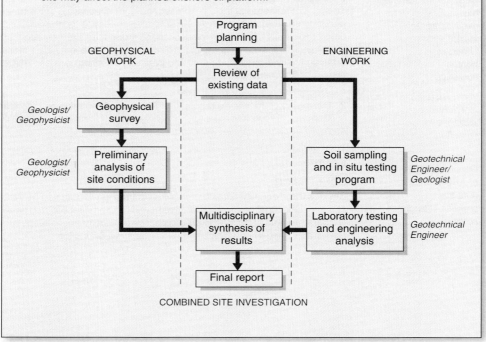

MODEL 6–4
A McDuff process description with a flowchart (both are included in an appendix to a report to a client)

Paragraph format
makes it difficult
for reader to
locate individual
steps.

MAKING TRAVEL ARRANGEMENTS
(Original Version)

When you're making travel arrangements, ask the person taking the trip to give you most of the details needed—dates, destinations, flight numbers, flight times, hotel requirements, rental car requirements, purpose of trip, and account number. Before proceeding, the first thing I do is confirm the flight information in the Official Airline Guide (OAG). You'll find the OAG on top of the credenza beside my typewriter. The next step is to call Turner Travel (566-0998). Although I've had great luck with all the people there, ask for Bonnie or Charlie—these two are most familiar with our firm. Turner Travel will handle reservations for flights, hotels, and rental cars. Remind them that we always use Avis midsize cars.

After you have confirmed the reservations information, fill out the McDuff travel form. Here's where you need to know the purpose of the trip and the traveler's McDuff account number. Blank forms are in the top drawer of my file cabinet in the folder labeled "Travel Forms—Blank." Once the form is complete, file the original in my "Travel Forms—Completed" folder, also in the top drawer of the file cabinet. Give the copy to the person taking the trip.

When you get the ticket in the mail from Turner Travel, check the flight information against the completed travel form. If everything checks out, give the ticket to the traveler. If there are errors, call Turner.

Also, when making any reservations for visitors to our office, call either the Warner Inn (566-7888) or the Hasker Hotel (567-9000). We have company accounts there, and they will bill us directly.

MODEL 6–5 *continues*

Instructions for making travel arrangements (McDuff's departed Baltimore travel coordinator left a narrative description of the procedure he followed [original version], which was then reformatted and edited [revised version])

MAKING TRAVEL ARRANGEMENTS
(Revised Version)

Arranging Travel for Employees

To make travel arrangements for employees, follow these instructions:

Step	Action
1.	Obtain the following information from the traveler:

    a. Dates
    b. Destinations
    c. Flight numbers
    d. Flight times
    e. Hotel requirements
    f. Rental car requirements
    g. Purpose of trip
    h. Account number

2.   Confirm flight information in the Official Airline Guide (OAG).
        Note: The OAG is on the credenza beside my typewriter.

3.   Call Turner Travel (566-0998) to make reservations.
        Note: Ask for Bonnie or Charlie.
        Note: For car rental, use Avis midsize cars.

4.   Complete the McDuff travel form.
        Note: Blank forms are in the folder labeled "Travel
        Forms—Blank," in the top drawer of my file cabinet.

5.   Make one copy of the completed travel form.

6.   Place the original form in the folder labeled "Travel
    Forms—Completed," in the top drawer of my file cabinet.

7.   Send the copy to the person taking the trip.

8.   Check the ticket and the completed travel form after the ticket
    arrives from Turner Travel.

9.   Do the ticket and the completed travel form agree?

    a. If *yes*, give the ticket to the traveler.
    b. If *no*, call Turner Travel.

Arranging Hotel Reservations for Visitors

    To make reservations for visitors, call the Warner Inn (566-7888) or the Hasker Hotel (567-9000). McDuff has company accounts there, and they will bill us.

Action steps all begin with "command" form of verb.

Letters are used to show long list of subpoints, for easy reference.

Notes are used to provide reader with *extra* information, separate from action of steps.

Though closely related, Steps 5–7 are best separated—for convenient reference by reader.

As noted in Guideline 8, two subpoints can show reader the *options* that exist.

MODEL 6–5
continued

Abstract
information in
ABC format
places
instructions in a
context.

Reader receives
purpose and
overview
information,
along with
reference to
attachment.

MEMORANDUM

DATE: June 23, 2002
TO: Employees Receiving New Message Recorders
FROM: Your Name, Purchasing Agent
SUBJECT: Instructions for new Message Recorders

INTRODUCTORY SUMMARY

We have just received the new phone message recorder you ordered. After processing, it will be delivered to your office within the week. The machine is one of the best on the market, but the instructions that accompany it are somewhat hard to follow. To help you begin using the recorder as soon as possible, I have simplified the instructions for setting up and operating the machine.

The illustration below labels the machine's parts. Following the illustration are seven easy steps you need to operate the recorder.

MODEL 6–6
continues
McDuff memo containing how-to instructions for a phone message recorder (rewritten from manufacturer's difficult-to-follow instructions)

Main tasks are indicated in heading and subheads.

For clarity, substeps under each main step have separate numbering.

Similar actions are separated into *two* different steps, to keep actions distinct.

Extra information is placed in note, *not* in step itself.

As with "notes," "results" should be *separated* from action in steps.

At point of decision, reader is given two clear options for action.

SETTING UP AND USING YOUR NEW RECORDER

If you devote about 15 minutes to these seven tasks, you can learn to operate your new message recorder.

1. Hooking Up Your Recorder
   a. Plug the recorder into any wall outlet using the *power cable.*
   b. Plug the *phone jack* into the *recorder outlet* located at the back of the unit.

2. Preparing Your Message
   a. Write down your:
      Greeting
      Name and department
      Time of return
   b. Write down the request you want to leave on the machine.
      NOTE: A sample request might be as follows: "Please leave your name, phone
              number, and a brief message after you hear the tone."

3. Recording Your Message
   a. Press the *ON* button.
      RESULT: The red *on* light will come on.
   b. Press and hold down the *RECORD* button, and keep holding it down for the entire
      time you record.
      RESULT: While the button is held down, the red *record/message* light will come on.
   c. Record your message directly into the *microphone.*
   d. Release the *RECORD* button when you are finished recording.
   e. Do you need to record the message again?
      If *yes*, repeat steps a through d (your previous message will be erased each time you
      record).
      If *no*, go on to the next step.

4. Turning On and Testing Your Recorder
   a. Press the *ON* button.
      RESULT: The red *on* light will come on.
   b. Press the *PLAY* button.
      RESULT: The red *in use* light will come on.
   c. Call in your message from another phone to make sure the unit is recording
      properly.

5. Playing Back Messages
   a. Look at the red *record/message* light to see if it is blinking.
      NOTE: The number of times the light blinks in succession indicates the number of
      messages you have received.

MODEL 6–6
continued

continues

b. Press the *REW* (rewind) button.

NOTE: The tape will stop automatically when it is completely rewound.

Because of potential damage to equipment, "caution" appears *before* steps to which it applies.

CAUTION: Do not press *PLAY* and *FF* (Fast Forward) at the same time! Doing so will break the tape. See the manufacturer's manual for process of replacing broken tape.

---

c. Press *PLAY* and listen to the messages.

d. Do you want to replay messages?

If *yes*, repeat steps b and c.

If *no*, go on to next step.

e. Do you want to skip ahead to other messages?

If *yes*, push the *FF* (fast forward) button.

If *no*, go on to next step.

6. Erasing Received Messages

a. Press and continue holding down the *PLAY* and *REW* (rewind) buttons at the same time.

b. Release the buttons when you hear a click.

NOTE: The tape automatically stops when the messages are erased.

7. Turning Off Your Recorder

a. Press the *OFF* button.

RESULT: All red lights will go off.

Conclusion of ABC format wraps up memo by telling readers what to do if they encounter problems.

CONCLUSION

These new recorders are fully guaranteed for three years, so please report any problems right away. Paul Hansey (ext. 765) will be glad to help fix the machine or return the machine to the manufacturer for repair. In particular, you need to report these problems to Paul:

- Lost or incomplete messages
- Interference or noise on line
- Faulty equipment
- Inability to record

MODEL 6–6

continued

Chapter 7

Letters, Memos, and Electronic Communication

Collaboration at Work

Choosing the Right Mode

General Instructions

Each chapter begins with a Collaboration at Work exercise that applies strategies for working in groups to chapter topics. The exercise assumes you (1) have been divided into groups of about three to six students, (2) will use group time inside or outside of class to complete the case, and (3) will produce an oral or written response. For guidelines about writing in groups, refer to pages 28–34.

Background for Assignment

One hundred years ago, business professionals had few opportunities for communication beyond the formal letter or meeting. But today the range of options is incredibly broad. On the one hand, we have to marvel at the choices for getting our message heard or read. On the other hand, the many ways to communicate present an embarrassment of riches that can be confusing.

In other words, when you have multiple communication options, you're challenged to match the right method with the right context—"right" in terms of what the reader wants and "right" in terms of the level of effort you should exert to suit the purpose. You may think this challenge only applies to your working life. However, it also can influence your life in college, as this exercise shows.

Group Assignment

Brainstorm with your group to list every means you have used to communicate with your college and university, from the time you applied to the present. Then for each communication option that follows, provide two or three situations for which the option would be the appropriate choice:

1. Letter that includes praise
2. Letter that describes a complaint
3. Letter that provides information
4. Letter that attempts to persuade
5. Telephone call
6. Fax transmission
7. Email
8. Memorandum
9. Personal meeting

M arie Stargill, McDuff's fire-science expert, just returned from a seminar that emphasized new techniques for preventing injuries from job-site fires. Within 24 hours of her return, she has already done three things:

1. Written her manager an email message over the office computer network
2. Sent a letter to a McDuff client suggesting use of fire-retardant gloves she learned about at the seminar
3. Sent the conference director a letter of appreciation about the meeting

Like Marie, you will write many letters, memos, and email in your career. In fact, you probably will write more of them than any other type of document.

Letters, memos, and email are short documents written to accomplish a limited purpose. Letters are directed outside your organization, memos are directed within your organization, and email can be directed to either an external or an internal audience. (Longer, more complicated letters and memos—called "letter reports" and "memo reports"—are covered in chapter 8.) Here are some working definitions:

> **Letter:** document that conveys information to a member of one organization from someone outside that same organization. Also called "correspondence," letters usually cover one major point and go on one page. This chapter classifies letters into these four groups, according to type of message: (1) positive, (2) negative, (3) neutral, and (4) sales.

> **Memorandum:** document written from a member of an organization to one or more members of the same organization. Abbreviated "memo," it usually covers just one main point and no more than a few. Readers prefer one-page memos.

> **Email:** document written usually in an informal style either to members of one's own organization or to an external audience. Characterized by the speed with which it is written and delivered, an email can include more formal attachments to be read and possibly printed by the audience.

As with other forms of technical writing, your ability to write good memos, letters, and email depends on a clear sense of purpose, thorough understanding of reader needs and close attention to correct formats. This chapter prepares you for this challenge by presenting sections that cover (1) general rules that apply to both letters and memos; (2) specific formats for positive letters, negative letters, neutral letters, sales letters, and memoranda; and (3) guidelines for email. Job letters and resumes are discussed in a separate chapter on the job search (chapter 14).

GENERAL GUIDELINES FOR LETTERS AND MEMOS

Letters convey your message to readers outside your organization, just as memos are an effective way to get things done within your own organization. By applying the guidelines in this chapter, you can master the craft of writing good letters and memos. You need to plan, draft, and revise each letter and memo as if your job depends on it—for it may.

Refer to Models 7–1 and 7–2 on pages 243 and 244–245 for McDuff examples that demonstrate the guidelines that follow. Later examples in this chapter show specific types of letters and additional memos.

● *Letter/Memo Guideline 1: Know Your Purpose*

Before beginning your draft, write down your purpose in one clear sentence. This approach forces you to sift through details to find a main reason for writing every letter or memo. This "purpose sentence" often becomes one of the first sentences in the document. Here are some samples:

- **Letter purpose sentence:** "As you requested yesterday, I'm sending samples of the new candy brands you are considering placing in McDuff's office vending machines."
- **Memo purpose sentence:** "This memo will explain McDuff's new policy for selecting rental cars on business trips."

Some purpose statements are implied; others are stated. An implied purpose statement occurs in the second paragraph of Model 7–1. That paragraph shows that the writer wishes both to respond to requests for McDuff brochures and, just as important, to seek the professor's help in soliciting good graduates for McDuff's Atlanta office. In a sense, one purpose leads into the other. In Model 7–2 you will find a more obvious purpose statement in the second sentence.

● *Letter/Memo Guideline 2: Know Your Readers*

Whom are you trying to inform or influence? The answer to this question affects the vocabulary you choose, the arguments you make, and the tone you adopt. Pay particular attention when a letter or memo will be read by more than one person. If these readers are from different technical levels or different levels within an organization, the challenge increases. A complex audience compels you to either (1) reduce the level of technicality to that which can be understood by all readers or (2) write different parts of the document for different readers.

Model 7–1 is directed to a professor with whom the writer wants to develop a reciprocal relationship—that is, George Lux gives free guest lectures in civil-engineering classes, hoping the professor in turn will help him inform potential job applicants about McDuff. Model 7–2 (pp. 244–245), directed to an in-house technical audience, contains fairly general information about the new technical editor. This information would apply to, and be understood by, all readers.

● *Letter/Memo Guideline 3: Follow Correct Format*

Most organizations adopt letter and memo formats that must be used uniformly by all employees. Here are the basic guidelines:

- **Letters:** There are three main letter formats—block, modified block, and simplified. Models 7–3, 7–4, and 7–5 on pages 246–248 show the basic page design of each; letter examples throughout the chapter use the three formats. As noted, you usually follow the preferred format of your own organization.

- **Memos:** With minor variations, all memos look much the same. The obligatory "date/to/from/subject" information hangs at the top left margin, in whatever order your organization requires. Model 7–6 on page 249 shows one basic format. These four lines allow you to dispense with lengthy introductory passages seen in more formal documents.

    Give the *subject line* special attention, for it telegraphs meaning to the audience immediately. In fact, readers use it to decide when, or if, they will read the complete memo. Be brief but also engage interest. For example, the subject line of the Model 7–2 memo could have been "Editing." Yet that brevity would have sacrificed reader interest. The actual subject line, "New Employee to Help with Technical Editing," conveys more information and shows readers that the contents of the memo will make their lives easier.

    Note that the sender signs his or her initials after or above the typed name in the "from" line.

- **Letters and Memos:** Several format conventions apply to letters and memos alike. Some of the more important features are listed here. See the models at the end of the chapter for details about spacing.

    Facsimile reference: Readers often need to know—for convenience and for the record—when memos or letters have been sent by fax. Type FAX TRANSMISSION or FACSIMILE before the "date/to/from/subject" lines for a memo and between the date and inside address for a letter. This fax line also can be used for other similar notations such as CONFIDENTIAL or PERSONAL or REGISTERED.

    Reference initials: If the document has been typed by someone other than the writer, place the typist's initials two lines below the signature block for letters and below the last paragraph for memos (example: jt). Some organizations prefer that the initials of the writer also be included, followed by those of the typist (example: GTY/jt).

    Enclosure notation: If attachments or enclosures accompany the letter or memo, type the singular or plural form of "Enclosure" or "Attachment" one or two lines below the reference initials. Some writers also list the item itself (example: Enclosure: Code of Ethics).

Copy notation: If the memo or letter has been sent to anyone other than the recipient, type "Copy" or "Copies" one or two lines below the enclosure notation, followed by the name(s) of the person or persons receiving copies (example: Copy: Preston Hinkley). Some organizations prefer the initials "c" (for copy), "cc" (for carbon copy, even though carbons hardly exist anymore), or "pc" (for photocopy). If you are sending a copy but do not want the original letter or memo to include a reference to that copy, write "bc" (for blind copy) and the person's name only on the copy—not on the original (example: bc: Rob McDuff). Note: only send blind copies when you are certain it is appropriate and ethical to do so.

Postscripts: Items marked "PS" or "P.S." can appear occasionally in letters and rarely in memos. They are considered by many readers to be symbols of poor planning, so use them with caution. If used, they appear as the last item on the document (below the copy notation) and can be typed or written in longhand.

Multiple-page headings: Each page after the first page often has a heading that includes the name of the person or company receiving the letter or memo, the date, and the page number. Some organizations may prefer an abbreviated form such as "Jones to Bingham, 2," without the date.

● *Letter/Memo Guideline 4: Follow the ABC Format for All Letters and Memos*

Letters and memos subscribe to the same three-part ABC (**A**bstract/**B**ody/**C**onclusion) format used throughout this book. This approach responds to each reader's need to know "What does this document have to do with me?" According to the ABC format, your letter or memo comprises these three main sections:

- **Abstract:** The abstract introduces the purpose and usually gives a summary of main points to follow. It includes one or two short paragraphs.
- **Body:** The body contains supporting details and thus makes up the largest part of a letter or memo. You can help your readers by using techniques like these:

  Deductive patterns for paragraphs: In this general-to-specific plan, your first sentence should state the point that will help the reader understand the rest of the paragraph. This pattern avoids burying important points in the middle or end of the paragraph, where they might be missed. Fast readers tend to focus on paragraph beginnings and expect to find crucial information there. Note how most paragraphs in Model 7–2 (pp. 244–245) follow this format.

  Personal names: If they know you, readers like to see their names in the body of the letter or memo. Your effort here shows concern for the reader's perspective, gives the letter a personal touch, and helps strengthen your personal relationship with the reader. (See the last paragraph in Model 7–1, page 243.) Of course, the same technique in direct mail can sometimes backfire, since it is an obvious ploy to create an artificially personal relationship.

Lists that break up the text: Listed points are a good strategy for highlighting details. Readers are especially attracted to groupings of three items, which create a certain rhythm, attract attention, and encourage recall. Use bullets, numbers, dashes, or other typographical techniques to signal the listed items. For example, the bulleted list in Model 7–1 draws attention to three important points about McDuff that the writer wants to emphasize.

Strongest points first or last: If your letter or memo presents support or makes an argument, include the most important points at the beginning and/or at the end—not in the middle. For example, Model 7–2 begins and ends with two crucial issues: the effect of poor editing on company productivity and the need to decide upon specific ways the new editor can improve McDuff writing.

Headings to divide information: Even one-page letters and memos sometimes benefit from the emphasis achieved by headings. The three headings in Model 7–2 quickly steer the reader to main parts of the document.

- **Conclusion:** Readers remember first what they read last. The final paragraph of your letter or memo should leave the reader with an important piece of information—for example, (1) a summary of the main idea or (2) a clear statement of what will happen next. The Model 7–1 letter makes an offer that will help to continue the reader's association with the university, while the Model 7–2 memo gives readers a specific task to accomplish before the next meeting.

- ### *Letter/Memo Guideline 5: Use the 3Cs Strategy for Persuasive Messages*

The ABC format provides a way to organize all letters and memos. Another pattern of organization for you to employ is the "3Cs strategy"—especially when your letter or memo has a persuasive objective. This strategy has three main goals:

- **Capture** interest with a good opener, which tells the reader what the letter or memo can do for him or her.
- **Convince** the reader with supporting points, all of which confirm the opening point that this document will make life easier.
- **Control** the closing, with a statement that puts you in the position of following up on the letter or memo *and* solidifies your relationship with the reader.

Although neither Model 7–1 nor Model 7–2 is overtly persuasive, each has an underlying persuasive purpose. Note how both employ the 3Cs strategy.

- ### *Letter/Memo Guideline 6: Stress the "You" Attitude*

As noted earlier, using the reader's name in the body helps convey interest. But your efforts to see things from the reader's perspective must go deeper than a name reference. For example, you should perform these tasks:

- **Anticipate questions** your reader might raise and then answer these questions. You can even follow an actual question ("And how will our new testing lab help your firm?") with an answer ("Now McDuff's labs can process samples in 24 hours").

- **Replace the pronouns "I," "me," and "we" with "you" and "your."** Of course, you have to use first-person pronouns at certain points in a letter, but many pronouns should be second-person. The technique is quite simple. You can change almost any sentence from writer-focused prose ("We feel that this new service will . . .") to reader-focused prose ("You'll find that this new service will . . .").

Model 7–1 shows this "you" attitude by emphasizing what McDuff and the writer himself can do for the professor and his students. Model 7–2 shows it by emphasizing that the new editor will make the readers' job easier.

Letter/Memo Guideline 7: Use Attachments for Details

Keep text brief by placing details in attachments, which readers can examine later, rather than bogging down the middle of the letter or memo. In this way, the supporting facts are available for future reference, without distracting readers from the main message. The memo in Model 7–2, for example, includes a list of possible job tasks for the new McDuff editor. The listing would only clutter the body of the memo, especially since its purpose is to stimulate discussion at the next meeting.

Letter/Memo Guideline 8: Be Diplomatic

Without a tactful tone, all your planning and drafting will be wasted. Choose words that will persuade and cajole, not demand. Be especially careful of memos written to subordinates. If you sound too authoritarian, your message may be ignored—even if it is clear that what you are suggesting will help the readers. Generally speaking, negative (or "bad news") letters often use the passive voice, whereas positive (or "good news") letters often use the active voice.

For example, the letter in Model 7–1 would fail in its purpose if it sounded too pushy and one-sided about McDuff's interest in hiring graduates. Similarly, the editing memo in Model 7–2 would be poorly received if it used stuffy, condescending wording such as "Be advised that starting next month, you are to make use of proofreading services provided in-house by . . ."

Letter/Memo Guideline 9: Edit Carefully

Because letters and memos are short, editing errors may be obvious to readers. Take special care to avoid the following errors:

- **Mechanics:**
  - Misspelled words of any kind, but especially the reader's name
  - Wrong job title (call the reader's office to double-check, if necessary)
  - Old address (again, call the reader's office to check)

- **Grammar:**
  - Subject-verb and pronoun nonagreement
  - Misused commas

- **Style:**
  - Stuffy phrases such as "per your request" and "enclosed herewith"

- Long sentences with more than one main and one dependent clause
- Presumptuous phrases such as "Thanking you in advance for . . ."
- Negative tone suggested by phrases such as "We cannot," "I won't," and "Please don't hesitate to"

The last point is crucial and gets more attention later in this chapter. Use the editing stage to rewrite any passage that could be phrased in a more positive tone. You must always keep the reader's goodwill, no matter what the message.

● *Letter/Memo Guideline 10: Respond Quickly*

A letter or memo that comes too late will fail in its purpose, no matter how well written. Mail letters within 48 hours of your contact with, or request from, the reader. Send memos in plenty of time for your reader to make the appropriate adjustments in schedule, behavior, and so forth. This rule applies, for example, to letters and memos written in response to these situations:

- You want to write a follow-up letter after meeting or talking with a client.
- A customer requests information about a product or service.
- You discover that there will be a delay in your supply of a product or service to a customer.
- You select a candidate to interview for a position.
- You announce a change in company policy.
- You set the time for a company meeting.

The first sentence in the Model 7–1 letter, for example, shows that George Lux writes the day after his guest lecture. This responsiveness will help secure the goodwill of the professor.

SPECIFIC GUIDELINES FOR LETTERS AND MEMOS

Letters are to your clients and vendors what memos are to your colleagues. They relay information quickly and keep business flowing. This section gives you specific guidelines for these documents:

- Letters with a positive message
- Letters with a negative message
- Letters with a neutral message
- Letters with a sales message
- Memoranda

To be sure, many documents are hybrid forms that combine these patterns. As a technical sales expert for McDuff, for example, you may be writing to answer a customer question about a new piece of equipment just purchased from McDuff's Equipment Development group. Your main task is to solve a problem caused by a confusing passage in the owner's manual. At the same time, however, your concern for the customer's satisfaction can pave the way for purchase of a second machine later in the year. Thus the letter has both a positive message and a sales message. This example also points to a common thread that weaves all four letter types together: the need to maintain the reader's goodwill toward you and your organization.

The next five sections present (1) a pattern for each type of correspondence, based on the ABC (**A**bstract/**B**ody/**C**onclusion) format used throughout this text, and (2) one or more brief case studies in which the pattern might be used at McDuff.

Positive Letters

Everyone likes to give good news; fortunately, you will often be in the position of providing it when you write. Here are some sample situations:

- Replying to a question about products or services
- Acknowledging that an order has been received
- Recommending a colleague for a promotion or job
- Responding favorably to a routine request
- Responding favorably to a complaint or an adjustment
- Hiring an employee

The trick is to recognize the good-news potential of many situations. This section gives you an all-purpose format for positive letters, followed by a case study from McDuff.

ABC Format for Positive Letters.   All positive letters follow one overriding rule. You must always:

> State good news immediately!

Any delay gives readers the chance to wonder whether the news will be good or bad, thus causing momentary confusion. Here is a complete outline for positive letters that corresponds to the ABC format:

ABC FORMAT: *Positive Letter*

Abstract
- Bridge between this letter and last communication with person
- Clear statement of good news you have to report

Body
- Supporting data for main point mentioned in abstract
- Clarification of any questions reader may have
- Qualification, if any, of the good news

Conclusion
- Statement of eagerness to continue relationship, complete project, etc.
- Clear statement, if appropriate, of what step should come next

McDuff Case Study for a Positive Letter. As a project manager at McDuff's Houston office, Nancy Slade has agreed to complete a foundation investigation for a large church about 300 miles away. There are cracks in the basement floor slab and doors that do not close, so her crew will need a day to analyze the problem (observing the site, measuring walls, digging soil borings, taking samples, and so on). She took this small job on the condition that she could schedule it around several larger (and more profitable) projects in the same area during mid-August.

Yesterday Nancy received a letter from the minister (speaking for the church committee), who requested that McDuff change the date. He was just asked by the regional headquarters to host a three-day conference at the church during the same time that McDuff was originally scheduled to complete the project.

After checking her project schedule, Nancy determines that she can reschedule the church job. Model 7–7 on page 250 shows her response to the minister.

Negative Letters

It would be nice if all your letters could be as positive as those just described. Unfortunately, the real world does not work like that. You will have many opportunities to display both tact and clarity in relating negative information. Here are a few cases:

- Explaining delays in projects or delivery of services
- Declining invitations or requests
- Registering complaints about products or services
- Refusing to make adjustments based on complaints
- Denying credit
- Giving bad news about employment or performance
- Explaining changes from original orders

This section gives you a format to follow in writing sensitive letters with negative information. Then it provides one application at McDuff.

ABC Format for Negative Letters. One main rule applies to all negative letters:

> Buffer the bad news, but still be clear.

Despite the bad news, you want to keep the reader's goodwill. Spend time at the beginning building your relationship with the reader by introducing less controversial information—*before* you zero in on the main message. Here is an overall pattern to apply in each negative letter:

ABC FORMAT: *Negative Letter*

Abstract

- Bridge between your letter and previous communication
- General statement of purpose or appreciation—in an effort to find common bond or area of agreement

Body

- Strong emphasis on what *can* be done, when possible
- Buffered yet clear statement of what cannot be done, with clear statement of reasons for negative news
- Facts that support your views

Conclusion

- Closing remarks that express interest in continued association
- Statement, if appropriate, of what will happen next

McDuff Case Study for a Negative Letter.   Reread the letter situation described in the section on positive letters. Now assume that instead of being able to comply with the minister's request, the writer is unable to complete the work on another date without changing the fee. This change would be necessary because Nancy would have to send a new crew 300 miles to the site, rather than using a crew already working on a nearby project.

Nancy knows the church is on a tight budget, but she also knows that McDuff would not be in business too long by working for free. Most importantly, since the church is asking for a change in the original agreement, she believes it is fair to request a change in the fee. Model 7–8 on page 251 is the letter she sends. Note her effort to buffer the negative news.

Neutral Letters

Some letters express neither positive nor negative news. They are simply the routine correspondence written every day to keep businesses and other organizations operating. Some situations follow:

- Requesting information about a product or service
- Inviting the reader to an event
- Responding to an invitation or routine request
- Placing orders
- Providing a transmittal letter for fax transmissions
- Sending solicited or unsolicited items through the mail

Use the following outline in writing your neutral letters. Also, refer to the McDuff examples that follow the outline.

ABC Format for Neutral Letters.   Because the reader usually has no personal stake in the news, neutral letters require less emphasis on tone and tact than other types. Yet they still require careful planning. In particular, always abide by this main rule:

> Be absolutely clear about your inquiry or response.

Neutral letters operate a bit like good-news letters. You need to make your point early, without giving the reader time to wonder about your message. Neutral letters vary greatly in specific organization patterns. The "umbrella plan" suggested here emphasizes the main criterion of clarity.

ABC FORMAT: *Neutral Letters*

Abstract
- Bridge or transition between letter and previous communication, if any
- Precise purpose of letter (request, invitation, response to invitation)

Body
- Details that support the purpose statement—for example,
  Description of item(s) requested
  Requirements related to the invitation
  Description of item(s) being sent

Conclusion
- Statement of appreciation
- Description of actions that should occur next

McDuff Case Studies for Neutral Letters.   Letters with neutral messages get written by the hundreds each week at McDuff. Here are four situations that would require a neutral letter; items 2 and 4 provide the context for the examples in Model 7–9 on page 252 and Model 7–10 on page 253:

1. Zach Bowers, a lab assistant, writes a laboratory supply company for information about a new unbreakable beaker to use in testing.

2. Faron Abdullah, president of the Student Government Association at River College, asks representatives of McDuff's St. Louis office to attend a career fair.

3. Donna Martinich, a geologist, responds to the request of a past client for a copy of a report done three years ago.

4. Farah Linkletter, a supply assistant with McDuff's San Francisco office, orders three new transits, making sure to emphasize that one is not to include a field case.

Sales Letters

Upon hearing the term *sales letter,* some people have visions of direct-mail requests for magazine subscriptions, vacation land, or diet plans. In this text, however, sales letters mean something quite different. They name all your correspondence with a customer—from the first contact letter through the last thank-you note. This list gives you some idea of the possibilities for sales letters:

- Starting a relationship ("I'll be calling you . . .")
- Following a phone call ("Good talking to you. . . . Can we meet to discuss your needs regarding . . .")
- Following a meeting ("You mentioned that you could use more information . . . so here's a brochure on . . .")
- Following completion of sale or project ("We enjoyed working with you on . . .")
- Seeking repeat business ("I'd like to know how the new machinery has been working . . .")

Notice that sales letters almost always work together with personal contacts, such as meetings and phone calls. Your goal is to build a continuing relationship with the customer. Consult the following outline when writing sales letters for any context; the McDuff example shows the outline in action.

ABC Format for Sales Letters.   The one main rule that governs all sales letters is as follows:

> Help readers solve their problems.

Customers are interested in your product or service only insofar as it can assist them. You must engage the readers' interest by showing that you understand their needs and can help fulfill them. Here is a plan for writing a successful sales letter. Note reference to the 3Cs (Capture/Convince/Control) strategy mentioned earlier in the chapter:

ABC FORMAT: *Sales Letters*

Abstract

(choose one or two to capture attention)

- Cite a surprising fact
- Announce a new product or service that client needs
- Ask a question
- Show understanding of a client's problem
- Show potential for solving a client's problem
- Present a testimonial
- Make a challenging claim
- Summarize results of a meeting
- Answer a question reader previously asked

Body

(choose one or two to convince the reader)

- Stress one main problem reader has concern about
- Stress one main selling point of your solution
- Emphasize what is unique about your solution
- Focus on value and quality, rather than price
- Put details in enclosures
- Briefly explain the value of any enclosures

Conclusion

(keep control of the next step in sales process)

- Leave the reader with one crucial point to remember
- Offer to call (first choice) or ask reader to call (last choice)

McDuff Case Study for a Sales Letter. McDuff provides customers with professional services and equipment, so sales letters have an important place in the firm. Barbara Feinstein is one employee who writes them almost every day. As a first-year employee with a degree in industrial hygiene, Barbara works in the newly formed asbestos-abatement group. Basically, she helps clients find out if there is any asbestos that needs to be removed from structures, recommends a plan for removal, and has the work done by another division of McDuff.

Here is one series of sales contacts that involves several letters. First, Barbara sent "cold call" sales letters to 100 schools and small businesses in the St. Louis area, suggesting that they might want to have their structures checked for unsafe levels of asbestos. The letter contained a reply card. After calling and then meeting with a number of the respondents, she sent individualized follow-up letters that answered questions that came up in discussions and provided additional information. After another series of phone calls and meetings with some of the potential customers, she negotiated contracts with five of the businesses and completed the projects. Then within a few months of completion, she sent a final letter proposing additional McDuff services and began the cycle again. Model 7–11 on page 254 provides a sample sales letter that Barbara used at the beginning of the cycle.

Memoranda

Memoranda (also called *memos*) may be the single most common type of writing in business today. You will write them to peers, subordinates, and superiors in your organization—from the first days of your career until you retire. Even if you work in an organization that uses email extensively, you still will compose print messages that convey your point with brevity, clarity, and tact.

Memos can contain all four types of messages discussed with respect to letters—positive, negative, neutral, and sales. Here are some situations that would require good memos in an organization.

Positive

- Announcing high bonuses for the fiscal year
- Commending an employee for performance on a project
- Informing employees about improved fringe benefits

Negative

- Reporting decreased quarterly revenues for the year
- Requesting closer attention to filling out time sheets
- Asking for volunteers to work on a holiday

Neutral

- Announcing a meeting
- Summarizing the results of a meeting with a client
- Explaining a new laboratory procedure

Sales

- Requesting funding for a training seminar
- Recommending another staff member for the proposals unit
- Suggesting changes in the performance evaluation system

Following is an ABC format for memos, along with several case studies from McDuff.

ABC Format for Memoranda.   Abide by this one main rule in every memo-writing situation:

> Be clear, brief, and tactful.

Because many activities are competing for their time, readers expect information to be related as quickly and clearly as possible. Yet be sure not to sacrifice tact and sensitivity as you strive to achieve conciseness. This ABC format will help you accomplish both goals:

ABC FORMAT: *Memoranda*

Abstract

- Clear statement of memo's purpose
- Outline of main parts of memo

Body

- Supporting points, with strong points at the beginning and/or end
- Frequent use of short paragraphs or listed items
- Absolute clarity about what memo has to do with reader
- Tactful presentation of any negative news
- Reference to attachments, when much detail is required

Conclusion

- Clear statement of what step should occur next
- Another effort to retain goodwill and cooperation of readers

McDuff Case Studies for Printed Memoranda.   At McDuff, memoranda are written to and from employees at all levels. To reflect this diversity, this section describes two different contexts for writing memoranda and shows accompanying examples.

In the first context, the lead secretary at McDuff's Baltimore office has chaired an office committee to improve efficiency in using the centralized word-processing center. The committee was formed when the branch manager realized that many technical staff members had not been trained to use the center. This lack of training led to sloppy habits and loss of productivity. Rather than issue a "dictum" from his office, the branch manager established a small committee to review the problem and issue guidelines to the office staff. The memorandum in Model 7–12 resulted from the committee's meetings.

In the second McDuff case, the St. Louis personnel director, Timothy Fu, must announce several changes in benefits to the entire staff. Some changes are good news in that they expand employee benefits, and others are bad news in that they further limit benefits. Timothy has the difficult task of imparting both types of information in one memo to a broad audience. The memorandum in Model 7–13 is a result of his efforts.

EMAIL

Electronic communication has become the preferred means of communication for many people in their professional lives and, often, in their personal lives. Some of us receive up to 50 to a 100 or more messages a day. Any medium so widely used deserves special attention in a chapter on informal options of technical communication. The purpose of this overview is to describe the appropriate use and style of email, provide an ABC format, list some basic guidelines for using email, and give several examples.

Appropriate Use and Style of Email

Email is an appropriate reflection of the speed at which we conduct business today. Indeed, it mirrors the pace of popular culture as well. Following are some of the obvious advantages that using email provides:

- It gets quickly to the intended receiver.
- Its arrival can be easily confirmed.
- Your reader can reply to your message quickly.
- It's cheap to use—once you have invested in the hardware and software.
- It permits cheap transmission of multiple copies.

Adding to the ease of transmission is the fact that email allows you to create mailing lists. One address label can be an "umbrella" for multiple recipients, saving you much time.

Of course, remember the flip side of this ease of use: Email is NOT private. Every time you send an email, remember that it may end up being read by "the world." Either by mistake or design, many supposedly private emails arrive in the in boxes of unintended recipients.

Email communication is often considered less formal and, therefore, less demanding in its format and structure than print-based messages like memoranda and letters. However, because email messages have become so pervasive a means of communication, you should consider constructing them as carefully as you would a memorandum or a letter.

Chapter 1, which mentions email in the context of group writing, shows how electronic mail helps you collaborate with others during the writing process—especially the planning stage. Interestingly, the email medium has produced a casual writing style similar to that of handwritten notes. It even has its own set of abbreviations and shortcut languages, which ranges so widely and changes so often that no list of abbreviations is included here. Following is an email message from one McDuff employee to another. Josh Bergen and Natalie Long are working together on a report wherein they must offer suggestions for designing an operator's control panel at a large dam. Josh has just learned about another control panel that McDuff designed and installed for a Russian nuclear power plant (see Project 6 in the color insert). Josh wrote this email message to draw Natalie's attention to the related McDuff project:

DATE: September 15, 2002
TO: Natalie Long
FROM: Josh Bergen
SUBJECT: Zanger Dam Project

Natalie—

I've got an idea that might save us A LOT of time on the Zanger Dam project. Check out the company project sheet on the Russian nuclear plant job done last year.

Operators of hi-tech dams and nuke plants seem to face the same hassles:

- confusing displays
- need to respond fast
- distractions

When either a dam or nuke operator makes a mistake, there's often big trouble. I think we'd save time—and our client's money—if we could go right to some of the technical experts used in the nuke job. At least as a starting place. Maybe we'd even make our deadline on this project. That would be a change, considering the schedule delays this month on other jobs.

What do you think about this idea? Let me know today, if possible.

This message displays some of the most common stylistic features of electronic mail.

ABC Format for Email

Simply understanding that email SHOULD have a format puts you ahead of many writers, who consider email a license to ramble on without structure. Yes, email is casual and quick, but that does not make it formless. In fact, the overriding rule for email is as follows:

> Don't send it *too* quickly!

By taking an extra minute to give your message structure—and to check its style and tone, as will be noted later—you'll have the best chance of sending an email that will be well received. The following three-part ABC format resembles that used for letters:

ABC FORMAT: *Email*

Abstract
- Casual, friendly greeting if justified by relationship
- Short, clear statement of purpose for writing
- List of main topics to be covered

Body
- Supporting information for points mentioned in abstract
- Use of short paragraphs that start with main ideas
- Use of headings and lists
- Use of abbreviations and jargon only when understood by all readers

Conclusion
- Summary of main point
- Clarity about action that comes next

Remember—your reader is confronted with many emails during the day. Further, the configurations of some computers make reading a screen harder on the eyes than reading print memos. So give each email a structure that will make it simple for your reader to find important information.

Guidelines for Email

Besides the appropriate style and ABC format, there are some additional guidelines that apply to email. Again, you are trying to strike the appropriate balance between speed of delivery, on the one hand, and quality of the communication to your reader, on the other.

● *Email Guideline 1: Begin with Standard Memorandum Format*

The message should contain an indication of the date, to, from, and subject information that a print-based memorandum would have. The exact wording and order will depend on the particular email system in your organization.

● *Email Guideline 2: Focus on One Main Subject in a Message*

State your subject in the "subject line" as briefly and clearly as possible. Then in the body of the message, begin with a succinct description of your subject.

● *Email Guideline 3: Use a Positive Conversational Style*

As mentioned earlier, a casual style is usually appropriate. Adopt a style that resembles how you would talk to the recipient on the phone. Sentence fragments and slang are acceptable, as long as they contribute to your objectives and are in good taste. Most important, avoid displaying a negative, angry tone. Don't be too quick to push the "send" button when an email would not produce a *constructive* exchange.

● *Email Guideline 4: Be Sure Your Message Indicates the Context to Which It Applies*

Tell your readers what the subject is and what prompts you to write your message. If you are replying to a message, be sure to include the previous message or summarize the message to which you are replying. Most email software packages include a copy of the message to which you are replying.

● *Email Guideline 5: Choose the Most Appropriate Method for Replying to a Message*

Short email messages may only require that you include a brief response at the beginning or end of the email to which you are responding. For complex, multitopic messages, however, you may wish to split your reply by commenting on each point individually (see Figure 7–1).

● *Email Guideline 6: Format Your Message Carefully*

Because email messages frequently replace more formal print-based documents, they should be organized and formatted so that the readers can easily locate the information you want to communicate.

- Use headings to identify important chunks of information.
- Use bulleted lists to display a series of information.
- Use sufficient "white space" to separate important chunks of information.
- Use "separators" to divide one piece of information from another.

Figure 7–2 illustrates an email message with headings, separators, and white space.

● *Email Guideline 7: Chunk Information for Easy Scanning*

Break the information into coherent chunks dealing with one specific topic, including all the details that a reader would need to get all of the essential information. Depending on the nature of the information, include specific topic, time, date, location, and necessary prerequisites and details.

● *Email Guideline 8: When Writing to Groups, Give Readers a Method to Abstain from Receiving Future Notices*

Email can easily become invasive and troublesome for recipients. You will gain favor—or at least not lose favor—if you are considerate and allow recipients to decide what email they wish to receive.

```
*********************************************************************************
X-Sender: mckinley@mail2.mcduff.com
Date: Tue, 10 Nov 2002 09:25:30 -0800
To: pcarmich@advantage.com
From: Mike McKinley <mckinley@mduff.com>
Subject: our recent visit
Mime-Version: 1.0


Dear Paul,

YOU WROTE:

>I hope that you had a good flight back home. I certainly enjoyed meeting you
and look forward to the possibility of working with you this coming spring on the
project that your firm, McDuff, may do for us.

REPLY:

The trip back was fine, but tiring. I enjoyed meeting you also and visiting with your
staff. I particularly enjoyed meeting Harold Black, for he will be very valuable in
developing the plans for the possible water purification plant.

YOU WROTE:

>If Advantage, Inc., does decide to build the water purification plant, we would be
very interested in having McDuff's Mary Stevens as the project manager.

REPLY:

That certainly will be a possibility; Mary is one of our best managers.

YOU WROTE:

>After you left, I called the city administration here in Murrayville. McDuff does not
need a business license for your work here, but, of course, you will need the
necessary construction permits.

REPLY:

Thanks for taking care of this matter—I had not thought of that. We will supply the
details to you for applying for the construction permits if you accept our proposal.
*********************************************************************************
```

FIGURE 7–1   An email message that separates different topics for reply

● *Email Guideline 9: When Writing to Groups, Suppress the Email Addresses of Recipients—Unless the Group Has Agreed to Let Addresses Be Known*

It is inappropriate to reveal the email addresses of group members to other group members. Use the "bcc" line to suppress group members' addresses.

```
*****************************************************************************************
X-Sender: mckinley@mail2.mcduff.com
Date: Tue, 10 Oct. 2002 09:25:30 -0800
To: Branch employees
From: Paul Carmichael <pcarmich@advantage.com>
Subject: October update
Mime-Version: 1.0
```

This is the October Electronic Update for Advantage, Inc. If you do not wish to receive this electronic update, send a message to

pcarmich@advantage.com

With the message in the subject line: Unsubscribe.

```
****************************
```
UPCOMING EVENTS
```
****************************
```

Project managers' meeting

October 21—project managers meeting (notice the change of location): Hereford building, room 209.

```
**************************************
```
November department meetings

All departments will have their planning and reporting meetings on November 18 at noon, with a joint lunch in the main dining room and breakout sessions at 12:30. Meetings should conclude at 2 p.m.

```
**************************************
```
December department meetings

NOTE CHANGE OF DATE: The December department meetings will be held on December 9 (second Wednesday), NOT December 16 (third Wednesday).
```
*****************************************************************************************
```

FIGURE 7–2   Long email message with use of appropriate headings, separators, and white space

● *Email Guideline 10: When Composing an Important Message, Consider Composing It on Your Word Processor*

Important email messages should be not only clear in format but also correct in mechanics. Since email software may not have a spelling checker, compose important messages on your word processor and use your spelling checker to check accuracy. Then either cut and paste it into an email message or attach it as a file.

For everything you want to know about email, see **http://www.learnthenet.com/ english/section/email.html** Also, see Model 7–14 for another email example.

Communication Challenge

"Ethics and Sales Letters"

Technical firms like McDuff are known for taking a conservative approach to marketing. For the last six months, however, the McDuff corporate marketing department has been considering more imaginative marketing strategies. This case study explains their approach, presents the draft of a sales letter that is part of the proposed strategy, and ends with questions and comments for discussion.

The Plan

Traditionally, technical consulting firms have done little if any advertising and sent few sales letters. Instead, they have just "been there" when clients needed services like ones McDuff performs, such as foundation design, construction management, equipment development, environmental remediation, and training. McDuff has been successful because it has built up a stable of clients that return regularly for continuing services. Lately, however, a more competitive economy has reduced the amount of repeat business. Now some McDuff corporate managers want to introduce more aggressive marketing techniques into the firm's long-term plan.

Specifically, the marketing staff is proposing a three-stage plan to corporate management and branch managers. Their focus will be government and private groups who own wetlands property being considered for commercial or recreational development. Eventually owners will be told by federal agencies that the property must be rigorously evaluated for environmental impact. This analysis will determine to what degree, if any, the land can be developed for parks, residential purposes, etc. McDuff would like to write owners about this requirement now, before the owners officially request environmental studies. Thus McDuff may beat competitors to the work. Here's the plan they propose:

1. Use public records to identify owners of wetland areas throughout the country.
2. Send a sales letter to all these owners.
3. Follow the letter with phone calls and meetings that will provide a more detailed account of McDuff's services.

The marketing staff has decided the sales letter must depart from the predictable, boring letters sent in the past. In fact, the staff has devised a "grabber": any recipient hiring McDuff to perform an environmental study before January 15 will receive a fee reduction of 10%, up to a maximum of $1,000. Although the writers do not want their letter to look too commercial, they *do* want to depart from letters of the past. They believe something is needed to get readers to act; a financial incentive seems appropriate.

Continued

The Letter

The marketing staff just sent out the following letter to McDuff's corporate officers and branch managers. Included with the letter was a list of prospective clients and a reference to the three-point marketing strategy noted in the preceding bulleted list. Readers are to call or email the marketing staff with their response.

September 15, 2002

Drummond Recreational Development, Inc.
P.O. Box 303
Wilmington, NC 20065

Reference: 100-Acre Parcel at Star-Gazer Swamp

Dear Mr. Drummond:

Would you like to save up to $1,000 on a study the government requires to evaluate the development potential of Star-Gazer Swamp? For a limited time, McDuff can save you this amount on an environmental evaluation of the site.

As you may know, the federal government requires that you complete an environmental assessment of property that may include wetlands areas important for biological diversity. McDuff has completed over 500 such studies during the last 10 years. In each case, clients have received information they needed to determine (1) possible commercial or recreational development of their property and (2) applicable federal restrictions. In making this evaluation, our team of technical experts has no equal among consulting firms throughout the country.

For a limited time, you have the opportunity to save 10%, up to a maximum of $1,000, on an environmental assessment—which will cost you full

CHAPTER SUMMARY

Letters and memoranda keep the machinery of business, industry, and government moving. Letters usually go to readers outside your organization, while memos go to readers inside. In both types of correspondence, abide by these rules:

1. Know your purpose.
2. Know your readers.
3. Follow correct format.
4. Follow the ABC format for all letters and memos.
5. Use the 3Cs strategy for persuasive messages.
6. Stress the "you" attitude.
7. Use attachments for details.
8. Be diplomatic.
9. Edit carefully.

price for any services contracted after January 15, 2003. Enclosed is a brochure that gives specifics about the features of our assessments and lists our previous clients.

I'll call you in a few days to see if your firm wants to take advantage of our offer for a significant reduction on an evaluation assessment. If you have questions before that time, please give me a call.

Sincerely,

Kathryn P. Phyzer

Kathryn P. Phyzer, Marketing Manager

Enclosure

Questions and Comments for Discussion

1. Explain how McDuff's letter does or does not satisfy this chapter's criteria for the content and style of an effective sales letter.
2. What kind of response do you think the marketing staff will get from the McDuff managers to whom they sent the draft?
3. Do you believe the financial incentive is *ethical*? Explain.
4. Do you think the financial incentive would be *effective*? Explain.
5. Generally speaking, what type of marketing strategy is either ethical or appropriate for technical firms like McDuff? Explain your answer. In your view, should marketing techniques for firms like McDuff differ from techniques used by retail stores, credit card companies, and other similar organizations? Explain.

Besides following these basics, you need to follow specific strategies for the four basic business letters, memos, and email. In letters with a positive message, the good news always goes first. In letters with a negative message, work on maintaining goodwill by placing a "buffer" before the bad news. Neutral letters, such as requests for information, should be absolutely clear in their message. Sales letters should show an interest in solving the readers' problems, more than an eagerness to sell a product or service. Memos should strive for brevity, clarity, and tact. Your relationship with both superiors and subordinates can depend in part on how well you write memoranda. Email messages give you the additional flexibility of adopting an informal, conversational style. Use email when speed and informality are desired.

www.prenhall.com/pfeiffer **Access the Technical Writing Web site by pointing your browser at www.prenhall.com/pfeiffer. Here, you will find additional resources to aid you in learning and applying the information from this chapter.**

ASSIGNMENTS

Follow these general guidelines for assignments here:

- Print or design a letterhead when necessary.
- Use whatever letter format your instructor requires.
- Invent addresses when necessary.
- Invent any extra information you may need for letters or memos, but do not change the information presented here.

Part 1: Letters

1. **Positive Letter—Job Offer.** Assume that you are the personnel director for McDuff's San Francisco office. Yesterday, you and your hiring committee decided to offer a job to Ashley Tasker, one of ten recent graduates you interviewed for an entry-level position as a lab technician. Write Ashley an offer letter and indicate a starting date (in two weeks), a specific salary, and the need for her to sign and return an acceptance letter immediately. In the interview, you outlined the company's benefit plan, but you are enclosing with your letter a detailed description of fringe benefits (such as health insurance, long-term disability insurance, retirement plan, and vacation policy). Although Ashley is your first choice for this position, you will offer the job to another top candidate if Ashley is unable to start in two weeks at the salary you stated in the letter.

2. **Positive Letter—Recommendation.**

   Option A. Select one of the model resumes in chapter 14. Assume you have known the writer as his or her professor, supervisor, colleague, etc. Now write a letter of recommendation that highlights what you see as the person's high points. Keep in mind the job objective on the resume. Have a specific job and reader in mind for this exercise.

   Option B. Pair up with a class colleague and share information about your respective academic backgrounds, work experiences, and career goals. Now write a letter that recommends your counterpart for either (a) a scholarship based on grades, need, and/or some other criterion; (b) a specific job; (c) an internship in an organization related to the person's career interest; or (d) an academic award. Give yourself a simulated role that would give you firsthand knowledge of the person's background.

   Option C. Assume that a professor in a class you have already completed—and a class that you found valuable—has asked you to write a letter of recommendation for his or her promotion package at the college. You are assuming that this individual, in assembling his or her promotion materials, is permitted to include reference letters from a variety of sources, such as fellow professors and former students. Your letter should only address what you know about the professor's abilities from your firsthand experience.

3. **Positive Letter—Favorable Response to Complaint.** The following letter was written in response to a complaint from the office manager at McDuff's Denver office. She wrote the manufacturer that the lunchroom toaster-oven broke down just three days after the warranty expired. Although she did not ask for a specific monetary adjustment, she did make clear her extreme dissatisfaction with the product. The manufacturer responded with the following letter. Be prepared to discuss what is right and what is wrong with the letter. Also, rewrite it using this chapter's guidelines.

This letter is in response to your August 3 complaint about the Justrite toaster-oven you purchased about six months ago for your lunchroom at McDuff, Inc. We understand that the "light-dark" adjustment switch for the toaster device broke shortly after the warranty expired.

Did you know that last year our toaster-oven was rated "best in its class" and "most reliable" by *Consumers Count* magazine? Indeed, we have received so few complaints about the product that a recent survey of selected purchasers revealed that 98.5% of first-time purchasers of our toaster-ovens are pleased that they chose our products and would buy another.

Please double-check your toaster-oven to make sure that the switch is broken—it may just be temporarily stuck. We rarely have had customers make this specific complaint about our product. But if the switch is in need of repair, return the entire appliance to us, and we will have it repaired free of charge or have a new replacement sent to you. We stand behind our product, since the warranty period only recently expired.

It is our sincere hope that you continue to be a satisfied customer of Justrite appliances.

4. **Negative Letter—Explanation of Project Delay.** You work for McDuff's Boston office. As project manager for the construction of a small strip shopping center, you have had delays about halfway through the project because of bad weather. Even worse, the forecast is for another week of heavy rain. Yesterday, just when you thought nothing else could go wrong, you discovered that your concrete supplier, Atlas Concrete, has a truck drivers' strike in progress. Because you still need half the concrete for the project, you have started searching for another supplier.

Your client, an investor/developer named Tanya Lee located in a city about 200 miles away, probably will be upset by any delays in construction, whether or not they are within your control. Write her a letter in which you explain weather and concrete problems. Try to ease her concern, especially because you will want additional jobs from her in the future.

5. **Negative Letter—Declining a Request.** Assume that you work at the McDuff office that completed the Sentry Dam (see Project 1 in the color insert in chapter 2). Word of your good work has spread to the state director of dams. He has asked you, as manager of the Sentry project, to deliver a 20-minute speech on dam safety to the annual meeting of county engineers. Unfortunately, you have already agreed to be at a project site in another state on that day, and you cannot reschedule the site visit. Write the director of dams—who is both a former and, you hope, a future client—and decline the request. Though you know he expressly wanted you to speak, offer to send a substitute from your office.

6. **Negative Letter—Request for Prompt Payment.** Recently your McDuff office completed the project described in Project 4 included in the color insert in chapter 2. As indicated on the sheet, the city of Rondo was quite satisfied with your work. You billed the city within a week after completion and requested payment within 30 days of receipt of the bill, as you do for most clients. Forty-five days elapsed without payment being received, so you sent a second bill. Now it has been three months since completion of the project, and you still have not been paid. You suspect that your bill got lost in the paperwork at city hall, for there was a change of mayoral administrations shortly after you finished the project. Yet two phone calls to the city's business office have brought no satisfaction—two different assistants told you they could not find the bill and that you should rebill the city. You are steamed but would like to keep the client's goodwill, if possible. Write another letter requesting payment.

7. **Negative Letter—Change in Project Scope and Schedule.** As a marketing executive at the Cleveland office, you oversee many of the large accounts held by the office. One important account is a company that owns and operates a dozen radio and television stations throughout the Midwest. On one recent project, McDuff engineers and technicians did the foundation investigation for, and supervised construction of, a new transmitting tower for a television station in Toledo. First, your staff members completed a foundation investigation, at which time they examined the soils and rock below grade at the site. On the basis of what they learned, McDuff ordered the tower and the guy wires that connect it to the ground. Once the construction crew actually began excavating for the foundation, however, they found mud that could not support the foundation for the tower. While unfortunate, it sometimes happens that actual soil conditions cannot be predicted by the preliminary study. Because of this discovery of mud, the tower must be shifted to another location on the site. As a result, the precut guy wires will be the wrong length for the new site, requiring McDuff to order wire extenders. The extenders will arrive in two weeks, delaying placement of the tower by that much time. All other parts of the project are on schedule, so far.

    Your client, Ms. Sharon West of Midwest Media Systems in Cleveland, doesn't understand much about soils and foundation work. But she does understand what construction delays mean to the profit margin of her firm's new television station as it attempts to compete with larger stations in Toledo. You need to console this important client, while informing her of this recent finding.

8. **Neutral Letter—Response to Request for Information.** As reservations clerk for the Best Central Inn in St. Louis, you just received a letter from Jerald Pelletier, an administrative secretary making arrangements for a meeting of McDuff managers from around the country. The group is considering holding its quarterly meeting in St. Louis in six months. Pelletier has asked you to send some brief information on hotel rates, conference facilities (meeting rooms), and availability. Send him some room rates for double and single rooms, and let him know that you have four conference rooms to rent out at $50 each per day. Also, tell him that at this time, the hotel rooms and conference rooms are available for the three days he mentioned.

9. **Neutral Letter—Request for Information.** For this assignment choose one of the seven projects in the color insert in chapter 2. Assume you are a potential client of McDuff and have happened onto this project, which is similar to one your organization may need completed in the near future. Although you are not yet at the stage where you want to receive proposals, you would like additional information about firms that might be interested in your project. Using this context and information in the project description, write the marketing manager of McDuff for more information. Be specific about what you want to receive.

10. **Sales Letter—to McDuff.** Select a product or service with which you are familiar because of home, work, or school experience. Now assume that you are responsible for marketing this product or service to Ms. Janis Black, purchasing agent for McDuff, Inc. In a phone conversation earlier today, she showed some initial interest in purchasing the product or service for her firm. In fact, you managed to set up a meeting for two weeks from today, after she returns from a business trip. Now you need to write a follow-up letter in which you summarize the phone conversation, confirm the meeting date and time, and offer some additional information about your product or service that will keep her interest.

    In selecting your product or service, you might want to review the information about McDuff, Inc., contained in chapter 2. Here are some sample products and services that Janis must routinely evaluate for use by the firm:

- Cleaning supplies, such as lavatory soap and paper towels
- Laboratory equipment, such as glass beakers and lab coats
- Office materials, such as notepads and pencils
- Office equipment, such as fax machines and pencil sharpeners
- Leased products, such as automobiles for managers and salespeople

11. **Sales Letter—from McDuff.** For this assignment, choose one of the seven project sheets included with the color insert. Although you *will* be using the sheets as a source of information in writing your letter, you will *not* be sending the sheets as an attachment.

    Option A. Assume you are a McDuff employee writing to a new client. This client has shown initial interest in the service or product described in the project sheet you're using for this assignment. You've had one phone conversation with the potential client. Now you hope your letter will lay the foundation for a personal meeting. Write a sales letter that briefly describes (1) the service or product on the project sheet and/or (2) your success on the specific project.

    Option B. Assume you are a McDuff employee writing to the same client for whom you completed the project described on the sheet. You have heard there may be similar work available with this client, so you are seeking repeat business. Write your follow-up letter, referring to the success on the project and seeking information about possible future work.

Part 2: Memoranda and Email

12. **Memo—Positive News.** As branch manager of McDuff's Atlanta office, you just learned from your accounting firm that last year's profits were even higher than previously expected. Apparently, several large construction jobs had not been counted in the first reporting of profits. You and your managers had already announced individual raises before you learned this good news. Now you want to write a memo that states that every branch employee will get a $500 across-the-board bonus, in addition to whatever individual raises have been announced for next year.

13. **Memo—Positive News.** Kevin Kehoe, an employee at the San Francisco office, is being considered for promotion to Manager of Technical Services. He has asked you to write a memo to the San Francisco branch manager on his behalf. Although you now work as a marketing expert at the corporate office, several years ago you worked directly for Kevin on the Ocean Exploration Program in Cameroon (see Project 2 in the color insert). Kevin has asked that your memo deal exclusively with his work on that program. Kevin was manager of the project; you believe that it was largely through his technical expertise, boundless energy, and organizational skills that the project was so successful. He developed the technical plan of work that led to the clear-cut set of findings. Write a memo that conveys this information to the branch manager considering Kevin for the promotion. Because the branch manager is new, he is not familiar with the project on which you and Kevin worked. Thus your memo may need to mention some details from Project 2 in the color insert.

14. **Memo—Negative News.** You are project manager of the construction-management group at McDuff's St. Paul office. The current policy in your office states that employees must pass a pre-employment drug screening before being hired. After that, there are no tests unless you or one of your job supervisors has reason to suspect that an employee is under the influence of drugs on the job.

Lately a number of clients have strongly suggested that you should have a *random* drug-screening policy for all employees in the construction-management group. They argue that the on-the-job risk to life and property is great enough to justify this periodic testing, without warning. You have consulted your branch manager, who likes the idea. You have also talked with the company's attorney, who assures you that such random testing should be legal, given the character of the group's work. After considerable thought, you decide to implement the policy in three weeks. Write a memo to all employees of your group and relate this news.

15. **Memo—Negative News.** A year ago you introduced a pilot program to McDuff's London office, where you are branch manager. The program gave up to half the office employees the choice to work four 10-hour days each week, as opposed to five 8-hour days. You wanted to offer this flexibility to workers who, for whatever reason, desired longer weekends. As branch manager, you made it clear at the time that you would evaluate the program at the end of the one-year pilot.

Having completed your review, you've decided that employees need to return to the old schedule. Your main reason is that having the office short of staff on Friday (when the four-day employees are gone) has proved awkward in dealing with current and prospective clients. On many occasions, clients have called to find that their company contact is not working that day. In addition, the secretarial and word processing staff that does work on Friday cannot keep up with the end-of-week workload. Whereas you originally thought a split schedule in the office would work, now you know it causes more confusion than it's worth. People in the office are constantly forgetting who is working what schedule, though the schedule is published. So, your memo to office employees must inform them of your decision to discontinue the pilot program and to return to a five-day workweek for the whole office. The change will take place in one month.

16. **Memo—Neutral Message.** As mailroom supervisor at McDuff's Baltimore office, you have a number of changes to announce to employees of the corporate office. Write a memo that clearly relates the following information. Deliveries and pickups of mail, which currently are at 8:30 A.M. and 3:00 P.M., will change to 9:00 A.M. and 3:30 P.M., starting in two weeks. Also, there will be an additional pickup at noon on Monday, Wednesday, and Friday. The mailroom will start picking up mail to go out by Federal Express or any other one-day carrier, rather than the sender having to wait for the carrier's representative to come to the sender's office. The sender must call the mailroom to request the pickup; and the carrier must be told by the sender to go to the mailroom to pick up the package. The memo should also remind employees that the mail does not go out on federal holidays, even though the mailroom continues to pick up mail from the offices on those days.

17. **Memo—Persuasive Message.** For this assignment, choose either (1) a good reference book or textbook in your field of study or (2) an excellent periodical in your field. The book or periodical should be one that could be useful to someone working in a profession, preferably one that you may want to enter yourself.

Now assume that you are an employee of an organization that would benefit by having this book or periodical in its staff library, customer waiting room, or perhaps as a reference book purchased for employees in your group. Write a one-page memo to your supervisor recommending the purchase. You might want to consider criteria such as those listed here:

- Relevance of information in the source to the job
- Level of material with respect to potential readers
- Cost of book or periodical as compared with its value

- Amount of probable use
- Important features of the book or periodical (such as bibliographies or special sections)

18. **Memo—Neutral Message.** Assume you work in the public relations area at McDuff's corporate office in Baltimore. Recently your office began designing and producing company project sheets, each of which uses one page to describe a specific project completed by McDuff and provides an accompanying graphic on the page. The color insert in this textbook shows seven examples of these project sheets. Write a memo to all 15 branch managers. Give them the information that follows: Beginning next month, project sheets will be written for every McDuff job that grosses $10,000 or more. Your office will have each sheet finished within 30 days of project completion and will then send each office the sheets for projects that were coordinated by that office. Then, as time permits, the public relations office will go back to significant previous projects, like the ones in the color insert, to do additional project sheets on previous work.

19. **Memo—Persuasive Message.** Assume you work at a McDuff office and have no undergraduate degree. You are not yet sure what degree program you want to enter, but you have decided to take one night course each term. Your McDuff office has agreed to pay 100 percent of your college expenses on two conditions. First, before taking each course, you must write a memo of request to your supervisor, justifying the value of the class to your specific job or to your future work with the company. Clearly, your boss wants to know that the course has specific application or that it will form the foundation for later courses. Second, you must receive a *C* or better in every class for which you want reimbursement.

   Write the persuasive memo just described. For the purposes of this assignment, choose one course that you actually have taken or are now taking. Yet in your simulated role for the assignment, write as if you have not taken the course.

20. **Email—Collaborative Project #1.** This assignment applies only if your campus offers students the use of electronic mail. Select one of the preceding memo assignments to complete as a group-writing project with two or three members of your class. Set up a plan of work that (1) involves the group in several face-to-face meetings and (2) requires each member of the group to send and receive at least one assignment-related email message to and from every other member of the group.

21. **Email—Collaborative Project #2.** Pair up with one or two members of your class. Assume your team has been asked to write a McDuff project sheet to add to the seven sheets in the color insert in chapter 2. Although you do not need to include a photo, hand in a description of what kind of photograph you would recommend for the sheet you are writing.

   After your group has one or two face-to-face meetings to agree on a topic and research agenda, conduct all further team communication by email. One member will be responsible for emailing a draft of the project sheet to the others. Then other members will make all comments and suggestions by email. Once the email communication is complete, print and submit the project sheet along with hard copy of all email correspondence within the group.

22. **Ethics Assignment.** In the "Communication Challenge" section of this chapter you were asked to examine the ethics of a particular sales letter. This assignment moves that ethical concern into the realm of electronic communication and assumes you have some familiarity with email. The project is best completed as a group assignment.

   Pooling the experience that members of your group have had with email, focus specifically on inappropriate or unethical behavior. Possible topics include the content of messages, tone of language, and the use of distribution lists. Now draft a simple "code of ethics" that could be distributed to members of any organization—such as McDuff, Inc.—whose members use email on a daily basis. See Assignment #7 in chapter 2 for an example of an actual code of ethics.

23. **International Communication Assignment.** Email messages can be sent around the world as easily as they can be sent to the next office. If you end up working for a company with international offices or clients, you probably will use email to conduct business.

Investigate the email conventions of one or more countries outside your own. Search for any ways that the format, content, or style of international email may differ from email in your country. Gather information by collecting hard copy of email messages sent from other countries, interviewing people who use international email, and/or consulting the library for information on international business communication. Write a memo to your instructor in which you (1) note differences you found and (2) explain why these differences exist. If possible, focus on any differences in culture that may affect email transactions.

McDuff, Inc.

12 Peachtree Street
Atlanta, GA 30056
(404) 555-7524

August 2, 2002

Professor Willard R. Burton, Ph.D.
Department of Civil Engineering
Southern University of Technology
Paris, GA 30007

Dear Professor Burton:

Thanks very much for your hospitality during my visit to your class yesterday. I appreciated the interest your students showed in my presentation on stress fractures in highway bridges. Their questions were very perceptive.

You may recall that several students requested further information on McDuff, so I have enclosed a dozen brochures for any students who may be interested. As you know, job openings for civil-engineering graduates have increased markedly in the last five years. Some of the best opportunities lie in these three areas of the discipline:

• Evaluation of environmental problems
• Renovation of the nation's infrastructure
• Management of construction projects

These areas are three of McDuff's main interests. As a result, we are always searching for top-notch graduates from solid departments like yours.

Again, I enjoyed my visit back to Southern last Friday, Professor Burton. Please call when you want additional guest lectures by me or other members of the McDuff staff.

Sincerely,

George F. Lux

George F. Lux, P.E.

Enclosures

Annotations (left margin):

Expresses appreciation *and* provides lead-in to body.

Responds to question that arose at class presentation.

Uses bulleted list to emphasize information of value to professor's students.

Adds unobtrusive reference to McDuff's needs.

Closes with offer to visit class again.

Includes reference to enclosures.

MODEL 7–1
McDuff sample letter

M McDuff, Inc.

MEMORANDUM

DATE: December 4, 2002
TO: Technical Staff
FROM: Ralph Simmons, Technical Manager RS
SUBJECT: New Employee to Help with Technical Editing

Last week we hired an editor to help you produce top-quality reports, proposals, and other documents. This memorandum gives you some background on this change, highlights the credentials of our new editor, and explains what the change will mean to you.

BACKGROUND

At September's staff meeting, many technical staff members noted the excessive time spent editing and proofreading. For example, some of you said that this final stage of writing takes from 15-30 percent of the billable time on an average report. Most important, editing often ends up being done by project managers—the employees with the highest billable time.

Despite these editing efforts, many errors still show up in documents that go out the door. Last month I asked a professional association, the Engineers Professional Society (EPS), to evaluate McDuff-Boston documents for editorial correctness. (EPS performs this service for members on a confidential basis.) The resulting report showed that our final reports and proposals need considerable editing work. Given your comments at September's meeting and the results of the EPS peer review, I began searching for a solution.

SOLUTION: IN-HOUSE EDITOR

To come to grips with this editing problem, the office just hired Ron Perez, an experienced technical editor. He'll start work January 3. For the last six years, Ron has worked as an editor at Jones Technical Services, a Toronto firm that does work similar to ours. Before that he completed a master's degree in technical writing at Sage University in Buffalo.

At next week's staff meeting, we'll discuss the best way to use Ron's skills to help us out. For now, he will be getting to know our work by reviewing recent reports and proposals. Also, the attached list of possible activities can serve as a springboard for our discussion.

CONCLUSION

By working together with Ron, we'll be able to improve the editorial quality of our documents, free up more of your time for technical tasks, and save the client and ourselves some money.

I look forward to meeting with you next week to discuss the best use of Ron's services.

Enclosure
Copy: Ron Perez

Uses informative subject line.

Gives purpose of memo and highlights contents.

Uses side headings for easy reading.

Shows that the change arose from *their* concerns.

Adds evidence from outside observer.

Gives important information about Ron in *first* sentence.

Establishes his credibility.

Refers to attachment.

Focuses on *benefit* of change to reader. Restates next action to occur.

MODEL 7–2
McDuff sample letter

POSSIBLE ACTIVITIES FOR IN-HOUSE EDITOR

1. Reviewing reports at all levels of production

2. Helping to coordinate the writing of proposals

3. Preparing a format manual for the word-processing operations and secretaries

4. Preparing a report/proposal guide for the technical staff

5. Teaching luncheon sessions on editing

6. Teaching writing seminars for the technical staff

7. Working with the graphics department to improve the page design of our documents

8. Helping to write and edit public-relations copy for the company

9. Visiting other offices to help produce consistency in the editing of documents throughout the company

MODEL 7–2
continued

Letterhead of your organization

Two or more blank lines (adjust space to center letter on page)

Date of letter

Two or more blank lines (adjust space to center letter on page)

Address of reader

One blank line

Greeting

One blank line

Paragraph: single-spaced (indenting optional)

One blank line

Paragraph: single-spaced (indenting optional)

One blank line

Paragraph: single-spaced (indenting optional)

One blank line

Complimentary close

Three blank lines (for signature)

Typed name and title

One blank line

Typist's initials (optional: Writer's initials before typist's initials)

Computer file # (if applicable)

One blank line (optional)

Enclosure notation

One blank line (optional)

Copy notation

MODEL 7–3
Block style for letters

Letterhead of your organization

Two or more blank lines (adjust space to center letter on page)

Date of letter

Two or more blank lines (adjust space to center letter on page)

Address of reader

One blank line

Greeting

One blank line

Paragraph: single-spaced, with first line indented 5 spaces

One blank line

Paragraph: single-spaced, with first line indented 5 spaces

One blank line

Paragraph: single-spaced, with first line indented 5 spaces

One blank line

Complimentary close

Three blank lines (for signature)

Typed name and title

One blank line

Typist's initials (optional: Writer's initials before typist's initials)

Computer file # (if applicable)

One blank line (optional)

Enclosure notation

One blank line (optional)

Copy notation

MODEL 7–4
Modified block style (with indented paragraphs) for letters

| Letterhead of organization |

Two or more blank lines (adjust space to center letter on page)

| Date of letter |

Two or more blank lines (adjust space to center letter on page)

| Address of reader |

Three blank lines

| Short subject line |

Three blank lines

| Paragraph: single-spaced, no indenting |

One blank line

| Paragraph: single-spaced, no indenting |

One blank line

| Paragraph: single-spaced, no indenting |

Five blank lines (for signature)

| Typed name and title |

One blank line (optional)

| Typist's initials (optional: Writer's initials before typist's initials) |

| Computer file # (if applicable) |

One blank line (optional)

| Enclosure notation |

One blank line (optional)

| Copy notation |

MODEL 7–5
Simplified style for letters

Facsimile reference

One or more blank lines

Date of memo

Reader's name (and position, if appropriate)

Writer's name (and position, if appropriate)

Subject of memo

One or more blank lines

Paragraph: Single-spaced (optional–first line indented)

One blank line

Paragraph: Single-spaced (optional–first line indented)

One blank line

Paragraph: Single-spaced (optional–first line indented)

One blank line

Typist's initials (optional–writer's initials before typist's initials)

One blank line

Enclosure notation

One blank line

Copy notation

MODEL 7–6
Memo style

M McDuff, Inc.

12 Post Street
Houston, Texas 77000
(713) 555-9781

July 23, 2002

The Reverend Mr. John C. Davidson
Maxwell Street Church
Canyon Valley, Texas 79195

Dear Reverend Davidson:

Thanks for your letter asking to reschedule the church project from mid-August to another, more convenient time. Yes, we'll be able to do the project on one of two possible dates in September, as explained below.

As you know, McDuff originally planned to fit your foundation investigation between two other projects planned for the Canyon Valley area. In making every effort to lessen church costs, we would be saving money by having a crew already on site in your area—rather than having to charge you mobilization costs to and from Canyon Valley.

As it happens, we have just agreed to perform another large project in the Canyon Valley area beginning on September 18. We would be glad to schedule your project either before or after that job. Specifically, we could be at the church site for our one-day field investigation on either September 17 or September 25, whichever date you prefer.

Please call me by September 2 to let me know your scheduling preference for the project. In the meantime, have a productive and enjoyable conference at the church next month.

Sincerely,

Nancy Slade

Nancy Slade, P.E.
Project Manager

NS/mh
File #34678

Mentions letter that prompted this response. Gives good news *immediately.*

Reminds reader of rationale for original schedule—cost *savings.* Offers two options—both save the church money.

Shows McDuff's flexibility.

Makes clear what should happen next.

MODEL 7–7
Positive letter in block style

McDuff, Inc.

12 Post Street
Houston, Texas 77000
(713) 555-1381

July 23, 2002

Provides "bridge" and compliments Davidson on conference.

The Reverend John C. Davidson
Maxwell Street Church
Canyon Valley, Texas 79195

Dear Reverend Davidson:

Thanks for your letter asking to reschedule the foundation project at your church from mid-August to late August, because of the regional conference. I am sure you are proud that Maxwell was chosen as the conference site.

Reminds him about original agreement—in tactful manner.

One reason for our original schedule, as you may recall, was to save the travel costs for a project crew going back and forth between Houston and Canyon Valley. Because McDuff has several other jobs in the area, we had planned not to charge you for travel.

Phrases negative message as *positively* as possible, giving rationale for necessary change.

We can reschedule the project, as you request, to a more convenient date in late August, but the change will increase project costs from $1,500 to $1,800 to cover travel. At this point, we just don't have any other projects scheduled in your area in late August that would help defray the additional expenses. Given our low profit margin on such jobs, that additional $300 would make the difference between our firm making or losing money on the foundation investigation at your church.

Makes it clear what will happen next. Ends on *positive* note.

I'll call you next week, Reverend Davidson, to select a new date that would be most suitable. McDuff welcomes its association with the Maxwell Street Church and looks forward to a successful project in late August.

Sincerely,

Nancy Slade

Nancy Slade, P.E.
Project Manager

NS/mh
File #34678

MODEL 7–8
Negative letter in modified block style (with indented paragraphs)

251

River
College

January 4, 2002

Mr. Timothy Fu, Personnel Director
McDuff, Inc.
127 Rainbow Lane
St. Louis, MO 63103

Dear Mr. Fu:

McDuff, Inc., has hired 35 graduates of River College since 1975. To help continue that tradition, we would like to invite you to the college's first Career Fair, to be held February 21, 2002, from 8 a.m. until noon.

Sponsored by the Student Government Association, the Career Fair gives juniors and seniors the opportunity to get to know more about a number of potential employers. We give special attention to organizations, like McDuff, that have already had success in hiring River College graduates. Indeed, we have already had a number of inquiries about whether your firm will be represented at the fair.

Participating in the Career Fair is simple. We will provide you with a booth where one or two McDuff representatives can talk with students that come by to ask about your firm's career opportunities. Feel free to bring along whatever brochures or other written information that would help our students learn more about McDuff's products and services.

I will call you next week, Mr. Fu, to give more details about the fair and offer a specific booth location. We at River College look forward to building on our already strong association with McDuff.

Sincerely,

Faron G. Abdullah

Faron G. Abdullah, President
Student Government Association

Copy: Gene Abrams, Placement Director

States purpose clearly.

Describes Career Fair and its importance.

Shows value of event to McDuff.

Gives clear instructions for participating.

States appreciation and indicates what will happen next.

56 New Lane
Bolt, Missouri
65101
(314) 555-0272

MODEL 7–9
Neutral letter (invitation) in block style

345 Underwood Street
Belforth, California 90706
(713) 555-9781

April 2, 2002

Faraday Supply Company
34 State Street
San Francisco, CA 94987

ORDER FOR FIELD TRANSITS

Provides bridge to previous contact.

Yesterday I called Ms. Gayle Nichols to ask what transits you had in current inventory. Having considered what you have in stock, I wish to order those listed below.

States purpose clearly.

Please send us these items:

Gives *exact* information needed by reader.

1. One Jordan #456 Transit, with special field case
2. One Smith-Beasley #101FR, with special field case
3. One Riggins #6NMG, without special field case

Emphasizes important details about the order.

Note that we *do* want the special field cases with the Jordan and Smith-Beasley units, but do not want the case with the Riggins unit.

States exactly what should happen next.

Please send the units and the bill to my attention. As always, we appreciate doing business with Faraday.

Farah Linkletter

Farah Linkletter
Supply Assistant

gh

MODEL 7–10
Neutral letter (placing order) in simplified style

McDuff, Inc.

127 Rainbow Lane
St. Louis, Missouri 63103
(314) 555-8175

Simplified style eliminates salutation and closing.

August 21, 2002

Mr. James Swartz, Safety Director
Jessup County School System
1111 Clay Street
Smiley, MO 64607

Uses subject line to gain attention.

Refers to previous successful work.

Leads in naturally to letter's subject (asbestos abatement).

NEW ASBESTOS-ABATEMENT SERVICE NOW AVAILABLE

We enjoyed working with you last year, James, to update your entire fire alarm system. Given the current concern in the country about another safety issue, asbestos, we wanted you to know that our staff now does abatement work.

As you know, many of the state's school systems were constructed during years when asbestos was used as a primary insulator. No one knew then, of course, that the material can cause illness and even premature death for those who work in buildings where asbestos was used in construction. Now we know that just a small portion of asbestos produces a major health hazard.

Comforts reader by showing how problem can be discovered and solved.

Reinforces relationship between writer's and reader's organizations.

Refers briefly to enclosures; stays in control by mentioning follow-up phone call.

Fortunately, there's a way to tell whether you have a problem: the asbestos survey. This procedure, done by our certified asbestos-abatement professionals, results in a report that tells whether or not your buildings are affected. And if we find asbestos, we can remove it for you.

Jessup showed real foresight in modernizing its alarm system last year, James. Your desire for a thorough job on that project was matched, as you know, by the approach we take to our business. Now we'd like to help give you the peace of mind that will come from knowing that either (1) there is no asbestos problem in your 35 structures or (2) you have removed the material.

The enclosed brochure outlines our asbestos services. I'll call you in a few days to see whether McDuff can help you out.

Barbara Feinstein

Barbara H. Feinstein
Certified Industrial Hygienist

BHF/sg

MODEL 7–11
Sales letter in simplified style

MEMORANDUM

DATE: August 1, 2002
TO: Technical Staff
FROM: Gini Preston, Chair, Word Processing Committee *GP*
SUBJECT: Word Processing Suggestions

The Word Processing Committee has met for 6 weeks to consider changes in McDuff's Word Processing Center. This memo highlights the recommendations that have been approved by management.

Please note these changes in your daily use of the company's Word Processing Center.

1. **Document status:** Documents will be designated either "rush" or "regular" status, depending on what you request. If at all possible, rush documents will be returned within four hours. Regular documents will be returned within one working day.
2. **Draft stages:** Both users and operators should make every effort to produce no more than three hard-copy drafts of any document. Typically, these would include:

   - **First typed draft** (typed from writer's handwritten or cut-and-paste copy)
   - **Second typed draft** (produced after user has made editing corrections on first-draft copy)
   - **Final typed draft** (produced after user makes final editing changes, after the proofreader makes a pass through the document, and after the operator incorporates final changes into the copy)

3. **New proofreader:** A company proofreader has been hired to improve the quality of our documents. This individual will have an office in the Word Processing Center and will review all documents produced by the word processing operators.

These changes will all take effect August 15. Your efforts to implement them will help improve the efficiency of the center, the quality of your documents, and the productivity of the company.

Feel free to call me at ext. 567 if you have any questions.

Copy: Rob McDuff

Gives brief purpose statement and overview of contents.

Provides lead-in to list.

Uses numbered list of three main changes reader needs to note.

Emphasizes need for limiting number of hard copies—uses list within list.

Makes it clear *when* change will take place.

MODEL 7–12
Memorandum: changes in procedures

MEMORANDUM

DATE: May 3, 2002
TO: All Employees of Cleveland Office
FROM: Timothy Fu, Personnel Director TF
SUBJECT: New Cost Containment Measure for Health Care

The next fiscal year will bring several changes in the company's fringe benefit plan. Later this month, you'll receive a complete report on all adjustments to go into effect July 1. For now, this memo will outline one major change in health care. Specifically, McDuff will adopt a cost-containment program called PAC—intended to help you and the company get more health care for the dollar.

WHAT IS PAC AND HOW DOES IT WORK?

Health costs have risen dramatically in the last 10 years. The immediate effect on McDuff has been major increases in insurance premiums. Both you and the company have shared this burden. This year McDuff will fight this inflationary trend by introducing a new cost-containment program called PAC—Pre-Admission Check.

Started by Healthco, our company medical supplier, PAC changes the procedure by which you and your dependents will be recommended for hospitalization. Except in emergencies, you or your physician will need to call the PAC hotline before admission to the hospital. The PAC medical staff will do the following:

1. Review the length of stay recommended by your physician, to make sure it conforms to general practice
2. Request a second opinion if the PAC staff believes that such an opinion is warranted
3. Approve final plans for hospitalization

If your physician recommends that you stay in the hospital beyond the length originally planned, he or she will call PAC for authorization.

WILL PAC AFFECT THE LEVEL OR QUALITY OF HEALTH CARE?

No. PAC will in no way restrict your health care or increase your personal costs. Quite the contrary, it may reduce total costs considerably, leading to a stabilization of the employee contributions to premiums next year. The goal is to make sure physicians give careful scrutiny to the lengths of hospital stays, staying within the norms associated with a particular illness unless there is good reason to do otherwise.

Programs like PAC have worked well for many other firms around the country; there is a track record of lowering costs and working efficiently with physicians and hospitals. Also, you will be glad to know that Healthco has the firm support of its member physicians on this program.

MEMO TO: all employees
May 3, 2002
Page 2

WHAT WILL HAPPEN NEXT?

As mentioned earlier, this change goes into effect with the beginning of the new fiscal year on July 1. Soon you will receive a report about this and other changes in benefits. If you have any questions before that time, please call the Corporate Benefits Department at ext. 678.

Margin notes:

Puts this memo in context of all benefits changes.

Emphasizes point of *agreement*—concerns about costs. Describes problem that led to need for change.

Gives overview of program.

Uses list to highlight three main elements of PAC.

Uses heading to focus on *main concern* of reader—*quality of care.*

Indicates that similar programs have worked well elsewhere.

Leaves reader with clear sense of next step.

MODEL 7–13
Memorandum: changes in benefits

TO: Lab, Marketing, and Administrative Staff in U.S. Offices
FROM: Janice Simmons, Benefits Manager
SUBJECT: Training Funds for Fiscal Year 2002
DATE: January 2, 2002

Happy New Year to all of you! I hope you had a good break. I'm writing to announce some guidelines for approved training for the next twelve months—including an increased reimbursement. Please read on to see how these changes affect all lab, marketing, and administrative staff.

1. **Lab Staff**
   Maximum Reimbursement: $3,000 (up from $2,000)
   Approval Process: Discuss with your manager 21 days before trip
   Trip Purpose: To improve lab procedures

2. **Marketing Staff**
   Maximum Reimbursement: $4,000 (up from $3,500)
   Approval Process: Discuss with your manager 21 days before trip
   Trip Purpose: To learn new sales techniques

3. **Administrative Staff**
   Maximum Reimbursement: $4,500 (up from $4,000)
   Approval Process: Discuss with your manager 21 days before trip
   Trip Purpose: To improve productivity of office procedures

In the past most employees have failed to make use of their maximum training allotment. I encourage all of you to seek training opportunities that fit the guidelines listed above. **Please note the required 21-day lead time in the approval process!**

Just send me an email if you have any questions about the procedure.

Janice

The following annotations appear in the left margin:

Begins with casual, friendly tone.

Includes clear purpose statement and 3 topics to be covered.

Supplies details about topics mentioned in first paragraph.

Uses list and parallel structure for easy reading.

Uses short paragraphs.

Concludes with reminder about an important part of the procedure.

Encourages them to contact her if there are questions.

MODEL 7–14
Simple email message

8

Informal Reports

Suggestions for High School Students

General Instructions

Each chapter begins with a Collaboration at Work exercise that applies strategies for working in groups to chapter topics. The exercise assumes you (1) have been divided into groups of about three to six students, (2) will use group time inside or outside of class to complete the case, and (3) will produce an oral or written response. For guidelines about writing in groups, refer to pages 28–34.

Background for Assignment

This chapter introduces you to eight types of short reports, one of which is focused on recommendations. When you are writing a recommendation report with a group of colleagues, agreeing on content can be a challenge. Recommendations involve opinions, and opinions often vary about what should be presented to the reader. As with other collaborative efforts, first you share information in a nonjudgmental way. Then you choose what should be included in the report based on your group discussions.

Group Assignment

Assume an association of colleges and universities has asked your group to help write a short report to be sent to high school students. The report's purpose is to assist students in selecting a college or university. Your group will prepare an outline for the body of the report by (1) choosing several headings that classify groupings of recommendations and (2) providing specific recommendations within each grouping. For example, one grouping might be "Support for Job Placement," with one recommendation in this grouping being "Request data on the job placement rates of graduates of the institution." After producing your outline, share the results with other groups in the class.

*A*lan Murphy, a salesperson for McDuff's St. Paul office, has a full day ahead. Besides having to make some sales calls in the morning, he must complete two short reports back in the office. The first is a short progress report to Brasstown Bearings, a company that recently hired McDuff to train its technical staff in effective sales techniques. (As noted in chapter 2, McDuff works in the field of training as well as in technical areas.) As manager of the project, Alan has overseen the efforts of three McDuff trainers for the last three weeks. According to the contract, he must send a progress report to Brasstown every three weeks during the project. Alan's second report is internal. His boss wants a short report recommending ways that McDuff can pursue more training projects like the Brasstown job.

Like Alan Murphy, you will spend much of your time writing informal reports in your career. Though short and easy to read like letters and memos, informal reports have more substance, are longer, and thus require more organization skills than correspondence. A working definition follows:

Informal report: this document contains about two to five pages of text, not including attachments. It has more substance than a simple letter or memo but less than a formal report. It can be directed to readers either outside or inside your organization. If *outside*, it may be called a letter report. If *inside*, it may be called a memo report. In either case, its purpose can be informative (to clarify or explain) or persuasive (to convince) or both.

This chapter has three sections. The first one shows you when to use informal reports in your career by describing some McDuff cases. The second provides 10 main writing guidelines that apply to both letter and memo reports. The third focuses on specific suggestions for writing five common types of informal reports.

At the end of the chapter are examples with marginal annotations. They will give you specific, real-life applications of the chapter's writing guidelines. As such, the models will help you complete chapter assignments and do actual reports on the job. During your career, you will write many types of informal reports other than those presented here. If you grasp this chapter's principles, however, you can adapt to other formats.

WHEN TO USE INFORMAL REPORTS

As noted in the definition just given, informal reports are clearly distinguished from both formal reports and routine letters and memos. Early in your career, however, you may have trouble deciding exactly where to draw the line. To help you decide, the two sections that follow briefly describe situations in which informal reports would be appropriate at McDuff.

Letter Reports at McDuff

Written to people outside your organization, letter reports use the format of a business letter because of their brevity. Yet they include more detail than a simple business letter. Here are some sample projects at McDuff that would require letter reports:

- **Training recommendation:** McDuff's corporate training staff recommends changes in the training program of a large construction company. Courses that are recommended include technical writing, interpersonal communication, and quality management.
- **Seafloor study:** McDuff's Nairobi staff writes a preliminary report on the stability of the seafloor where an oil rig might be located off the coast of Africa. This preliminary study includes only a survey of information on file about the site. The final report, involving fieldwork, will be longer and more formal.
- **Marketing report:** A marketing specialist at the corporate office completes a study on "New Markets Beyond 2003." The report has been solicited by a professional marketing association to which McDuff belongs.
- **Asbestos project:** McDuff's Houston staff reports to a suburban school board about possible asbestos contamination of an old elementary school. After two days on-site, the crew of two technicians determined that the structure had no asbestos in its walls, plumbing, floors, or storerooms.
- **Environmental study:** McDuff's San Francisco staff reports to the local Sierra Club chapter on possible environmental effects of an entertainment park, proposed for a rural area where eagles often nest. The project involved one site visit, interviews with a biologist, and some brief library research.
- **Equipment design project:** McDuff's equipment-development staff reports to a manufacturer on tentative designs for a computer-controlled device to cut plastic drainage pipe. The project involved several days' drafting work.

As these examples show, letter reports are the best format for projects with a limited scope. Also, this informal format is a good sales strategy when dealing with customers greatly concerned about the cost of your work. When reading letter reports, they realize—consciously or subconsciously—that these documents cost them less money than formal reports. Your use of letter reports for small jobs shows a sensitivity to their budget and may help gain their repeat work. See Model 8–1 on pages 285–286 for a letter report based on a small project at McDuff.

Memo Reports at McDuff

Memo reports are the informal reports that go back and forth among McDuff's own employees. Though in memorandum format, they include more technical detail and are longer than routine memos. These situations at McDuff show the varied contexts of memo reports:

- **Need for testing equipment:** Joan Watson, a lab technician in the Denver office, evaluates a new piece of chemical testing equipment for her department manager, Wes Powell. Powell discusses the report with his manager.

- **Personnel problem:** Werner Hoffman, a field engineer in the Munich office, writes to his project manager, Hans Schulman, about disciplinary problems with a field hand. Hoffman discusses the report with his manager and with the personnel manager.
- **Need for drafting tables:** Susan Gindle, an equipment-development technician in Baltimore, writes a report to the equipment-development manager, Ralph Peak. Gindle recommends the company purchase five drafting tables from Simulon, Inc., as opposed to similar tables from Sonet, Inc. Peak will discuss the report with the company's vice president for research and with the finance officer.
- **Progress in hiring minorities:** Scott Sampson, personnel manager, reports to Lynn Redmond, vice president of human resources, on the company's initial efforts to hire more minorities. Redmond will discuss Sampson's progress report with the company president and with all office managers.
- **Report on training session:** Pamela Martin, a field engineer in St. Louis, reports to her project manager, Mel Baron, on a one-week course she took in Omaha on new techniques for removing asbestos from buildings. Baron circulates the report to his office manager, Ramsey Pitt. Then Pitt sends copies to the manager of every company office, since asbestos projects are becoming common throughout the firm.

These five reports would require enough detail to justify writing memo reports, rather than simple memos. As for audience, each report would go directly to, or at least be discussed with, readers at *high* levels within the company. That means good memo reports can help advance your career. Model 8–2 on pages 287–288 provides an annotated example of a memo report about proposed computer software at McDuff.

GENERAL GUIDELINES FOR INFORMAL REPORTS

The following are 10 guidelines that focus mainly on report format.

● *Informal Report Guideline 1: Plan Well Before You Write*

Like other chapters in this book, this section emphasizes the importance of the planning process. Complete the Planning Form at the end of the book for each assignment in this chapter, as well as for informal reports you write in your career. Before you begin writing a draft, use the Planning Form to record specific information about these points:

- The document's purpose
- The variety of readers who will receive the document
- The needs and expectations of readers, particularly decision-makers
- An outline of the main points to be covered in the body

● *Informal Report Guideline 2: Use Letter or Memo Format*

Model 8–1 shows that letter reports follow about the same format as typical business letters (see chapter 7). For example, both are produced on letterhead and both often include the reader's name, the date, and the page number on all pages after the first. Yet the format of letter reports differs from that of letters in these respects:

- The greeting is sometimes left out or replaced by an attention line, especially when your letter report will go to many readers in an organization.
- A report title often comes immediately after the inside address. It identifies the specific project covered in the report. You may have to use several lines because the project title should be described fully, in the same words that the reader would use.
- Spacing between lines might be single, one-and-one-half, or double, depending on the reader's preference.

Model 8–2 shows the typical format for a memo report. Like most memos, it includes "date/to/from/subject" information at the top and has the reader's name, date, and page number on every page after the first. Also, both memos and memo reports have a subject line that should engage interest, give readers their first quick look at your topic, and be both specific and concise—for example, "Fracture Problems with Molds 43-D and 42-G" is preferable to "Problems with Molds."

There are, however, some format differences between memos and memo reports. Memo reports are longer and tend to contain more headings than routine memos. Also, spacing between lines varies from company to company, though one-and-one-half or double spacing is most common.

Informal Report Guideline 3: Make Text Visually Appealing

Your letter or memo report must compete with other documents for each reader's attention. Here are three visual devices that help get attention, maintain interest, and highlight important information:

- Bulleted points, for short lists like this one
- Numbered points, for lists that are longer or that include a list of ordered steps
- Frequent use of headings and subheadings

Headings are particularly useful in memo and letter reports. As Models 8–1 and 8–2 (pp. 285–288) show, they give readers much-needed visual breaks. Since informal reports have no table of contents, headings also help readers locate information quickly. (Chapter 4 gives more detail on headings and other features of page design.)

Informal Report Guideline 4: Use the ABC Format for Organization

Headings and lists attract attention, but these alone will not keep readers interested. You also need to organize information effectively. Most technical documents, including informal reports, follow what this book calls the ABC format. This approach to organization includes three parts: (1) **A**bstract, (2) **B**ody, and (3) **C**onclusion.

- **Abstract:** Start with a capsule version of the information most needed by decision-makers.
- **Body:** Give details in the body of the report, where technical readers are most likely to linger awhile to examine supporting evidence.
- **Conclusion:** Reserve the end of the report for a description or list of findings, conclusions, or recommendations.

"Abstract," "body," and "conclusion" are only generic terms. They indicate the *types* of information included at the beginning, middle, and end of your reports— *not* necessarily the exact headings you will use. The next four guidelines give details on the ABC format as applied to memo and letter reports.

● *Informal Report Guideline 5: Call the Abstract an Introductory Summary*

Abstracts should give readers a summary, the "big picture." This text suggests that in informal reports, you label this overview as "Introductory Summary," a term that gives the reader a good idea of what the section contains. (You do have the option of leaving off a heading label. In this case, your first few paragraphs would contain the introductory summary information, followed by the first body heading of the report.)

In letter reports, the introductory summary comes immediately after the title. In memo reports, it comes after the subject line. Note that informal reports do not require long, drawn-out beginnings; just one or two paragraphs in this first section give readers three essential pieces of information:

1. **Purpose** for the report—why are you writing it?
2. **Scope** statement—what range of information does the report contain?
3. **Summary** of essentials—what main information does the reader most want or need to know?

● *Informal Report Guideline 6: Put Important Details in the Body*

The body section provides details needed to expand upon the outline presented in the introductory summary. If your report goes to a diverse audience, managers often read the quick overview in the introductory summary and then skip to conclusions and recommendations. Technical readers, on the other hand, may look first to the body section(s), where they expect to find supporting details presented in a logical fashion. In other words, here is your chance to make your case and to explain points thoroughly.

Yet the discussion section is no place to ramble. Details must be organized so well and put forth so logically that the reader feels compelled to read on. Here are three main suggestions for organization:

- **Use headings generously.** Each time you change a major or minor point, consider whether a heading change would help the reader. Informal reports should include at least one heading per page.
- **Precede subheadings with a lead-in passage.** Here you mention the subsections to follow, before you launch into the first subheading. (For example, "This section covers these three phases of the field study: clearing the site, collecting samples, and classifying samples.") This passage does for the entire section exactly what the introductory summary does for the entire report—it sets the scene for what is to come by providing a "road map."
- **Move from general to specific in paragraphs.** Start each paragraph with a topic sentence that includes your main point. Then give supporting details. This approach always keeps your most important information at the beginnings of paragraphs, where readers tend to focus first while reading.

Another important consideration in organizing the report discussion is the way you handle facts versus opinions.

● *Informal Report Guideline 7: Separate Fact from Opinion*

Some informal reports contain strong points of view. Others contain only subtle statements of opinion, if any. In either case, you must avoid any confusion about what constitutes fact or opinion. The safest approach in the report discussion is to move logically from findings to your conclusions and, finally, to your recommendations. Since these terms are often confused, here are some working definitions:

- **Findings:** Facts you uncover (for example, you observed severe cracks in the foundations of two adjacent homes in a subdivision).
- **Conclusions:** Ideas or beliefs you develop based on your findings (for example, you conclude that foundation cracks occurred because the two homes were built on soft fill, where original soil had been replaced by construction scraps). Opinion is clearly a part of conclusions.
- **Recommendations:** Suggestions or action items based on your conclusions (for example, you recommend that the foundation slab be supported by adding concrete posts below it). Recommendations are almost exclusively made up of opinions.

● *Informal Report Guideline 8: Focus Attention in Your Conclusion*

Letter and memo reports end with a section labeled "Conclusion," "Closing," "Conclusions," or "Conclusions and Recommendations." Choose the wording that best fits the content of your report. In all cases, this section gives details about your major findings, your conclusions, and, if called for, your recommendations. People often remember best what they read last, so think hard about what you place at the end of a report.

The precise amount of detail in your conclusion depends on which of these two options you choose for your particular report:

Option 1: If your major conclusions or recommendations have already been stated in the discussion, then you only need to restate them briefly to reinforce their importance (see Model 8–2, pp. 287–288).

Option 2: If the discussion leads up to, but has not covered, these conclusions or recommendations, then you may want to give more detail in this final section (see Model 8–1, pp. 285–286).

As in Models 8–1 and 8–2, lists often are mixed with paragraphs in the conclusion. Use such lists if you believe they will help readers remember your main points.

● *Informal Report Guideline 9: Use Attachments for Less Important Details*

The trend today is to avoid lengthy text in informal reports. Yet technical detail is often needed for support. One solution to this dilemma is to replace as much report text as possible with clearly labeled attachments that could include these items:

- **Tables and figures:** Illustrations in informal reports usually appear in attachments unless it is crucial that one is within the text. Memo and letter reports are so short that attached illustrations are easily accessible.
- **Costs:** It is best to list costs on a separate sheet. First, you do not want to bury important financial information within paragraphs. Second, readers often need to circulate cost information, and a separate cost attachment is easy to photocopy and send.

● *Informal Report Guideline 10: Edit Carefully*

Many readers judge you on how well you edit a report. A few spelling errors or some careless punctuation makes you seem unprofessional. Your career and your firm's future can depend on your ability to write final drafts carefully. Chapter 15 and the Handbook at the end of this text give detailed information about editing. For now, remember these basic guidelines:

- Keep most sentences short and simple.
- Proofread several times for mechanical errors such as misspellings (particularly personal names).
- Triple-check all cost figures for accuracy.
- Make sure all attachments are included, are mentioned in the text, and are accurate.
- Check the format and wording of all headings and subheadings.
- Ask a colleague to check over the report.

These 10 guidelines help memo and letter reports accomplish their objectives. Remember—both your supervisors and your clients will judge you as much on communication skills as they do on technical ability. Consider each report to be part of your resume.

SPECIFIC GUIDELINES FOR FIVE INFORMAL REPORTS

Report types vary from company to company. The ones described here are only a sampling of what you will be asked to write on the job. These five types were chosen because they are common, they can be written as either memo reports or letter reports, and they incorporate the writing patterns described in chapters 5 and 6. If you master these five informal reports, you can probably handle other types that come your way.

The sections that follow include an ABC format for each report being discussed, some brief case studies from McDuff, and report models based on the cases. Remember to consult chapters 5 and 6 if you need to review general patterns of organization used in short and long reports, such as cause-effect and technical description.

Note: Remember that informal reports were defined earlier as having *less substance* than the formal reports discussed in the next chapter. Thus the following five report types also could be constructed as long, formal reports—assuming the report content was more substantial and detailed than that included in an informal report.

Problem Analyses

Every organization faces both routine and complex problems. The routine ones often get handled without much paperwork; they are discussed and then solved. But other problems often need to be described in reports, particularly if they involve many people, are difficult to solve, or have been brewing for a long time. Use this working definition of a report that analyzes a problem:

> **Problem analysis:** this informal report presents readers with a detailed description of problems in areas such as personnel, equipment, products, services, and so forth. Its main goal is to provide *objective* information so that the readers can choose the next step. Any opinions must be well supported by facts.

Problem analyses, which can be either internal or external documents, should follow the pattern of organization described here.

ABC Format for Problem Analyses.   Like other informal reports, problem analyses fit the simple ABC (**A**bstract/**B**ody/**C**onclusion) format recommended throughout this text. The three sections contain some or all of the following information, depending on the specific report. Note that solutions to problems are not mentioned; this chapter deals separately with (1) problem analyses, whose main focus is problems, and (2) recommendation reports, whose main focus is solutions. Of course, be aware that during your career, you will be called on to write reports that combine both types.

*ABC FORMAT: *Problem Analysis*

Abstract

- Purpose of report
- Capsule summary of problems covered in report discussion

Body

- Background on source of problems
- Well-organized description of the problems observed
- Data that support your observations
- Consequences of the problems

Conclusion

- Brief restatement of main problems (unless report is so short that such restatement would seem repetitious)
- Degree of urgency required in handling problems
- Suggested next step

*The content and order of items within each of the three sections can vary.

McDuff Case Study for a Problem Analysis. Model 8–3 on pages 289–290 presents a sample problem analysis that follows this chapter's guidelines. Harold Marshal, a longtime McDuff employee, supervises all technical work aboard the *Seeker II,* a boat that McDuff leases during the summer. Staffed with several technicians and engineers, the boat is used to collect and test soil samples from the ocean floor. Different clients purchase these data, such as oil companies that need to place oil rigs safely and telecommunications companies that need to lay cable.

After a summer on the *Seeker II,* Harold has severe reservations about the safety and technical adequacy of the boat. Yet he knows that his supervisor, Jan Stillwright, will require detailed support of any complaints before she seriously considers negotiating a new boat contract next season. Given this critical audience, Harold focuses on specific problems that affect (1) the safety of the crew, (2) the accuracy of the technical work performed, and (3) the morale of the crew. He believes that this pragmatic approach, rather than an emotional appeal, will best persuade his boss that the problem is serious.

Recommendation Reports

Most problem analyses contain both facts and opinions. You as the writer must make special efforts to separate the two, for this reason: Most readers want the opportunity to draw their *own* conclusions about the problem. Also, support all opinions with facts. This guideline holds true especially with recommendation reports, for they include more personal views than other report types. Use the following working definition for recommendation reports:

> **Recommendation report:** this informal report presents readers with specific suggestions that affect personnel, equipment, procedures, products, services, and so on. Although the report's main purpose is to persuade, every recommendation must be supported by objective data.

Recommendation reports can be either internal or external documents. Both follow the ABC format suggested here.

ABC Format for Recommendation Reports. Problem analyses and recommendation reports sometimes overlap in content. You may recommend solutions in a problem analysis, just as you may analyze problems in a recommendation report. The ABC format assumes that you want to mention the problem briefly before proceeding to discuss solutions.

McDuff Case Study for a Recommendation Report. Model 8–1 (pp. 285–286) shows a typical recommendation report written at McDuff. The reader is a client oil firm about to place an oil rig at an offshore site in the Gulf of Mexico. Given the potential for risk to human life and to the environment, Big Muddy Oil wants to take every precaution. Therefore, it has hired McDuff to determine whether the preferred site is safe.

*ABC FORMAT: *Recommendation Report*

Abstract

- Purpose of report
- Brief reference to problem to which recommendations respond
- Capsule summary of recommendations covered in report discussion

Body

- Details about problem, if necessary
- Description of recommendations (though some short reports place recommendations in the conclusion)
- Data that support recommendations (with reference to attachments, if any)
- Main benefits of recommendations
- Any possible drawbacks

Conclusion

- Brief statement or restatement of main recommendations (optional)
- The main benefit of recommended change
- Your offer to help with next step

*The content and order of items within each of the three sections can vary.

This example presents an important problem you may face in writing recommendation reports. Occasionally you may be asked to deliver recommendations sooner than you would prefer if you were working under ideal circumstances. In such cases, assume the cautious approach taken by Bartley Hopkins, the McDuff writer of Model 8–1. That is, make sure to state that your report is preliminary and based on incomplete data. This approach is even more important in situations like this case study in which there is risk to human life. Take pains to qualify your recommendations so that they cannot possibly be misunderstood by your audience.

Equipment Evaluations

Every organization uses some kind of equipment, and someone has to help buy, maintain, or replace it. Because companies put so much money into this part of their business, evaluating equipment has become an important activity. Here is a working definition of evaluation reports:

Equipment evaluation: this informal report provides objective data about how equipment has, or has not, functioned. The report may cover topics such as machinery, tools, vehicles, office supplies, computer hardware, and computer software.

Like a problem analysis, an equipment evaluation may just focus on problems. Or like a recommendation report, it may go on to suggest a change in equipment. Whatever its focus, an equipment evaluation *must* provide a well-documented review of the exact manner in which equipment has performed. Follow this ABC format in evaluating equipment.

ABC Format for Equipment Evaluations.   Equipment evaluations that are informal reports should include some or all of the points listed here. Remember that in this type of report, the discussion must include evaluation criteria most important to the *readers,* not you.

---

*ABC FORMAT: *Equipment Evaluation*

Abstract
- Purpose of report
- Capsule summary of what your report says about the equipment

Body
- Thorough description of the equipment being evaluated
- Well-organized critique, either analyzing the parts of one piece of equipment or contrasting several pieces of similar equipment according to selected criteria
- Additional supporting data, with reference to any attachments

Conclusion
- Brief restatement of major findings, conclusions, or recommendations

*The content and order of items within each of the three sections can vary.

McDuff Case Study for an Equipment Evaluation.   Like other firms, McDuff relies on word processing for almost all internal and external documents. Model 8–2 (pp. 287–288) contains an evaluation of a new word processing package used on a trial basis. Melanie Frank, office manager in San Francisco, conducted the trial in her office and wrote the report to the branch manager, Hank Worley. Note that she analyzes each of the software's five main features. Then she ends with a recommendation, much like a recommendation report.

Pay special attention to the tone and argumentative structure of this example. Frank shows restraint in her enthusiasm, knowing that facts will be more convincing than opinions. Indeed, every claim about Best Choice software is supported either by evidence from her trial or by a logical explanation. For example, her praise of the file-management feature is supported by the experience of a field engineer who used the system for three days. And her statement about the well-written user's guide is supported by the few calls made to the Best Choice support center during the trial.

Progress/Periodic Reports

Some short reports are intended to cover activities that occurred during a specific period of time. They can be directed inside or outside your organization and are defined in this way:

> **Progress report:** this informal report provides your manager or client with details about work on a specific project. Often you agree at the beginning of a project to submit a certain number of progress reports at certain intervals.

> **Periodic report:** this informal report, usually directed within your own organization, summarizes your work on diverse tasks over a specific time period. For example, as supervisor of company publications, you may be asked to submit periodic reports each month on McDuff's new brochures, public-relations releases, and product flyers.

Progress and periodic reports contain mostly objective data. Yet both of them, especially progress reports, sometimes may be written in a persuasive manner. (See chapter 5 for techniques of argumentation.) After all, you are trying to put forth the best case for the work you have completed. The next section provides an ABC format for these two report types.

ABC Format for Progress/Periodic Reports. Whether internal or external, progress and periodic reports follow a basic ABC format. They contain some or all of the parts shown in the box on the next page.

McDuff Case Study for a Progress Report. As Model 8–4 on pages 291–292 indicates, Scott Sampson, McDuff's personnel manager, is in the midst of an internal project being conducted for Kerry Camp, vice president of domestic operations. Sampson's goal is to find ways to improve the company's training for technical employees. Having completed two of three phases, he is reporting his progress to Camp. Note that Sampson organizes the body sections by task. This arrangement helps to focus the reader's attention on the two main accomplishments—the successful phone interviews and the potentially useful survey. If, instead, Sampson had

---

*ABC FORMAT: *Progress/Periodic Report*

Abstract

- Purpose of report
- Capsule summary of main project(s)
- Main progress to date or since last report

Body

- Description of work completed since last report, organized either by task or by time or by both
- Clear reference to any dead ends that may have taken considerable time but yielded no results
- Explanation of delays or incomplete work
- Description of work remaining on project(s), organized either by task or by time or by both
- Reference to attachments that may contain more specific information

Conclusion

- Brief restatement of work since last reporting period
- Expression of confidence, or concern, about overall work on project(s)
- Indication of your willingness to make any adjustments the reader may want to suggest

---

*The content and order of items within each of the three sections can vary.

completed many smaller tasks, he may have wanted to organize the body of the report by time, not tasks.

Also note that Sampson adopts a persuasive tone at the end of the report. That is, he uses his solid progress as a way to emphasize the importance of the project. In this sense, he is "selling" the project to his "internal customer," Kerry Camp, who ultimately will be in the position to make decisions about the future of technical training at McDuff.

McDuff Case Study for a Periodic Report. Model 8–5 on page 293 shows the rather routine nature of most periodic reports. Here Nancy Fairbanks is simply submitting her usual monthly report. The greatest challenge in such reports is to classify, divide, and label information in such a way that readers can quickly find what they need. Fairbanks selected the kind of substantive headings that help the reader locate information (for example, "Jones Fill Project" and "Performance Reviews").

Lab Reports

College students write lab reports for courses in science, engineering, psychology, and other subjects. Yet this report type also exists in technical organizations such as hospitals, engineering firms, and computer companies. Perhaps more than any other type of informal report, the lab report varies in format from organization to organization (and from instructor to instructor, in the case of college courses). This chapter will present a format to use when no other instructions have been given. A working definition follows:

> **Lab report:** this informal report describes work done in any laboratory—with emphasis on topics such as purpose of the work, procedures, equipment, problems, results, and implications. It may be directed to someone inside or outside your own organization. Also, it may stand on its own or it may become part of a larger report that uses the laboratory work as supporting detail.

The next section shows a typical ABC format for lab reports, with the types of information that might appear in the three main sections.

ABC Format for Lab Reports. Whether simple or complicated, lab reports usually contain some or all of these parts:

---

*ABC FORMAT: *Lab Report*

Abstract
- Purpose of report
- Capsule summary of results

Body
- Purpose or hypothesis of lab work
- Equipment needed
- Procedures or methods used in the lab test
- Unusual problems or occurrences
- Results of the test with reference to your expectations (results may appear in conclusion, instead)

Conclusion
- Statement or restatement of main results
- Implications of lab test for further work

---

*The content and order of items within each of the three sections can vary.

Communication Challenge

"A Nonprofit Job: Good Deed or Questionable Ethics?"

It all started with an innocent conversation. Velora Nescon, a project manager at McDuff's Houston branch, had lunch with her old college friend, Sibyl Sanders. As principal of Houston's Downtown Academy, Sibyl talked about her effort to keep the new private academy financially healthy and academically strong. Now she was busy with an expansion. What follows is a description of a problem she faced with the expansion, the help that Velora tried to offer, and some ethical and procedural questions raised by the situation.

A Working Lunch

Houston's Downtown Academy was established as a private elementary and middle school for bright inner-city kids who couldn't afford other private schools. Five years ago, business leaders had donated a renovated building for the new school and had raised money for a scholarship account. Now the academy had a respectable enrollment, a good academic reputation, and excellent morale among faculty and staff. All in all, the future looked good for the school. Recently a benefactor gave the school a piece of land adjacent to its campus, where a recreational area would be built.

In her conversation with Velora, Sibyl noted that although the school had some money to begin construction, the budget would be tight. At this point, Velora reminded her friend that McDuff, Inc., offered some of the technical services the project might require. She added that she could ask her branch manager if McDuff might handle the job just for cost, as its way of contributing to the growth of the school. Velora genuinely wanted to help her friend with this worthwhile venture, but in truth she also saw an opportunity to keep her technical staff busy during a slack period. By the end of lunch Sibyl and Velora had reached tentative agreement on McDuff doing the property study required before the land could be developed. Then back at the office, Velora got her boss to agree to allowing the job to be done for cost.

Velora's Lucky Find

The project involved a soil and environmental study. McDuff was to drill borings to determine what foundations would be needed for small structures in the recreation area. Also, soil samples would be taken from the site to check for contamination. These tasks were routine.

Before sending out her crew, Velora mentioned the project to a colleague, George Lightfoot, who thought he remembered doing some soil borings at the same location a couple years ago. Upon checking his files, George found a report that included two borings paid for by Ace Enterprises, a firm that had considered buying the property. Later Ace backed out of the purchase for reasons unrelated to the report. George loaned Velora the report and suggested she ask Ace for permission to use it, since she was trying to save the Downtown Academy money on the project. Velora thanked George for the report and said she'd call Ace.

Velora wanted to get on this no-profit job right away, while her crew wasn't busy, so she tried all afternoon to contact Ace Enterprises. There was no listing for the firm in Houston, and the McDuff librarian found no address in a quick search of her regional files. Velora assumed the firm had gone out of business, changed names, or left town. That being the case, she decided to move ahead in using information in the report—thus saving the Downtown Academy money for soil borings that would have been done. She assumed that Ace, if it still existed, would not mind contributing information for a nonprofit job like this one.

McDuff's Fieldwork

The next day the McDuff crew gathered soil samples from the surface and from shallow borings dug with a hand auger. Results of the lab tests on the samples showed there was an underground storage tank on the property. It had been used for kerosene, which had leaked into the surrounding soil. Both the tank and the soil would have to be removed.

Again trying to save her friend some money, Velora had the small tank and soil removed by a McDuff subcontractor working nearby later that week. The crew had been working for another McDuff client most of the day. Because that client had to pay for a full day's use of the crew and crane anyway, Velora didn't charge the Downtown Academy for the two hours it took to remove the tank and soil.

Questions and Comments for Discussion

Within a week, Velora handed Sibyl a complete report showing her fieldwork, lab tests, and conclusions. Sibyl was overjoyed that the study had come in even under the zero-profit budget she and Velora had first discussed. In thanking Velora, she assured her that McDuff would be a serious contender for the *profit* contracts that the Downtown Academy was sure to have in the future.

1. Do you think there are any ethical problems raised by this case study? If so, what are they and how would you have dealt with them? If not, explain your views.
2. Specifically, how do Sibyl's and Velora's actions either satisfy or violate the Equal Consideration of Interests principle described in chapter 2 of this text?
3. Putting aside the ethical issue, do you think Velora followed wise procedures in her handling of the Downtown Academy project? Why or why not?

McDuff Case Study for a Lab Report. Model 8–6 on pages 294–295 shows a McDuff lab report that is not part of a larger document. In this case, the client sent McDuff some soils taken from borings made into the earth. McDuff has analyzed the samples in its company laboratory and then drawn some conclusions about the kind of rock from which the samples were taken. The report writer, a geologist named Joseph Rappaport, uses the body of the report to provide background information, lab materials, procedures, and problems encountered. Note that the report body uses process analysis, a main pattern of organization covered in chapter 6.

CHAPTER SUMMARY

This chapter deals exclusively with the short, informal reports you will write through-out your career. On the job, you will write them for readers inside your organization (as **memo reports**) and outside your organization (as **letter reports**). In both cases, follow these 10 basic guidelines:

1. Plan well before you write.
2. Use letter or memo format.
3. Make text visually appealing.
4. Use the ABC format for organizing information.
5. Start with an introductory summary.
6. Put detailed support in the body.
7. Separate fact from opinion.
8. Focus attention in your conclusion.
9. Use attachments for details.
10. Edit carefully.

Although letter and memo reports come in many varieties, this chapter conveys only five common types: problem analyses, recommendation reports, equipment evaluations, progress/periodic reports, and lab reports. Each follows its own type of three-part ABC format for organizing information.

Access the Technical Writing Web site by pointing your browser at www.prenhall.com/pfeiffer. Here, you will find additional resources to aid you in learning and applying the information from this chapter.

ASSIGNMENTS

This chapter includes both short and long assignments. The short assignments in Part 1 are designed to be used for in-class exercises and short homework assignments. The assignments in Part 2 generally require more time to complete.

Part 1: Short Assignments

1. **Problem Analysis—Critiquing a Report.** Using the guidelines in this chapter, analyze the level of effectiveness of the following McDuff problem analysis.

April 16, 2002

Mr. Jay Henderson
Christ Church
10 Smith Dr.
Jar, Georgia 30060

**PROBLEM ANALYSIS
NEW CHURCH BUILDING SITE**

Introductory Summary

Last week your church hired our firm to study problems caused by the recent incorporation of the church's new building site into the city limits. Having reviewed the city's planning and zoning requirements, we have found some problems with your original site design—which initially was designed to meet the county's requirements only. My report focuses on problems with four areas on the site:

1. Landscaping screen
2. Church sign
3. Detention pond
4. Fire truck access

Attached to this report is a site plan to illustrate these problems as you review the report. The plan was drawn from an aerial viewpoint.

Landscaping Screen

The city zoning code requires a landscaping screen along the west property line, as shown on the attached site illustration sheet. The former design does not call for a screen in this area. The screen will act as a natural barrier between the church parking lot and the private residence adjoining the church property. The code requires that the trees for this screen be a minimum height of 8 feet with a height maturity level of at least 20 feet. The trees should be an aesthetically pleasing barrier for all parties, including the resident on the adjoining property.

Church Sign

After the site was incorporated into the city, the Department of Transportation decided to widen Woodstock Road and increase the setback to 50 feet, as illustrated on our site plan. With this change, the original location of the sign fell into the road setback. Its new location must be out of the setback and moved closer to the new church building.

Detention Pond

The city's civil engineers reviewed the original site drawing and found that the detention pond was too small. If the detention pond is not increased, rainwater may build up and overflow into the building, causing a considerable amount of flood damage to property in the building and to the building itself. There is a sufficient amount of land in the rear of the site to enlarge and deepen the pond to handle all expected rainfall.

Fire Truck Access

On the original site plan, the slope of the ground along the back side of the new building is so steep that a city fire truck would not be able to gain access to the rear of the building in the event of a fire. This area is shown on our site illustration around the north and east sides of the building. The zoning office enforces a code that is required by the fire marshal's office. This code states that all buildings within the city limits must provide a flat and unobstructed access path around the buildings. If the access is not provided, the safety of the church building and its members would be in jeopardy.

Conclusion

The just-stated problems are significant, yet they can be solved with minimal additional cost to the church. Once the problems are remedied and documented, the revised site plan must be approved by the zoning board before a building permit can be issued to the contractor.

I look forward to meeting with you and the church building committee next week to discuss any features of this study and its ramifications.

Sincerely,

Thomas K. Jones

Thomas K. Jones
Senior Landscape Engineer

Enclosure

2. **Problem Analysis—Group Project.** Divide into three- or four-person teams, as your instructor directs. In your group, share information about any problems that team members have encountered with services or facilities at the college or university you attend. Then select a problem substantive enough to be described in a short report. As a group, write a problem analysis in the format put forth in this chapter. Assume that your group represents a McDuff technical team that has been hired to investigate, and then write a series of reports on, problems at the school. Your report is one in the series. Select as your audience the appropriate administrators at the college or university.

3. **Recommendation Report—Critiquing a Report.** Using the guidelines in this chapter, analyze the level of effectiveness of the following McDuff recommendation report.

April 20, 2002

Kenman Aircraft Company
76 Jonesboro Road
Sinman, Colorado 87885

Attention: Mr. Ben Randall, Facilities Manager

EMERGENCY EXIT STUDY

Introductory Summary

As you requested, I have just completed a study of the emergency exits in your accounting office at the plant. My study indicates that you have two main problems: (1) easier access to exits is needed and (2) more exit signs and better visibility of these signs are needed. This report contains recommendations for rearranging the floor plan and improving signage.

Problems with Current Floor Plan

Two main problems cause the accounting office to fail to meet the county's guidelines for access to fire exits. First, the file cabinets on the north wall of the office are partially blocking the Reynolds Lane exit. Second, the office photocopier partially blocks the exit to the east hallway. In the first case, the file cabinets are so heavy that they could not be moved by one person. In the second case, the photocopier could be rolled out of the way only by a very strong individual. Obviously, both situations are unacceptable and violate the current code.

The other problem is signage. The Reynolds Lane exit has an exit sign, but it is not easily seen. The east hallway exit has no sign at all. In addition, the rest of the office lacks any maps that show people the location of the two fire exits.

Recommendations for Solving Exit Problem

Fortunately, the existing problems can be corrected with only minor cost to the company. The following recommendations should be implemented immediately upon your receipt of this report.
1. Move the file cabinets on the north wall to the east wall so that they no longer block the Reynolds Street exit.
2. Relocate the photocopier to the office supply room or the cubicle adjacent to it.
3. Remove the undersized exit sign from the Reynolds Street exit.
4. Purchase and install two county-approved exit signs above the two fire exits.
5. Draw up an emergency plan map and post a copy in every cubicle within the accounting office.

When you implement these recommendations, you will be in accordance with the county's current fire regulations.

Conclusion

I strongly suggest that my recommendations be put into action as soon as possible. By doing so you will greatly reduce the risk to your employees and your associated liability.

If you have any questions or need additional information, please call me at your convenience.

Sincerely,

Howard B. Manwell

Howard B. Manwell
Field Engineer

4. **Writing a Recommendation Report.** Divide into groups of three or four students, as your instructor directs. Consider your group to be a technical team from McDuff. Assume that the facilities director of your college or university has hired your team to recommend changes that would improve your classroom. Write a group report that includes the recommendations agreed to by your group. For example, you may want to consider structural changes of any kind, additions of equipment, changes in the type and arrangement of seating, and so forth.

5. **Writing an Equipment Evaluation.** Assume you are a supervisor at McDuff's Equipment Development shop, located in Baltimore. The Procurement Office routinely asks you to write evaluations of new pieces of equipment being used in the shop. Such evaluations help the director of procurement, Brenda Seymour, decide on future purchases.

Write a brief memo to Seymour, evaluating the Brakoh cordless drills that your staff began using in the shop about a year ago. The Brakoh brand replaced a more expensive brand that the shop had used for the 10 previous years. In the last few months, your technicians have reported that the cheaper models have been falling apart after six or eight months of use. Information coming to you suggests that there are two main problems: (1) a grinding noise can be heard in the housing, resulting in the failure of the chuck (the piece that holds the drill bit) to rotate; and (2) drilling time between rechargings tends to decrease as the drill gets older. You believe that the manufacturer, in order to cut the cost of the drill, has substituted poorly made components in high-wear locations. For example, the gears responsible for turning the chuck are made of plastic. With a little wear, the gears tend to slip, which produces the grinding sound and the rotation failure. As for the recharging problem, the power cell just seems to hold less charge than the previous drill. In summary, the drill has broken down four times faster than the other model, causing many repair bills and a loss in productivity.

In writing your memo, remember that in this case your main job is to provide information to the director of procurement, *not* to make recommendations one way or the other. After receiving your memo, she probably will complete a cost analysis to determine if the problems with the cheaper drill outweigh the advantage of the initial cost savings.

Part 2: Longer Assignments—Individual or Group Work

While planning some of these assignments—especially assignment 6—you may need to review information in chapter 2 about McDuff, Inc. Also, for each assignment you should complete a copy of the Planning Form (included at the end of the book). These assignments can be completed as individual or group projects.

6. **Report Based on Color Insert.** The color insert included in this book contains summaries of seven McDuff projects, in various topic areas. These summaries were written for marketing purposes, after the jobs were completed.

Using the information on one of these sheets, write a brief informal report that summarizes the project for the client. If necessary, add details that are not on the sheet. *Caution:* Remember that marketing sheets may *not* be organized as reports are organized. Consider your purpose and audience carefully before writing.

7. **Report Based on Internet "Surfing."** Use the Internet to collect actual information, and/or a list of sources that may contain information, about a topic that relates to your academic major. Then write an informal equipment evaluation in which you analyze (1) the ease with which the Internet allowed you to collect information on your topic and (2) the quality of the sources or information you received. Your audience is your instructor, who will let you know the degree of knowledge you can assume he or she has on this topic.

8. **Problem Analysis.** Assume you are a McDuff field engineer working at the construction site of a nuclear power plant in Jentsen, Missouri. For the past three weeks, your job has been to observe the construction of a water cooling tower, a large cylindrical structure. As consultants to the plant's construction firm, you and your McDuff crew were hired to make sure that work proceeds properly and on schedule. As the field engineer, you are supposed to report any problems in writing to your project manager, John Raines, back at your St. Louis office. Then he will contact the construction firm's office, if necessary.

Write a short problem analysis in the form of a memo report to Raines. (Follow the guidelines in the "Problem Analyses" section of this chapter.) Take the following randomly organized information and present it in a clear, well-organized fashion. If you wish, add information of your own that might fit the context.

- Three cement pourings for the tower wall were delayed an hour each on April 21 because of light rain.
- Cement-truck drivers need to slow down while driving through the site. Other workers complain about the excessive dust raised by the trucks.
- Mary Powell, a McDuff safety inspector on the crew, has cited 12 workers for not wearing their hard hats.
- You just heard from one subcontractor, Allis Wire, Inc., that there will be a two-day delay in delivering some steel reinforcing wires that go into the concrete walls. That delay will throw off next week's schedule. Last Monday's hard rain and flooding kept everyone home that day.
- It is probably time once again to get all the subcontractors together to discuss safety at the tower site. Recently two field hands had bad cuts from machinery.
- Although there have not been any major thefts at the site, some miscellaneous boards and masonry pieces are missing each day—probably because nearby residents (doing small home projects) think that whatever they find at the site has been discarded. Are additional "No trespassing" signs needed?
- Construction is only two days behind schedule, despite the problems that have occurred.

9. **Problem Analysis.** As a landscape engineer for McDuff, one of your jobs is to examine problems associated with the design of walkways, the location of trees and garden beds, the grading of land around buildings, and any other topographical features. Assume that you have been hired by a specific college, community, or company with which you are familiar. Your objective is to evaluate one or more landscaping problems at the site.

   Write an informal report that describes the problem(s) in detail. (Follow the guidelines in the "Problem Analyses" section of this chapter.) Be specific about how the problem affects people—the employees, inhabitants, students, etc. Here are some sample problems that could be evaluated:

   - Poorly landscaped entrance to a major subdivision
   - Muddy, unpaved walkway between dormitories and academic buildings on a college campus
   - Unpaved parking lot far from main campus buildings
   - Soil runoff into the streets from several steep, muddy subdivision lots that have not yet been sold
   - City tennis courts with poor drainage
   - Lack of adequate flowers or bushes around a new office building
   - Need for a landscaped common area within a subdivision or campus
   - Need to save some large trees that may be doomed because of proposed construction

10. **Recommendation Report.** For this paper, choose a design problem at your college or company. Now put yourself in the position of a McDuff employee hired by your school or company to recommend solutions to the problem.

    Your ideas must be in the form of a report that gives one or more recommendations resulting from your study. (Consult writing guidelines in the "Recommendation Reports" section of this chapter.) Assume that the problem is well enough understood to require only a brief summary, before launching into your recommendations. Because this is a short report, it may not contain many technical details for implementing your recommendations. Also, you need to choose a limited-enough topic so that it can be covered in a short memo report. Here are some sample topics:

- Poor ventilation in an office or a classroom, such as one with sealed windows
- Inadequate space for quick exits during emergencies
- Poor visibility in a large auditorium
- Poor acoustics in a large classroom or training room
- Lack of, or improper placement of, lighting
- Energy inefficiency caused by structural flaws, such as poor insulation or high ceilings
- Rooms or walkways that are not handicap-accessible
- Failure to take advantage of solar heating
- Inefficient heating or air-conditioning systems

11. **Recommendation Report.** This project will require some research. Assume that your college plans either to embark on a major recycling effort or to expand a recycling program that has already started. Put yourself in the role of a McDuff environmental scientist or technician who has been asked to recommend these recycling changes.

First, do some research about recycling programs that have worked in other organizations. A good place to start would be periodical indexes such as *Readers' Guide to Periodicals* or the *Environment Index*, which will lead you to some magazine articles of interest. Choose to discuss one or more recoverable resources such as paper, aluminum, cardboard, plastic, or glass bottles. Be specific about how your recommendations can be implemented by the organization or audience about which you are writing. (Consult the guidelines in the "Recommendation Reports" section of this chapter.)

12. **Equipment Evaluation.** For six months you have driven a new Ford 150 company truck at remote job sites. As lead field hand for McDuff's Boston office, you have been asked to write an evaluation of the vehicle for Brenda Seymour, director of procurement at the corporate office in Baltimore. Seymour will use your report to decide whether to recommend ordering five more F-150s for other offices. She has told you that you need to discuss only major positive or negative features, not every detail. If she needs more information after reading your report, she will let you know.

Consider the list below to be your random notes. Use all this information to write a memo report that evaluates the truck. Make sure to follow the guidelines in this chapter.

- My 150 has been very reliable—it never failed to start, even during subzero ice storms last winter.
- The 4.6 liter small V-8, has provided plenty of power to handle any hauling I have done. No need to order the more expensive and less fuel-efficient 5.4 liter V-8.
- Have been to 18 job sites with the truck, from marshes in Maine to mountains in New Hampshire. Have put about 12,000 miles on it, on all kinds of roads and in all conditions.
- Tires that came with the truck did not work well in muddy locations, even with four-wheel drive. Suggest we buy all-terrain tires for future vehicles. Continue to order four-wheel drive—it is necessary at over half our job sites.
- The short bed (6 ft) did not provide enough hauling room, once I put my toolbox across the truck bed near the back window. Suggest company buy long-bed trucks with the added 2 ft of room.
- Given what I know now, I give the truck a good to excellent rating.
- Automatic transmission worked great. Am told by other owners that the automatic is better than the manual for construction jobs because the manual tends to burn out

clutches, especially when the truck needs to be "rocked" back and forth to get out of mud holes. My automatic has taken a lot of abuse without problems.

- Have had some problems with front-end handling on rough roads. Suggest that future trucks be ordered with special handling package, which includes two shock absorbers—not just one—on each front wheel.
- Have had no major repairs, just the regular maintenance checks at the dealer.
- There was one recall from the manufacturer concerning an exhaust pipe hanger that might bend, but the dealer fixed the problem in 20 minutes.
- Really need to have another six months to see how well truck holds up.

13. **Equipment Evaluation.** McDuff, Inc., has decided to make a bulk purchase of 20 typewriters or word processors. (For the purposes of this assignment, choose a typewriter or word processor with which you are familiar.) The machinery will go in a new department being set up in several months.

   Assume that McDuff now uses five different types of machines (again, remember you can choose typewriters or word processing systems). In the interests of a fair comparison/contrast, Brenda Seymour, director of McDuff's corporate Procurement Department, has asked you and several other employees to evaluate the effectiveness of your own system. Write Seymour a memo report that includes your evaluation. (Consult guidelines in the "Equipment Evaluations" section of this chapter.) She will use the data and opinions in all the equipment evaluations she receives to make her choice for the bulk purchase. Your criteria for evaluation might include topics such as one or more of these:

   - Physical design of the equipment
   - Ease with which system can be learned
   - Quality of the written instructions
   - Frequency and cost of maintenance
   - Availability of appropriate software
   - Length of coverage of warranty
   - Nearness to a service center
   - Reputation of the manufacturer

14. **Progress or Periodic Report.** Assume that you have worked as a field hand at McDuff's Atlanta office for 15 years. Because of your reliability, good judgment, and intelligence, the company is paying for your enrollment at a local college. Also, you get half time off, with pay. Because of its investment in you, McDuff expects you to report periodically on your college work. Choose one of these two options for this assignment:

   Progress report: Select a major project you are now completing in any college course. Following the guidelines in the "Progress/Periodic Reports" section of this chapter, write a progress report on this project. Direct the memo report to the Atlanta office's manager of engineering, Wade Simkins. Sample topics might include a major paper, laboratory experiment, field project, or design studio.

   Periodic report: Assume that McDuff requires that you submit periodic reports on your schooling every few weeks. Following the guidelines in the "Progress/Periodic Reports" section of this chapter, write a periodic report on your recent course work (completed or ongoing classes or both). Direct the memo report to the manager of engineering, Wade Simkins. Organize the report by class, and then give specific updates on each one.

15. **Lab Report.** For this assignment, you must be taking a lab course now or have taken such a course recently. As in assignment 14, assume you work as a field hand with McDuff's Atlanta office. The company is sponsoring your schooling and has requested that you report on a specific college lab.

Following the guidelines in this chapter's "Lab Reports" section, write a report to Wade Simkins, manager of engineering. The quality of your report may affect whether or not McDuff continues to fund your schooling. Be specific about the goals, procedures, and results of your laboratory—just as you would in an actual college lab report.

16. **Ethics Assignment.** Reread Model 8–3, a problem analysis concerning *Seeker II*, a ship leased by McDuff for its offshore jobs. Jan Stillwright receives the report just after she has signed a contract to complete a highly profitable, one-week assignment for one of McDuff's best clients. This client urgently needs some geological data from the floor of the Gulf of Mexico, so that a bid for a construction project can be submitted to the government. Jan immediately contacts other leasing operations but discovers that no other drilling ships are available when her project must begin—the day after tomorrow. At this point Jan emails Harold Marshall that she will definitely address all concerns with the boat owner after this urgent one-week study. Moreover, she says that she will (1) give overtime pay to the crew for the upcoming one-week trip and (2) assign McDuff's top safety officer to the trip so that she can observe, record, and validate the points made in Harold's report. What are the ethical implications, if any, of Jan's response? If you were Harold, how would you respond?

17. **Informal Report—International Context.** Investigate features such as style, format, structure, and organization of short reports written in another country. For this assignment, it would be best to interview someone who does business in another country and, if possible, to get an actual report that you can submit. Write a memo report to your instructor that presents the results of your study.

12 Post Street
Houston, Texas 77000
(713) 555-9781

April 22, 2002

Big Muddy Oil Company, Inc.
12 Rankin St.
Abilene, TX 79224

ATTENTION: Mr. James Smith, Engineering Manager

SHARK PASS STUDY
BLOCK 15, AREA 43-B
GULF OF MEXICO

INTRODUCTORY SUMMARY

You recently asked our firm to complete a preliminary soils investigation at an offshore rig site. This report presents the tentative results of our study, including major conclusions and recommendations. A longer, formal report will follow at the end of the project.

On the basis of what we have learned so far, it is our opinion that you can safely place an oil platform at the Shark Pass site. To limit the chance of a rig leg punching into the seafloor, however, we suggest you follow the recommendations in this report.

WORK AT THE PROJECT SITE

On April 16 and 17, 2002, McDuff's engineers and technicians worked at the Block 15 site in the Shark Pass region of the gulf. Using McDuff's leased drill ship, *Seeker II*, as a base of operations, our crew performed these main tasks:

- Seismic survey of the project study area
- Two soil borings of 40 feet each

Both seismic data and soil samples were brought to our Houston office the next day for laboratory analysis.

LABORATORY ANALYSIS

On April 18 and 19, our lab staff examined the soil samples, completed bearing capacity tests, and evaluated seismic data. Here are the results of that analysis.

Soil Layers

Our initial evaluation of the soil samples reveals a 7-9 ft layer of weak clay starting a few feet below the seafloor. Other than that layer, the composition of the soils seems fairly typical of other sites nearby.

Margin notes (left column):

Includes specific title.

Uses *optional* heading for abstract part of ABC format.

Draws attention to *main point* of report.

Gives on-site details of project—dates, location, tasks.

Uses *lead-in* to subsections that follow.

Highlights most important point about soil layer—that is the *weak clay.*

MODEL 8–1
Recommendation report (letter format)

continues

Bearing Capacity

We used the most reliable procedure available, the XYZ method, to determine the soil's bearing capacity (that is, its ability to withstand the weight of a loaded oil rig). That method required that we apply the following formula:

$$Q = cNv + tY, \text{ where}$$

Q	=	ultimate bearing capacity
c	=	average cohesive shear strength
Nv	=	the dimensionless bearing capacity factor
t	=	footing displacement
Y	=	weight of the soil unit

The final bearing capacity figure will be submitted in the final report, after we repeat the tests.

Seafloor Surface

By pulling our underwater seismometer back and forth across the project site, we developed a seismic "map" of the seafloor surface. That map seems typical of the flat floor expected in that area of the gulf. The only exception is the presence of what appears to be a small sunken boat. This wreck, however, is not in the immediate area of the proposed platform site.

CONCLUSIONS AND RECOMMENDATIONS

Based on our analysis, we conclude that there is only a slight risk of instability at the site. Though unlikely, it is possible that a rig leg could punch through the seafloor, either during or after loading. We base this opinion on (1) the existence of the weak clay layer, noted earlier, and (2) the marginal bearing capacity.

Nevertheless, we believe you can still place your platform if you follow careful rig-loading procedures. Specifically, take these precautions to reduce your risk:

1. Load the rig in 10-ton increments, waiting one hour between loadings.
2. Allow the rig to stand 24 hours after the loading and before placement of workers on board.
3. Have a soils specialist observe the entire loading process, to assist with any emergency decisions if problems arise.

As noted at the outset, these conclusions and recommendations are based on preliminary data and analysis. We will complete our final study in three weeks and submit a formal report shortly thereafter.

McDuff, Inc., enjoyed working once again for Big Muddy Oil at its Gulf of Mexico lease holdings. I will phone you this week to see if you have any questions about our study. If you need information before then, please give me a call.

Sincerely,

Bartley Hopkins

Bartley Hopkins, Project Manager
McDuff, Inc.

hg

Notes *why* this method was chosen (that is, reliability).

Explains both *how* the mapping procedure was done and *what results* it produced.

Leads off section with major conclusion, for emphasis.

Restates points (made in body) that support conclusion.

Uses list to emphasize recommendations to *reduce risk.*

Again mentions tentative nature of information, to prevent misuse of report.

Maintains control and shows initiative by offering to *call* client.

MODEL 8–1
continued

M McDuff, Inc.

MEMORANDUM

DATE: July 26, 2002
TO: Hank Worley, Project Manager
FROM: Melanie Frank, Office Manager *MF*
SUBJECT: Evaluation of Best Choice Software

INTRODUCTORY SUMMARY

When the office purchased one copy of Best Choice Software last month, you suggested I send you an evaluation after 30 days' use. Having now used Best Choice for a month, I have concluded that it meets all our performance expectations. This memo presents our evaluation of the main features of Best Choice.

HOW BEST CHOICE HAS HELPED US

Best Choice provides five primary features: word processing, file management, spreadsheet, graphics, and a user's guide. Here is my critique of all five.

Word Processing

The system contains an excellent word-processing package that the engineers as well as the secretaries have been able to learn easily. This package can handle both our routine correspondence and the lengthy reports that our group generates. Of particular help is the system's 90,000-word dictionary, which can be updated at any time. The spelling correction feature has already saved much effort that was previously devoted to mechanical editing.

File Management

The file-manager function allows the user to enter information and then to manipulate it quickly. During one three-day site visit, for example, a field engineer recorded a series of problems observed in the field. Then she rearranged the data to highlight specific points I asked her to study, such as I-beam welds and concrete cracks.

Spreadsheet

Like the system's word-processing package, the spreadsheet is efficient and quickly learned. Because Best Choice is a multipurpose software package, spreadsheet data can be incorporated into letter or report format. In other words, spreadsheet information can be merged with our document format to create a final draft for submission to clients or supervisors, with a real savings in time. For example, the memo I sent you last week on budget projections for field equipment took me only an hour to complete; last quarter, the identical project took four hours.

Uses optional first heading for abstract section of ABC format. Gives background, main points, and scope statement.

Notes five main points to be covered.

Begins paragraph with most important point. Supports claim with evidence.

Uses specific example to document opinion.

Gives simple explanation of how spreadsheet works.

MODEL 8–2
Equipment evaluation (memo format)

continues

Shows relevance of graphics to current work.

Graphics

The graphics package permits visuals to be drawn from the data contained in the spreadsheet. For example, a pie chart that shows the breakdown of a project budget can be created easily by merging spreadsheet data with the graphics software. With visuals becoming such an important part of reports, we have used this feature of Best Choice quite frequently.

Supplies strong supporting statistic.

User's Guide

Eight employees in my group have now used the Best Choice user's guide. All have found it well laid out and thorough. Perhaps the best indication of this fact is that in 30 days of daily use, we have placed only three calls to the Best Choice customer-service number.

Wraps up report by restating main points.

CONCLUSION

Best Choice seems to contain just the right combination of tools to help us do our job, both in the field and in the office. These are the system's main benefits:

- Versatility—it has diverse functions
- Simplicity—it is easy to master

Offers follow-up effort.

The people in our group have been very pleased with the package during this 30-day trial. If you like, we would be glad to evaluate Best Choice for a longer period.

MODEL 8–2
continued

288

MEMORANDUM

DATE: October 15, 2002
TO: Jan Stillwright, Vice President of Research and Training
FROM: Harold Marshal, Technical Supervisor HW
SUBJECT: Boat Problems During Summer Season

INTRODUCTORY SUMMARY

We have just completed a one-month project aboard the leased ship, *Seeker II*, in the Pacific Ocean. All work went just about as planned, with very few delays caused by weather or equipment failure.

However, there were some boat problems that need to be solved before we lease *Seeker II* again this season. This report highlights the problems so that they can be brought to the owner's attention. My comments focus on four areas of the boat: drill rig, engineering lab, main engine, and crew quarters.

DRILL RIG

Thus far the rig has operated without incident. Yet on one occasion, I noticed that the elevator for lifting pipe up the derrick swung too close to the derrick itself. A quick gust of wind or a sudden increase in sea height caused these shifts. If the elevator were to hit the derrick, causing the elevator door to open, pipe sections might fall to the deck below.

I believe the whole rig assembly needs to be checked over by someone knowledgeable about its design. Before we put men near that rig again, we need to know that their safety would not be jeopardized by the possibility of falling pipe.

ENGINEERING LAB

Quite frankly, it is a tribute to our technicians that they were able to complete all lab tests with *Seeker II's* limited facilities. Several weeks into the voyage, these four main problems became apparent:

1. Ceiling leaks
2. Poor water pressure in the cleanup sink
3. Leaks around the window near the electronics corner
4. Two broken outlet plugs

Although we were able to devise a solution to the window leaks, the other problems stayed with us for the entire trip.

MODEL 8–3
Problem analysis (memo format)

continues

Margin notes:

Gives abstract (or *summary*) in first paragraph.

Provides *capsule listing* of problems discussed in report.

Opens with *most important point*—then qualifies it. Explains problem in *layperson's language*, indicating possible consequences.

Uses *listing* to draw attention to four main lab problems on board.

Uses *simple language* to describe technical problems.

Closes section with opinion that flows from facts presented.

Gives lead-in to three sections that follow.

Describes three problem areas in great detail—knowing the owner will want facts to support complaints.

Briefly restates problem, with emphasis on *safety* and *profits*.

Ends with specific recommendation.

MAIN ENGINE

On this trip, we had three valve failures on three different cylinder heads. From our experience on other ships, it is very unusual to have one valve fail, let alone three. Fortunately for us, these failures occurred between projects, so we did not lose time on a job. And fortunately for the owner, the broken valve parts did not destroy the engine's expensive turbocharger.

Only an expert will be able to tell whether these engine problems were flukes or if the entire motor needs to be rebuilt. In my opinion, the most prudent course of action is to have the engine checked over carefully before the next voyage.

CREW QUARTERS

When 15 men live in one room for three months, it is important that basic facilities work. On *Seeker II* we experienced problems with the bedroom, bathroom, and laundry room that caused some tension.

Bedroom

Three of the top bunks had such poor springs that the occupants sank 6 to 12 in. toward the bottom bunks. More important, five of the bunks are not structurally sound enough to keep from swaying in medium to high seas. Finally, most of the locker handles are either broken or about to break.

Bathroom

Poor pressure in three of the commodes made them almost unusable during the last two weeks. Our amateur repairs did not solve the problem, so I think the plumbing leading to the holding tank might be defective.

Laundry Room

We discovered early that the filtering system could not screen the large amount of rust in the old 10,000-gallon tank. Consequently, undergarments and other white clothes turned a yellow-red color and were ruined.

CONCLUSION

As noted at the outset, none of these problems kept us from accomplishing the major goals of this voyage. But they did make the trip much more uncomfortable than it had to be. Moreover, in the case of the rig and engine problems, we were fortunate that injuries and downtime did not occur.

I strongly urge that the owner be asked to correct these deficiencies before we consider using *Seeker II* for additional projects this season.

MODEL 8–3
continued

MEMORANDUM

DATE: June 11, 2002
TO: Kerry Camp, Vice President of Domestic Operations
FROM: Scott Sampson, Manager of Personnel *SS*
SUBJECT: Progress Report on Training Project

INTRODUCTORY SUMMARY

On May 21 you asked that I study ways our firm can improve training for technical employees in all domestic offices. We agreed that the project would take about six or seven weeks and involve three phases:

Phase 1: Make phone inquiries to competing firms
Phase 2: Send a survey to our technical people
Phase 3: Interview a cross section of our technical employees

I have now completed Phase 1 and part of Phase 2. My observation thus far is that the project will offer many new directions to consider for our technical training program.

WORK COMPLETED

In the first week of the project, I had extensive phone conversations with people at three competing firms about their training programs. Then in the second week, I wrote and sent out a training survey to all technical employees in McDuff's domestic offices.

Phone Interviews

I contacted three firms for whom we have done similar favors in the past: Simkins Consultants, Judd & Associates, and ABG Engineering. Here is a summary of my conversations:

1. Simkins Consultants
   Talked with Harry Roland, training director, on May 23. Harry said that his firm has most success with internal training seminars. Each technical person completes several one- or two-day seminars every year. These courses are conducted by in-house experts or external consultants, depending on the specialty.

2. Judd & Associates
   Talked with Jan Tyler, manager of engineering, on May 24. Jan said that Judd, like Simkins, depends mostly on internal seminars. But Judd spreads these seminars over one or two weeks, rather than teaching intensive courses in one or two days. Judd also offers short "technical awareness" sessions at the lunch hour every two weeks. In-house technical experts give informal presentations on some aspect of their research or fieldwork.

Summarizes project, to refresh reader's memory and establish common ground.

Gives overview of report.

Summarizes two main tasks, as lead-in to subsections.

Organizes this section by the companies consulted.

Creates parallel form in organization of all three points.

MODEL 8–4
Progress report (memo format)

continues

3. ABG Engineering

Talked with Newt Mosely, personnel coordinator, on May 27. According to Newt, ABG's training program is much as it was two decades ago. Most technical people at high levels go to one seminar a year, usually sponsored by professional societies or local colleges. Other technical people get little training beyond what is provided on the job. In-house training has not worked well, mainly because of schedule conflicts with engineering jobs.

Internal Survey

After completing the phone interviews noted, I began the survey phase of the project. Last week, I finished writing the survey, had it reproduced, and sent it with cover memo to all 450 technical employees in domestic offices. The deadline for returning it to me is June 17.

WORK PLANNED

With phone interviews finished and the survey mailed, I foresee the following schedule for completing the project:

June 17:	Surveys returned
June 18-21:	Surveys evaluated
June 24-28:	Trips taken to all domestic offices to interview a cross section of technical employees
July 3:	Submission of final project report to you

CONCLUSION

My interviews with competitors gave me a good feel for what technical training might be appropriate for our staff. Now I am hoping for a high-percentage return on the internal survey. That phase will prepare a good foundation for my on-site interviews later this month. I believe this major corporate effort will upgrade our technical training considerably.

I would be glad to hear any suggestions you may have about my work on the rest of the project. For example, please call if you have any particular questions you want asked during the on-site interviews (ext. 348).

Gives important details about the survey.

Organizes section chronologically, making sure to stay within a six- or seven-week schedule.

Looks to future tasks.

Emphasizes major benefit, to "sell" the project internally.

Indicates flexibility and encourages response from reader.

MODEL 8–4
continued

DATE: August 2, 2002
TO: Ralph Buzby, Manager of Engineering
FROM: Nancy Fairbanks, Project Manager *NF*
SUBJECT: Activity Report for July 2002

> *Begins with overview of entire report.*

     July has been a busy month in our group. Besides starting and finishing many smaller jobs, we completed the Jones Fill project. Also, the John Lewis Dam borings began just a week ago. Finally, I did some marketing work and several performance reviews.

SMALL PROJECTS

> *Gives summary of small projects.*
>
> *Uses list to highlight main types.*
>
> *Indicates reasons for delays.*

     Last month, my group completed nine small projects, each with a budget under $20,000 and each lasting only a few days. These jobs were in three main areas:

1. Surveying subdivisions—five jobs
2. Taking samples from toxic sites—two jobs
3. Doing nearby soil borings—two jobs

All nine were completed within budget. Eight of the nine projects were completed on time. The Campbell County survey, however, was delayed for a day because of storms on July 12.

JONES FILL PROJECT

     Our written report on this 12-month job was finally submitted to Trunk Engineering, Inc., on July 23. The delay was caused by Trunk's decision to change the scope of the project again. The firm wanted another soil boring, which we completed on July 22.

JOHN LEWIS DAM PROJECT

> *Again, gives* reasons *for delay.*
>
> *Supports section with* specifics*—for example, the exact number of meetings.*

     As you know, we had hoped to start work at the dam site last month. However, the client decided to make a lot of design changes that had to be approved by subcontractors. The final approval to start came just last week; thus our first day on site was July 29.

MARKETING

     During July, my main marketing effort was to meet with some previous clients, acquainting them with some of our new services. I met with eight different clients at their offices, with two meetings occurring on each of these dates: July 15, 16, 22, and 23. There's a good possibility that several of these meetings will lead to additional waste-management work in the next few months.

PERFORMANCE REVIEWS

     As we discussed last month, I fell behind on my staff's performance reviews in June. In July, I completed the three delayed reviews, as well as the four that were due in July. Copies of the paperwork were sent to your office and to the Personnel Department on July 19. This brings us up to date on all performance reviews.

> *Lays foundation for* next *meeting.*

Ralph Buzby
August 2, 2002
Page 2

CONCLUSION

     July was a busy month in almost all phases of my job. Because of this pace, I haven't had time to work on the in-house training course you asked me to develop. In fact, I'm concerned that time I devote to that project will take me away from my ongoing client jobs. At our next meeting, perhaps we should brainstorm about some solutions to this problem.

MODEL 8–5
Periodic report (memo format)

M McDuff, Inc.

105 Halsey Street
Baltimore, Maryland 21212
(301) 555-7588

December 12, 2002

Mr. Andrew Hawkes
Monson Coal Company
2139 Lasiter Dr.
Baltimore, MD 21222

**LABORATORY REPORT
BOREHOLE FOSSIL SAMPLES
BRAINTREE CREEK SITE, WEST VIRGINIA**

INTRODUCTORY SUMMARY

Last week you sent us six fossilized samples from the Braintree Creek site. Having analyzed the samples in our lab, we believe they suggest the presence of coal-bearing rock. As you requested, this report will give a summary of the materials and procedures we used in this project, along with any problems we had.

As you know, our methodology in this kind of job is to identify microfossils in the samples, estimate the age of the rock by when the microfossils existed, and then make assumptions about whether the surrounding rock might contain coal.

LAB MATERIALS

Our lab analysis relies on only one piece of specialized equipment: a Piketon electron microscope. Besides the Piketon, we use a simple 400-power manual microscope. Other equipment is similar to that included in any basic geology lab, such as filtering screens and burners.

LAB PROCEDURE

Once we receive a sample, we first try to identify the exact kinds of microfossils that the rocks contain. Our specific lab procedure for your samples consisted of these two steps:

Step 1

We used a 400-power microscope to visually classify the microfossils that were present. Upon inspection of the samples, we concluded that there were two main types of microfossils: nannoplankton and foraminifera.

Margin notes:

Gives *overview* of results.

Outlines *procedure* to be detailed in following paragraph.

Describes main equipment, in layperson's language.

Breaks down procedure into easy-to-read "chunks."

MODEL 8–6
Lab report (letter format)

Provides smooth transitions.

Step 2

Next, we had to extract the microfossils from the core samples you provided. We used two different techniques:

Nannoplankton Extraction Technique

Itemizes steps because of their importance in procedure.

a. Selected a pebble-size piece of the sample
b. Thoroughly crushed the piece under water
c. Used a dropper to remove some of the material that floats to the surface (it contains the nannoplankton)
d. Dried the nannoplankton-water combination
e. Placed the nannoplankton on a slide

Uses *parallel form* in describing this process.

Foraminifera Extraction Technique

a. Boiled a small portion of the sample
b. Used a microscreen to remove clay and other unwanted material
c. Dried remaining material (foraminifera)
d. Placed foraminifera on slide

PROBLEMS ENCOUNTERED

Does not bury sampling error— gives it proper treatment.

The entire lab procedure went as planned. The only problem was minor and occurred when we removed one of the samples from the container in which it was shipped. As the bag was taken from the shipping box, it broke open. The sample shattered when it fell onto the lab table. Fortunately, we had an extra sample from the same location.

CONCLUSION

Ends with wrap-up that reinforces main point of report.

Judging by the types of fossils present in the sample, they come from rock of an age that might contain coal. This conclusion is based on limited testing, so we suggest you test more samples at the site. We would be glad to help you with this additional sampling and testing.

Offers follow-up services.

I will call you this week to discuss our study and any possible follow-up you may wish us to do.

Sincerely,

Joseph Rappaport

Joseph Rappaport
Senior Geologist

MODEL 8–6
continued

Chapter 9

Formal Reports

Critiquing an Annual Report

General Instructions

Each chapter begins with a Collaboration at Work exercise that applies strategies for working in groups to chapter topics. The exercise assumes you (1) have been divided into groups of about three to six students, (2) will use group time inside or outside of class to complete the case, and (3) will produce an oral or written response. For guidelines about writing in groups, refer to pages 28–34.

Background for Assignment

Many companies prepare annual reports that summarize activities during the previous year. Readers include stockholders, employees, and investment groups, among others. Besides providing useful data about the company, annual reports often provide good examples of the following:

- How complex and diverse information can be organized
- How words can be selected to accentuate the positive, even if negative information is being reported
- How page design can be used to create an engaging, readable document

CHAPTER OUTLINE

Continued

Annual reports are sent to stockholders. However, they also may be available in libraries and on the Internet. For example, at *www.pearson.com* you will find the annual report for Pearson Education, the parent company of Prentice Hall, the publisher of this textbook.

Group Assignment

You may complete this assignment with a hard copy annual report provided by your instructor or secured from your library. Or you can complete it by locating an annual report on the World Wide Web, such as the Pearson Education annual report already mentioned. Meet with your group to review the print or online version of the report. Develop responses to the following questions:

1. How is information organized? (Give examples)
2. Does the report follow or not follow the ABC format? (Explain)
3. Who is the audience for the report? (Support your conclusion)

*T*oby West, McDuff's director of marketing, was given an interesting assignment two months ago. Rob McDuff asked him to take a long, hard look at the company's clients. Are they satisfied with the service they received? Do they routinely reward McDuff with additional work? Are there any features of the company, its employees, or its services that frustrate them? What do they want to see changed? In other words, Toby West was asked to step back from daily events and evaluate the company's level of service. He attacked the project in four stages:

1. He designed and sent out a survey to all recent and current clients.
2. He followed up on some of the returned surveys with phone and personal interviews.
3. He evaluated the data he collected.
4. He wrote a *formal report* on the results of his study. Besides going to all corporate and branch managers, West's report later served as a basis for some in-house training sessions called "Quality at McDuff."

Like Toby West, you will write a number of long, formal reports during your career. Most will be written collaboratively with colleagues; others will be just your creations. All of them will require major efforts at planning, organizing, drafting, and revising. Though informal reports are the most common report in business writing, formal reports become a larger part of your writing as you move along in your career. This text uses the following working definition:

> **Formal report:** this report covers complex projects and is directed to readers at different technical levels. Although not defined by length, a formal report usually contains at least 6 to 10 pages of text, not including appendices. It can be directed to readers either inside or outside your organization. Often bound, it usually includes these separate parts: (1) cover/title page, (2) letter/memo of transmittal, (3) table of contents, (4) list of illustrations, (5) executive summary, (6) introduction, (7) discussion sections, and (8) conclusions and recommendations. (Appendices often appear after the report text.)

To prepare you to write excellent formal reports, this chapter includes three main sections. The first section briefly describes four situations that would require formal reports—two are in-house (to readers *within* McDuff) and two are external (to readers *outside* McDuff). The second section provides guidelines for writing the main parts of a long report. The third section includes a complete long report from McDuff that follows this chapter's guidelines.

WHEN TO USE FORMAL REPORTS

Like most people, you probably associate formal reports with important projects. What else would justify all that time and effort? In comparison to informal reports, formal reports usually (1) cover more complicated projects and (2) are longer than their informal counterparts.

While complexity of subject matter and length are the main differences between formal and informal reports, sometimes there is another distinction. Formal reports may have a more diverse set of readers. In this case, readers who want just a quick overview can turn to the executive summary at the beginning or the conclusions and recommendations at the end. Technical readers who want to check your facts and figures can turn to discussion sections or appendices. And all readers can flip to the table of contents for a quick outline of what sections the report contains. You need to think about the needs of all these readers as you plan and write your formal reports.

The intended audience for formal reports can be inside or outside, though the latter is more common in formal reports. Here are four situations at McDuff for which formal reports would be appropriate:

- **Salary study and recommendations (internal):** Mary Kennelworth, a supervisor at McDuff's San Francisco office, has just completed a study of technicians' salaries among McDuff's competitors on the West Coast. What prompted the study was the problem she has had hiring technicians to assist environmental engineers and geologists. Lately some top applicants have been choosing other firms. Because the salary scales of her office are set by McDuff's corporate headquarters, she wants to give the main office some data showing that San Francisco starting salaries should be higher. Mary decides to submit a formal report, complete with data and recommendations for adjustments. Her main audience includes the branch manager and the vice president of human resources (the company's top decision-maker about salaries and other personnel matters).

- **Analysis of marketing problems (internal):** For several years, Jim Springer, engineering manager at McDuff's Houston office, has watched profits decline in onshore soils work. (In this type of work, engineers and technicians investigate the geological and surface features of a construction site and then recommend foundation designs and construction practices.) One problem has been the "soft" construction market in parts of Texas. But the slump in work has continued despite the recent surge in construction. In other words, some other company is getting the work. Jim and his staff have analyzed past marketing errors, with a view toward developing a new strategy for gaining new clients and winning back old ones. He plans to present the problem analysis and preliminary suggestions in a formal report. The main readers will be his manager, the corporate marketing manager in Baltimore, and managers at other domestic offices who have positions that correspond to his.

- **Waste-management survey (external):** For the last several years, the city of Belton, Georgia, has noticed increased fish kills on the Channel River, which flows through the city and serves as the city's main source of drinking water. Pollution has always been fairly well monitored on the river, so city officials are puzzled by the kills. McDuff's Atlanta office was hired to analyze the problem and present its opinion about the cause. A team of chemists, environmental engineers, and field technicians has just completed a study and will present its formal report. The audience will be quite diverse—from the technical experts in the city's water department to the members of a special citizens' panel representing the residents.

- **Collapse of oil rig (external):** A 10-year-old rig in the North Sea recently collapsed during a mild storm. Several rig workers died, and several million dollars' worth of equipment was lost. Also, the accident created an oil spill that destroyed a significant amount of fish and wildlife before it was finally contained. McDuff's London office was hired to examine the cause of the collapse of this structure, which supposedly was able to withstand hurricanes. After three months of on-site analysis and laboratory work, McDuff's experts are ready to submit their report. It will be read by corporate managers of the firm, agencies of the Norwegian government, and members of several major wildlife organizations. Also, it will be used as the basis for some articles in magazines and newspapers throughout the world.

As these four situations show, formal reports are among the most difficult on-the-job writing assignments you face in your career. You will write some yourself; you will write others as a member of a team of technical and professional people. In all cases, you need to (1) understand your purpose, (2) grasp the needs of your readers, and (3) design a report that responds to these needs. Guidelines in the next two sections will help you meet these goals.

STRATEGY FOR ORGANIZING FORMAL REPORTS

You will encounter different report formats in your career, depending on your profession and your specific employer. Whatever format you choose, however, there is a universal approach to good organization that always applies. This approach is based on these main principles, discussed in detail in chapter 3:

Principle 1:   Write different parts for different readers.
Principle 2:   Place important information first.
Principle 3:   Repeat key points when necessary.

These principles apply to formal reports even more than they do to short documents, for these reasons:

1. A formal report often has a very mixed audience—from laypersons to highly technical specialists to executives.
2. The majority of readers of long reports focus on specific sections that interest them most, reading selectively each time they pick up the report.
3. Few readers have time to wade through a lot of introductory information before reaching the main point. They will get easily frustrated if you do not place important information first.

This chapter responds to these facts about readers of formal reports by following the ABC format (for **A**bstract, **B**ody, **C**onclusion). As noted in chapter 3, the three main rules are that you should (1) start with an abstract for decision-makers, (2) put supporting details in the body, and (3) use the conclusion to produce action. This simple ABC format should be evident in *all* formal reports, despite their complexity. Here is how the particular sections of formal reports fit within the ABC format:

ABC FORMAT: *Formal Report*

Abstract

- Cover/title page
- Letter or memo of transmittal
- Table of contents
- List of illustrations
- Executive summary
- Introduction

Body

- Discussion sections
- [Appendices—appear after text but support Body section]

Conclusion

- Conclusions
- Recommendations

Several features of this structure deserve special mention. First, note that the generic abstract section includes five different parts of the report that help give readers a capsule version of the entire report. As you will learn shortly, the executive summary is by far the most important section for providing this "big picture" of the report. Second, appendices are placed within the body part of the outline, even though sequentially they come at the end of the report. The reason for this outline placement is that both appendices and body sections provide supporting details for the report. Third, remember that the generic conclusion section in the ABC format can contain conclusions or recommendations or both, depending on the nature of the report.

Before moving to a discussion of the specific sections that make up the ABC format, take note of the use of main headings in complex formal reports. (See Figure 4–7, pp. 113–115.) Much like chapter titles, these headings are often centered, in full caps, in bold type, and oversized. Also, they usually begin a new page. In this way, each major section of the formal report seems to exist on its own. Then you have three remaining heading levels for use within each section.

GUIDELINES FOR THE EIGHT PARTS OF FORMAL REPORTS

What follows is a description of eight parts of the formal report:

1. Cover/title page
2. Letter of transmittal
3. Table of contents
4. List of illustrations
5. Executive summary
6. Introduction
7. Discussion sections
8. Conclusions and recommendations

What could be considered a ninth part, appendices, is mentioned in the context of the discussion section. Also, note that formal reports may contain work-cited pages or bibliographies, which could be grouped as the first appendix. See chapter 13 for guidelines on formats.

As for pagination of a formal report, many styles abound. Here are some guidelines for one commonly used pattern that will be acceptable unless you have been instructed to use another:

- Use lowercase roman numerals for some or all of the front matter that precedes—and includes—the table of contents.
- Use arabic numbers for items that follow the table of contents (all of which are listed *in* the table of contents).
- Continue the arabic numbering for appendices if they are relatively short. Long sets of appendices sometimes have their own internal numbering (A-1, 2, 3. . . .; B-1, 2, 3. . . .).

See Model 9–9 (326–341) for an example of this approach to pagination and for an example of guidelines that follow.

Cover/Title Page

Formal reports are normally bound, usually with a cover used for all reports in the writer's organization. (Reports prepared for college courses, however, are often placed in a three-tab folder, the outside of which serves as the report cover.) Since the cover is the first item seen by the reader, it should be attractive and informative. Usually it contains the same four pieces of information mentioned in the following list with regard to the title page; sometimes it may have only one or two of these items.

Inside the cover is the title page, which should include these four pieces of information:

- Project title (exactly as it appears on the letter/memo of transmittal)
- Your client's name ("Prepared for . . .")
- Your name and/or the name of your organization ("Prepared by . . .")
- Date of submission

To make your title page or cover distinctive, you might want to place a simple illustration on it. Do not clutter the page, of course. Use a visual only if it reinforces a main point and if it can be simply and tastefully done. For example, assume that McDuff, Inc., has submitted a formal report to a coastal city in California, concluding that an industrial park can be built near the city's bird sanctuary without harming the habitat—if stringent guidelines are followed. The report writer decides to place the picture of a bird on the title page, punctuating the report's point about the industrial park, as in Model 9–1 on page 318.

Letter/Memo of Transmittal

Letters or memos of transmittal are like an appetizer—they give the readers a taste of what is ahead. If your formal report is to readers outside your own organization, write a letter of transmittal. If it is to readers inside your organization, write a memo of transmittal. Models 9–2 and 9–3 on pages 319–320 show examples of both. Follow these guidelines for constructing this part of your report:

● *Transmittal Guideline 1: Place Letter/Memo Immediately after Title Page*

This placement means that the letter/memo is bound with the document, to keep it from becoming separated. Some organizations paper-clip this letter or memo to the front of the report, making it a cover letter or memo. In so doing, however, they risk having it become separated from the report.

● *Transmittal Guideline 2: Include a Major Point from Report*

Remember that readers are heavily influenced by what they read first in reports. Therefore, take advantage of the position of this section by including a major finding, conclusion, or recommendation from the report—besides supplying necessary transmittal information.

● *Transmittal Guideline 3: Follow Letter and Memo Conventions*

Like other letters and memos, letters and memos of transmittal should be easy to read, inviting readers into the rest of the report. Keep introductory and concluding paragraphs relatively short—no more than three to five lines each. Also, write in a conversational style free of technical jargon and stuffy phrases such as "per your request" or "enclosed herewith." See the models at the end of chapter 7 for more details concerning letter/memo format. For now, here are some highlights about the mechanics of format:

Letters and Memos

- Use single spacing and ragged-edge copy, even if the rest of the report is double-spaced and right-justified.
- Use only one page.

Letters

- Include company project number with the letter date.
- Correctly spell the reader's name.
- Be sure the inside address includes the mailing address to appear on the envelope.
- Use the reader's last name ("Dear Mr. Jamison:") in the salutation or attention line because of the formality of the report—unless your close association with the reader would make it more appropriate to use first names ("Dear Bill:").
- Usually include a project title, as with letter reports. It is treated like a main heading. Use concise wording that matches wording on the title page.
- Use "Sincerely" as your closing.
- Include a line to indicate those who will receive copies of the report ("cc" for carbon copy, "pc" for photocopy, or just "c" or "copy" for copy).

Memos

- Give a clear description of the project in the subject line of the memo, including a project number if there is one.
- Include a distribution list to indicate those who will receive copies.

Table of Contents

Your contents page acts as an outline. Many readers go there right away to grasp the structure of the report, and then return again and again to locate report sections of most interest to them. Guidelines follow for assembling this important component of your report. See Model 9–4 on page 321 for an example.

- ### Table of Contents Guideline 1: Make It Very Readable

The table of contents must be pleasing to the eye so that readers can find sections quickly and see their relationship to each other. Be sure to:

- Space items well on the page
- Use indenting to draw attention to subheadings
- Include page numbers for every heading and subheading, unless there are many headings in a relatively short report, in which case you can delete page numbers for all of the lowest-level headings listed in the table of contents

- ### Table of Contents Guideline 2: Use the Contents Page to Reveal Report Emphases

Choose the wording of headings and subheadings with care. Be specific yet concise so that each heading listed in the table of contents gives the reader a good indication of what the section contains.

Readers associate the importance of report sections with the number of headings and subheadings listed in the table of contents. If, for example, a discussion section called "Description of the Problem" contains many more heading breakdowns than other sections, you are telling the reader that the section is more important. When possible, it is best to have about the same number of breakdowns for report sections of about the same importance. In short, the table of contents should be balanced.

● *Table of Contents Guideline 3: Consider*
  Leaving Out Low-Level Headings

In very long reports, you may want to unclutter the table of contents by removing low-level headings. As always, the needs of the readers are the most important criterion to use in making this decision. If you think readers need access to all levels of headings on the contents page, keep them there. If you think they would prefer a simple contents page instead of a comprehensive one, delete all the lowest-level headings from the table of contents (see Model 9–5 on page 322 compared to Model 9–4 on page 321).

● *Table of Contents Guideline 4: List Appendices*

Appendices include items such as tables of data or descriptions of procedures that are inserted at the end of the report. Typically, they are listed at the end of the table of contents. They may be paged with arabic numerals, like the rest of the report. However, sometimes no page numbers are given in the table of contents, since many appendices contain "off-the-shelf" material and are thus individually paged (for example, Appendix A might be paged A-1, A-2, A-3, etc.). Tabs on the edges of pages can help the reader locate these sections.

● *Table of Contents Guideline 5: Use Parallel Form in All Entries*

All headings in one section, and sometimes even all headings and subheadings in the report, have parallel grammatical form. Readers find mixed forms distracting. For example, "Subgrade Preparation" and "Fill Placement" are parallel, in that they are both the same type of phrase. However, if you were to switch the wording of the first item to "Preparing the Subgrade" or "How to Prepare the Subgrade," parallel structure would be lost.

● *Table of Contents Guideline 6: Proofread Carefully*

The table of contents is one of the last report sections to be assembled; thus it often contains errors. Wrong page numbers and incorrect heading wording are two common mistakes. Another is the failure to show the correct relationship of headings and subheadings. Obviously, errors in the table of contents can confuse the reader and prove embarrassing to the writer. Proofread the section carefully.

List of Illustrations

Illustrations within the body of the report are usually listed on a separate page right after the table of contents. When there are few illustrations, another option is to list them at the bottom of the table of contents page rather than on a separate page. In either case, this list should include the *number, title,* and *page number* of every table and figure within the body of the report. If there are many illustrations, separate the list into tables and figures. See the examples in Model 9–6 on page 323 and in Model 9–9 on pages 326–341. (For more information on illustrations, see chapter 11.)

Executive Summary

No formal report would be complete without an executive summary. This short section provides decision-makers with a capsule version of the report. Consider it a stand-alone section that should be free of technical jargon. See Model 9–7 on page 324 for an example. Follow these basic guidelines in preparing this important section of your formal reports:

● *Executive Summary Guideline 1: Put It on One Page*

The best reason to hold the summary to one page is that most readers expect and prefer this length. It is a comfort to know that somewhere within a long report there is one page to which one can turn for an easy-to-read overview. Moreover, a one-page length permits easy distribution at meetings. When the executive summary begins to crowd your page, it is acceptable to switch to single spacing if such a change helps keep the summary on one page—even though the rest of the report may be space-and-a-half or double-spaced.

Some extremely long formal reports may require that you write an executive summary of several pages or longer. In this case, you still need to provide the reader with a section that summarizes the report in less than a page. The answer to this dilemma is to write a brief "abstract," which is placed right before the executive summary. Consider the abstract to be a condensed version of the executive summary, directed to the highest-level decision-makers. (See chapter 13 for further discussion of abstracts.)

● *Executive Summary Guideline 2: Avoid Technical Jargon*

Include only that level of technical language the decision-makers will comprehend. It makes no sense to talk over the heads of the most important readers.

● *Executive Summary Guideline 3: Include Only Important Conclusions and Recommendations*

The executive summary mentions only major points of the report. An exhaustive list of findings, conclusions, and recommendations can come later at the end of the report. If you have trouble deciding what is most important, put yourself in the position of the readers. What information is most essential for them? If you want to leave them with one, two, or three points about the report, what would these points be? *That* is the information that belongs in the executive summary.

● *Executive Summary Guideline 4: Avoid References to the Report Body*

Avoid the tendency to say that the report provides additional information. It is understood that the executive summary is only a generalized account of the report's contents. References to later sections do not provide the busy reader with further understanding.

An exception is those instances when you are discussing issues that involve danger or liability. Here you may need to add qualifiers in your summary—for example, "As noted in this report, further study will be necessary." Such statements protect you and the client in the event the executive summary is removed from the report and used as a separate stand-alone document.

● *Executive Summary Guideline 5: Use Paragraph Format*

Whereas lists are often appropriate for body sections of a report, they can give executive summaries a fragmented effect. Instead, the best summaries create unity with a series of fairly short paragraphs that flow together well. Within a paragraph, there can be a short listing of a few points for emphasis (see Model 9–7 on page 324), but the listing should not be the main structural element of the summary.

Occasionally, you may be convinced that the paragraph approach is not desirable. For example, a project may involve a series of isolated topics that would not mesh into unified paragraphs. In this case, use a modified list. Start the summary with a brief introductory paragraph, followed by a numbered list of three to nine points. Each numbered point should include a brief explanation (for example, "3. *Sewer Construction:* We believe that seepage influx can be controlled by. . . . 4. *Geologic Fault Evaluation:* We found no evidence of surficial. . . .").

● *Executive Summary Guideline 6: Write the Executive Summary Last*

Only after finishing the report do you have the perspective to write a summary. Approach the task this way. First, sit back and review the report from beginning to end. Then ask yourself, "What would my readers really need to know if they had only a minute or two to read?" The answer to that question becomes the core of your executive summary.

Introduction

View this section as your chance to prepare both technical and nontechnical readers for the discussion ahead. You do not need to summarize the report, for your executive summary has accomplished that goal. Instead, give information on the report's purpose, scope, and format, as well as a project description. Follow these basic guidelines, which are reflected in the Model 9–8 example on page 325:

● *Introduction Guideline 1: State Your Purpose and Lead-in to Subsections*

The purpose statement for the document should appear immediately after the main introduction heading ("This report presents McDuff's foundation design recommendations for the new Hilltop Building in Franklin, Maine"). Follow it with a sentence that mentions the introduction subdivisions to follow ("This introduction provides a description of the project site and explains the scope of activities we conducted").

● *Introduction Guideline 2: Include a Project Description*

Here you need to be precise about the project. Depending on the type of project, you may be describing a physical setting, a set of problems that prompted the report study, or some other data. The information may have been provided to you or you may have collected it yourself. Accuracy in this section will help prevent any later misunderstandings between you and the reader. (When the project description would be too long for the introduction, sometimes it is placed in the body of the report.)

● *Introduction Guideline 3: Include Scope Information*

This section must outline the precise objectives of the study. Include all necessary details, using bulleted or numbered lists when appropriate. Your listing or description should parallel the order of the information presented in the body of the report. Like the project description, this subsection must be accurate in every detail. Careful, thorough writing here can prevent later misunderstanding about the tasks you were hired to perform.

● *Introduction Guideline 4: Consider Including Information on Report Format*

Often the scope section lists information as it is presented in the report. If this is not the case, end the introduction with a short subsection on the report format. Here you can give readers a brief preview of the main sections that follow. In effect, the section acts as a condensed table of contents and may list the report's major sections and appendices.

Discussion Sections

Discussion sections compose the longest part of formal reports. In general, they are written for the most technically oriented members of your audience. You can focus on facts and opinions, demonstrating the technical expertise that the reader expects from you. General guidelines for writing the report discussion are listed here. For a complete example of the discussion component, see the formal-report example in Model 9–9 on pages 326–341.

● *Discussion Guideline 1: Move from Facts to Opinions*

As you have learned, the ABC format requires that you start your formal report with a summary of the most important information. That is, you skip right to essential conclusions and recommendations the reader needs and wants to know. Once into the discussion section, however, you back up and adopt a strategy that parallels the stages of the technical project itself. You begin with hard data and move toward conclusions and recommendations (that is, those parts that involve more opinion).

One way to view the discussion is that it should follow the order of a typical technical project, which usually involves these stages:

First, you collect data (samples, interviews, records, etc.).

Second, you subject these data to verification or testing (lab tests or computer analyses, for example).

Third, you analyze all the information, using your professional experience and skills to form conclusions (or convictions based on the data).

Fourth, you develop recommendations that flow directly from the conclusions you have formed.

Thus, the body of your report gives technical readers the same movement from fact toward opinion that you experience during the project itself. There are two reasons for this approach, one ethical and the other practical. First, as a professional, you are obligated to draw clear distinctions between what you have observed and what you have concluded or recommended. Second, your reports will usually be more persuasive if you give readers the chance to draw conclusions for themselves. If you move carefully through the four-stage process just described, readers will be more likely to reach the same conclusions that you have drawn.

● *Discussion Guideline 2: Use Frequent Headings and Subheadings*

Headings give readers handles by which to grasp the content of your report. They are especially needed in the report body, which presents technical details. Your readers will view headings, collectively, as a sort of outline by which they can make their way easily through the report.

● *Discussion Guideline 3: Use Listings to Break Up Long Paragraphs*

Long paragraphs full of technical details irritate readers. Use paragraphs for brief explanations, not for descriptions of processes or other details that could be listed.

● *Discussion Guideline 4: Use Illustrations for Clarification and Persuasion*

A simple table or figure can sometimes be just the right complement to a technical discussion in the text. Incorporate illustrations into the report body to make technical information accessible and easier to digest.

● *Discussion Guideline 5: Place Excessive Detail in Appendices*

Today's trend is to place cumbersome detail in appendices that are attached to formal reports, rather than weighing down the discussion with this detail. In other words, you give readers access to supporting information without cluttering up the text of the formal report. Of course, you need to refer to appendices in the body of the report and label appendices clearly so that readers can locate them easily.

Tabbed sheets are a good way to make all report sections, including appendices, very accessible to the reader. Consider starting each section with a tabbed sheet so that the reader can "thumb" to it easily.

Conclusions and Recommendations

This final section of the report should give readers a place to turn to for a *comprehensive* description—sometimes in the form of a *listing*—of all conclusions and recommendations. The points may or may not have been mentioned in the body of the report, depending on the length and complexity of the document. Conclusions, on the one hand, are convictions or beliefs based on the findings of your study. Recommendations, on the other hand, are actions you are suggesting based on your conclusions. For example, your conclusion may be that there is a dangerous level of toxic chemicals in a town's water supply. Your recommendation may be that the toxic site near the reservoir should be immediately cleaned.

What distinguishes this last section of the report text from the executive summary is the level of detail and the audience. The conclusions and recommendations section provides an *exhaustive* list of conclusions and recommendations for technical and management readers. The executive summary provides a *selected* list or description of the most important conclusions and recommendations for decision-makers, who may not have technical knowledge.

In other words, view the conclusions and recommendations section as an expanded version of the executive summary. It usually assumes one of these three headings, depending of course on the content:

1. Conclusions
2. Recommendations
3. Conclusions and Recommendations

Another option for reports that contain many conclusions and recommendations is to separate this last section into two sections: (1) Conclusions and (2) Recommendations.

FORMAL REPORT EXAMPLE

Model 9–9 (pp. 326–341) provides a long, formal technical report from McDuff, Inc. It contains the main sections discussed previously, including the list of illustrations. Marginal annotations indicate how the model reflects proper use of this chapter's guidelines for format and organization.

The report results from a study that McDuff completed for the city of Winslow, Georgia. Members of the audience come from both technical and nontechnical backgrounds. Some are full-time professionals hired by the city, whereas others are part-time, nonpaid citizens appointed by the mayor to explore environmental problems. The paid professionals include engineers, environmental specialists, accountants, city planners, managers, lawyers, real estate experts, and public-relations specialists. The part-time appointees include citizens who work in a variety of blue-collar and white-collar professions or who are homemakers.

Communication Challenge

"The Ethics of Clients Reviewing Report Drafts"

Last week Hank Wallace of McDuff's Kenya office completed the draft of an ocean exploration project for the Republic of Cameroon (see Project #2 on the color insert in chapter 2). As is routine with major reports, Hank showed the client a draft before the final draft was submitted. For the first time in his career, he was asked by the client to make changes he thinks are difficult to justify by project data. What follows is an explanation of why McDuff shares report drafts with some clients, some background on the ocean exploration project, and questions and comments for discussion.

Sharing Drafts with Clients

In McDuff's business, some of the firm's reports must be submitted both to the paying client and to regulatory agencies of the government. This dual audience has created a review procedure common in the industry. Client firms have an opportunity to review a draft and make suggestions before both they and the regulatory agencies are sent final drafts.

For example, a U.S. mining company hired McDuff to examine a Siberian site to determine if gold reserves could be mined without damaging the delicate permafrost surface of the tundra. Because the Russian government regulates development of the region, it received a final copy of the report. But before the final copy was submitted to the government, McDuff shared a draft with engineers and executives from the mining company. These client representatives questioned several technical assumptions McDuff made about the site, but McDuff had adequate justifications for its work. In the end, McDuff made no change in its original draft recommendation—that is, that further study was needed before mining is permitted in the permafrost region.

In another case, however, a client's review of a report on a dam in the Midwest prompted McDuff to adjust its report before submission of the final draft to the state's Department of Natural Resources, which regulates high-hazard dams. The owners of the dam—who paid for the study—convinced McDuff that the report should emphasize the fact that poor installation of a guardrail over the dam created a drainage problem. When heavy rains came, soil washed out an embankment near the dam's spillway. The first draft had failed to mention that the state's transportation group bore some responsibility for the dam's problems.

The Ocean Exploration Report Review

The draft review of the ocean exploration report did not go as smoothly as the two reviews just described. Major differences of opinion were evident between the McDuff project manager and the client, Worldwide Energy, Inc.

Continued

As indicated on the project sheet in the chapter 2 color insert, McDuff engineers developed conclusions and recommendations for the Cameroon coastal site. For the most part, they found that the offshore environment where they did the study would be too environmentally sensitive to drill offshore wells or run pipelines. There were two locations where a pipeline might be safely placed, but even in this case some environmental damage was likely. When Worldwide Energy got the draft report, the company asked for a meeting.

At the meeting the following week, McDuff engineers reviewed their findings, conclusions, and recommendations with the client. Ultimately, McDuff managers were asked to change the wording in the report to present a more favorable view of oil exploration at the site because, in the client's opinion, McDuff was being too conservative in its conclusions. If McDuff would just adjust some wording so as not to emphasize what is, after all, only *possible* environmental damage, then the Cameroon government might be provided the support it needs to develop this potentially rich oil field. Cameroon, the client argued, needs oil revenues to improve its economy and assist poor farmers with the transition to a modern economy. McDuff is not being asked to alter the facts—only to adjust the tone of the language.

Back at the office, the McDuff project manager met with the branch manager and later with corporate staff via teleconference. The project manager presented the facts of the project and a summary of the meeting. To all present, it was clear that the relationship with a long-term client was at stake.

Questions and Comments for Discussion

1. How should McDuff, Inc., respond to the client's request?
2. Generally, do you think McDuff's procedure for reviewing report drafts with clients is *ethically* sound? Support your answer.
3. If you answered "yes" to question 2, do you have any suggestions to improve the *procedure* for this client report review? That is, how might the process be adjusted to reduce the potential for misunderstandings and abuse?
4. If you answered "no" to question 2, is there any circumstance in which you would support the review of a report draft by a client before final submission to the client and its regulatory agency?
5. It is often said, in this text and elsewhere, that collaboration is essential in the workplace. Describe the kinds of on-the-job situations wherein you think collaboration between writer and reader would be useful, appropriate, and ethical.

CHAPTER SUMMARY

In your career you will write formal reports for large, complex projects—either inside or outside your organization. In both cases, you will be sending the report to people with different technical backgrounds. This complex audience will respond best to reports that subscribe to the ABC format, for it organizes information so that different readers can read different sections of the report. Although long-report formats vary according to company and profession, most will have these eight basic parts: cover/title page, letter/memo of transmittal, table of contents, list of illustrations, executive summary, introduction, discussion sections, and conclusions and recommendations. Follow the specific guidelines in this chapter for these sections. The annotated model can serve as your reference.

Access the Technical Writing Web site by pointing your browser at www.prenhall.com/pfeiffer. Here, you will find additional resources to aid you in learning and applying the information from this chapter.

ASSIGNMENTS

Part 1 assignments ask you to evaluate a whole report, to write an individual section, or to evaluate an individual section. Part 2 assignments ask you to write complete formal reports. Remember to submit Planning Forms with the Part 2 assignments.

Part 1: Short Assignments

These assignments can be completed by individuals or by teams. If you are instructed to use teams, first review guidelines in chapter 1 on group writing.

1. **Evaluation—A Formal Report.** Use Model 9–9 on pages 326–341, the complete formal report example in this chapter, for this assignment. The audience for the report is described in the chapter section entitled "Formal Report Example." Although the writer directed the report to a mixed technical and nontechnical audience, some sections clearly are more technical than others.

   ● Evaluate the likely audience for each section of the report.
   ● Discuss ways that the writer did, or did not, address the needs of specific audience types.
   ● Offer suggestions for improving the manner in which the report meets the needs of its intended audience.

2. **Evaluation—A Formal Report.** Locate a formal report written by a private firm or government agency *OR* use a long report provided by your instructor. Determine the degree to which the example follows the guidelines in this chapter. Depending on the instructions given by your teacher, choose between these options:

   ● Present your findings orally *or* in writing.
   ● Select part of the report *or* all of the report.

3. **Executive Summary.** Choose one of the seven project sheets included in the color insert in chapter 2. Write a brief executive summary for the project. If necessary, provide additional information or transitional wording not included on the sheet, but do not change the nature of the information already provided.

4. **Evaluation—An Introduction.** Review the chapter guidelines for writing an effective introduction to a formal report. Then evaluate the degree to which the following example follows or does not follow the guidelines presented.

INTRODUCTION

McDuff, Inc., has completed a three-week study of the manufacturing and servicing processes at King Radio Company. As requested, we have developed a blueprint for ways in which Computer-Aided Testing (CAT) can be used to improve the company's productivity and quality.

Project Description

M. Dan Mahoney familiarized our project team with the problems that prompted this study of computer-aided testing. According to Mr. Mahoney, the main areas of concern are as follows:

- Too many units on the production line are failing postproduction testing and thus returning to the repair line.
- Production bottlenecks are occurring throughout the plant because of the testing difficulties.
- Technicians in the servicing center are having trouble repairing faulty units because of their complexity.
- Customers' complaints have been increasing, both for new units under warranty and repaired units.

Scope

From May 3–5, 2002, McDuff, Inc., had a three-person team of experts working at the King Radio Company plant. This team interviewed many personnel, observed all the production processes, and acquired data needed to develop recommendations.

Upon returning to the McDuff office, team members met to share their observations and develop the master plan included in this report.

Report Format

This report is largely organized around the two ways that CAT can improve operation at the King Radio plant. Based on the detailed examination of the plant's problems in this regard, the report covers two areas for improvement and ends with a section that lists main conclusions and recommendations. The main report sections are as follows:

- Production and Servicing Problems at King Radio
- CAT and the Manufacturing Process
- CAT and the Servicing Process
- Major Conclusions and Recommendations

The report ends with two appendices. Appendix A offers detailed information on several pieces of equipment we recommend that you purchase. Appendix B provides three recent articles from the journal *CAT Today*. All three deal with the application of CAT to production and service problems similar to those you are experiencing.

Part 2: Longer Assignments

This section contains assignments for writing entire formal reports. Remember to complete the Planning Form for each assignment.

These assignments can be written by individual writers or by group-writing teams. If your instructor has made this a team assignment, review the chapter 1 guidelines on group writing.

5. **Research-Based Formal Report.** Complete the following procedure for writing a research-based report:

   ● Conduct either a computer-assisted search or a traditional library search on a general topic in a field that interests you. Do some preliminary reading to screen possible specific topics.

   ● Choose three to five specific topics that would require further research and for which you can locate information.

   ● Work with your instructor to select the one topic that would best fit this assignment, given your interests and the criteria set forth here.

   ● Develop a simulated context for the report topic, whereby you select a *purpose* for the report, a specific *audience* to whom it could be addressed (as if it were a "real" report), and a specific *role* for you as a writer.

     For example, assume you have selected "earth-sheltered homes" as your topic. You might be writing a report to the manager of a local design firm on the features and construction techniques of such structures. As a newly hired engineer or designer, you are presenting information so that your manager can decide whether the firm might want to begin building and marketing such homes. This report might present only data, or it could present data and recommendations.

   ● Write the report according to the format guidelines in this chapter and in consideration of the specific context you have chosen.

   ● Document your sources appropriately (see chapter 13).

6. **Work-Based Formal Report.** This assignment is based on the work experience that you may have had in the past or that you may be experiencing now.

   ● Choose 5 or 10 report topics that are based on your current or past work experience. For example, you could choose "warehouse design" if you stock parts, "check-out procedure" if you work behind the counter at a video-rental store, "report-production procedures" if you work as a secretary at an engineering firm, etc. In other words, find a subject that you know about or about which you can find more information, especially through interviews.

   ● Work with your instructor to select the one topic that holds out the best possibilities for a successful report, on the basis of the criteria given here.

   ● Develop a context for the report in which you give yourself a *role* in the company where you work(ed). This role should be one in which you would actually write a formal in-house or external report about the topic you have chosen, but the role does not have to be the exact one you had or have. Then select a precise *purpose* for which you might be writing the report and finally a set of *readers* that might read such a report within or outside the organization. Your report can be a presentation of data and conclusions *or* a presentation of data, conclusions, and recommendations.

   ● Follow the guidelines included in this chapter for format and organization.

7. **School-Based Formal Report.** This assignment can be completed as an individual project or as a group project. As an individual project, it will rely on observations you have made during the time you spent at a high school, college, or university—either the one where you are taking this course or another you attended previously. As a group project, it will rely on either (1) group members from diverse majors using their varied backgrounds to examine a common campus problem, or (2) group members majoring in the same field or working in the same department exploring a problem they have in common.

Whether you write an individual or a group report, follow this general procedure:

- Assemble a list of 5 or 10 problems that you have observed at your school. These problems might concern (1) the physical campus (as in poor design of parking lots or inadequate lab space), (2) the curriculum (as in the need to update certain courses), (3) extracurricular activities (as in the need for more cultural or athletic events), or (4) difficulties with campus support services (as in red tape during registration).
- Work with your instructor to choose the one topic for which you can find the most information and for which you can develop the context described here.
- Collect information in whatever ways seem useful—for example, site observations, surveys to students, follow-up phone calls, or interviews.
- Submit progress reports at intervals requested by your instructor. (Consult guidelines in the "Progress/Periodic Reports" section of chapter 8.)
- Consider your *role* to be the one that you, in fact, have—a student or a group of students at the school. Then select as your *reader(s)* the school officials who would actually be in charge of solving the problem you have identified. (You may or may not end up sending the report. Follow the advice of your instructor in this matter.) The *purpose* of this report will be to explain, in great detail, all aspects of the problem *and* to form conclusions as to its cause. If it seems appropriate, you may take one further step to suggest recommendations for a solution—*if* your research has taken you this far. In any case, detail and also tact are important criteria.

8. **McDuff-Based Formal Report.** For this assignment you will place yourself into a role of your choosing at McDuff, Inc. Use the following procedure, which may be modified by your instructor:

- Review the section at the beginning of this chapter that lists McDuff cases for formal reports to get a sense of when formal reports are used at companies like McDuff.
- Review the McDuff information in chapter 2, especially with regard to the kinds of jobs people hold at the company and the kinds of projects that are undertaken.
- Choose a specific job that *you* could assume at McDuff—based on your academic background, your work experience, or your career interests.
- Choose a specific project that (a) could conceivably be completed at McDuff by someone in the role you have chosen, (b) would result in a formal report directed either inside or outside the company, and (c) would be addressed to a complex audience at two or three of the levels indicated on the Planning Form at the end of the book.
- Be sure you have access to information that will be used in this simulated report—for example, from work experience, from a term paper or class project in another course, or from your interviews of individuals already in the field. (For this assignment, you may want to talk with a professional, such as a recent graduate in your major.)
- Prepare a copy of the Planning Form at the end of the book for your instructor's approval—*before* proceeding further with the project.
- Complete the formal report, following the guidelines in this chapter.

9. **Ethics Assignment.** Illustrations on cover pages of formal reports are one strategy for attracting the readers' attention to the document. Note the use of an illustration in Model 9–1, along with the rationale on page 303. Do you think Model 9–1 uses its graphics in an ethically sound way to engage the reader with the report? Why or why not? How do you determine whether a cover page illustration is an appropriate persuasive tool, on the one hand, or an inappropriate attempt to manipulate the reader, on the other? Give hypothetical examples.

10. **International Communication Assignment.** This assignment requires that you gain information about writing long technical reports that are designed for readers outside the United States (or outside the country where you are taking this course, if it is not the United States). The suggestions you develop can relate to either (1) reports written in English that will be read in English or (2) reports written in English that will be translated into another language.

Specifically, write a report that provides a wide range of recommendations for writing long technical documents to a specific international audience. Cover as many writing-related issues as possible—organization, format, page design, and style.

To gather information for this assignment, you will need to locate someone who works for an international firm, who deals with international clients, or who has in some other way acquired information about the needs of international readers of technical documents. Possible sources include (1) your institution's alumni office, which may be able to provide names of graduates or employers of graduates, (2) friends or colleagues, (3) individuals contacted through web sites of international organizations, and (4) local chambers of commerce and other organizations that promote international trade.

Oceanside's New Industrial Park

Prepared for: City Council
                          Oceanside, California

Prepared by: McDuff, Inc.
                          San Francisco, California

Date: March 3, 2002

MODEL 9–1
Title page with illustration (see explanation on page 303)

12 Post Street
Houston, Texas 77000
(713) 555-9781

Report #82-651
July 19, 2002

Belton Oil Corporation
P.O. Box 301
Huff, Texas 77704

Attention: Mr. Paul A. Jones

GEOTECHNICAL INVESTIGATION
DREDGE DISPOSAL AREA F
BELTON OIL REFINERY
HUFF, TEXAS

This is the second volume of a three-volume report on our geotechnical investigation concerning dredge materials at your Huff refinery. This study was authorized by Term Contract No. 604 and Term Contract Release No. 20-6 dated May 6, 2002.

This report includes our findings and recommendations for Dredge Disposal Area F. Preliminary results were discussed with Mr. Jones on July 16, 2002. We consider the soil conditions at the site suitable for limited dike enlargements. However, we recommend that an embankment test section be constructed and monitored before dike design is finalized.

We appreciate the opportunity to work with you on this project. We look forward to assisting you with the final design and providing materials-testing services.

Sincerely,

George Fursten

George H. Fursten
Geotechnical/Environmental Engineer

GHF/dnn

MODEL 9–2
Letter of transmittal

MEMORANDUM

DATE: March 18, 2002
TO: Lynn Redmond, Vice President of Human Resources
FROM: Abe Andrews, Personnel Assistant   *aa*
SUBJECT: Report on Flextime Pilot Program at Boston Office

As you requested, I have examined the results of the six-month pilot program to introduce flextime to the Boston office. This report presents my data and conclusions about the use of flexible work schedules.

To determine the results of the pilot program, I asked all employees to complete a written survey. Then I followed up by interviewing every fifth person off an alphabetical list of office personnel. Overall, it appears that flextime has met with clear approval by employees at all levels. Productivity has increased and morale has soared. This report uses the survey and interview data to suggest why these results have occurred and where we might go from here.

I enjoyed working on this personnel study because of its potential impact on the way McDuff conducts business. Please give me a call if you would like additional details about the study.

MODEL 9–3
Memo of transmittal

TABLE OF CONTENTS

MODEL 9–4
Table of contents (all subheadings included)

TABLE OF CONTENTS

MODEL 9–5
Table of contents (third-level subheadings omitted)

LIST OF ILLUSTRATIONS

MODEL 9–6
List of illustrations—formal report

Gives brief
background of
project.

Keeps verbs in
active voice, for
clarity and brevity.

Uses short list to
emphasize major
findings.

Emphasizes major
conclusion in
separate
paragraph. (Note
that this major
point *could* have
been placed after
first paragraph, for
a different effect.)

EXECUTIVE SUMMARY

Quarterly monitoring of groundwater showed the presence of nickel in Well M–17 at the Hennessey Electric facility in Jones, Georgia. Nickel was not detected in any other wells on the site. Hennessey then retained McDuff's environmental group to determine the source of the nickel.

The project consisted of four main parts. First, we collected and tested 20 soil samples within a 50-yard radius of the well. Second, we collected groundwater samples from the well itself. Third, we removed the stainless steel well screen and casing and submitted them for metallurgical analysis. Finally, we installed a replacement screen and casing built with teflon.

The findings from this project are as follows:

- The soil samples contained *no* nickel.
- We found *significant* corrosion and pitting in the stainless steel screen and casing that we removed.
- We detected *no* nickel in water samples retrieved from the well after replacement of the screen and casing.

Our study concluded that the source of the nickel in the groundwater was corrosion on the stainless steel casing and screen.

MODEL 9–7
Executive summary–formal report

324

INTRODUCTION

This document examines the need for a McDuff, Inc., <u>Human Resources Manual</u>. Such a manual would apply to all U.S. offices of the firm. As background for your reading of this report, I have included (1) a brief description of the project, (2) the scope of my activities during the study, and (3) an overview of the report format.

Project Description

Three months ago, Rob McDuff met with the senior staff to discuss diverse human resources issues, such as performance appraisals and fringe benefits. After several meetings, the group agreed that the company greatly needed a manual to give guidance to managers and their employees. Shortly thereafter, I was asked to study and then report on three main topics: (1) the points that should be included in a manual, (2) the schedule for completing the document, and (3) the number of employees that should be involved in writing and reviewing policies.

Scope of Activities

This project involved seeking information from many McDuff employees and completing some outside research. Specifically, the project scope involved:

- Sending a survey to employees at every level at every domestic office
- Tabulating the results of the survey
- Interviewing some of the survey respondents
- Completing library research on the topic of human resource manuals
- Developing conclusions and recommendations that were based on the research completed

Report Format

To fulfill the report's purpose of examining the need for a McDuff <u>Human Resources Manual</u>, this report includes these main sections:

Section 1: Research Methods
Section 2: Findings of the Survey and Interviews
Section 3: Findings of the Library Research
Section 4: Conclusions and Recommendations

Appendices at the end of the text contain the survey form, interview questions, sample survey responses, and several journal articles of most use in my research.

Margin annotations:

Gives purpose of report and overview of introduction (as lead-in).

Describes the task the writer was given.

Denotes the major activities that were accomplished.

Provides reader with a preview of main sections to follow (as a sort of "mini" table of contents).

MODEL 9–8
Introduction—formal report

STUDY OF WILDWOOD CREEK

WINSLOW, GEORGIA

Prepared for:

The City of Winslow

Prepared by:

Christopher S. Rice, Hydro/Environmental Engineer
McDuff, Inc.

November 30, 2002

Uses graphic on title page to reinforce theme of environmental protection.

MODEL 9–9
Formal report

12 Peachtree Street
Atlanta, GA 30056
(404) 555-7524

McDuff Project #99-119
November 30, 2002

Adopt-a-Stream Program
City of Winslow
300 Lawrence Street
Winslow, Georgia 30000

Attention : Ms. Elaine Sykes, Director

**STUDY OF WILDWOOD CREEK
WINSLOW, GEORGIA**

We have completed our seven-month project on the pollution study of Wildwood Creek. This project was authorized on May 18, 2002. We performed the study in accordance with our original proposal No. 14-P72, dated April 24, 2002.

This report mentions all completed tests and discusses the test results. Wildwood Creek scored well on many of the tests, but we are concerned about several problems—such as the level of phosphates in the stream. The few problems we observed during our study have led us to recommend that several additional tests should be completed.

Thank you for the opportunity to complete this project. We look forward to working with you on further tests for Wildwood Creek and other waterways in Winslow.

Sincerely,

Christopher S. Rice, P.E.

Christopher S. Rice, P.E.
Hydro/Environmental Engineer

In the left margin:

Lists project title as it appears on title page.

Gives brief statement of project information.

Provides major point from report.

MODEL 9–9
continued

continues

TABLE OF CONTENTS

Uses white space, indenting, and bold to accent organization of report.

i

MODEL 9–9
continued

328

Includes illustration titles as they appear in text.

LIST OF ILLUSTRATIONS

MODEL 9–9
continued

continues

EXECUTIVE SUMMARY

The City of Winslow hired McDuff, Inc., to perform a pollution study of Wildwood Creek. The section of the creek that was studied is a one-mile-long area in Burns Nature Park, from Newell College to U.S. Highway 42. The study lasted seven months.

McDuff completed 13 tests on four different test dates. Wildwood scored fairly well on many of the tests, but there were some problem areas. For example, high levels of phosphates were uncovered in the water. The phosphates were derived either from fertilizer or from animal and plant matter and waste. Also uncovered were small amounts of undesirable water organisms that are tolerant to pollutants and can survive in harsh environments.

McDuff recommends that (1) the tests done in this study be conducted two more times, through spring, 2003, (2) other environmental tests be conducted, as listed in the conclusions and recommendations section, and (3) a voluntary cleanup of the creek be scheduled. With these steps, we can better analyze the environmental integrity of Wildwood Creek.

Summarizes purpose and scope of report.

Describes major findings and conclusions.

Includes main recommendation from report text.

MODEL 9–9
continued

INTRODUCTION

McDuff, Inc., has completed a follow-up to a study completed in 1993 by Ware County on the health of Wildwood Creek. This introduction describes the project site, scope of our study, and format for this report.

PROJECT DESCRIPTION

By law, all states must clean up their waterways. The State of Georgia shares this responsibility with its counties. Ware County has certain waterways that are threatened and must be cleaned. Wildwood Creek is one of the more endangered waterways. The portion of the creek that was studied for this report is a one-mile stretch in the Burns Nature Park between Newell College and U.S. Highway 42.

SCOPE OF STUDY

The purpose of this project was to determine whether the health of the creek has changed since the previous study in 1993. Both physical and chemical tests were completed. The nine physical tests were as follows:

- Air temperature
- Water temperature
- Water flow
- Water appearance
- Habitat description
- Algae appearance
- Algae location
- Visible litter
- Bug count

The four chemical tests were as follows:

- pH
- Dissolved oxygen (DO)
- Turbidity
- Phosphate

REPORT FORMAT

This report includes three main sections:

1. Field Investigation: a complete discussion of all the tests that were performed for the project
2. Test Comparison: charts of the test results and comparisons
3. Conclusions and Recommendations

Gives lead-in to Introduction.

Briefly describes project.

Uses bulleted list to emphasize scope of activities.

Provides "map" of main sections in report.

MODEL 9–9
continued

continues

FIELD INVESTIGATION

Wildwood Creek has been cited repeatedly for environmental violations in the pollution of its water. Many factors can generate pollution and affect the overall health of the creek. In 1993, the creek was studied in the context of a study of all water systems in Ware County. Wildwood Creek was determined to be one of the more threatened creeks in the county.

The city needed to learn if much has changed in the past nine years, so McDuff was hired to perform a variety of tests on the creek. Our effort involved a more in-depth study than that done in 1993. Tests were conducted four times over a seven-month period. The 1993 study lasted only one day.

The field investigation included two categories of tests: physical tests and chemical tests.

PHYSICAL TESTS

The physical tests covered a broad range of environmental features. This section will discuss the importance of the tests and some major findings. The Test Comparison section on page 9 includes a table that lists results of the tests and the completion dates. The test types were as follows: air temperature, water temperature, water flow, water appearance, habitat description, algae appearance, algae location, visible litter, and bug count.

Air Temperature

The temperature of the air surrounding the creek will affect life in the water. Unusual air temperature for the seasons will determine if life can grow in or out of the water.

Three of the four tests were performed in the warmer months. Only one was completed on a cool day. The difference in temperature from the warmest to coolest day was 10.5°C, an acceptable range.

Water Temperature

The temperature of the water determines which species will be present. Also affected are the feeding, reproduction, and metabolism of these species. If there are one or two weeks of high temperature, the stream is unsuitable for most species. If water temperature changes more than 1° to 2°C in 24 hours, thermal stress and shock can occur, killing much of the life in the creek.

During our study, the temperature of the water averaged 1°C cooler than the temperature of the air. The water temperature did not get above 23°C or below 13°C. These ranges are acceptable by law.

Amplifies information presented later in report.

MODEL 9–9
continued

Water Flow

The flow of the water will influence the type of life in the stream. Periods of high flow can cause erosion to occur on the banks and sediment to cover the streambed. Low water flow can decrease the living space and deplete the oxygen supply.

The flow of water was at the correct level for the times of year the tests were done—except for June, which had a high rainfall. With continual rain and sudden flash floods, the creek was almost too dangerous for the study to be performed that month.

In fact, in June we witnessed the aftermath of one flash flood. Figure 1 shows the creek with an average flow of water, and Figure 2 shows the creek during the flood. The water's average depth is 10 inches. During the flash flood, the water level rose and fell 10 feet in about one hour. Much dirt and debris were washing into the creek, while some small fish were left on dry land as the water receded.

Incorporates graphic into page of text.

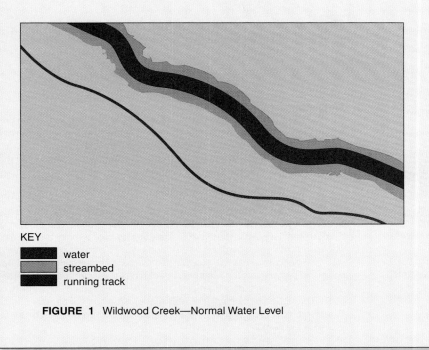

KEY

- ■ water
- ▨ streambed
- ■ running track

FIGURE 1 Wildwood Creek—Normal Water Level

MODEL 9–9
continued

continues

KEY

- water
- streambed
- running track

FIGURE 2   Wildwood Creek—Flash Flood Water Level

Water Appearance

The color of the water gives a quick but fairly accurate view of the health of the creek. If the water is brown or dirty, then silt or human waste may be present. Black areas of water may contain oil or other chemical products.

On each of the four test days, the water was always clear. Thus the appearance of the creek water was considered excellent.

Habitat Description

The habitat description concerns the appearance of the stream and its surroundings. An important criterion is the number of pools and the number of ripples—that is, points where water flows quickly over a rocky area. Both pools and ripples provide good locations for fish and other stream creatures to live and breed.

In describing habitat, McDuff also evaluates the amount of sediment at the bottom of the stream. Too much sediment tends to cover up areas where aquatic life lays eggs and hides them from predators. We also evaluate the stability of the stream banks; a stable bank indicates that erosion has not damaged the habitat. Finally, we observe the amount of stream cover. Such vegetation helps keep soil in place on the banks.

Elaborates on importance of information shown in Table 1. Description parallels five items in table.

MODEL 9–9
continued

Wildwood Creek tested fairly well for habitat. The number of pools and ripples was about average for such creeks. Stream deposits and stream bank stability were average to good, and stream cover was good to excellent. For more detail about test results, see the chart in the Test Comparison section on page 9.

Algae Appearance and Location

Algae is naturally present in any creek. The amount of algae can be a warning of pollution in the water. If algae is growing out of control, disproportionate amounts of nutrients such as nitrogen or phosphate could be present. These chemicals could come from fertilizer washed into the creek. Excessive amounts of algae cause the oxygen level to drop when they die and decompose.

During the four studies, algae was everywhere, but it was especially heavy on the rocks in the ripples of the creek. The algae was always brown and sometimes hairy.

Visible Litter

Litter can affect the habitat of a creek. While some litter has chemicals that can pollute the water, other litter can cover nesting areas and suffocate small animals. Whether the litter is harmful or not, it is always an eyesore.

On all four test dates, the litter we saw was heavy and ranged from tires to plastic bags. Some of the same trash that was at the site on the first visit was still there seven months later.

Bug Count

The bug count is a procedure that begins by washing dirt and water onto a screen. As water drains, the dirt with organisms is left on the screen. The bugs are removed and classified. Generally, the lower the bug count, the higher the pollution levels. Bug counts were considered low to average.

Two types of aquatic worms were discovered every time during our count, but in relatively small amounts. In addition, the worms we observed are very tolerant of pollution and can live in most conditions. Finally, we observed only two crayfish, animals that are somewhat sensitive to pollution.

CHEMICAL TESTS

While physical tests cover areas seen with the naked eye, chemical tests can uncover pollutants that are not so recognizable. Certain chemicals can wipe out all life in a creek. Other chemicals can cause an overabundance of one life-form, which in turn could kill more sensitive animals.

A chart of results of chemical tests is included in the Test Comparison section on page 9. The chemical tests that McDuff performed were pH, dissolved oxygen (DO), turbidity, and phosphate.

Gives specific details that support the report's conclusions and recommendations, which come later.

MODEL 9–9
continued

continues

pH

The pH test is a measure of active hydrogen ions in a sample. The range of the pH test is 0-14. If the sample is in the range of 0-7.0, it is acidic; but if the sample is in the range of 7.0-14, it is basic. By law, the pH of a water sample must be within the range of 6.0 to 8.5.

For the tests we completed, the water sample was always 7.0, which is very good for a creek.

Dissolved Oxygen (DO)

Normally, oxygen dissolves readily into water from surface air. Once dissolved, it diffuses slowly in the water and is distributed throughout the creek. The amount of DO depends on different circumstances. Oxygen is always highest in choppy water, just after noon, and in cooler temperatures.

In many streams, the level of DO can become critically low during the summer months. When the temperature is warm, organisms are highly active and consume the oxygen supply. If the amount of DO drops below 3.0 ppm (parts per million), the area can become stressful for the organisms. An amount of oxygen that is 2.0 ppm or below will not support fish. DO that is 5.0 ppm to 6.0 ppm is usually required for growth and activity of organisms in the water.

According to the Water Quality Criteria for Georgia, average daily amounts of DO should be 5.0 ppm with a minimum of 4.0 ppm. Wildwood Creek scored well on this test. The average amount of DO in the water was 6.9 ppm, with the highest amount being 9.0 ppm on November 11, 2002.

Turbidity

Turbidity is the discoloration of water due to sediment, microscopic organisms, and other matter. One major factor of turbidity is the level of rainfall before a test.

Three of our tests were performed on clear days with little rainfall. On these dates, the turbidity of Wildwood Creek was always 1.0, the best that creek water can score on the test. The fourth test, which scored worse, occurred during a rainy period.

Phosphate

Phosphorus occurs naturally as phosphates—for example, orthophosphates and organically bound phosphates. Orthophosphates are phosphates that are formed in fertilizer, while organically bound phosphates can form in plant and animal matter and waste.

Phosphate levels higher than .03 ppm contribute to an increase in plant growth. If phosphate levels are above 0.1 ppm, plants may be stimulated to grow out of control. The phosphate level of Wildwood was always 0.5 ppm, considerably higher than is desirable.

MODEL 9–9
continued

TEST COMPARISON

There was little change from each of the four test dates. The only tests that varied greatly from one test to another were air temperature, water temperature, water flow, and DO. On the basis of these results, it would appear that Wildwood Creek is a relatively stable environment.

Brings together test results for easy reference.

TABLE 1 Physical Tests

TEST DATES	5/26/02	6/25/02	9/24/02	11/19/02
Air Temperature in °C	21.5	23.0	24.0	13.5
Water Temperature in °C	20.0	22.0	23.0	13.0
Water Flow	Normal	High	Normal	Normal
Water Appearance	Clear	Clear	Clear	Clear
Habitat Description				
Number of Pools	2.0	3.0	2.0	5.0
Number of Ripples	1.0	2.0	2.0	2.0
Amount of Sediment Deposit	Average	Average	Good	Average
Stream Bank Stability	Average	Good	Good	Good
Stream Cover	Excellent	Good	Excellent	Good
Algae Appearance	Brown	Brown/hairy	Brown	Brown
Algae Location	Everywhere	Everywhere	Attached	Everywhere
Visible Litter	Heavy	Heavy	Heavy	Heavy
Bug Count	Low	Average	Low	Average

TABLE 2 Chemical Tests

Test	5/26/02	6/25/02	9/24/02	11/19/02
pH	7.0	7.0	7.0	7.0
Dissolved Oxygen (DO)	6.8	6.0	5.6	9.0
Turbidity	1.0	3.0	1.0	1.0
Phosphate	.50	.50	.50	.50

MODEL 9–9
continued

continues

CONCLUSIONS AND RECOMMENDATIONS

This section includes the major conclusions and recommendations from our study of Wildwood Creek.

CONCLUSIONS

Draws conclusions that flow from data in body of report.

Generally, we were pleased with the health of the stream bank and its floodplain. The area studied has large amounts of vegetation along the stream, and the banks seem to be sturdy. The floodplain has been turned into a park, which handles floods in a natural way. Floodwater in this area comes in contact with vegetation and some dirt. Floodwater also drains quickly, which keeps sediment from building up in the creek.

Uses paragraph format instead of lists because of lengthy explanations needed.

However, we are concerned with the number and types of animals uncovered in our bug counts. Only two bug types were discovered, and these were types quite tolerant to pollutants. The time of year these tests were performed could affect the discovery of some animals. However, the low count still should be considered a possible warning sign about water quality. Phosphate levels were also high and probably are the cause of the large amount of algae.

We believe something in the water is keeping sensitive animals from developing. One factor that affects the number of animals discovered is the pollutant problems in the past (see Appendix A). The creek may still be in a redevelopment stage, thus explaining the small numbers of animals.

RECOMMENDATIONS

On the basis of these conclusions, we recommend the following actions for Wildwood Creek:

Gives numbered list of recommendations for easy reference.

1. Conduct the current tests two more times, through spring 2003. Spring is the time of year that most aquatic insects are hatched. If sensitive organisms are found then, the health of the creek could be considered to have improved.
2. Add testing for nitrogen. With the phosphate level being so high, nitrogen might also be present. If it is, then fertilizer could be in the water.
3. Add testing for human waste. Some contamination may still be occurring.
4. Add testing for metals, such as mercury, that can pollute the water.
5. Add testing for runoff water from drainage pipes that flow into the creek.
6. Schedule a volunteer cleanup of the creek.

With a full year of study and additional tests, the problems of Wildwood Creek can be better understood.

MODEL 9–9
continued

APPENDIX A

Background on Wildwood Creek

Wildwood Creek begins from tributaries on the northeast side of the city of Winslow. From this point, the creek flows southwest to the Chattahoochee River. Winslow Wastewater Treatment Plant has severely polluted the creek in the past with discharge of wastewater directly into the creek. Wildwood became so contaminated that signs warning of excessive pollution were posted along the creek to alert the public.

Today, all known wastewater discharge has been removed. The stream's condition has dramatically improved, but nonpoint contamination sources continue to lower the creek's water quality. Nonpoint contamination includes sewer breaks, chemical dumping, and storm sewers.

Another problem for Wildwood Creek is siltration. Rainfall combines with bank erosion and habitat destruction to wash excess dirt into the creek. This harsh action destroys most of the macroinvertebrates. At the present time, Wildwood Creek may be one of the more threatened creeks in Ware County.

MODEL 9–9
continued

continues

APPENDIX B

Water Quality Criteria for Georgia

All waterways in Georgia are classified in one of the following categories: fishing, recreation, drinking, and wild and scenic. Different protection levels apply to the different uses. For example, the protection level for dissolved oxygen is stricter in drinking water than fishing water. All water is supposed to be free from all types of waste and sewage that can settle and form sludge deposits.

In Ware County, all waterways are classified as "fishing," according to Chapter 391-3-6.03 of "Water Use Classifications and Water Quality Standards" in the Georgia Department of Natural Resources *Rules and Regulations for Water Quality Control.* The only exception is the Chattahoochee River, which is classified as "drinking water supply" and "recreational."

MODEL 9–9
continued

APPENDIX C

Map 6
Location of City of Winslow
Parks and Recreation Facilities

LEGEND

1) Birney Street Park
2) Custer Park
3) Nelson Park
4) Newell College
5) Indian Bluff
6) West View Park
7) Elmwood Park
8) Austin Heights
9) Riverview Park
10) Lewis Park
11) Burns Nature Park

N BY:S.C. SCOTT
CITY OF WINSLOW, GA
PUBLIC WORKS ENGR./DRAFT.
NO SCALE

NORTH

DEPARTMENT of PLANNING
and DEVELOPMENT

THE CITY OF WINSLOW
1945

MODEL 9–9
continued

Chapter 10

Proposals and Feasibility Studies

Proposing Changes in Security

General Instructions

Each chapter begins with a Collaboration at Work exercise that applies strategies for working in groups to chapter topics. The exercise assumes you (1) have been divided into groups of about three to six students, (2) will use group time inside or outside of class to complete the case, and (3) will produce an oral or written response. For guidelines about writing in groups, refer to pages 28–34.

Background for Assignment

Assume your school is reviewing all issues related to (1) the safety of students, faculty, and staff and (2) the security of equipment. This review has not been triggered by any particular event or problem; it is simply a periodic evaluation of conditions on campus. One step in the process has been to request that five consulting firms visit campus, spend a day on a field investigation, and then submit a proposal that lists specific work to be done and the cost of the work. Your school will choose one firm to do the job.

Group Assignment

Assume your group is one of five consulting firms submitting a proposal. As a preliminary step, team members have toured the campus, recorded observations, and collected initial ideas to propose. After discussing your observations, agree on three to five main changes to propose to the school's administration. (As an alternative, your group could focus on proposing three to five changes in one particular activity, building, or area of campus.)

Gini Harris, a structural technician at McDuff's Houston office, has written many reports in her two years at the company. Last month, however, she had the chance to write her first major proposal. Anchor Productions, a Hollywood film company, asked McDuff for a proposal to build a full-scale replica of the Alamo. In an upcoming film about the Texas battle, most scenes will be filmed in this replica to be built on leased land west of San Antonio. To write her proposal for Anchor, Gini had to seek advice from a variety of technical specialists, as well as architectural historians.

As it happens, Gini wrote a successful proposal that got the job. The work went so well that she asked her supervisor, Ken Blair, if McDuff could start a technical group to work just on historical restoration/replication projects around the country. Intrigued by the idea, Blair asked Harris to prepare a *feasibility study* that would focus on start-up costs, involvement of McDuff offices, hiring needs, and potential profit during the first five years. If the feasibility study showed promise, Blair would present the idea to McDuff's corporate management.

Although you probably won't write proposals to build an Alamo, you will write proposals and feasibility studies in your own field during your career. These modes of writing are crucial to most organizations. Indeed, many companies rely on them, especially proposals, for their very survival. Proposals and feasibility studies are defined as follows:

Proposals: documents written to convince your readers to adopt an idea, a product, or a service. They can be directed to colleagues inside your own organization (*in-house proposals*), to clients outside your organization (*sales proposals*), or to organizations that fund research and other activities (*grant proposals*).

In all three cases, proposals can be presented in either a short, simple format (*informal proposal*) or a longer, more complicated format (*formal proposal*). Also, proposals can be either requested by the reader (*solicited*) or submitted without a request (*unsolicited*).

Feasibility studies: documents written to show the *practicality* of a proposed policy, product, service, or other change within an organization. Often prompted by ideas suggested in a *proposal*, they examine details such as costs, alternatives, and likely effects. Though they must reflect the objectivity of a report, most feasibility studies also try to convince readers either (1) to adopt or reject the one idea discussed or (2) to adopt one of several alternatives presented in the study.

Feasibility studies can be *in-house* (written to decision-makers in your own organization) or *external* (requested by clients from outside your organization).

There are four main sections in this chapter. The first gives specific situations in which you might write proposals and feasibility studies at McDuff. The second and third sections discuss informal and formal proposal formats, while also denoting differences between in-house and sales versions of these formats. The fourth section covers feasibility studies. All chapter guidelines are followed by annotated examples to use as models for your own writing.

PROPOSALS AND FEASIBILITY STUDIES AT MCDUFF

As the Alamo example shows, proposals and feasibility studies often work together. The proposal may suggest a topic upon which a feasibility study is then written. The flowchart in Figure 10–1 shows another possible communication cycle that would involve both a proposal and a feasibility study. Note that the diagram includes the term *RFP*, which stands for "request for proposal":

> **Request for proposal (RFP):** documents sometimes sent out by organizations that want to receive proposals for a product or service. The RFP gives guidelines on (1) what the proposal should cover, (2) when it should be submitted, and (3) to whom it should be sent. As writer, you should follow the RFP religiously in planning and drafting your proposal.
>
> RFPs generally are *not* used in these situations:
>
> - When the proposal is solicited from within your own organization
> - When the proposal is requested less formally, as through a letter, phone call, or memo
> - When the proposal is unsolicited, meaning that you are writing it without a request from the person who will read it

The sections that follow describe additional situations in which proposals and feasibility studies would be written at McDuff. Reading through these brief cases will show you the varied contexts for persuasive writing within just one company.

McDuff Proposals

Like many organizations, McDuff depends on (1) in-house proposals to breathe new life into its internal operations, (2) sales proposals to request work from clients, and (3) occasional grant proposals to seek research funds from outside organizations. Proposals are a main activity in healthy, growing organizations like McDuff.

Of the five cases described here, the first and second are internal, requiring in-house proposals; the third and fourth are external, requiring sales proposals; and the fifth is external, requiring a grant proposal.

- **In-house proposal for computer drafting equipment:** Meg Stevens, a graphics expert at the Denver office, writes an in-house proposal to the drafting manager, Elvin Lipkowsky, in which she proposes that the company purchase a new computer-drafting system. Her proposal includes a schedule whereby the department can shift entirely to computer drafting in the next five years.

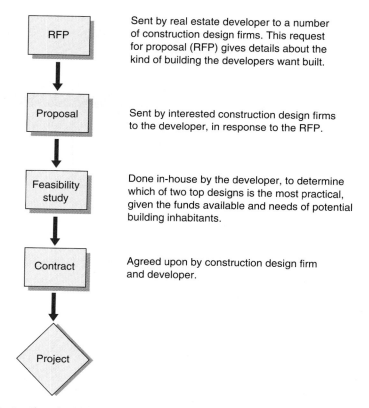

RFP

Sent by real estate developer to a number of construction design firms. This request for proposal (RFP) gives details about the kind of building the developers want built.

Proposal

Sent by interested construction design firms to the developer, in response to the RFP.

Feasibility study

Done in-house by the developer, to determine which of two top designs is the most practical, given the funds available and needs of potential building inhabitants.

Contract

Agreed upon by construction design firm and developer.

Project

FIGURE 10–1　Flowchart showing the main documents involved in one possible construction project

- **In-house proposal for retaining legal counsel:** Jake Washington, an employment specialist in the Human Resources Department in the Baltimore office, writes an in-house proposal to Lynn Redmond, vice president of human resources. In it he proposes that the company retain legal counsel on a half-time basis (20 hours a week). In his position at McDuff, Jake uses outside legal advice in dealing with new hiring laws, unemployment compensation cases, affirmative action regulations, and occasional lawsuits by employees who have been fired. He is proposing that the firm retain regular half-time counsel, rather than dealing with different lawyers as is done now.
- **Sales proposal for asbestos removal:** Jane Wiltshire, asbestos department manager at McDuff's St. Paul office, regularly talks with owners of buildings that may contain asbestos. After an initial discussion with the head minister of First Street Church, she writes an informal sales proposal in which she offers McDuff's services in performing an asbestos survey of the church building. Specifically, she explains how McDuff will examine the structure for possible asbestos, gives a schedule for completing the survey and writing the final report, and proposes a lump-sum price for the project.

- **Sales proposal for work on wind turbine project:** A utility company in California plans to build 10 giant turbines in a desert valley in the southern part of the state. The "free" power that is generated will help offset the large increases in fuel costs for the company's other plants. Although the firm has selected a turbine design and purchased the units, it needs to decide where to place them and what kind of foundations to use. Thus it sent out a request for proposal to companies that have experience with foundation and environmental engineering. Louis Bergen, engineering manager at McDuff's San Francisco office, writes a proposal that offers to test the soils at the site, pinpoint the best locations for the heavy turbines, and design the most effective foundations.
- **Grant proposal for new equipment design:** Oilarus, Ltd., a British oil company, sometimes gives research and development funds to small companies. Such funding usually goes toward development of new technology or products in the field of petroleum engineering. Angela Issam, who works in McDuff's Equipment Development group, decides to apply for some of the funding. Her proposed project, if successful, would provide a new piece of oil drilling safety equipment that would reduce the chance of offshore oil spills at production sites.

McDuff Feasibility Studies

The next two examples show that feasibility studies often flow from proposals. They can be internal, to managers who need facts before making a decision, or they can be external, to clients who request a service.

- **In-house feasibility study about legal counsel:** Lynn Redmond, the vice president of human resources, recently received Jake Washington's proposal that McDuff retain half-time legal counsel (see second case in previous section). The idea interests her, but she is not convinced of its practicality. She calls a meeting with Jake Washington and Scott Sampson, the personnel manager, who also uses legal counsel on a part-time basis. Because Lynn will have to sell her boss on this idea, she asks Jake and Scott to write a feasibility study on it.

  This study, unlike the proposal, must include a detailed comparison of the present mode of operating versus the proposed strategy of retaining a lawyer for 20 hours per week on a regular basis. The study must examine criteria such as current costs versus projected costs, current level of satisfaction versus projected level of satisfaction, and current level of services provided versus projected level of services provided. Lynn asks that the report include a clear recommendation, made on the basis of the data.
- **External feasibility study on plant site:** Tarnak, Inc., a large furniture manufacturing company in North Carolina, has decided to build a new plant in northern Georgia. After getting proposals from many cities that want the plant, the company has narrowed its choices to three spots that are about equal in cost of living, access to workers, standard of living, construction costs, transportation facilities, and access to raw materials. Yet the firm has not studied any site with respect to waste management.

Specifically, Tarnak needs to know which of the three cities is best prepared to handle the solid and chemical wastes from the plant in a safe and economical manner. It hires McDuff's Atlanta office to write a feasibility study. The first goal of the study is to determine what sites, if any, meet Tarnak's criteria for waste management. If the first objective yields more than one site, McDuff then must compare the sites and recommend the best one.

GUIDELINES FOR INFORMAL PROPOSALS

Like informal reports, informal proposals are short documents that cover projects with a limited scope. But how short is short? And just what does "limited" mean? Following are some specific guidelines that you can use if your employer or client has not provided others:

Use Informal Proposals When:

- The text of the proposal (excluding attachments) is no more than 5 pages
- The size of the proposed project is such that a long, formal proposal would appear to be inappropriate
- The client has expressed a preference for a leaner, less formal document

Use Formal Proposals When:

- The text of the proposal (excluding attachments) is over 5 pages
- The size and importance of the project is such that a formal proposal would be appropriate
- The client has expressed a preference for a more formal document

These two formats can be used for proposals that are either *in-house* (to readers within your own organization) or *external* (to readers outside your organization). The rest of this section provides writing guidelines and an annotated model for informal proposals, the type of persuasive writing you will do most often in your career.

Informal proposals have two formats: (1) memos (for in-house proposals) and (2) letters (for external proposals). The guidelines recommended apply to both. With some variations, they are essentially the guidelines suggested in chapter 8 for informal reports. The formats are much the same, though the content and tone are different. Reports explain, whereas proposals persuade.

● *Informal Proposal Guideline 1: Plan Well Before You Write*

Complete the Planning Form at the end of the book for all proposal assignments in class, as well as for proposals you write on the job. Carefully consider your purpose, audience, and organization. Two factors make this task especially difficult in sales-proposal writing:

1. You may know nothing more about the client than what is written on the RFP.
2. Proposals almost always are on a tight schedule that limits your planning time.

Despite these limitations, try to find out exactly who will be making the decision about your proposal. Many clients will tell you if you give them a call. In fact, they may be pleased that you care enough about the project to target the audience. Once you identify the decision-makers, spend time brainstorming about their needs *before* you begin writing. Proposals that betray an ignorance of client needs often do so because the writer began writing too soon about the product or service.

● *Informal Proposal Guideline 2: Use Letter or Memo Format*

Letter proposals, such as the example shown in Model 10–1 on pages 372–374 basically follow the format of routine business letters (see chapter 7). This casual style gives readers the immediate impression that your document will be "approachable"—that is, easy to get through and limited in scope. Memo proposals, such as the example shown in Model 10–2 on pages 375–376, follow the format of an internal memorandum (see chapter 7). Here are a few highlights:

- Line spacing is usually single, but it may be $1\frac{1}{2}$ or double, depending on the reader's or company's preference.
- The recipient's name, date, and page number appear on sheets after the first.
- Most readers prefer an uneven or "ragged-edge" right margin, as opposed to an even or "right-justified" margin.

Your subject line in a memo proposal (if you choose to use one) gives readers the first impression of the proposal's purpose. Choose concise yet accurate wording. Furthermore, the wording must match that which you have used in the proposal text. See Model 10–2 for wording that gives the appropriate information and tries to engage the reader's interest.

● *Informal Proposal Guideline 3: Make Text Visually Appealing*

The page design of informal proposals must draw readers into the document. Remember—you are trying to *sell* a product, a service, or an idea. If the layout is unappealing, then you will lose readers before they even get to your message. Also, remember that your proposal may be competing with others. Put yourself in the place of the reader who is wondering which one to pick up first. How the text looks on the page can make a big difference. Here are a few techniques to follow to help make your proposal visually appealing:

- Use lists (with bullets or numbered points) to highlight main ideas.
- Follow your readers' preferences as to font size, type, line spacing, and so forth. Proposals written in the preferred format of the reader will gain a competitive edge.
- Use headings and subheadings to break up blocks of text.

These and other techniques help to reveal the proposal's structure and lead readers through the informal proposal. Given that there is no table of contents, you need to take advantage of such strategies.

● *Informal Proposal Guideline 4:*
   Use the ABC Format for Organization

The ABC format used throughout this book also applies, of course, to informal proposals.

ABC FORMAT: *Informal Proposal*

Abstract

- Gives the summary or "big picture" for those who will make decisions about your proposal. Usually includes some kind of "hook" or "grabber"—a point that will interest the audience in reading further.

Body

- Gives the details about exactly what you are proposing to do.

Conclusion

- Drives home the main benefit and makes clear the next step.

Note: Beginning and ending sections should be easy to read and stress just a few points. They provide a short "buffer" on both ends of the longer, more technical body section in the middle. The next four guidelines give more specific advice for writing the main parts of an informal proposal.

● *Informal Proposal Guideline 5: Use the Heading "Introductory Summary" for the Generic Abstract Section*

Here you capture the client's attention with a capsule summary of the entire proposal. This one- or two-paragraph starting section permits space only for what the reader *really* needs to know at the outset, such as the following:

- Purpose of proposal
- Reader's main *need*
- Main *features* you offer and related *benefits*
- Overview of proposal sections to follow

As Models 10–1 and 10–2 show, the introductory summary appears immediately after the subject line of the memo proposal or salutation of the letter proposal. As with informal reports, you have the option of actually labeling the section "Introductory Summary" or simply leaving off the heading. In either case, keep this overview very brief. Answer the one question clients are thinking: "Why should we hire this firm instead of another?" If you find yourself starting to give too much detail, move background information into the first section of the discussion.

● *Informal Proposal Guideline 6: Put Important Details in the Body*

The discussion of your proposal should address these basic questions:

1. What problem are you trying to solve, and why?
2. What are the technical details of your approach?
3. Who will do the work, and with what?
4. When will it be done?
5. How much will it cost?

Discussion formats vary from proposal to proposal, but here are some sections commonly used to respond to these questions:

    1. Description of problem or project and its significance. Give a precise technical description, along with any assumptions that you have made on the basis of previous contact with the reader. Explain the importance or significance of the problem, especially to the reader of the proposal.

    2. Proposed solution or approach. Describe the specific tasks you propose in a manner that is clear and well organized. If you are presenting several options, discuss each one separately—making it as easy as possible for the reader to compare and contrast information.

    3. Personnel. If the proposal involves people performing tasks, it may be appropriate to explain qualifications of participants.

    4. Schedule. Even the simplest proposals usually require some sort of information about the schedule for delivering goods, performing tasks, and so on. Be both clear and realistic in this portion of the proposal. Use graphics when appropriate (see chapter 11 for guidelines on Gantt and milestone charts).

    5. Costs. Place complete cost information in the body of the proposal unless you have a table that would be more appropriately placed in an attachment. Above all, do not bury dollar figures in paragraph format. Instead, highlight these figures with indented or bulleted lists, or at least place them at the beginnings of paragraphs. Because your reader will be looking for cost data, it is to your advantage to make that information easy to find. Finally, be certain to include *all* costs—materials, equipment, personnel, salaries, etc.

● *Informal Proposal Guideline 7: Give Special
Attention to Establishing Need in the Body*

A common complaint about proposals is that writers fail to establish the need for what is being proposed. As any good salesperson knows, customers must feel that they need your product, service, or idea before they can be convinced to purchase or support it. In other words, do not simply try to dazzle readers with the good sense and quality of what you are proposing. Instead, lay the groundwork for acceptance by first showing the readers that a strong need exists.

    Establishing need is most crucial in unsolicited proposals, of course, when readers may not be psychologically prepared to accept a change that will cost them money. Even in proposals that have been solicited, however, you should give some attention to restating the basic needs of the readers. If nothing else, this special attention shows your understanding of the problem.

● *Informal Proposal Guideline 8: Focus Attention in Your Conclusion*

Called "conclusion" or "closing," this section gives you the opportunity to control the readers' last impression. It also helps you avoid the awkwardness of ending proposals with the statement of costs, which is usually the last section in the discussion. In this closing section you can:

- Emphasize a main benefit or feature of your proposal
- Restate your interest in doing the work
- Indicate what should happen next

Regarding the last point, sometimes you may ask readers to call if they have questions. In other situations, however, it is appropriate to say that you will follow up the proposal with a phone call. This approach leaves you in control of the next step.

Incidentally, for informal sales proposals, there is a special technique that can push the proposal one step closer to approval. After the signature section, place an "acceptance block." As shown in the following example, this item makes it as easy as a signature for the reader to accept your proposal, rather than his or her having to write a return letter.

ACCEPTED BY LMN DEVELOPMENT, INC.

By: _____

Title: _____

Date: _____

● *Informal Proposal Guideline 9: Use Attachments for Less Important Details*

Remember that the text of informal proposals is usually less than 10 pages. That being the case, you may have to put supporting data or illustrations in attachments that follow the conclusion. Cost and schedule information, in particular, is best placed at the end in well-labeled sections.

Make sure that the proposal text includes clear references to these visuals. If you have more than one attachment, give each one a letter and a title (for example, "Attachment A: Project Costs"). If you have only one attachment, include the title but no letter (for example, "Attachment: Resumes").

● *Informal Proposal Guideline 10: Edit Carefully*

In the rush of completing proposals, some writers fail to edit carefully. That is a big mistake. Make sure to build in enough time for a series of editing passes, preferably by different readers. There are two reasons why proposals of all kinds deserve this special attention.

1. They can be considered contracts in a court of law. If you make editing mistakes that alter meaning (such as an incorrect price figure), you could be bound to the error.

2. Proposals often present readers with their first impression of you. If the document is sloppy, they can make assumptions about your professional abilities as well.

GUIDELINES FOR FORMAL PROPOSALS

Sometimes the complexity of the proposal may be such that a formal response is best. As always, the final decision about format should depend upon the needs of your readers. Ask yourself questions like these in deciding whether to write an informal or a formal proposal:

- Is there too much detail for a letter or memo?
- Is a table of contents needed so that sections can be found quickly?
- Will the professional look of a formal document lend support to the cause?
- Are there so many attachments that a series of lengthy appendices would be useful?
- Are there many different readers with varying needs, such that there should be different sections for different people?

If you answer yes to one or more of these questions, give careful consideration to writing a formal proposal. This long format is most common in external sales proposals; however, important in-house proposals may sometimes require the same approach—especially in large organizations in which you may be writing to unknown persons in distant departments. Both in-house and sales examples follow the writing guidelines given next.

Formal proposals can be long and complex, so this part of the chapter treats each proposal section separately—from title page through conclusion. Two points will become evident as you use these guidelines. First, formal reports and formal proposals are a lot alike. A quick look at the last chapter will show you the similarities in format. Second, a formal proposal—like *all* technical writing described in this text—follows the basic ABC format described in chapter 3. Specifically, the parts of the formal proposal fit the pattern as shown in the box on the following page.

As you read through and apply these guidelines, refer to Model 10–3 on pages 377–385 and Model 10–4 on pages 386–406 for annotated examples of the formal proposal. Note that each major section in the model proposals starts on a new page. Another alternative is to run most sections together, changing pages only at the end of the letter of transmittal and executive summary. Minor format variations abound, of course, but this chapter's guidelines will stand you in good stead throughout your career.

Cover/Title Page

Like formal reports, formal proposals usually are bound documents with a cover. The cover will include one or more of the items listed here for inclusion on the title page.

ABC FORMAT: *Formal Proposal*

Abstract
- Cover/Title Page
- Letter of Transmittal
- Table of Contents
- List of Illustrations
- Executive Summary
- Introduction

Body
- Technical information
- Management information
- Cost information
- (Appendices—appear after text but support Body section)

Conclusion
- Conclusion

Just as important, the cover should be designed to attract the reader's interest—with good page layout and perhaps even a graphic. Remember—proposals are sales documents. No one *has* to read them.

Inside the cover is the title page, which contains these four pieces of information:

- **Project title,** sometimes preceded by "Proposal for" or similar wording
- **Your reader's name** (sometimes preceded by "Prepared for . . .")
- **Your name or the name of your organization** spelled out in full (sometimes preceded by "Prepared by . . .")
- **Date of submission**

The title page gives clients their first impression of you. For that reason, consider using some tasteful graphics to make the proposal stand out from those of your competitors. For sales proposals, a particularly persuasive technique is to place the logo of the *client's company* on the cover or title page. In this way, you imply your interest in linking up with that firm *and* your interest in satisfying its needs, rather than simply selling your products or services.

Letter/Memo of Transmittal

Internal proposals have memos of transmittal; external sales proposals have letters of transmittal. These letters or memos must grab the reader's interest. The guidelines for format and organization presented here will help you write attention-getting prose. In particular, note that the letter or memo should be in single-spaced, ragged-right-edge format, even if the rest of the proposal is double-spaced copy with right-justified margins.

For details of letter and memo format, see chapter 7. The guidelines for the letter/memo of transmittal for formal reports also apply to formal proposals (see Transmittal Guideline 3 in chapter 9). For now, here are some highlights of format and content that apply especially to letters and memos of transmittal:

1. Use short beginning and ending paragraphs (about three to five lines each).
2. Use a conversational style, with little or no technical jargon. Avoid stuffy phrases such as "per your request."
3. Use the first paragraph for introductory information, mentioning what your proposal responds to (for example, a formal RFP, a conversation with the client, or your perception of a need).
4. Use the middle of the letter to emphasize one main benefit of your proposal, though the executive summary and proposal proper will mention benefits in detail. Stress what you can do to solve a problem, using the words "you" and "your" as much as possible (rather than "I" and "we").
5. Use the last paragraph to retain control by orchestrating the next step in the proposal process. When appropriate, indicate that you will call the client soon to follow up on the proposal.
6. Follow one of the letter formats described in chapter 7. Following are some exceptions, additions, or restrictions:

   - Use single-spaced, ragged-right-edge copy, which will make your letter stand out from the proposal proper.
   - Keep the letter on one page—a two-page letter loses that crisp, concise impact you want a letter to make.
   - Place the company proposal number (if there is one) at the top, above the date. Exact placement of both number and date depends on your letter style.
   - Include the client's company name or personal name on the first line of the inside address, followed by the mailing address that will be on the envelope. If you use a company name, place an "attention" line below the inside address. Include the full name (and title, if appropriate) of your contact person at the client firm. If you use a personal name, follow the last line of the inside address with a conventional greeting ("Dear Mr. Adams:").
   - (Optional) Include the project title below the attention line, using the exact wording that appears on the title page.
   - Close with "Sincerely" and your name at the bottom of the page. Also include your company affiliation.

Table of Contents

Create a very readable table of contents by spacing items well on the page. List all proposal sections, subsections, and their page references. At the end, list any appendices that may accompany the proposal.

Given the tight schedule on which most proposals are produced, errors can be introduced at the last minute because of additions or revisions. Therefore, take time to proofread the table of contents carefully. In particular, make sure to follow these guidelines:

- Wording of headings should match within the proposal text.
- Page references should be correct.
- All headings of the same order should be parallel in grammatical form.

List of Illustrations

When there are many illustrations, the list of illustrations appears on a separate page after the table of contents. When there are few entries, however, the illustrations may be listed at the end of the table of contents page. In either case, the list should include the number, title, and page number of every illustration appearing in the body of the text. (If there is only one illustration, a number need not be included.) You may divide the list into tables and figures if many of both appear in your proposal.

Executive Summary

Executive summaries are the most frequently read parts of proposals. That fact should govern the time and energy you put into their preparation. Often read by decision-makers in an organization, the summary should present a concise one-page overview of the proposal's most important points. It should also accomplish these objectives:

- Avoid technical language
- Be as self-contained as possible
- Make brief mention of the problem, proposed solution, and cost
- Emphasize the main benefits of your proposal

Start the summary with one or two sentences that command readers' attention and engage their interest. Then focus on just a few main selling points (three to five is best). You might even want to highlight these benefits with indented lead-ins such as "Benefit 1, Benefit 2." When possible, use the statement of benefits to emphasize what is unique about your company or your approach so that your proposal will attract special attention. Finally, remember to write the summary after you have completed the rest of the proposal. Only at this point do you have the perspective to sit back and develop a reader-oriented overview.

Introduction

The introduction provides background information for both nontechnical and technical readers. Although the content will vary from proposal to proposal, some general guidelines apply. Basically, you should include information on the (1) purpose, (2) description of the problem to which you are responding, (3) scope of the proposed study, and (4) format of the proposal. (A *lengthy* problem or project statement should be placed in the first discussion section of the proposal, not in the introduction.)

- Use subheadings if the introduction goes over a page. In this case, begin the section with a lead-in sentence or two that mention the sections to follow.
- Start with a purpose statement that concisely states the reason you are writing the proposal.
- Include a description of the problem or need to which your proposal is responding. Use language directly from the request for proposal or other document the reader may have given you so that there is no misunderstanding. For longer problem or need descriptions, adopt the alternative approach of including a separate needs section or problem description after the introduction.
- Include a scope section in which you briefly describe the range of proposed activities covered in the proposal, along with any research or preproposal tasks that have already been completed.
- Include a proposal format section if you feel the reader would benefit from a listing of the major proposal sections that will follow.

Discussion Sections

Aim the discussion or body toward readers who need supporting information. Traditionally, the discussion of a formal sales proposal contains three basic types of information: (1) technical, (2) management, and (3) cost. Here are some general guidelines for presenting each type. Remember that the exact wording of headings and subheadings will vary, depending on proposal content.

1. **Technical Sections**
   - Respond thoroughly to the client's concerns, as expressed in writing or meetings.
   - Follow whatever organization plan that can be inferred from the request for proposal.
   - Use frequent subheadings with specific wording.
   - Back up all claims with facts.

2. **Management Sections**
   - Describe who will do the work.
   - Explain when the work will be done.
   - Display schedule information graphically.
   - Highlight personnel qualifications (but put resumes in appendices).

3. **Cost Section**
   - Make costs extremely easy to find.
   - Use formal or informal tables when possible.
   - Emphasize value received for costs.
   - Be clear about add-on costs or options.
   - Always total your costs.

Conclusion

Formal proposals should always end with a section labeled "Conclusion" or "Closing." This final section of the text gives you the chance to restate a main benefit, summarize the work to be done, and assure clients that you plan to work with them closely to satisfy their needs. Just as importantly, this brief section helps you end on a positive note. You come back full circle to what you stressed at the beginning of the document—benefits to the client and the importance of a strong personal relationship. (Without the conclusion, the client's last impression would be made by the cost section in the discussion.)

Appendices

Because formal proposals are so long, readers sometimes have trouble locating information they need. Headings help, but they are not the whole answer. Another way you can help readers is by transferring technical details from the proposal text into appendices. The proposal still will contain detail—for technical readers who want it—but detail will not intrude into the text. This technique can save you or your employer considerable time by permitting you to develop standard appendices (also called *boilerplate*) to be used in later proposals.

Any supporting information can be placed in appendices, but here are some common items included there:

- Resumes
- Organization charts
- Company histories
- Detailed schedule charts
- Contracts
- Cost tables
- Detailed options for technical work
- Summaries of related projects already completed
- Questionnaire samples

Boilerplate is often taken right off the shelf and thus is not paged in sequence with your text. Instead, it is best to use individual paging within each appendix. For example, pages in an Appendix B would be numbered B-1, B-2, B-3.

GUIDELINES FOR FEASIBILITY STUDIES

Much like recommendation reports (see chapter 8), feasibility studies guide readers toward a certain line of action. Another similarity is that both report types can be either in-house or external. Yet most feasibility studies have these five distinctive features that justify their being considered separately here:

1. They *always* are solicited by the reader, usually for the purpose of deciding on the best course of action.
2. They *always* assume one of these two patterns of organization:
   - An analysis of the advantages and disadvantages of one course of action, product, or idea
   - A comparison of two or more courses of action, products, or ideas
3. They *always* are intended to help managers and other decision-makers vote for or against an idea or select among several alternatives.
4. They *usually* "nudge" (as opposed to "urge" or "push") the reader toward a decision. That is, they are supposed to be written in such a way that the facts speak for themselves.
5. They *often* are preceded in time by a proposal.

In some ways, feasibility studies could be viewed as a cross between technical reports and proposals. As a writer you are expected to deal with the topic objectively and honestly, yet you are also expected to express your point of view. The *American Heritage Dictionary* defines "feasible" in this way: "capable of being accomplished or brought about." Thus a feasibility study determines if some course of action is practical. For example, it may be *desirable* for a student to quit work and return to college full-time. Yet if that same student has hefty car and apartment payments, the only *feasible* alternative may be part-time course work.

The following guidelines will help you prepare the kinds of feasibility studies requested by your boss (if the study is in-house) or by your client (if the study is external). In either case, your study may be used as the basis for a major decision. Refer to Model 10–5 on pages 407–408 as you read and apply these guidelines to your own writing.

● *Feasibility Study Guideline 1: Choose Format Carefully*

In deciding whether to use the format of an informal (letter or memo) or formal document, use the same criteria mentioned earlier in the chapter with regard to proposals. As always, the central questions concern your readers:

- What format will give them easiest access to the data, conclusions, and recommendations of your study?
- Are there enough pages to suggest need of a table of contents (that is, a formal report)?
- What is the format preference of your readers?
- What has been the format of previous feasibility studies written for the same organization?

● *Feasibility Study Guideline 2: Use the ABC Structure*

Like other forms of technical writing, good feasibility studies have the basic three-part structure of **A**bstract, **B**ody, and **C**onclusion—though the exact headings you choose may vary from report to report. Here is an overview of what the study would include:

ABC FORMAT: *Feasibility Study*

Abstract

- Capsule summary of information for the most important readers (that is, the decision-makers)

Body

- Details that support whatever conclusions and recommendations the study contains, working logically from fact toward opinion

Conclusion

- Wrap-up in which you state conclusions and recommendations resulting from study

---

The following guidelines examine specific sections of feasibility studies, along with details of content and tone. The focus here is on what would be included in an informal document. See chapter 9 for guidelines that could be applied to a formal feasibility study.

● *Feasibility Study Guideline 3: Call Your Abstract an Introductory Summary*

This section provides information that the most important readers would want if they were in a rush to read your study. With that criterion in mind, consider including these items:

- Brief statement about who has authorized the study and for what purpose
- Brief mention of the criteria used during the evaluation
- Brief reference to your recommendation

The last item is important, for it saves readers the frustration of having to wade through the whole document in search of the answers to the questions "Is this a practical idea?" or "Which alternative is best?" It is best to mention the recommendation up front, giving readers a frame through which to see the entire report.

● *Feasibility Study Guideline 4: Organize the Body Well*

More than anything else, readers of feasibility studies expect an unbiased presentation. That means the midsection of your report must clearly and logically work from facts toward recommendations. Here is one approach that works:

1. **Describe evaluation criteria used during your study,** if readers need more detail than was presented in the introductory summary.
2. **Describe exactly WHAT was evaluated and HOW,** especially if you are comparing several items.
3. **Choose criteria that are most meaningful to the readers,** such as:
   - Cost
   - Practicality of implementing idea
   - Changes that may be needed in personnel
   - Effect on growth of organization
   - Effects on day-to-day operations

        Of course, exact criteria will depend upon the precise topic you are investigating.
4. **Discuss both advantages and disadvantages** when you are evaluating just one item. Move from advantages to disadvantages. The conclusion will allow you to come back around to supporting points.
5. **Follow organization guidelines for comparisons** when evaluating several alternatives (see chapter 5). You can discuss one item at a time OR you can discuss one criterion at a time.

● *Feasibility Study Guideline 5: Use the Conclusion*
 for Detailed Conclusions and Recommendations

Here you get the opportunity to state (or restate) the conclusions evident from data you have presented in the discussion. First state conclusions, and then state your recommendations. Use listings for three or more points, to make this last section of the study as easy as possible to read.

● *Feasibility Study Guideline 6: Use Graphics for Comparisons*

When comparing several items, you need to consider most readers' preference for tabulated information. Tables can appear either in the discussion section or in attachments. In both cases, follow graphics guidelines explained in chapter 11.

● *Feasibility Study Guideline 7: Offer to Meet with the Readers*

Most readers have many questions after reading a feasibility study, even if that study has been quite thorough. You score points for eagerness and professionalism if you anticipate needs and express your willingness to meet with readers later. Such meetings give you another opportunity to demonstrate your understanding of the topic.

Communication Challenge

"The Black Forest Proposal: Good Marketing or Bad Business?"

To strengthen its proposals, McDuff hired a new proposal writer at the corporate office in Baltimore. Ben Sadler came well recommended, having both advertising and marketing experience with technical firms. After he finished his first major proposal at McDuff, he had some disagreements during an internal review of his draft by one of the firm's technical experts. What follows is background information on the project, an overview of decisions Ben made in writing the proposal, and some questions and comments related to the review.

Background on Black Forest Project

Rob McDuff and his staff hired Ben Sadler for a specific reason. Although business was going fairly well, they felt the company needed new direction and energy in its marketing. Staff members seemed to be taking for granted that clients would always return and that word-of-mouth would keep new clients coming through the doors. Company leaders knew such an attitude was dangerous. Because they wanted to venture into new types of work, they believed the time was right for a new marketing expert. Ben Sadler seemed to be the catalyst the firm needed.

Just after arriving, Ben learned about a request for a proposal recently issued for a large construction job—building a new university campus in southern Germany. The project at Black Forest University included four buildings, for a total of $35 million in construction. It would become the centerpiece of the new campus. McDuff had thus far done no major construction work in Germany, nor had it done much work at colleges and universities anywhere in the world. Yet Ben felt the firm had the technical tools and the personnel to be a contender. After convincing his immediate boss, the vice president for business and marketing, that the proposal was worth writing, Ben went to work.

Proposal Strategy

Although Ben had research assistance from McDuff employees in the United States and Germany, he wrote the draft himself. Here are parts of the writing and marketing strategy he planned to use:

- *Experience:* He emphasized the large construction jobs McDuff had done for other types of government-related agencies in Germany and around the world. Although McDuff had done no major college or university construction, Ben felt that including work for government agencies would be an adequate substitute.
- *Technical Experts:* Although McDuff's London and Munich offices had no experts to coordinate large-scale construction, Ben knew the company could bring in experts from the United States. Admittedly, it would be more expen-

sive to import talent. Plus these high-salary individuals would be able to visit the site only periodically. Yet this arrangement satisfied the minimum technical requirements in the industry. Ben felt comfortable including the experts' resumes in the proposal and highlighting their experience, without mentioning the fact that they did not work out of the German office. That fact seemed to him to be an internal matter.

- **Costs:** When Ben calculated the tentative cost for the project, he was surprised that the figure was so high. The extra personnel costs previously noted were apparently part of the problem. But Ben also thought some costs may have been overstated because McDuff was not used to bidding on such jobs in Germany. (McDuff's corporate accounting manager had done the tentative cost estimate.) Believing the figures were inflated, Ben cut about 10 percent from personnel costs mentioned in the draft. He thought McDuff would perform more efficiently than the accounting manager had estimated.

- **Proposal Strategy:** Ben wanted to come on strong in the executive summary with what he saw as the benefits McDuff could offer. He focused on three main selling points: (a) availability of a nearby office and lab in Munich for project coordination, (b) experience with other large construction jobs in the United States, and (c) McDuff's history of good working relationships in Germany and the rest of Europe.

- **Personal Contacts:** Ben happened to have a close friend at a former firm (not a competitor of McDuff's) who went to college with an official now on the Black Forest University board of directors. Ben wanted to use this "friend of a friend" connection to get a meeting with the board member, perhaps to find out more about the project. He might then get the chance to give part of the "McDuff story" that was not revealed in the proposal.

Questions and Comments for Discussion

After Ben completed the draft, it was reviewed first by J. R. Link, one of the top technical experts at the Baltimore branch and an old-timer with McDuff. J. R. met with Ben and expressed reservations about the project and about the proposal. The questions that follow reflect their conversation, as well as some other concerns about the proposal:

1. J. R. Link first wondered why Sadler proposed on the job in the first place. Given that such proposals cost McDuff $10,000 or $15,000 to write, why did he bother with the Black Forest job when McDuff doesn't have experience on large construction jobs in Germany? Most projects out of the Munich office are environmental studies. Wouldn't it be a long shot to get the work? And shouldn't McDuff managers have some ethical concern about trying to get a job when they know they don't have the experience other competing firms probably have (or at least *should* have)?

    What is your view of the practical and ethical concerns raised by J. R.?

    Continued

2. J. R. also questioned whether Ben was being deceptive about the way in which technical experts would be provided for the project. If the resumes were to be included in the proposal, shouldn't McDuff also mention that these experts reside in the United States? This matter didn't seem to be an "internal" one, as Ben stated.

   What's your view of the way Ben handled the issue of outside experts from the United States? How much of this sort of information has to be put forth in a competitive proposal?

3. Ben had an honest disagreement about the calculation of costs by the accounting manager. He may have been right or wrong in his reservations about the accountant's estimate. Putting this point aside, was it procedurally correct for him, as project manager, to make changes in the costs submitted by an advisor? Why or why not? Was it ethical?

4. Ben chose a "direct" approach to proposal content by placing main selling points first (in the executive summary). Do you think this strategy is appropriate in all cultures? Why or why not? (If possible, do some research on technical communication in Germany, Japan, or China before answering the question.)

5. As noted earlier, Ben decided to pursue a connection he had on the board of directors of Black Forest University. Is this strategy ethical? Would it work? What are some possible results of such a strategy?

6. Have two students in the class conduct a role-play of the conversation between J. R. Link and Ben Sadler. The students can use the information just presented and any additional points that conceivably could be put forth by these two men, considering the sketches provided of them.

CHAPTER SUMMARY

Proposals and feasibility studies stand out as documents that aim to *convince* readers. In the case of proposals, you are writing to convince someone inside or outside your organization to adopt an idea, a product, or a service. In the case of feasibility studies, you are marshalling facts to support the practicality of one approach to a problem—sometimes in comparison with other approaches. Both documents can be either informal or formal, depending on length, complexity, or reader preference.

This chapter includes lists of writing guidelines for informal proposals, formal proposals, and feasibility studies. For informal proposals, follow these basic guidelines:

1. Plan well before you write.
2. Use letter or memo format.
3. Make text visually appealing.
4. Use the ABC format for organization.
5. Use the heading "Introductory Summary" for the generic abstract section.
6. Put important details in the body.
7. Give special attention to establishing need in the body.
8. Focus attention in your conclusion.

9. Use attachments for less important details.
10. Edit carefully.

In formal proposals, abide by the same general format presented in chapter 9 for formal reports. To be sure, formal proposals have a different tone and substance because of their more persuasive purpose. Yet they do have the same basic parts, with minor variations: cover/title page, letter/memo of transmittal, table of contents, list of illustrations, executive summary, introduction, discussion sections, conclusion, and appendices.

Feasibility studies demonstrate that an idea is or is not practical. Also, they may compare several alternatives. Follow these basic writing guidelines:

1. Choose format carefully.
2. Use the ABC structure.
3. Call your abstract an introductory summary.
4. Organize the body well.
5. Use the conclusion for detailed conclusions and recommendations.
6. Use graphics for comparisons.
7. Offer to meet with the readers.

Access the Technical Writing Web site by pointing your browser at www.prenhall.com/pfeiffer. Here, you will find additional resources to aid you in learning and applying the information from this chapter.

ASSIGNMENTS

The assignments in Part 1 and Part 2 can be completed either as individual projects or as group-writing projects. If your instructor assigns group projects, review the information in chapter 1 on group writing.

Part 1: Short Assignments

These short assignments require either that you write parts of informal or formal proposals *or* that you evaluate the effectiveness of an informal proposal included here.

1. **Introductory Summary.** For this assignment, select one of the seven projects in the color insert. Now assume that you were responsible for writing the proposal that resulted in the project. In other words, work backwards from the project to the informal proposal that McDuff used to get the work. Write a short introductory summary for the original proposal. Focus on the main reason you think the client would have for hiring McDuff. If necessary, invent additional information to complete this assignment successfully.

2. **Needs Section.** As this chapter suggests, informal proposals—especially those that are unsolicited—must make a special effort to establish the need for the product or service being proposed. Assume that you are writing an informal proposal to suggest a change in procedures or equipment at your college. Keep the proposal limited to a small change; you may even see a need in the classroom where you attend class (audiovisual equipment? lighting? heating or air systems? aesthetics? soundproofing?). Write the needs section that would appear in the body of the informal proposal.

3. **Conclusion or Closing.** For this assignment, as with assignment 1, select a project from the color insert. Assume that you were the McDuff employee responsible for writing the

informal proposal that resulted in the work described in the project. Write an effective conclusion or closing for the proposal.

4. **Evaluation—Informal Proposal.** Review the informal proposal that follows, submitted by MainAlert Security Systems to the McDuff, Inc., office in Atlanta. Evaluate the effectiveness of every section of the proposal.

200 Roswell Road
Marietta, Georgia 30062
(770) 555-2000

September 15, 2002

Mr. Bob Montrose
Operations Manager
McDuff, Inc.
3295 Peachtree Road
Atlanta, Georgia 30324

Dear Bob,

Thank you for giving MainAlert Security Systems an opportunity to submit a proposal for installation of an alarm system at your new office. The tour of your nearly completed office in Atlanta last week showed me all I need to know to provide you with burglary and fire protection. After reading this proposal, I think you will agree with me that my plan for your security system is perfectly suited to your needs.

This proposal describes the burglary and fire protection system I've designed for you. This proposal also describes various features of the alarm system that should be of great value. To provide you with a comprehensive description of my plan, I have assembled this proposal in several main sections:

1. Burglary Protection System
2. Fire Protection System
3. Arm/Disarm Monitoring
4. Installation Schedule
5. Installation and Monitoring Costs

BURGLARY PROTECTION SYSTEM

The burglary protection system would consist of a 46-zone MainAlert alarm control set, perimeter protection devices, and interior protection devices. The alarm system would have a strobe light and a siren to alert anyone nearby of a burglary in progress. Our system also includes a two-line dialer to alert our central station personnel of alarm and trouble conditions.

Alarm Control Set

The MainAlert alarm control set offers many features that make it well suited for your purposes. Some of these features are as follows:

1. Customer-programmable keypad codes
2. Customer-programmable entry/exit delays
3. Zone bypass option
4. Automatic reset feature
5. Point-to-point annunciation

I would like to explain the point-to-point annunciation feature, since the terminology is not as self-explanatory as the other features are. Point-to-point annunciation is a feature that enables the keypad to display the zone number of the point of protection that caused the alarm. This feature also transmits alarm-point information to our central station. Having alarm-point information available for you and the police can help prevent an unexpected confrontation with a burglar.

Interior and Perimeter Protection

The alarm system I have designed for you uses both interior and perimeter protection. For the interior protection, I plan to use motion detectors in the hallways. The perimeter protection will use glass-break detectors on the windows and door contacts on the doors.

There are some good reasons for using both interior and perimeter protection:

1. Interior and perimeter protection used together provide you with two lines of defense against intrusion.
2. A temporarily bypassed point of protection will not leave your office vulnerable to an undetected intrusion.
3. An employee who may be working late can still enjoy the security of the perimeter protection while leaving the interior protection off.

Although some people select only perimeter protection, it is becoming more common to add interior protection for the reasons I have given. Interior motion detection, placed at carefully selected locations, is a wise investment.

Local Alarm Signaling

The local alarm-signaling equipment consists of a 40-watt siren and a powerful strobe light. The siren and strobe will get the attention of any passerby and unnerve the most brazen burglar.

Remote Alarm Signaling

Remote alarm signaling is performed by a two-line dialer that alerts our central station to alarm and trouble conditions. The dialer uses two telephone lines so that a second line is available if one of the lines is out. Any two existing phone lines in your office can be used for the alarm system. Phone lines dedicated for alarm use are not required.

FIRE PROTECTION SYSTEM

My plan for the fire protection system includes the following equipment:

1. Ten-zone fire alarm panel
2. Eight smoke detectors
3. Water flow switch
4. Water cutoff switch
5. Four Klaxon horns

The ten-zone fire alarm panel will monitor one detection device per zone. Because each smoke detector, the water flow switch, and the water cutoff switch have a separate zone, the source of a fire alarm can be determined immediately.

To provide adequate local fire alarm signaling, this system is designed with four horns. Remote signaling for the fire alarm system is provided by the MainAlert control panel. The fire alarm would report alarm and trouble conditions to the MainAlert control panel. The MainAlert alarm control panel would, in turn, report fire alarm and fire trouble signals to our central station. The MainAlert alarm panel would not have to be set to transmit fire alarm and fire trouble signals to our central station.

Continued

ARM/DISARM MONITORING

Since 20 of your employees would have alarm codes, it is important to keep track of who enters and leaves the office outside of office hours. When an employee would arm or disarm the alarm system, the alarm would send a closing or opening signal to our central station. The central station would keep a record of the employee's identity and the time the signal was received. With the arm/disarm monitoring service, our central station would send you opening/closing reports on a semi-monthly basis.

INSTALLATION SCHEDULE

Given the size of your new office, our personnel could install your alarm in three days. We could start the day after we receive approval from you. The building is now complete enough for us to start anytime. If you would prefer for the construction to be completed before we start, that would not present any problems for us. To give you an idea of how the alarm system would be laid out, I have included an attachment to this proposal showing the locations of the alarm devices.

INSTALLATION AND MONITORING COSTS

Installation and monitoring costs for your burglary and fire alarm systems as I have described them in this proposal will be as follows:

- $8,200 for installation of all equipment
- $75 a month for monitoring of burglary, fire, and opening/closing signals under a two-year monitoring agreement

The $8,200 figure covers the installation of all the equipment I have mentioned in this proposal. The $75-a-month monitoring fee also includes opening/closing reports.

CONCLUSION

The MainAlert control panel, as the heart of your alarm system, is an excellent electronic security value. The MainAlert control panel is unsurpassed in its ability to report alarm status information to our central station. The perimeter and interior protection offers complete building coverage that will give you peace of mind.

The fire alarm system monitors both sprinkler flow and smoke conditions. The fire alarm system I have designed for you can provide sufficient warning to allow the fire department to save your building from catastrophic damage.

The arm/disarm reporting can help you keep track of employees who come and go outside of office hours. It's not always apparent how valuable this service can be until you need the information it can provide.

I'll call you early next week, Bob, in case you have any questions about this proposal. We will be able to start the installation as soon as you return a copy of this letter with your signature in the acceptance block.

Sincerely,

Anne Rodriguez Evans

Anne Rodriguez Evans
Commercial Sales

Enc.

ACCEPTED by McDuff, Inc.

By: _____

Title: _____

Date: _____

ALARM SYSTEM LAYOUT FOR
McDUFF INC. – ATLANTA, GA

LEGEND

BAC	Burglar alarm control	K	Klaxon horn
FAC	Fire alarm control	KP	Keypad
		M	Motion detector
		S	Siren

WC Water cutoff switch
WF Water flow switch
GB Glass-break detector
DC Door contact

Part 2: Longer Assignments

For each of these assignments except number 8, complete a copy of the Planning Form included at the end of the book.

   5. **Informal Proposal—McDuff.** Choose Option A or Option B. Remember that informal proposals should be fairly limited in scope, given their length and format.

Option A: In-House

- Use your past or present work experience to write a memo proposal suggesting a change at McDuff, Inc. Possible topic areas include changes in operating procedures, revisions to company policies, additions to the work force, alterations of the physical plant, or purchase of products or services.
- Place yourself in the role of an employee of McDuff. The proposal may be solicited or unsolicited, whatever best fits your situation.
- Make sure that your proposal topic is limited enough in scope to be covered fully in an informal proposal with memo format.
- Choose at least two levels of readers who could conceivably be decision-makers about a proposal such as the one you are writing—for example, branch or corporate managers. Review chapter 2 if necessary.

Option B: Sales

- Select a product or service (1) with which you are reasonably familiar (on the basis of your work experience, research, or other interests) and (2) that could conceivably be purchased by a company like McDuff.
- Put yourself in the role of someone representing the company that makes the product or provides the service.
- Write an informal sales proposal in which you propose purchase of the product or service by a representative of McDuff.

6. **Formal Proposal—McDuff.** Choose Option A, B, or C. Make sure that your topic is more complex than the one you would choose for the preceding informal-proposal assignments.

Option A: Community-Related

- Write a formal proposal in which you propose a change in (1) the services offered by a city or town (for example, mass transit or waste management) or (2) the structure or design of a building, garden, parking lot, shopping area, school, or other civic property.
- Select a topic that is reasonably complex and yet one about which you can locate information.
- Place yourself in the role of an outside consultant with a division of McDuff, Inc., who is proposing the change.
- Choose either an unsolicited or a solicited context.
- Write to an audience that could actually be the readers. Do enough research to identify at least two levels of audience.

Option B: School-Related

- Write a proposal in which you propose a change in some feature of a school you attend or have attended.
- Choose from topics such as operating procedures, personnel, curricula, activities, and physical plant.
- Select an audience that would actually make decisions on such a proposal.
- Give yourself the role of an outside consultant working for McDuff, Inc.

Option C: Work-Related

- Write a proposal in which you, as a representative of McDuff, propose purchase of a product or service by another firm.

- Choose a topic about which you have work experience, research knowledge, or keen interest—and one that could conceivably be offered in one of McDuff's project areas. Make sure you have good sources of information.
- Choose either a solicited or an unsolicited context.

7. **Feasibility Study—McDuff.**

- Choose any one of the proposal assignments that you completed as part of the preceding assignments. (Or for this assignment, you can use a proposal completed by one of your classmates.)
- Take yourself out of the role of proposal writer. Instead, consider yourself to be someone assigned (or hired) to complete the task of evaluating the practicality of the proposal, after it has been received.
- If appropriate, choose several alternatives to evaluate.

8. **Ethics Assignment.** Reread this chapter's "Communication Challenge" and its "Questions and Comments for Discussion." Put yourself in the position of Rob McDuff, company president. You've just had a phone call from J. R. Link, your old friend and colleague, who proceeds to relate his concerns about Ben Sadler. Put bluntly, J. R. thinks you should fire Sadler immediately before he does serious damage to the firm. You indicate you'll think about it, talk to Sadler, and then call J. R. to tell him your decision. If you were Rob McDuff, how would you handle the conversation with Sadler? (Remember why you hired him.) Would you fire him, praise him, or seek to moderate or redirect his efforts? Explain your position in detail in an essay.

9. **Informal or Formal Proposal—International Context.**

- Assume you are a consultant asked to propose a one-week training course to one of McDuff's offices outside the United States. (See the map on the color insert in chapter 2 for a list of these six locations.) Most or all seminar participants will be residents native to the country you choose—not U.S. citizens working overseas.
- Choose a seminar topic familiar to you—for example, from college courses, work experience, or hobbies—or one that you are willing to learn about quickly through some study.
- Research work habits, learning preferences, social customs, and other relevant topics concerning the country where the McDuff office you have chosen is located.
- Write McDuff, Inc., an informal or a formal proposal that reflects your understanding of the topic, your study of the country, and your grasp of the proposal-writing techniques presented in this chapter.

Optional Group Approach: If this assignment is done by groups within your class, assume that members of your team work for a company proposing training seminars at McDuff offices around the world. Each group member has responsibility for one of McDuff's non-U.S. offices.

Different sections of the proposal will be written by different group members, who may be proposing the same seminar for all offices or different seminars. Whatever the case, the document as a whole should be unified in structure, format, and tone. It will be read by (1) the vice president for international operations at the corporate office, (2) the vice president for research and training at the corporate office, and (3) all six branch managers in Venezuela, England, Saudi Arabia, Kenya, Germany, and Japan.

**Professional
Documentation, Inc.**

3450 Jones Mill Road, Neming, Georgia 30092
(404) 555-8438

January 15, 2002

Mr. David Barker
Technical Communication Manager
Real Big Professional Software
P.O. Box 123456
Atlanta, Georgia 30339

Dear David:

> *Shows understanding of client's main concern—scheduling.*

I enjoyed meeting with you and learning about your new General Ledger software product. Because you require a March release, I can understand why you want to choose an approach to documentation and get the project started.

> *Asserts ability to meet scheduling need.*

This proposal describes a strategy for completing the documentation in the 10 weeks between now and your March deadline. Included are these main sections:

> *Gives helpful overview of sections to follow.*

1. Selection of the Best Format
2. Adoption of a Publication Plan
3. Control of Costs
4. Conclusion

SELECTION OF THE BEST FORMAT

I think your customers will be best served by a combined installation and user's guide. It uses a functional approach to show how General Ledger works. My assessment results from these completed steps:

> *Uses list to itemize important points—that is, the basis for his assessment.*

- Interviews with support staffers responsible for providing technical support to customers using the company's other accounting products
- Interviews with programmers developing General Ledger, who have an intimate knowledge of how it works
- Conversations with you that clarified your organization's general expectations for the documentation

The assessment is also based on my experience developing documentation for other products. I strive to use clear, concise prose and ample white space to provide a visually appealing text. The text will be enhanced and supplemented with graphics depicting General Ledger's feature screens. The screens themselves will be captured directly

MODEL 10–1
Letter proposal

from the program and inserted into the text by your staff using your in-house publishing system.

This approach will yield a thorough and easy-to-use document that will allow your customers to take full advantage of General Ledger's many innovative features.

As you know, writing documentation is a cooperative effort. Each member of the General Ledger product team will play key roles during the development process. To keep us all on track, I have put together a publication plan that shows how the project will progress from beginning to end.

ADOPTION OF A PUBLICATION PLAN

The publication plan shows how we can have the documentation ready for General Ledger's March unveiling. The four major steps are described here.

Define the Project

Much of this work has already been accomplished as a result of doing the research for this proposal. As a preliminary step, we will meet and review the project's scope and priority within the organization. We will detail the resources that will be available to complete the project. Most important, we will look at expectations: management's, yours, and the customers'.

Develop a Schedule

This step is the key to the publication process and ensures a common understanding of what has to be done and in what period of time. It has three basic steps:

- We define the tasks that are part of the project.
- We define the resources we have available to deal with the identified task.
- We assign tasks to the most appropriate individuals.

Manage the Project

What is good project management? In this plan, good management is essentially good communications. In the first three steps, we define the information that project members must have to understand how the document will be produced and their roles in that process. Ongoing management of the project will be a matter of keeping the channels of communication open.

Perform a Postmortem

The last step is an evaluation of the effectiveness of the publication plan. It provides the opportunity for us to learn how to do future documentation better. It is important to look back at what went right and what went wrong during a project and to share this information with the others. You will get a complete postmortem report from me after the project is completed.

Continues emphasis on benefits to reader.

Leads in smoothly to next section.

Starts with *overview* of sections to follow.

Organizes paragraph around three *main points*.

Uses bulleted list for primary *steps* in project.

Introduces section with *question* to attract attention to passage.

MODEL 10–1
continued

continues

Throughout the project, this management system will guide us in completing General Ledger's documentation on time and within budget.

CONTROL OF COSTS

Good documentation helps to sell software. By working smart, we can develop documentation that will enhance General Ledger's appeal, and we can do it at a reasonable cost.

My experience in this area and the management system described here will reduce waste and duplication of effort, two factors that affect cost. This savings means I can bring the project in within the 200-hour cap you mentioned.

This estimate assumes that three of the program's four main features are in a complete, or "fixed," state and that the fourth main feature is about 50 percent complete. This estimate also assumes that all programming will be finished by March 5, which will allow time to put the guide through final review and production.

CONCLUSION

The functional approach, which describes a product in terms of its operations, is the documentation format that will best serve General Ledger customers. Your goal of having the documentation ready by March will be aided by adopting a four-step publication plan. The plan will define the strategy for writing the documentation and will help keep costs down.

I'll call you in a few days, David, to answer any questions you might have about this proposal. I can begin work on the documentation as soon as you sign the acceptance block and return a copy of this letter to me.

Sincerely,

Steven Nickels

Steven Nickels
Documentation Specialist

Enclosure

ACCEPTED by Real Big Professional Software

By: _____

Title: _____

Date: _____

Shows interest in *following through.*

Places benefit in heading.

Shows he can meet project criteria—but also *clarifies* **the** *assumptions* **he is making.**

Returns to *main concern* **of reader— scheduling.**

Retains *control* **of next step.**

Includes acceptance block to simplify approval process.

MODEL 10–1
continued

DATE: October 4, 2002
TO: Gary Lane
FROM: Jeff Bilstrom _JB_
SUBJECT: Creation of Logo for Montrose Service Center

Part of my job as director of public relations is to get the Montrose name firmly entrenched in the minds of metro Atlanta residents. Having recently reviewed the contacts we have with the public, I believe we are sending a confusing message about the many services we offer retired citizens in this area.

To remedy the problem, I propose we adopt a logo to serve as an umbrella for all services and agencies supported by the Montrose Service Center. This proposal gives details about the problem and the proposed solution, including costs.

The Problem

The lack of a logo presents a number of problems related to marketing the center's services and informing the public. Here are a few:

- The letterhead mentions the organization's name in small type, with none of the impact that an accompanying logo would have.
- The current brochure needs the flair that could be provided by a logo on the cover page, rather than just the page of text and headings that we now have.
- Our 14 vehicles are difficult to identify because there is only the lettered organization name on the sides without any readily identifiable graphic.
- The sign in front of our campus, a main piece of free advertising, could better spread the word about Montrose if it contained a catchy logo.
- Other signs around campus could display the logo, as a way of reinforcing our identity and labeling buildings.

It's clear that without a logo, the Montrose Service Center misses an excellent opportunity to educate the public about its services.

The Solution

I believe a professionally designed logo could give the Montrose Service Center a more distinct identity. Helping to tie together all branches of our operation, it would give the public an easy-to-recognize symbol. As a result, there would be a stronger awareness of the center on the part of potential users and financial contributors.

Gives concise view of problem—_and_ his proposed solution.

Includes effective lead-in.

Uses bulleted list to highlight main difficulties posed by current situation.

Ends section with good _transition_ to next section.

Starts with _main point_—need for logo.

MODEL 10–2
Memo proposal

continues

The new logo could be used immediately to do the following:

Focuses on *benefits* of proposed change.

- Design and print letterhead, envelopes, business cards, and a new brochure.
- Develop a decal for all company vehicles that would identify them as belonging to Montrose.
- Develop new signs for the entire campus, to include a new sign for the entrance to the campus, one sign at the entrance to the Blane Workshop, and one sign at the entrance to the Administration Building.

Cost

Emphasizes *benefit* of possible price break.

Developing a new logo can be quite expensive. However, I have been able to get the name of a well-respected graphic artist in Atlanta who is willing to donate his services in the creation of a new logo. All that we must do is give him some general guidelines to follow and then choose among eight to ten rough sketches. Once a decision is made, the artist will provide a camera-ready copy of the new logo.

Uses *listing* to clarify costs.

• Design charge	$0.00
• Charge for new letterhead, envelopes, business cards, and brochures (min. order)	545.65
• Decal for vehicles 14 @ $50.00 + 4%	728.00
• Signs for campus	415.28
Total Cost	$1,688.93

Conclusion

Closes with major benefit to reader and urge to action.

As the retirement population of Atlanta increases in the next few years, there will be a much greater need for the services of the Montrose Service Center. Because of that need, it's in our best interests to keep this growing market informed about the organization.

Keeps control of next step.

I'll stop by later this week to discuss any questions you might have about this proposal.

MODEL 10–2
continued

PROPOSAL FOR SUPPLYING TEAK CAM CLEAT SPACERS

Prepared by
Totally Teak, Inc.

Prepared for
John L. Riggini
Bosun's Locker Marine Supply

August 22, 2002

MODEL 10–3
Formal proposal

continues

Totally Teak, Inc.
6543 Amster Avenue, N.W.
Atlanta, Georgia 30308
(404) 555-9425

In this example, the letter of transmittal appears immediately following the title page. It can also appear before the title page (see Model 10-4).

Establishes *link* with previous client contact.

Stresses two main benefits.

August 22, 2002

John L. Riggini, President
Bosun's Locker Marine Supply
38 Oakdale Parkway
Norcross, OH 43293

Dear Mr. Riggini:

I enjoyed talking with you last week about inventory needs at the 10 Bosun's stores. In response to your interest in our products, I'm submitting this proposal to supply your store with our Teak Cam Cleat Spacers.

This proposal outlines the benefits of adding Teak Cam Cleat Spacers to your line of sailing accessories. The potential for high sales volume stems from the fact that the product satisfies two main criteria for any boat owner:

1. It enhances the appearance of the boat.
2. It makes the boat easier to handle.

Your store managers will share my enthusiasm for this product when they see the response of their customers.

Says he will *call* (rather than asking client to call).

I'll give you a call next week to answer any questions you have about this proposal.

Sincerely,

William G. Rugg

William G. Rugg
President
Totally Teak, Inc.

WR/rr

MODEL 10–3
continued

378

TABLE OF CONTENTS

LIST OF ILLUSTRATIONS

Organizes entire proposal around *benefits*.

MODEL 10–3
continued

continues

Briefly mentions
main *need* to
which proposal
responds.

Reinforces main
points mentioned
in *letter (selective
repetition* of
crucial information
is acceptable).

EXECUTIVE SUMMARY

This proposal outlines features of a custom-made accessory designed for today's sailors—whether they be racers, cruisers, or single-handed skippers. The product, Teak Cam Cleat Spacers, has been developed for use primarily on the Catalina 22, a boat owned by many customers of the 10 Bosun's stores. However, it can also be used on other sailboats in the same class.

The predictable success of Teak Cam Cleat Spacers is based on two important questions asked by today's sailboat owners:

- Will the accessory enhance the boat's appearance?
- Will it make the boat easier to handle and, therefore, more enjoyable to sail?

This proposal answers both questions with a resounding affirmative by describing the benefits of teak spacers to thousands of people in your territory who own boats for which the product is designed. This potential market, along with the product's high profit margin, will make Teak Cam Cleat Spacers a good addition to your line of sailing accessories.

MODEL 10–3
continued

INTRODUCTION

The purpose of this proposal is to show that Teak Cam Cleat Spacers will be a practical addition to the product line at the Bosun's Locker Marine Supply stores. This introduction highlights the need for the product, as well as the scope and format of the proposal.

Background

Sailing has gained much popularity in recent years. The high number of inland impoundment lakes, as well as the vitality of boating on the Great Lakes, has spread the popularity of the sport. With this increased interest, more and more sailors have become customers for a variety of boating accessories.

What kinds of accessories will these sailors be looking for? Accessories that (1) enhance the appearance of their sailboats and (2) make their sailboats easier to handle and, consequently, more enjoyable to sail. With these customer criteria in mind, it is easy to understand the running joke among boat owners (and a profitable joke among marine supply dealers): "A boat is just a hole in the water that you pour your money into."

The development of this particular product originated from our designers' firsthand sailing experiences on the Catalina 22 and knowledge obtained during manufacture (and testing) of the first prototype. In addition, we conducted a survey of owners of boats in this general class. The results showed that winch and cam cleat designs are major concerns.

Proposal Scope and Format

The proposal focuses on the main advantages that Teak Cam Cleat Spacers will provide your customers. These six sections follow:

1. Practicality
2. Suitability for a Variety of Sailors
3. High-Quality Construction and Appearance
4. Dealer Benefits
5. Sizable Potential Market
6. Affordable Price

Side annotations (left margin):

Makes clear the proposal's purpose.

Gives lead-in about section to follow.

Establishes *need* for product.

Shows his understanding of *need* (personal experience of designing owner survey).

Gives list of sections to follow, to reinforce organization of proposal.

MODEL 10–3
continued

continues

Uses main
heading that
*engages reader's
interest.*

Phrases each side
heading in
"benefit-centered"
language.

FEATURES AND BENEFITS

Teak Cam Cleat Spacers offer Bosun's Locker Marine Supply the best of both worlds. On the one hand, the product solves a nagging problem for sailors. On the other, it offers your store managers a good opportunity for profitability. Described here are six main benefits for you to consider.

Practicality

This product is both functional and practical. When installed in the typical arrangement shown in the figure below, the Teak Cam Cleat Spacer raises the height of the cam cleat, thereby reducing the angle between the deck and the sheet as it feeds downward from the winch. As a result of this increased height, a crewmember is able to cleat a sheet with one hand instead of two.

SIDE VIEWS OF CLEATING ARRANGEMENT

3

MODEL 10–3
continued

Such an arrangement allows a skipper to maintain steerage of the boat, keeping one hand on the helm while cleating the sheet with the other. Securing a sheet in this manner can be done more quickly and securely. Also, this installation reduces the likelihood of a sheet "popping out" of the cam cleat during a sudden gust of wind.

Suitability for a Variety of Sailors

For the racer, cruiser, and single-handed sailor, sailing enjoyment is increased as sheets and lines become easier to handle and more secure. In a tight racing situation, these benefits can be a deciding factor. The sudden loss of sail tension at the wrong moment as a result of a sheet popping out of the cam cleat could make the difference in a close race.

A cruising sailor is primarily concerned with relaxation and pleasure. A skipper in this situation wants to reduce his or her workload as much as possible. In the instance of a sheet popping loose, the sudden chaos of a sail flapping wildly interrupts an otherwise tranquil atmosphere. Teak Cam Cleat Spacers reduce the chance of this happening.

A cruising sailor often has guests aboard. In this situation, as well as in a race, the skipper wants to maintain a high level of seamanship, especially where the control of the boat and the trim of its sails are concerned.

The single-handed sailor derives the greatest benefit from installing Teak Cam Cleat Spacers. Without crew nearby to assist with handling lines or sheets, anything that makes work easier for this skipper is welcome.

High-Quality Construction and Appearance

The teakwood frame from which this product is manufactured is well suited for use around water, since teak will not rot. It also looks nice when oiled or varnished.

The deck of most sailboats is made primarily of fiberglass. The appearance of such a boat can be significantly enhanced by the addition of some teak brightwork.

Each spacer is individually handcrafted by Totally Teak, Inc., to guarantee a consistent level of high quality.

Dealer Benefits

Teak Cam Cleat Spacers make a valuable addition to the dealer's product line. They complement existing sailing accessories as well as provide the customer with the convenience of a readily available prefabricated product.

A customer who comes in to buy a cam cleat is a ready prospect for the companion spacer. Such a customer will likely want to buy mounting hardware as well.

With this unique teak product readily available, a dealer can save the customer the time and trouble of fabricating makeshift spacers.

Explains *exactly* how product will work.

Moves logically from *racing* to *cruising* to *single-handed sailing*— all to show usefulness of product.

Stresses *quality*.

Appeals to self-interest of *individual* Bosun's dealers.

MODEL 10–3
continued

continues

Includes number
of owners to
emphasize
potential sales.

Keeps price
information *short*
and *clear.*

Sizable Potential Market

These Teak Cam Cleat Spacers are designed with a large and growing potential market in mind. They are custom-made for the Catalina 22, one of the most popular sailboats in use today. Over 13,000 of these sailboats have been manufactured to date. These spacers are also well suited for other similar-class sailboats.

Affordable Price

The Teak Cam Cleat Spacers made by Totally Teak, Inc., wholesale for $3.95/pair. Suggested retail is $6.95/pair. This low price is easy on the skipper's wallet and should help this product move well. And, of course, the obviously high profit margin should provide an incentive to your store managers.

MODEL 10–3
continued

CONCLUSION

Why should a marine supply dealer consider carrying Teak Cam Cleat Spacers? This product satisfies two common criteria of sailboat owners today: it enhances the appearance of any sailboat, and it makes the boat easier to handle. The potential success of this product is based on its ability to meet these criteria and the following features and benefits:

1. It is practical, allowing quick, one-handed cleating.

2. It is ideally suited for a variety of sailors, whether they are racing, cruising, or sailing single-handedly.

3. It is a high-quality, handcrafted product that enhances the appearance of any sailboat.

4. It is a product that benefits the dealer by making a valuable addition to her or his product. It complements existing sail accessories and satisfies a customer need.

5. It is geared toward a sizable potential market. Today there are thousands of sailboats in the class for which this accessory is designed.

6. It is affordably priced and provides a good profit margin.

Links list of *benefits* with order of same in discussion— drives home advantages of product to *user* and *dealer.*

MODEL 10–3
continued

Refers briefly to
problem.

Emphasizes
Hydrotech's main
selling point: an
innovative
cleaning system.

Stays in control of
proposal process
by saying he
will call.

Hydrotech Diving and Salvage, Inc.
Industrial Complex
Belle Chase, Louisiana 70433

February 24, 2002

Ms. Susan Bard Jackson
Standard Shipping International
Fleet Drive
New York, New York 10019

Dear Ms. Jackson:

Your February 1, 2002, RFP details the need for improved hull maintenance on
your galaxy-class oil tankers. Traditional approaches to hull cleaning, as you know,
have become costly and inefficient. This proposal offers Hydrotech's innovative and
affordable solution to Standard Shipping's maintenance needs.

Recent technology has spurred development of mobile hull maintenance systems
that can be used on anchored vessels almost anywhere. This equipment, used
regularly, will greatly reduce the accumulation of marine growth. As a result, your
ships will cruise faster and more efficiently.

I'll give you a call next week to answer any questions you may have about how
Hydrotech can improve hull maintenance at Standard Shipping.

Sincerely,

Stephen B. Wilson

Stephen B. Wilson
Underwater Maintenance Supervisor

sf

MODEL 10–4
Formal proposal

Title mentions benefit— increased efficiency.

INCREASED EFFICIENCY FOR STANDARD SHIPPING INTERNATIONAL

Prepared by

Hydrotech Diving and Salvage, Inc.

for

Susan Bard Jackson

Fleet Maintenance Manager

Standard Shipping International

February 24, 2002

MODEL 10–4
continued

continues

Uses white space
and indentation to
make contents
page highly
readable.

TABLE OF CONTENTS

MODEL 10–4
continued

LIST OF ILLUSTRATIONS

TABLES

FIGURES

MODEL 10–4
continued

continues

EXECUTIVE SUMMARY

Provides overview
of problem . . .

Marine growth on hulls can reduce the speed of ships, increase fuel consumption, and cause more frequent hull repairs. The galaxy-class oil tanker has a particularly large hull surface area below the waterline. If marine growth is allowed to accumulate, excessive drag will occur between the hull and water.

. . . and solution.

The removal of marine growth usually is expensive and time-consuming. Hydrotech, however, offers a hull maintenance system that provides a more convenient, economical way to clean and preserve oil tanker hulls. The Seasled Hull Maintenance System uses a diver-operated, self-propelled, scrubbing and painting device that can be brought directly to your anchored vessels. The primary benefits to your company include the following:

Focuses on main
benefits to client.

1. Bottom cleaning and preservation completed in about 100 hours, 30 hours less than most conventional cleaning systems, and

Uses style that
appeals to busy
management
readers—short
paragraphs,
numbered points,
no technical
jargon.

2. Hull maintenance done at times and locations convenient for you.

Because the Seasled System reduces the cost of each cleaning, you can now afford annual hull maintenance and semiannual inspections. This regular attention will contribute significantly to the average cruising speed and fuel efficiency of your ships.

Summarizes
entire proposal.

MODEL 10–4
continued

INTRODUCTION

Recent technology has made many contributions to the petroleum shipping industry. A new method of hull cleaning and repair has been developed that offers many advantages to oil tanker operators. We believe this new method will contribute significantly to the success and prosperity of Standard Shipping International. Maintaining clean hulls will increase ship cruising speed and fuel efficiency.

Purpose

This proposal will describe the benefits you receive using the new Seasled Hull Maintenance System. Your company will save time and money by using this service regularly for all galaxy-class tankers.

Description of Hull Maintenance Problem

Standard Shipping International operates a fleet of six galaxy-class oil tankers engaged in worldwide transporting of petroleum. Your vessels travel to oceans and ports around the world, where many forms of marine growth collect on ships' hulls. Because the galaxy-class tanker has a large hull surface area below the waterline, even light marine growth will cause excessive drag between the hull and water.

Repeats background information for readers who start with introduction.

Shows understanding of problem mentioned in RFP but doesn't belabor it, since client already knows problem exists.

MODEL 10–4
continued

continues

Furthermore, if marine growth is allowed to accumulate for an extended period, the hull will deteriorate. Marine growth results in slower ship speed, increased fuel consumption, and more frequent major hull repair.

Scope

Shows that Hydrotech has done its homework in preparing proposal.

This proposal reflects our thorough research and over ten years' experience on hull maintenance and cleaning for tankers. Besides relying on our own experience, we have consulted experts at firms such as the following:

1. Yamamoto Shipbuilding Service Consultants
    Kure, Japan
2. Marine Corrosion Consultants
    Belle Chase, Louisiana
3. Ocean Science Center
    Key West, Florida

The result has been the development of the Seasled System. This proposal explains how using Seasled reduces cleaning time, while maintaining safety-conscious crew procedures. We will also cover features of scheduling as well as the backgrounds of key people who will work on the program. Then you will find a cost estimate, per ship, for both maintenance and inspection procedures.

MODEL 10–4
continued

Proposal Format

For ease of reference, this proposal is divided into these three major sections:

Refers to remaining section so readers don't have to consult contents.

Gives brief overview of subsections to follow.

1. Seasled: How It Will Work for Standard Shipping—which describes the equipment, crew, and procedures

2. Schedule and Qualifications—which provides information on schedules, services, and personnel

3. Reduced Costs Using Seasled—which provides information on the cost of our hull-cleaning service

MODEL 10–4
continued

continues

SEASLED: HOW IT WILL WORK

FOR STANDARD SHIPPING

This section describes the equipment and procedure used for Seasled hull maintenance of the galaxy-class tanker. The main services proposed are semiannual inspections and annual cleanings (which include spot painting of bare or badly worn sections).

Time-Saving Equipment

Recent advances in underwater technology have yielded specialized equipment that has revolutionized the ship-maintenance industry. The Seasled Hull Maintenance System is a self-propelled, diver-operated device capable of cleaning 400 square feet of hull surface per hour. (See Figures 1, 2, and 3 for top, bottom, and side views of this device.)

Shows that Hydrotech has done its homework in preparing proposal.

MODEL 10–4
continued

Figure 1   Seasled: Top view

Shield

Steering
Wheel

Umbilical
Cable

Comments and
figures show
that Seasled
incorporates
latest technology
and that
Hydrotech
helped develop it.

    Hydrotech Diving and Salvage, Inc., helped develop the Seasled System and has used it successfully for the past two years. The system is quite reliable, as well as extremely efficient in removing marine growth.

MODEL 10–4
continued

continues

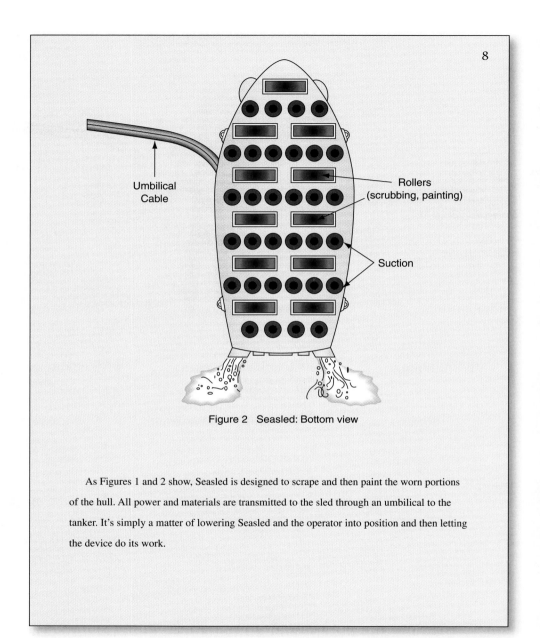

Umbilical Cable

Rollers (scrubbing, painting)

Suction

Figure 2   Seasled: Bottom view

As Figures 1 and 2 show, Seasled is designed to scrape and then paint the worn portions of the hull. All power and materials are transmitted to the sled through an umbilical to the tanker. It's simply a matter of lowering Seasled and the operator into position and then letting the device do its work.

MODEL 10–4
continued

Figure 3   Seasled: Side view

Labels in figure: Umbilical Cable, Steering Wheel, Shield, HYDROTECH Seasled, Suction, Bladder, Rollers (scrubbing, painting)

Mobility That Makes Sense

Mobility is the real advantage of Seasled. Any scheduled layover of your tankers can be transformed from lost shipping time to convenient maintenance time.

In the past, tankship operators were often required to take vessels hundreds of miles to deep-water ports where expensive pier space was needed to perform regular hull maintenance. Now our mobile hull-cleaning system can be transported to anchored vessels at locations that are more convenient for you.

You'll also be glad to know that the equipment is easily transferred from small craft to tanker deck, using the cargo lifting davits you already have on board.

Explicitly states difference between results from old and new systems.

MODEL 10–4
continued

continues

Furthermore, the hull cleaning and paint repair can be conducted while the ship is at anchor in calm water.

Diving Crew and Procedure

Describes procedures with short paragraphs and lists.

Our diving crews consist of highly trained professionals who ensure high standards of safety and productivity. A typical diving crew for hull maintenance includes the following members:

1. Two experienced diving supervisors
2. Six certified commercial divers
3. Four vocationally trained tenders
4. Two equipment technicians

These workers are on call 24 hours a day and will supervise the transport of the equipment to the desired location. Once aboard your ship, the system is assembled and operable in less than 12 hours.

Our personnel work 12-hour shifts around the clock with a diving supervisor on duty at all times. The hull-cleaning and paint-repair operation is conducted as follows:

1. Two divers, operating the sleds, start on the port and starboard sides of the bow and work toward the stern.

MODEL 10–4
continued

2. Each diver is monitored and assisted from topside by a tender.

3. Two standby divers are on deck and ready to assist the working diver in case of an emergency.

4. An equipment technician monitors and maintains the system.

During the entire operation, a surface-supplied life-support system and standard diving equipment are used by the diving crew.

Diver Rotation

The working divers are relieved every four hours, and the rotation proceeds as follows:

Emphasizes use of safe techniques.

1. The off-going divers are off duty for 12 hours following their dive.

2. The on-coming divers are on call eight hours before their dive.

3. The on-coming divers act as standby divers for the four hours immediately preceding their dive.

Completion Time

Table 1 lists individual tasks in the left column and the maximum completion time for each task in the right column. As you can see, the maximum completion time for an average galaxy-class tanker is 101 hours—far less than other methods.

MODEL 10–4
continued

continues

TABLE 1

JOB COMPLETION TIME:

CLEANING THE GALAXY HULL WITH SEASLED

Task	Completion Time
System Assembly	12 Hours
Hull Cleaning	75 Hours
Paint Repair	8 Hours
System Disassembly	6 Hours
Total Maximum Completion Time	101 Hours

The completion time was calculated for an empty galaxy-class tanker with a draft of 30 feet and 60,000 square feet of hull surface below the waterline.

Uses table for easy access to job-time data.

MODEL 10–4
continued

SCHEDULE AND QUALIFICATIONS

This section provides a recommended maintenance schedule and a description of services and personnel available to Standard Shipping International.

Schedules That Save Money

As you know, each ship should have annual cleanings and semiannual inspections to ensure that heavy growth does not develop. The thicker the encrustation, the harder it is to remove. Thus a regular maintenance schedule is needed to keep your ships operating at peak performance and to help us complete each job on time.

We realize that your ships run on irregular schedules due to the constantly changing requirements of your customers. Thus here are the two main features of our scheduling that respond to your needs:

1. Hydrotech will keep a record of each ship's annual cleaning and semiannual inspections and contact you to arrange a periodic rendezvous for hull cleaning.

2. If you need unscheduled work, Hydrotech *guarantees* that it can mobilize a crew to leave Hydrotech's headquarters within 72 hours of your call to us. This quick response time provides the flexibility you need to keep your fleet seaworthy and profitable.

Subheadings stress benefits to clients.

Numbered points emphasize how Hydrotech will meet important scheduling requirements.

MODEL 10–4
continued

continues

Introduces related services so client knows Hydrotech is a full-service firm despite its small size.

Variety of Services Available to You

Hydrotech offers many other maintenance and emergency services. Those that are frequently required include the following:

1. Removal of obstructions from propellers and shafts
2. Sea suction and discharge clearing
3. Recovery of lost equipment
4. Emergency damage-control repairs
5. Assistance to vessels that have run aground
6. Ship salvage operations

As you know, problems such as these are all too common in the shipping industry. Response time often determines whether schedules will be met, whether equipment can be recovered, or even if lives are saved. At Hydrotech, we respond to emergency calls with speed and dedication. You will receive the highest priority in an emergency.

Assures that Standard Shipping wil receive priority service.

Professionals That Make the Difference

Since it was started, Hydrotech has recruited professionals with experience in all phases of diving, underwater construction, and ship repair. Our personnel have the education and experience necessary to ensure safe, efficient, and professional performance for our customers. Some of our key personnel available to Standard Shipping include the following individuals:

List stresses experience of firm's personnel.

- William Baily, Jr., President and Chief Executive Officer, Hydrotech—formerly, Commander, Naval Undersea Systems Center, U.S.N. (Retired)
- Joseph Smith, Senior Vice-President, Hydrotech—formerly, Commander, Naval Ship Repair Facility, U.S.N. (Retired)

MODEL 10–4
continued

- John Delong, Diving Personnel Superintendent, Hydrotech—formerly, Master Diver, Naval Diving and Salvage Facility, U.S.N. (Retired)
- Samuel Johnson, Operations Officer, Hydrotech—formerly, Commander, Naval Undersea Construction Center, U.S.N. (Retired)

All crew members are carefully selected by our personnel department, under the supervision of Mr. Delong. The following list gives requirements for specific positions:

Lists criteria to further support claims of company's professionalism.

1. Diving Supervisors—selected from our senior divers or former U.S.N. Master Divers
2. Divers—selected from our senior tenders or former U.S.N. First Class Divers
3. Tenders—graduates of certified commercial diving schools; must complete our in-house training program
4. Equipment Technicians—trade-school graduates with at least two years' experience in our equipment repair shops

Our selection and promotion policy is time-consuming. But this policy pays off for you in terms of high productivity and safety records, which are among the best in the diving industry. We will accept nothing less.

MODEL 10–4
continued

continues

REDUCED COSTS USING SEASLED

As noted earlier, the Seasled System introduces increased efficiency, and thus reduced costs, to hull-cleaning projects. The charge for hull cleaning and paint repair for your vessels will vary by location, due to transportation costs. For your convenience, we have assigned flat rates for five geographic areas:

- Region 1-Southeastern U.S. (includes Georgia, Florida, Louisiana, and Texas)
- Region 2-the northern quadrant of the Western hemisphere (except that area designated Southeastern U.S.)
- Region 3-the southern quadrant of the Western hemisphere
- Region 4-the northern quadrant of the Eastern hemisphere
- Region 5-the southern quadrant of the Eastern hemisphere

Table 2 lists the charges for hull maintenance and semiannual inspection *per ship*. These costs are guaranteed through 2004 and include all transportation charges. As mentioned in your RFP, Standard Shipping will provide room and board for our crews on site.

Introduces innovative "flat rate" fee structure to simplify cost estimating for client.

Displays maintenance and inspection costs in table.

Guarantees costs through specific date to make budget more predictable.

MODEL 10–4
continued

TABLE 2

HULL MAINTENANCE AND INSPECTION CHARGES PER SHIP

Service	Regions				
	Region #1	Region #2	Region #3	Region #4	Region #5
Maintenance	$25,000	$30,000	$35,000	$40,000	$40,000
Inspection	$ 5,000	$ 7,000	$ 9,000	$11,000	$11,000

Our portable hull-maintenance system rates apply only to empty galaxy-class tankers with a draft of about 30 feet. Special rates can be quoted for loaded tankers on an individual basis.

MODEL 10–4
continued

continues

CONCLUSION

Returns to client's major concerns— regular schedules and cost savings.

Hull cleaning on oil tankers used to be expensive and time-consuming. Ships had to be taken hundreds of miles to deep-water ports, where expensive pier mooring was needed to perform underwater hull maintenance.

A new hull-maintenance system, which can be brought to your anchored vessels, is now available. The Seasled Hull Maintenance System uses a diver-operated, self-propelled, scrubbing and painting device. This portable system can clean the bottom of an empty galaxy-class oil tanker and paint damaged areas in about four days.

Our diving crew can rendezvous with your ships at times and locations that fit your schedule, and turn lost time into money-saving maintenance time. To keep your ships operating at maximum efficiency, we recommend annual hull cleaning and semiannual inspections of all surfaces beneath the waterline.

We look forward to tailoring a hull-maintenance schedule to fulfill your individual requirements.

MODEL 10–4
continued

McDuff, Inc.

MEMORANDUM

DATE: July 22, 2002
TO: Greg Bass
FROM: Mike Tran *MT*
SUBJECT: Replacement of In-House File Server

INTRODUCTORY SUMMARY

Gives context for feasibility study.

The purpose of this feasibility study is to determine if the NTR PC905 would make a practical replacement for our in-house file server. As we agreed in our weekly staff meeting, our current file-serving computer is damaged beyond repair and must be replaced by the end of the week. This study shows that the NTR PC905 is a suitable replacement that we can purchase within our budget and install by Friday afternoon.

FEASIBILITY CRITERIA

There are three major criteria that I addressed. First, the computer we buy must be able to perform the tasks of a file-serving computer on our in-house network. Second, it must be priced within our $4,000 budget for the project. Third, it must be delivered and installed by Friday afternoon.

Performance

As a file server, the computer we buy must be able to satisfy these criteria:

- Store all program used by network computers
- Store the source code and customer-specific files for Xtracheck
- Provide fast transfer of files between computers while serving as host to the network
- Serve as the printing station for the network laser printer

The NTR PC905 comes with a 120MB hard drive. This capacity will provide an adequate amount of storage for all programs that will reside on the file server. Our requirements are for 30MB of storage for programs used by network computers and 35MB of storage for source code and customer-specific programs. The 120MB drive will leave us with 55MB of storage for future growth and work space.

The PC905 can transfer files and execute programs across our network. It can perform these tasks at speeds up to five times faster than our current file server. Productivity should increase because the time spent waiting for transfer will decrease.

Summarizes conclusion of report.

Pinpoints three criteria to be discussed.

Shows how NTR PC905 will fulfill performance criteria.

Only covers advantages because there are *no* disadvantages to buying PC905

MODEL 10–5
Feasibility study (one alternative)

continues

The computer we choose as the file server must also serve as the printing station for our network laser printer. The PC905 is compatible with our Hewy Packer laser printer. It also has 2.0MB more memory than our current server. As a result, it can store larger documents in memory and print them with greater speed.

Budget

The budget for the new file server is $4,000. The cost of the PC905 is as follows:

PC905 with 120MB Hard Drive	$2,910
Keyboard	112
Monitor	159
Total	$3,181

No new network boards need to be purchased because we can use those that are in the current server. We also have all additional hardware and cables that will be required for installation. Thus the PC905 can be purchased for $800 under budget.

Time Frame

Our sales representative at NTR guarantees that we can have delivery of the system by Friday morning. Given this assurance, we can have the system in operation by Friday afternoon.

Additional Benefits

We are currently using NTR PCs at our customer sites. I am very familiar with the setup and installation of these machines. By purchasing a brand of computer currently in use, we will not have to worry about additional time spent learning new installation and operation procedures. In addition, we know that all our software is fully compatible with NTR products.

The warranty on the PC905 is for one year. After the warranty period, the equipment is covered by the service plan that we have for all our other computers and printers.

CONCLUSION

I recommend that we purchase the NTR PC905 as the replacement computer for our file server. It meets or exceeds all criteria for performance, price, and installation.

MODEL 10–5
continued

Makes costs easy to find with simple table.

Highlights major goal—quick installation.

Ends with "extras"—that is, benefits not among major criteria but still useful.

Restates significant point already noted in introductory summary.

Part Three

Chapter 11

Graphics

Critiquing an Annual Report

General Instructions

Each chapter begins with a Collaboration at Work exercise that applies strategies for working in groups to chapter topics. The exercise assumes you (1) have been divided into groups of about three to six students, (2) will use group time inside or outside of class to complete the case, and (3) will produce an oral or written response. For guidelines about writing in groups, refer to pages 28–34.

Background for Assignment

While planning and writing, you make two main decisions about the use of graphics—first, when they should be used; and second, what types to select. This chapter will help you make such decisions. Yet you already possess the quality that is most useful in your study of graphics: common sense. Whether consciously or subconsciously, most of us tend to seek answers to basic questions like the following when we read a document:

1. Is there an appropriate mix of text and graphics?
2. Are the graphics really useful or are they just visual "fluff"?
3. Can information in the graphic be understood right away?
4. Was the correct type of graphic selected for the context?
5. Do any of the graphics include errors, such as in proportion?

Your answers to these questions often determine whether you continue reading a document—or at least whether you enjoy the experience.

Group Assignment

Choose one document that includes a variety of graphics—newspaper, magazine, report, textbook, catalog, web site, etc. Using the questions previously listed, work with your group to evaluate the use of graphics in all or part of the document. Whether you think a graphic is successful or not, give specific reasons to support your analysis.

Technology has radically changed the world of graphics. Now, almost anyone with a computer and the right software can quickly produce illustrations that used to take hours to construct. As a result, today there are sophisticated graphics in every medium—newspapers, magazines, television, and, of course, technical communication.

Because readers *expect* graphics to accompany text, you as a technical professional must respond to this need. Well-designed and well-placed graphics will keep your documents competitive. Your graphics do not have to be fancy, however. Nor is it true that adding graphics will necessarily improve a document. Readers are impressed by visuals *only* when they are well done and appropriate.

Because so many different types of graphics are available, technical professionals must first understand the basics of graphics before they apply sophisticated techniques. To emphasize these basics, this chapter (1) defines some common graphics terms, (2) explains the main reasons to use graphics, (3) covers the use of special fonts, (4) suggests guidelines for using color, (5) gives some general suggestions for incorporating graphics into text, (6) lists specific guidelines for eight common graphics, and (7) shows you how to avoid graphics misuse.

TERMS IN GRAPHICS

Terminology for graphics is not uniform in the professions. That fact can lead to some confusion. For the purposes of this chapter, however, some common definitions are adopted and listed here:

- **Graphics:** This generic term refers to any nontextual portion of documents or oral presentations. It can be used in two ways: (1) to designate the field (for example, "Graphics is an area in which he showed great interest") or (2) to name individual graphical items ("She placed three graphics in her report").
- **Illustrations, visual aids:** Used synonymously with "graphics," these terms also can refer to all nontextual parts of a document. The term *visual aids,* however, often is limited to the context of oral presentations.
- **Tables and figures:** These terms name the two subsets of graphics.
  Tables refers to illustrations that place numbers or words in columns or rows or both.
  Figures refers to all graphics other than tables. Examples include charts (pie, bar, line, flow, and organization), engineering drawings, maps, and photographs.
- **Charts, graphs:** A subset of "figures," these synonymous terms refer to a type of graphic that displays data in visual form—as with bars, pie shapes, or lines on graphs. Chart is the term used most often in this text.

- **Technical drawing:** Another subset of "figures," a technical drawing is a representation of a physical object. Such illustrations can be drawn from many perspectives and can include "exploded" views.

Of course, you may see other graphics terms. For example, some technical companies use the word *plates* for figures. Be sure to know the terms your readers understand and the types of graphics they use.

REASONS FOR USING SPECIAL FONTS, COLOR, AND GRAPHICS

Although the technology for producing graphics is constantly changing, the rationale for using them remains the same. Before exploring specific types of illustrations, this section covers some reasons why readers might choose special fonts, color, and graphics to accompany text.

● *Reason 1: Special Fonts, Color, and Graphics Simplify Ideas*

Your readers usually know less about the subject than you. Font styles, color, and graphics can help them cut through technical details and grasp basic ideas. For example, if a "Danger" symbol styled with the boldface "Impact" font style is given a bright orange background, the message is much more powerful than a "buried" warning message that looks just like the text around it. Also, a simple illustration of a laboratory instrument, such as a Bunsen burner, can make the description of a lab procedure much easier to understand. In a more complex example, Figure 11–1 uses a group of four different charts to convey the one main point—that McDuff's new Equipment Development group lags behind the company's other profit centers. A quick look at the charts tells the story of the group's difficulties much better than would several hundred words of text. Of course, the text itself must still focus the reader's attention on the relevant information.

● *Reason 2: Graphics Reinforce Ideas*

When a point needs emphasis, you can use a graphic element like a special font style or color, or you can create a graphic. For example, if you need to get instant reader recognition of section or chapter breaks, you might use a distinctive font style or color to capture the reader's attention. Also, you might draw a map to show where computer terminals will be located within a building or use a pie chart to show how a budget will be spent. You might even include a drawing that indicates how to operate a VCR. Each of these solutions would reinforce points made in the accompanying text.

● *Reason 3: Graphics Create Interest*

Tools such as font styles, color, and graphics are "grabbers." They can be used to engage readers' interest. If your customers have three reports on their desks and must quickly decide which one to read first, they probably will pick up the one with a distinctive look—one that complements the text. It may be something as simple as (1) a map outline of the state, county, or city where you will be doing a project, (2) a

2001 McDuff Sales (for 7 project areas)

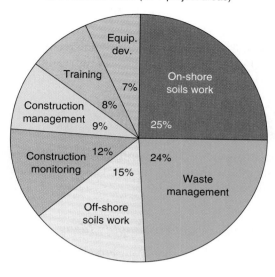

Equipment Development Group
1999-2001 Profits

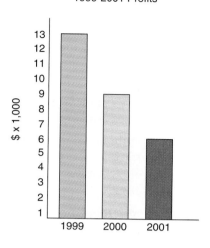

Equipment Development Group
$ Value of Contracts

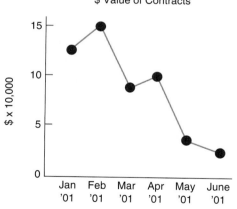

Equipment Development Group
Billable vs. Non-billable Time
(average employee/based on 40-hr week)

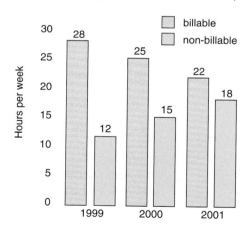

FIGURE 11–1   Graphics used to simplify ideas

picture of the product or service you are providing, or (3) a symbol of the purpose of your writing project.

Figure 11–2 shows how an outside consultant used a well-known Leonardo da Vinci drawing to attract attention to his McDuff proposal. The drawing helps to (1) add a classical touch to the cover, (2) focus on the human side of employee

Improving Productivity at McDuff, Inc.
An Innovative Approach to Employee Testing

Prepared for Robert S. McDuff
President, McDuff, Inc.
by
James H. Stephens
Infinite Vision, Inc.

FEBRUARY 22, 2002

FIGURE 11–2 Graphics used to create interest

testing, and (3) associate the innovation of Infinite Vision, Inc., with the creativity of da Vinci. Also, by using the font style and color preference of the client, the writer allows the reader to "take ownership" of the proposal before ever reading the contents.

● *Reason 4: Graphics Are Universal*

Some people wrongly associate the growing importance of graphics with today's reliance on television and other popular media—as if graphics pander to less-intellectual instincts. While media such as television and the Web obviously rely on visual effects, graphics have been mankind's universal language since cave drawings. Although those who sell or advertise products and services have long known the power of images, writers of technical documents have only recently started to merge the force of graphics with their text.

USING FONTS

Chapter 4 gave you an overview of two basic points about fonts to consider when designing documents: *font types and sizes of type*. The following section provides more details to help you meet the creative challenge of designing documents such as manuals, proposals, annual reports, and promotional materials.

Font Types

You already may have identified and grouped fonts into two categories—*sans serif* ("without feet or tails") and *serif* ("with feet or tails")—as you have explored the selection of fonts available to you on your desktop publishing software. Perhaps you have also noticed that many other font types exist, like those shown below:

- *Freehand* and *Freestyle Script*—appear to be handwritten
- *Vivaldi* and *Wedding Text*—seem to have an austere, more formal appearance
- *Parisian*—reflects a high fashion silhouette
- **Kids**—looks uneven, even awkward, as though it was printed by a child.

Your choice of fonts may determine whether your documents do or do not get read.

General Guidelines

How do you know which fonts to choose? Although there are no hard-and-fast rules to follow, keep these five guidelines in mind to make the task easier:

Font Style Guidelines

Consider:

1. the reader's or company's preferences
2. the need for clarity
3. the space available
4. the purpose of your document
5. the tone you want to convey

● *Font Style Guideline 1: Consider the Reader's or Company's Preferences*

Give font style the same consideration as you give your message. If your reader has clear preferences, by all means adhere to them. If, on the other hand, your audience is receptive to new ideas and images, you can become creative in selecting fonts.

● *Font Style Guideline 2: Consider the Need for Clarity*

All technical writers recognize the need for their messages to be clear. As you select a font, ask yourself questions such as the following:

- Am I using this type font for captions or for long passages of text?
- Does the material I've written contain technical terms or formulae, or was it written for a general audience?
- Will the font style enhance or detract from the readability of the material?

Readability of a font style partly depends on whether it uses sans serif characters or serif characters. Sans serif font styles ("without the feet or tails") and serif font styles ("with feet or tails") are shown in 18 pt. in Figure 11–3 in the word "Readability."

Sans serif font styles are more effective when used for headers, numeric data, and transparencies for these reasons: characters appear cleaner, and more white space exists between the letters. Serif font styles are more appropriate for straight text. The feet (or "tails") of serif font styles seem to lead the readers' eyes across the text in a natural flow, making it easier to scan the page. The result is better readability.

● *Font Style Guideline 3: Consider the Space Available*

Although all font styles are measured vertically using a scale of 72 points to an inch, they vary horizontally. If you are writing a long formal proposal, the space available might not be important. Adding or deleting one page might not matter. However, if you are writing the text for a software package, you might need to select a font style that is clear but that will—for this special assignment—fit into a 3″ × 6″ field.

Refer to Figure 11–3. What kinds of differences do you see? Wider letters? Thicker strokes? More or less white space (kerning) between the characters? More or less white space (leading) between rows of typed material? Each of these criteria varies from one font style to another.

In some cases, you may decide to use the same font style. Then you can make minor adjustments in line length by reducing the kerning (white space) between the characters to fit all your information on one page or to prevent the carryover of one word to the next page.

In other situations, you may want to adjust the leading (white space). In this way, the lines are closed up by one "point" of space, and one more line can fit on the page. Unless your organization uses a mandatory style sheet, these choices will be up to you.

SANS SERIF FONT STYLES	
Arial	Readability
Bauhaus Md	Readability
Basic Sans SF	Readability
Bernhard Fashion BT	Readability
Futura Md BT	**Readability**
Impact	**Readability**

Serif Font Styles	
Caslon Bd BT	Readability
Americana XBdCn BT	Readability
Poster Bodni BT	**Readability**
GarmdITC Bk BT	Readability
Kuenstler script B.T	Readability

FIGURE 11–3   Font styles

● *Font Style Guideline 4: Consider the Purpose of the Document*

Before making font style decisions, evaluate how different font styles may reinforce your document's purpose. Ask yourself these questions:

- Will the document be referred to frequently?
- Does the document present financial or statistical data that must be easily read and comprehended?
- Must it be eye-catching enough to make the audience eager to read it?
- Is it a routine document that needs only to be read once, handled, and filed?

Considering the above questions will help you choose a font style that serves the purpose of your entire document.

● *Font Style Guideline 5: Consider the Tone You Want to Convey*

Finally, give careful thought to which font style will reinforce the tone of your message.

For example, assume you have worked diligently to develop an annual report reflecting serious growth problems for the company and for the company's industry. The document is formal in tone; however, it is also a no-nonsense business document. A font type such as *Lydian Csv BT* is formal; however, it does not even remotely present the tone required for this annual report. In this instance, you may want to use Arial or Avant Garde because of its crisp, sans serif image.

In another example, you have been asked to invite management and hourly employees to a retirement party given for a mid-level manager. The tone, you correctly assess, will be informal, warm, and hospitable. As you scroll down the list of font styles, you find several that seem appropriate—**Dom Casual BT**, ZapfHumnst BT, and **Ad Lib**—and you wonder which, if any, will convey the right image. The first one looks interesting and casual, but it appears too small. The second one looks too impersonal. The third, though sans serif, appears casual, large, and powerful—like someone is shouting, *"Come on In!"*

In addition to the diversity of font styles available for your use, desktop publishing packages today offer you an opportunity to reconfigure your text into arcs, waves, slopes, and even circles. You can outline, shade, print vertically, and do much more whenever it is appropriate. That is the key: Your font styles must be *appropriate* for the tone you are trying to convey.

USING COLOR

Just as people expect color in photographs, movies, advertisements, and TV, color has emerged as a necessity in technical communications. This tool adds excitement to, and stimulates interest in, your documents. Most writers have software that lets them simply "click" on the colors they think are appropriate for a document and, when the job is completed, send them to a color printer. However, their color choices—and even the decision to use color—may be inappropriate for a number of reasons.

Because color is so costly and requires a longer time to print, you need to know more about its effective use. In this segment we will discuss what you must consider before you use color, including ways you can use it effectively.

The Cost and Time of Using Color

The cost of using color in technical communications is much like the cost of buying a luxury automobile. You receive many benefits, but you must be ready to pay for them. Before you spend your time and energy on using color, you'll want to consider answers to some important questions.

- **Has a budget been established for the job? If so, what does it include?** Whether we like it or not, budgets control our creative efforts. You won't want to invest massive amounts of time in thinking through your color strategy if you know up front that color is too costly. The seemingly high cost, though, may be justified by an anticipated return on investment.

  Developing color documents is usually a less significant cost factor than printing the materials. Most printer software gives you the option to print four-color documents. However, you will want to design your color graphics with clear, crisp images that retain their impact *even if they cannot be printed in color.*

- **What is the distribution?** You'll want to ask yourself these questions about distribution: Will the document be sent to potential customers or key executives in your organization? How many other people will receive copies? Must all recipients have color copies?

  The number of color copies required may be directly related to the hierarchy of receivers. If only five or six copies are needed and the pages are few, you may be able to print the documents on your desktop printer. If more copies are needed, you may need to use a printer to mass produce your documents. Four-color printing costs per document increase dramatically as volumes increase because (1) the printer must make a separate "master" for each page and ink color used (*Yellow, Magenta, Cyan, and Black*) and (2) the printer then "runs" each color for each page separately.

- **Does the company have printing/reproduction facilities?** Large companies frequently plan for high-volume reproduction of reports, advertising, training, and documentation. If they have the capability, you may receive more rapid turn-around times and benefit from lower costs.

- **Are other options available?** If you have a tight budget and still want to use color to stimulate interest in your document, you may want to consider (1) using colored paper to distinguish one section from another, (2) printing one or two copies in color for the most important readers and printing the balance in black and white, and (3) using one color with black to add some interest (perhaps the color most readily identified with your company's image).

Developing a Color Style Sheet

Once you decide to use color in your documents, you'll see a difference in how people react to your messages. If your company doesn't already have a color style sheet—which provides guidelines for the use of color—develop your own so you can use color uniformly and consistently from one project to another. Avoid overusing color. Too many colors, like too much data, will distort your message, confuse your audience, cost more, and take more time to produce. To develop a style sheet, ask yourself these four questions:

1. **How can I use color to help my audience read and retain the document's message?** You can define the levels of importance in a document by using a different color for different headings, and lighter or darker tones of a color to distinguish between subheadings or between major and minor concepts within a heading.

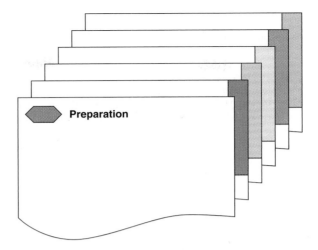

FIGURE 11–4 Using colored icons

For instance, in Figure 11–4 a report that contains five different sections uses colored icons that match section tabs. This technique adds color without giving up the readability of black font styles on white paper.

2. **How can I use color to attract attention to important data?** Use *one color* to highlight significant points or to provide a specific function. For example, if you consistently use color to frame tables or graphics, you create a visual cue that prompts your reader to look at the graphics. Or, you may want to use one color to *consistently* emphasize key words, phrases, or specific actions. In so doing, you build *continuity* in your document and enhance readability.

3. **Will my audience be able to see the differences in color?** About 10% of the male population has some degree of red/green color blindness. Therefore, be careful about using red and green together in your documents.

4. **Will the document be universally distributed in color, or will one or more sections be printed in black and white?** If your document will be reproduced—either all, or in part—in black and white, you need to place greater emphasis on textural differences as well as use distinctive shading and tinting. Most importantly, do not use blue images; unless shaded considerably, they will not reproduce on a photocopier. You must do everything you can to sharpen the images you want your audience to read.

After you answer questions about the general use of color, you can begin the process of selecting and combining colors to enhance your message and your graphics.

Color Terms

Most of us are familiar with the spectrum of colors seen when light passes through a prism or when rainbows appear in the sky. In most instances, your desktop publishing software displays a full color array that resembles a color wheel. However, you may not be familiar with some common color-related terms such as *hues, primary, secondary, tertiary,* and *complementary*. Let us look at them briefly (see Figure 11–5):

- *Hues:* Hues are the intense "true" colors we see on a "color wheel" that displays the relationships and intensities of colors.
- *Primary Colors:* Red, blue, and yellow are considered **primary** colors. We have learned ways to combine primary colors and make other colors that also appear on the color wheel through light, pigments, or inks.
- *Secondary Colors:* As children we often learn that when we mix equal parts of red and blue pigments, the result is purple. Blue and yellow become green, and yellow and red become orange. Because purple, green, and orange are mixed from the primary colors, they are often referred to as **secondary** colors.
- *Tertiary Colors:* Tertiary colors are colors mixed by a combination of a primary color and a secondary color as shown in Figure 11–5. They are also considered "hues" because of their intensity.
- *Complementary Colors:* Complementary colors are those that appear *immediately opposite* one another on the color wheel. For example, in Figure 11–5, you can see that red and green are opposite each other—as are blue and orange and yellow and violet. Because complementary colors are opposites, they are frequently used to give the best contrast between colors.

Guidelines for Using Color

This section includes four guidelines to help you select colors for your documents:

- ### *Color Selection Guideline 1: Use Colors Your Audience Will Associate with Your Topic in a Positive Way*

Take time to research your reader's reactions to colors. In particular, if your document has an international audience, make certain you know what colors are considered appropriate or inappropriate in specific cultures. Take time to do your research before you add color.

- ### *Color Selection Guideline 2: Use Colors That Enhance Your Company's Logo*

By using colors that complement and reinforce your company's logo, you are more likely to get immediate acceptance of your ideas.

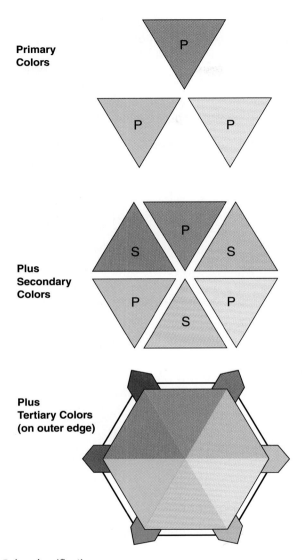

FIGURE 11–5 Color classifications

● *Color Selection Guideline 3: Use Dark or Textured Backgrounds Infrequently*

For years, advertisers thought the most productive newspaper ads were those with solid black backgrounds and large, white letters. When writers learned more about the use of white space, the myth of "reverse" messages was shattered. Although dark or textured backgrounds can be effective, like other design tools they may be overdone. Use them when you want to focus your reader's attention on a graphic by using darker tones as a frame.

If your documents will be printed in black, white, and grey, try using texture for only one or two elements of your graphic. It will help you avoid having a cluttered, busy appearance. If you decide to use shading to mask out—but not obliterate—important data, don't use shading any darker than a 10–25% screen.

Finally, if you really want your message to be surrounded with a dark color, remember you'll also need to use a larger, bolder font style and a lighter, brighter hue, or the information you want so much to be noticed will fade into the background.

● *Color Selection Guideline 4: Learn How Any Color Will Be Affected by the Medium It Will Be Printed on and Make Adjustments Accordingly*

Textures and colors of paper stock can drastically alter the appearance of your document. It is best to test the final colors *before* you use them in a final draft. Ask your peers, your boss, and your public relations department for their opinions and suggestions. A little planning can help you strengthen your document.

GENERAL GUIDELINES FOR GRAPHICS

A few basic guidelines apply to all graphics. Keep these fundamentals in mind as you move from one type of illustration to another.

● *Graphics Guideline 1: Determine the Purpose of the Graphic*

Graphics, like text, should only be used if they serve a purpose. Ask yourself:

- What kind of information does your audience need to better understand the scope, problem, or solution?
- What type of graphic can be used to present the data in the most interesting, informative way?
- Are any special symbols, colors, or font styles needed to reinforce the data?

● *Graphics Guideline 2: Evaluate the Accuracy and Validity of the Data*

Unless the information you plan to include in your document is accurate, you run the risk of presenting information that could damage your credibility as well as the credibility of the document. Remember, one false or inappropriate statistic tends to cast doubt on the balance of the data. Do the following:

- Check the accuracy of information
- Make sure the source is reputable
- Ensure that data are not distorted by flawed scales or images

● *Graphics Guideline 3: Refer to All Graphics in the Text*

With a few exceptions—such as cover illustrations used to grab attention—graphics should be accompanied by clear references within your text. Specifically, you should follow these rules:

- Include the graphic number in arabic, not roman, when you are using more than one graphic.
- Include the title, and sometimes the page number, if either is needed for clarity or emphasis.
- Incorporate the reference smoothly into text wording.

Here are two ways to phrase and position a graphics reference. In Example 1, there is the additional emphasis of the graphics title, whereas in Example 2, the title is left out. Also, note that you can draw more attention to the graphic by placing the reference at the start of the sentence in a separate clause. Or you can relegate the reference to a parenthetical expression at the end or middle of the passage. Choose the option that best suits your purposes.

- **Example 1:** In the past five years, 56 businesses in the county have started in-house recycling programs. The result has been a dramatic shift in the amount of property the county has bought for new waste sites, as shown in Figure 5 ("Landfill Purchases, 1985–1990").
- **Example 2:** As shown in Figure 5, the county has purchased much less land for landfills during the last five years. This dramatic reduction results from the fact that 56 businesses have started in-house recycling programs.

Graphics Guideline 4: Think About Where to Put Graphics

In most cases, locate a graphic close to the text in which it is mentioned. This immediate reinforcement of text by an illustration gives graphics their greatest strength. Variations of this option, as well as several other possibilities, are presented here:

- **Same page as text reference:** A simple visual, such as an informal table, should go on the same page as the text reference if you think it too small for a separate page.
- **Page opposite text reference:** A complex graphic, such as a long table, that accompanies a specific page of text can go on the page opposite the text—that is, on the opposite page of a two-page spread. Usually this option is exercised *only* in documents that are printed on both sides of the paper throughout.
- **Page following first text reference:** Most text graphics appear on the page after the first reference. If the graphic is referred to throughout the text, it can be repeated at later points. (*Note:* Readers prefer to have graphics positioned exactly where they need them, rather than their having to refer to another part of the document.)
- **Attachments or appendices:** Graphics can go at the end of the document in two cases: first, if the text contains so many references to the graphic that placement in a central location, such as an appendix, would make it more accessible; and second, if the graphic contains less important supporting material that would only interrupt the text.

Graphics Guideline 5: Position Graphics Vertically When Possible

Readers prefer graphics they can view without having to turn the document sideways. However, if the table or figure cannot fit vertically on a standard 8 1/2″ × 11″

page, either use a foldout or place the graphic horizontally on the page. In the latter case, position the illustration so that the top is on the left margin. (In other words, the page must be turned clockwise to be viewed.)

● *Graphics Guideline 6: Avoid Clutter*

Let simplicity be your guide. Readers go to graphics for relief from, or reinforcement of, the text. They do not want to be bombarded by visual clutter. Omit information that is not relevant to your purpose, while still making the illustration clear and self-contained. Also, use enough white space so that the readers' eyes are drawn to the graphic. The final section of this chapter discusses graphics clutter in more detail.

● *Graphics Guideline 7: Provide Titles, Notes, Keys, and Source Data*

Graphics should be as self-contained and self-explanatory as possible. Moreover, they must include any borrowed information. Follow these basic rules for format and acknowledgement of sources:

- **Title:** Follow the graphic number with a short, precise title—either on the line below the number *or* on the same line after a colon (for example, "Figure 3: Salary Scales").
- **Tables:** The number and title go at the top. (As noted in Table Guideline 1 on page 427, one exception is informal tables. They have no table number or title.)
- **Figures:** The number and title can go either above or below the illustration. Center titles or place them flush with the left margin.
- **Notes for explanation:** When introductory information for the graphic is needed, place a note directly underneath the title *or* at the bottom of the graphic.
- **Keys or legends for simplicity:** If a graphic needs many labels, consider using a legend or key, which lists the labels and corresponding symbols on the graphic. For example, a pie chart might have the letters *A, B, C, D,* and *E* printed on the pie pieces, while a legend at the top, bottom, or side of the figure would list what the letters represent.
- **Source information at the bottom:** You have a moral, and sometimes legal, obligation to cite the person, organization, or publication from which you borrowed information for the figure. Either (1) precede the description with the word "Source" and a colon, or (2) if you borrowed just part of a graphic, introduce the citation with "Adapted from."

Besides citing the source, it is sometimes necessary to request permission to use copyrighted or proprietary information, depending on your use and the amount you are borrowing. (A prominent exception is most information provided by the federal government. Most government publications are not copyrighted.) Consult a reference librarian for details about seeking permission.

SPECIFIC GUIDELINES FOR EIGHT GRAPHICS

Illustrations come in many forms. Almost any nontextual part of your document can be placed under the umbrella term *graphic*. Among the many types, these eight are often used in technical writing: (1) tables, (2) pie charts, (3) bar charts, (4) line charts, (5) schedule charts, (6) flowcharts, (7) organization charts, and (8) technical drawings. This section of the chapter highlights their different purposes and gives guidelines for using each type.

Tables

Tables present readers with raw data, usually in the form of numbers but sometimes in the form of words. Tables are classified as either informal or formal:

- **Informal tables:** limited data arranged in the form of either rows or columns
- **Formal tables:** data arranged in a grid, always with both horizontal rows and vertical columns

These five guidelines will help you design and position tables within the text of your documents.

● *Table Guideline 1: Use Informal Tables as Extensions of Text*

Informal tables are usually merged with the text on a page, rather than isolated on a separate page or attachment. As the following example shows, an informal table usually has (1) no table number or title, and (2) few if any headings for rows or columns. Also, it is not included in the list of illustrations in a formal document.

Our project in Alberta, Canada, will involve engineers, technicians, and salespeople from three other offices, in these numbers:	
San Francisco Office	45
St. Louis Office	34
London Office	6
Total	85

● *Table Guideline 2: Use Formal Tables for Complex Data Separated from Text*

Formal tables may appear on the page of text that includes the table reference, on the page following the first text reference, or in an attachment or appendix. In every case, you should do the following:

1. Extract important data from the table and highlight them in the text
2. Make every formal table as clear and visually appealing as possible by doing the following:

   - Use color to designate positive or negative totals, increases, or decreases or very important points.
   - Use colored or dark grey borders to frame the information and call attention to the table.
   - Use grey screens (no denser than 10–25%) to subordinate less important data that appear on the table.
   - Avoid excessive use of heavy horizontal and vertical crossed lines that create a "grid" look; instead, consider using white lines on a light grey background or light grey lines on a white background to lead the eye across the page.

See Figure 11–6 for examples of three tables that use different designs to present the same information.

● Table Guideline 3: Use Plenty of White Space

Used around and within tables, white space guides the eye through a table much better than do black lines. Avoid putting densely drawn black boxes around tables. Instead, leave one inch more of white space than you would normally leave around text and let it act as a frame.

● Table Guideline 4: Follow Usual Conventions for Dividing and Explaining Data

Figure 11–7 shows a typical formal table. It satisfies the overriding goal of being clear and self-contained. To achieve that objective in your tables, follow these guidelines:

1. **Titles and numberings:** Give a title to each formal table, and place title and number above the table. Number each table if the document contains two or more tables.
2. **Headings:** Create short, clear headings for all columns and rows.
3. **Abbreviations:** Include in the headings any necessary abbreviations or symbols, such as lb or %. Spell out abbreviations and define terms in a key or footnote if any reader may need such assistance.
4. **Numbers:** Round off numbers when possible, for ease of reading. Also, align multidigit numbers on the right edge, or at the decimal when shown.
5. **Notes:** Place any necessary explanatory headnotes either between the title and the table (if the notes are short) *or* at the bottom of the table.
6. **Footnotes:** Place any necessary footnotes below the table.
7. **Sources:** Place any necessary source references below the footnotes.
8. **Caps:** Use uppercase and lowercase letters, rather than full caps.

Table : Comparison of 1980–1994 Airline Operating Costs

Percent Distribution of Cash Operating Expenses	1980	1985	1990	1994
Labor	34.9	32.6	34.8	36
Interest	3.5	2.5	2.6	3
Fuel	22.3	17.6	12.7	11.7
Passenger Food	3.2	3.6	3.6	3.5
Advertising and Promotion	2.3	2.1	1.7	1.6
Landing Fees	1.7	1.9	2.1	2.2
Passenger Traffic Commissions	7.4	9.8	11.3	10
All Other Expenses	24.7	29.9	31.2	32
TOTAL	100	100	100	100

Source: 1995 Statistical Abstract of the United States
No. 1040, Airline Cost Indexes: 1980–1994

Table : Comparison of 1980–1994 Airline Operating Costs

Percent Distribution of Cash Operating Expenses	1980	1985	1990	1994
Labor	34.9	32.6	34.8	36
Interest	3.5	2.5	2.6	3
Fuel	22.3	17.6	12.7	11.7
Passenger Food	3.2	3.6	3.6	3.5
Advertising and Promotion	2.3	2.1	1.7	1.6
Landing Fees	1.7	1.9	2.1	2.2
Passenger Traffic Commissions	7.4	9.8	11.3	10
All Other Expenses	24.7	29.9	31.2	32
TOTAL	100	100	100	100

Source: Statistical Abstract of the United States
No. 1040, Airline Cost Indexes: 1980 to 1994

Table : Comparison of 1980–1994 Airline Operating Costs

Percent Distribution of Cash Operating Expenses	1980	1985	1990	1994
Labor	34.9	32.6	34.8	36
Interest	3.5	2.5	2.6	3
Fuel	22.3	17.6	12.7	11.7
Passenger Food	3.2	3.6	3.6	3.5
Advertising and Promotion	2.3	2.1	1.7	1.6
Landing Fees	1.7	1.9	2.1	2.2
Passenger Traffic Commissions	7.4	9.8	11.3	10
All Other Expenses	24.7	29.9	31.2	32
TOTAL	100	100	100	100

Source: 1995 Statistical Abstract of the United States
No. 1040, Airline Cost Indexes: 1980–1994

FIGURE 11–6 Use of shading and color highlights in 3 identical tables
Adapted from Table No. 1040, Statistical Abstract of the United States 1996, p 646

TABLE 22: McDuff's Employee Retirement Fund			
Investment Type	Book Value	Market Value	% of Total Market Value
Temporary Securities	$ 434,084	434,084	5.9%
Bonds	3,679,081	3,842,056	52.4
Common Stocks	2,508,146	3,039,350	41.4
Mortgages	18,063	18,063	.3
Real Estate	1,939	1,939	nil
Totals	$6,641,313	$7,335,492	100.0%

Note: This table contrasts the book value versus the market value of the McDuff Employee Retirement Fund, as of December 31, 2001.
Source: McDuff's accounting firm of Bumble and Bumble, Inc.

FIGURE 11–7   Example of formal table

● *Table Guideline 5: Pay Special Attention to Cost Data*

Most readers prefer to have complicated financial information placed in tabular form. Given the importance of such data, edit cost tables with great care. Devote extra attention to these two issues:

- Placement of decimals in costs
- Correct totals of figures

Documents like proposals can be considered contracts in some courts of law, so there is no room for error in relating costs.

Pie Charts

Familiar to most readers, pie charts show approximate relationships between the parts and the whole. Their simple circles with clear labels can provide comforting simplicity within even the most complicated report. Yet the simple form keeps them from being useful when you need to reveal detailed information or changes over time. Here are specific guidelines for constructing pie charts.

● *Pie Chart Guideline 1: Use No More than 6 or 7 Divisions*

To make pie charts work well, limit the number of pie pieces to no more than 6 or 7. In fact, the fewer segments the better. This approach lets the reader grasp major relationships without having to wade through the clutter of tiny divisions that are difficult to label and read. In Figure 11–8, for example, McDuff's client can readily see that most of the project staff will come from the three McDuff offices closest to the project site.

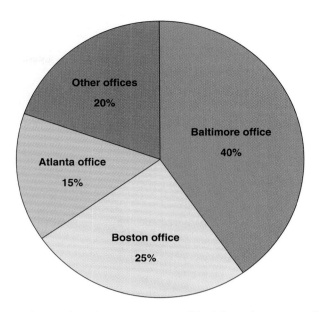

FIGURE 11–8 Pie chart with as few pieces as possible. (Chart shows McDuff workforce breakdown for an offshore Atlantic project. McDuff can draw most project workers from its East Coast offices.)

● *Pie Chart Guideline 2: Move Clockwise from 12:00, from Largest to Smallest Wedge*

Readers prefer pie charts oriented like a clock—with the first wedge starting at 12:00. Move from the largest to the smallest wedge to provide a convenient organizing principle.

Make exceptions to this design only for good reason. In Figure 11–8, for example, the last wedge represents a greater percentage than the previous wedge. In this way, it does not break up the sequence the writer wants to establish by grouping the three McDuff offices with the three largest percentages of project workers.

Another exception over which you have no control is your use of software packages that will not permit you to begin your pie at 12:00.

● *Pie Chart Guideline 3: Use Pie Charts Especially for Percentages and Money*

Pie charts catch the readers' eyes best when they represent items divisible by 100—as with percentages and dollars. Figure 11–8 shows percentages; Figure 11–9 shows money. Using the pie chart for money breakdowns is made even more appropriate by the coinlike shape of the chart. *In every case, make sure your percentages or cents add up to 100.*

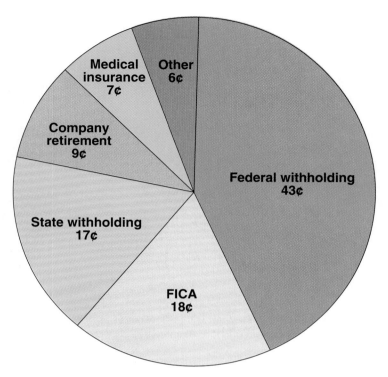

FIGURE 11–9   Pie chart showing money breakdown for average deductions from a McDuff paycheck

● *Pie Chart Guideline 4: Be Creative, But Stay Simple*

Figure 11–10 shows a number of options for designing pie charts, including:

1. Shading a wedge
2. Removing a wedge from the main pie

Be aware that some desktop publishing programs automatically format your pie charts using complex backgrounds and shading. However, they are often difficult to read and may distort the pieces of the pie. Remember to keep your pie charts simple and clean looking.

● *Pie Chart Guideline 5: Draw and Label Carefully*

The most common pie chart errors are (1) wedge sizes that do not correspond correctly to percentages or money amounts and (2) pie sizes that are too small to accommodate the information placed in them. Here are some suggestions for avoiding these mistakes:

● **Pie size:** Make sure the chart occupies enough of the page. On a standard 8 1/2″ × 11″ sheet with only one pie chart, your circle should be from 3″ to 6″ in diameter—large enough not to be dwarfed by labels and small enough to leave sufficient white space in the margins.

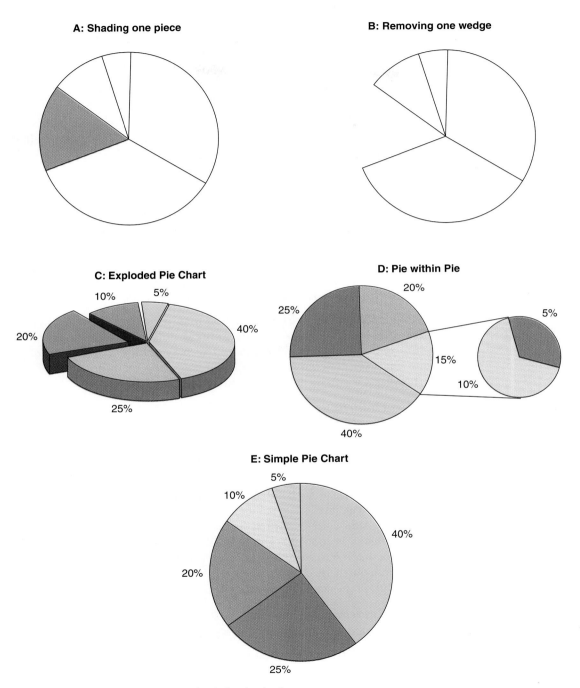

FIGURE 11–10 Techniques for emphasis in pie charts

- **Labels:** Place the wedge labels either inside the pie or outside, depending on the number of wedges, the number of wedge labels, or the length of the labels. Choose the option that produces the cleanest-looking chart.
- **Conversion of percentages:** If you are drawing the pie chart by hand, not using a computer program, use a protractor or similar device. One percent of the pie equals 3.6 degrees ($3.6 \times 100\% = 360$ degrees in a circle). With that formula as your guide, you can convert percentages or cents to degrees.

Remember, however, that a pie chart does not reveal fine distinctions very well; it is best used for showing larger differences.

Bar Charts

Like pie charts, bar charts are easily recognized, for they are seen every day in newspapers and magazines. Unlike pie charts, however, bar charts can accommodate a good deal of technical detail. Comparisons are provided by means of two or more bars running either horizontally or vertically on the page. Follow these five guidelines to create effective bar charts.

● *Bar Chart Guideline 1: Use a Limited Number of Bars*

Though bar charts can show more information than pie charts, both types of illustrations have their limits. Bar charts begin to break down when there are so many bars that information is not easily grasped. The maximum bar number can vary according to chart size, of course. Figure 11–11 shows several multibar charts. The impact of the charts is enhanced by the limited number of bars.

● *Bar Chart Guideline 2: Show Comparisons Clearly*

Bar lengths should be varied enough to show comparisons quickly and clearly. Avoid using bars that are too close in length, for then readers must study the chart before understanding it. Such a chart lacks immediate visual impact.

Also, avoid the opposite tendency of using bar charts to show data that are much different in magnitude. To relate such differences, some writers resort to the dubious technique of inserting "break lines" (two parallel lines) on an axis to reflect breaks in scale (see Figure 11–12). Although this approach at least reminds readers of the breaks, it is still deceptive. For example, note that Figure 11–12 provides no *visual* demonstration of the relationship between 50 and 2800. The reader must think about these differences before making sense out of the chart. In other words, the use of hash marks runs counter to a main goal of graphics—creating an immediate and accurate visual impact.

● *Bar Chart Guideline 3: Keep Bar Widths Equal and Adjust Space Between Bars Carefully*

While bar length varies, bar width must remain constant. As for distance between the bars, following are three options (along with examples in Figure 11–13):

- **Option A: Use no space** when there are close comparisons or many bars, so that differences are easier to grasp.

Sales

Operating Profits

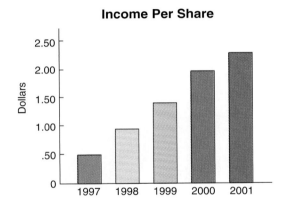

Income Per Share

FIGURE 11–11   Bar charts

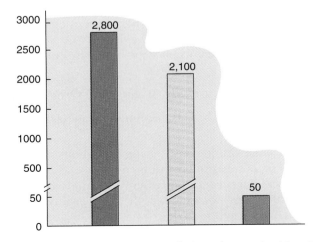

FIGURE 11–12   Hash marks on bar charts—a technique that can lead to misunderstanding
Adapted from William S. Pfeiffer, *Proposal Writing* (Columbus, Ohio: Merrill, 1989), 147.

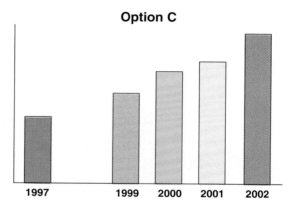

FIGURE 11–13 Bar chart variations
Adapted from William S. Pfeiffer, *Proposal Writing* (Columbus, Ohio: Merrill, 1989), 148.

- **Option B: Use equal space, but less than bar width** when bar height differences are great enough to be seen in spite of the distance between bars.
- **Option C: Use variable space** when gaps between some bars are needed to reflect gaps in the data.

● *Bar Chart Guideline 4: Carefully Arrange the Order of Bars*

The arrangement of bars is what reveals meaning to readers. Here are two common approaches:

- **Sequential:** used when the progress of the bars shows a trend—for example, McDuff's increasing number of environmental projects in the last five years
- **Ascending or descending order:** used when you want to make a point by the rising or falling of the bars—for example, the 1998 profits of McDuff's six international offices, from lowest to highest

● *Bar Chart Guideline 5: Be Creative*

Figure 11–14 shows two bar chart variations that help display multiple trends. The *segmented bars* in Option A produce four types of information: the total sales (A + B + C)

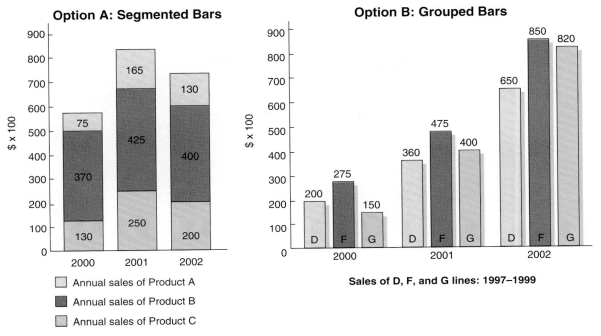

FIGURE 11-14 Bar chart variations for multiple trends
Adapted from William S. Pfeiffer, *Proposal Writing* (Columbus, Ohio: Merrill, 1989), 150.

and the individual sales for A, B, and C. The *grouped bars* in Option B show the individual sales trends for D, F, and G, along with a comparison of all three by year. Note that the amounts are written above the bars to highlight comparisons. Figure 11–15 uses different data to show a bar chart that groups individual *and* total sales figures by year.

Line Charts

Line charts are a common graphic. Almost every newspaper contains a few charts covering topics such as stock trends, car prices, or weather. More than other graphics, line charts telegraph complex trends immediately.

They work by using vertical and horizontal axes to reflect quantities of two different variables. The vertical (or *y*) axis usually plots the dependent variable; the horizontal (or *x*) axis usually plots the independent variable. (The dependent variable is affected by changes in the independent variable.) Lines then connect points that have been plotted on the chart. When drawing line charts, follow these five main guidelines:

● *Line Chart Guideline 1: Use Line Charts for Trends*

Readers are affected by the direction and angle of the chart's line(s), so take advantage of this persuasive potential. In Figure 11–16, for example, the writer wants to show the feasibility of adopting a new medical plan for McDuff. Including a line chart in the study gives immediate emphasis to the most important issue—the effect the new plan would have on stabilizing the firm's medical costs.

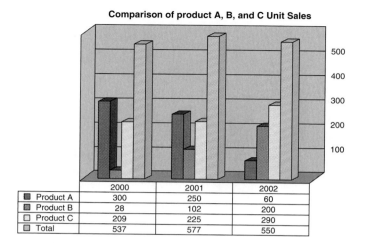

FIGURE 11–15   Grouped bar chart with data

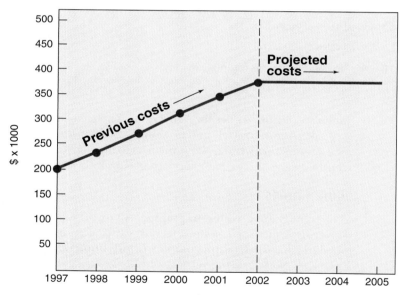

FIGURE 11–16   Line chart used to show effect of proposed medical plan on McDuff health costs

● *Line Chart Guideline 2: Locate Line Charts with Care*

Given their strong impact, line charts can be especially useful as attention-grabbers. Consider placing them (1) on cover pages (to engage reader interest in the document), (2) at the beginning of sections that describe trends, and (3) in conclusions (to reinforce a major point of your document).

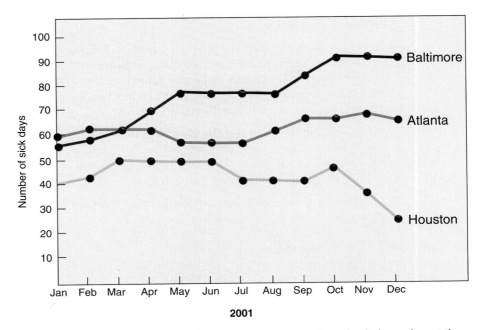

FIGURE 11–17 Line chart using multiple lines to show number of sick days taken at three McDuff offices: 2001

● *Line Chart Guideline 3: Strive for Accuracy and Clarity*

Like bar charts, line charts can be misused or just poorly constructed. Be sure that the line or lines on the graph truly reflect the data from which you have drawn. Also, select a scale that does not mislead readers with visual gimmicks. Here are some specific suggestions to keep your line charts accurate and clear:

- Start all scales from zero to eliminate the possible confusion of breaks in amounts (see Bar Chart Guideline 2).
- Select a vertical-to-horizontal ratio for axis lengths that is pleasing to the eye (three vertical to four horizontal is common).
- Make chart lines as thick as, or thicker than, the axis lines.
- Use shading under the line when it will make the chart more readable.

● *Line Chart Guideline 4: Do Not Place Numbers on the Chart Itself*

Line charts derive their main effect from the simplicity of lines that show trends. Avoid cluttering the chart with a lot of numbers that only detract from the visual impact.

● *Line Chart Guideline 5: Use Multiple Lines with Care*

Like bar charts, line charts can show multiple trends. Simply add another line or two. If you place too many lines on one chart, however, you run the risk of confusing the reader with too much data. Use no more than four or five lines on a single chart (see Figure 11–17).

Schedule Charts

Many documents, especially proposals and feasibility studies, include a special kind of chart that shows readers when certain activities will be accomplished. This kind of chart usually highlights tasks and times already mentioned in the text. Often called a milestone or Gantt chart (after Henry Laurence Gantt, 1861–1919), it usually includes these parts (see Figure 11–18):

- **Vertical axis,** which lists the various parts of the project, in sequential order
- **Horizontal axis,** which registers the appropriate time units
- **Horizontal bar lines** (Gantt) or separate markers (milestone), which show the starting and ending times for each task

Follow these basic guidelines for constructing effective schedule charts in your proposals, feasibility studies, or other documents.

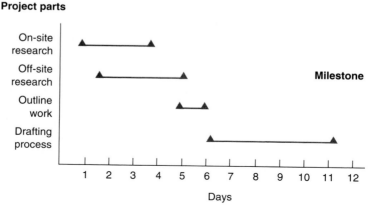

FIGURE 11–18 Gantt and milestone schedule charts

● *Schedule Chart Guideline 1: Include Only Main Activities*

Keep readers focused on no more than 10 or 15 main activities. If more detail is needed, construct a series of schedule charts linked to the main "overview" chart.

● *Schedule Chart Guideline 2: List Activities in Sequence, Starting at the Top of the Chart*

As shown in Figure 11–18, the convention is to list activities from the top to the bottom of the vertical axis. Thus the readers' eyes move from the top left to the bottom right of the page, the most natural flow for most readers of English.

● *Schedule Chart Guideline 3: Run Labels in the Same Direction*

If readers have to turn the chart sideways to read labels, they may lose interest.

● *Schedule Chart Guideline 4: Create New Formats When Needed*

Figure 11–18 shows only two common types of schedule charts; you should devise your own hybrid form when it suits your purposes. Your goal is to find the simplest format for telling your reader when a product will be delivered, a service completed, and so forth. Figure 11–19 includes one such variation.

● *Schedule Chart Guideline 5: Be Realistic about the Schedule*

Schedule charts can come back to haunt you if you do not include feasible deadlines. As you set dates for activities, be realistic about the likely time something can be accomplished. Your managers and clients understand delays caused by weather, equipment breakdowns, and other unforeseen events. However, they will be less charitable about schedule errors that result from sloppy planning.

Flowcharts

Flowcharts tell a story about a process, usually by stringing together a series of boxes and other shapes that represent separate activities (see Figure 11–20). Because they

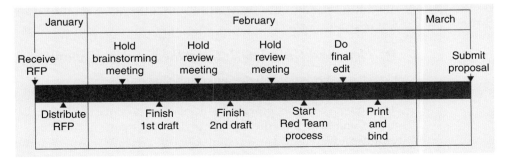

FIGURE 11–19 Schedule chart variation

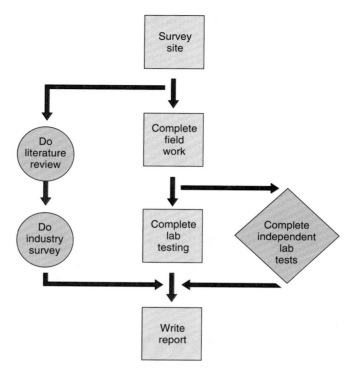

FIGURE 11–20   Flowchart for basic McDuff project
Adapted from William S. Pfeiffer, *Proposal Writing* (Columbus, Ohio: Merrill, 1989), 155.

have a reputation for being hard to read, you need to take extra care in designing them. These five guidelines will help.

● *Flowchart Guideline 1: Present Only Overviews*

Readers usually want flowcharts to give them only a capsule version of the process, not all the details. Reserve your list of particulars for the text or the appendices, where readers expect it.

● *Flowchart Guideline 2: Limit the Number of Shapes*

Flowcharts rely on rectangles and other shapes to relate a process—in effect, to tell a story. Different shapes represent different types of activities. This variety helps in describing a complex process, but it can also produce confusion. For the sake of clarity and simplicity, limit the number of different shapes in your flowcharts. Figure 11–21 includes a flowchart that is complex but still readable. Note that the writer has modified geometric shapes to match what they represent (for example, note the tractor-feed holes on the sides of the "printout" block).

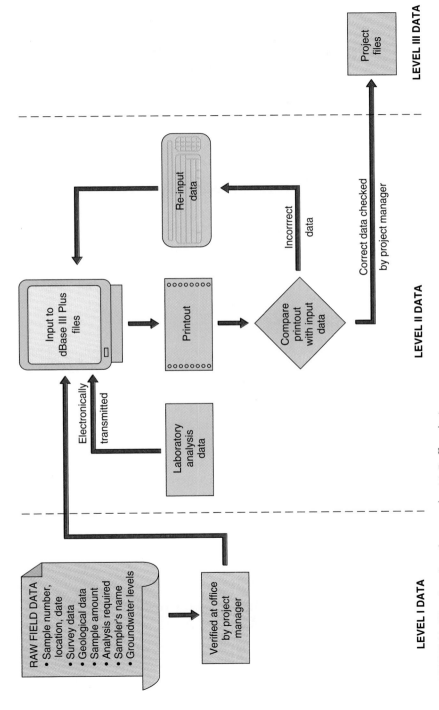

FIGURE 11–21 Flowchart for complex McDuff project

● *Flowchart Guideline 3: Provide a Legend When Necessary*

Simple flowcharts often need no legend. The few shapes on the chart may already be labeled by their specific steps. When charts get more complex, however, include a legend that identifies the meaning of each shape used.

● *Flowchart Guideline 4: Run the Sequence from Top to Bottom or from Left to Right*

Long flowcharts like the one in Figure 11–21 may cover the page with several columns or rows. Yet they should always show some degree of uniformity by assuming either a basically vertical or horizontal direction.

● *Flowchart Guideline 5: Label All Shapes Clearly*

Besides a legend that defines meanings of different shapes, the chart usually includes a label for each individual shape or step. Follow one of these approaches:

- Place the label inside the shape.
- Place the label immediately outside the shape.
- Put a number in each shape and place a legend for all numbers in another location (preferably on the same page).

Organization Charts

Organization charts reveal the structure of a company or other organization—the people, positions, or work units. The challenge in producing this graphic is to make sure that the arrangement of information accurately reflects the organization.

● *Organization Chart Guideline 1: Use the Linear "Boxes" Approach to Emphasize High-Level Positions*

This traditional format uses rectangles connected by lines to represent some or all of the positions in an organization (see Figure 11–22). Because high-level positions usually appear at the top of the chart, where the attention of most readers is focused, this design tends to emphasize upper management.

● *Organization Chart Guideline 2: Connect Boxes with Solid or Dotted Lines*

Solid lines show direct reporting relationships; dotted lines show indirect or staff relationships (see Figure 11–22).

● *Organization Chart Guideline 3: Use a Circular Design to Emphasize Mid- and Low-Level Positions*

This arrangement of concentric circles gives more visibility to workers outside upper management. These are often the technical workers most deeply involved in the details of a project. For example, Figure 11–23 draws attention to the project engineers perched on the chart's outer ring.

FIGURE 11–22 Basic organization chart
Adapted from William S. Pfeiffer, *Proposal Writing* (Columbus, Ohio: Merrill, 1989), 157.

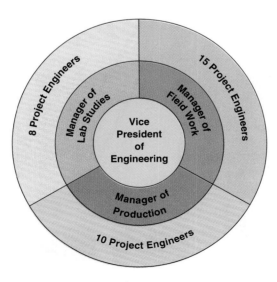

FIGURE 11–23 Concentric organization chart
Adapted from William S. Pfeiffer, *Proposal Writing* (Columbus, Ohio: Merrill, 1989), 159.

● *Organization Chart Guideline 4: Use Varied Shapes Carefully*

Like flowcharts, organization charts can use different shapes to indicate different levels or types of jobs. However, beware of introducing more complexity than you need. Use more than one shape only if you are convinced this approach is needed to convey meaning to the reader.

● *Organization Chart Guideline 5: Be Creative*

When standard forms will not work, create new ones. For example, Figure 11–24 uses an organization chart as the vehicle for showing the lines of responsibility in a specific project.

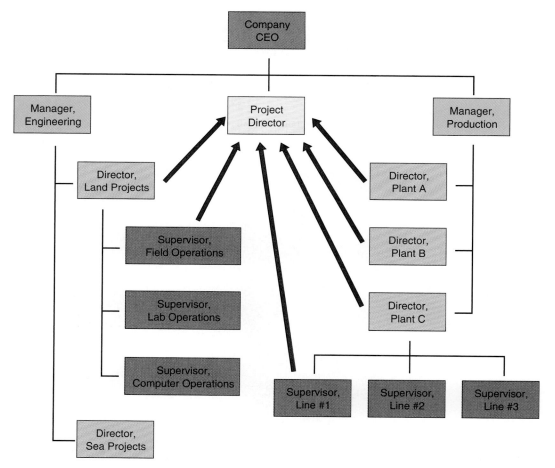

FIGURE 11–24   Organization chart focusing on project (indicates individuals most involved with upcoming project)

Adapted from William S. Pfeiffer, *Proposal Writing* (Columbus, Ohio: Merrill, 1989), 159.

Technical Drawings

Technical drawings are important tools of companies that produce or use technical products. These drawings can accompany documents such as instructions, reports, sales orders, proposals, brochures, and posters. They are preferred over photographs when specific views are more important than photographic detail. Whereas all drawings used to be produced mainly by hand, now they are usually created by CAD (computer-assisted design) systems. Follow these guidelines for producing technical drawings that complement your text.

● *Drawing Guideline 1: Choose the Right Amount of Detail*

Keep drawings as simple as possible. Use only that level of detail that serves the purpose of your document and satisfies your readers' needs. For example, Figure 11–25 will be used in a McDuff brochure on maintaining home heating systems. Its intention is to focus on one part of the thermostat—the lever and attached roller. Completed on a CAD system, this drawing presents an exploded view so that the location of the arm can be easily seen.

● *Drawing Guideline 2: Label Parts Well*

A common complaint of drawings is that parts included in the illustration are not carefully or clearly labeled. Place labels on every part you want your reader to see. (Conversely, you can choose *not* to label those parts that are irrelevant to your purpose.)

FIGURE 11–25   Technical drawing (exploded view) of home thermostat

When you label parts, use a typeface large enough for easy reading. Also, arrange labels so that (1) they are as easy as possible for your reader to locate and (2) they do not detract from the importance of the drawing itself. The simple labeling in Figure 11–25 fulfills these objectives.

● *Drawing Guideline 3: Choose the Most Appropriate View*

As already noted, illustrations—unlike photographs—permit you to choose the level of detail needed. In addition, drawings offer you a number of options for perspective or view:

- **Exterior view** shows surface features with either a two- or three-dimensional appearance—see Figure 11–26
- **Cross-section view** shows a "slice" of the object so that interiors can be viewed—see Figure 11–27
- **Exploded view** shows relationship of parts to each other by "exploding" the mechanism—see Figure 11–25
- **Cutaway view** shows inner workings of object by removing part of the exterior—see Figure 11–28.

FIGURE 11–26 Technical drawing using CAD system

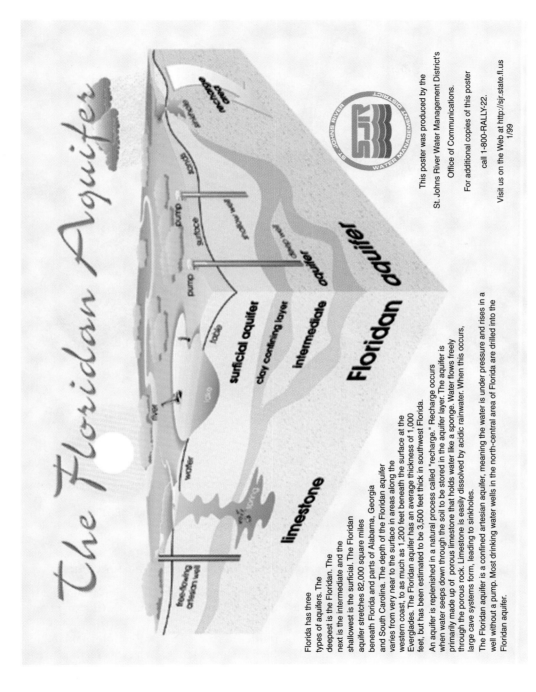

FIGURE 11–27 The 3-D aspect of the central graphic in the *Aquifer* poster makes the layered components easily understood

From Barbara Bartholomew, and Michael Bates, et al., "STC 1997–1998 International Competitions: Best of Show and Distinguished Technical Communication Winners," *Technical Communication*, 3rd quarter, 1998, p. 300. Courtesy of St. Johns River Water Management District.

FIGURE 11–28 The air brush and cutaway effects of the *Multi-turn actuator* make it easy to identify the relationships of parts

From Barbara Bartholomew, and Michael Bates, et al., "STC 1997–1998 International Competitions: Best of Show and Distinguished Technical Communication Winners," *Technical Communication*, 3rd quarter, 1998, p. 300. Courtesy of Michael Llewellyn, who produced the drawing.

● *Drawing Guideline 4: Use Legends When There Are Many Parts*

In complex drawings, avoid cluttering the illustration with many labels. Figure 11–26, for example, places all labels in one easy-to-find spot, rather than leaving them on the drawing.

However, other drawings can be rendered in such a way that internal labeling is the best option, as in Figure 11–27, which includes labels on a cross-section. Winner of a 1998 "Distinguished Technical Communication" award from the Society for Technical Communication (STC), this poster was cited by the STC for being "professionally executed," for "making technical information understandable," and for "being visually attractive." The 3-D view permits even nontechnical viewers to grasp details of a complex geological structure.

Finally, remember some drawings may have a purpose that requires no labels in any location. For example, Figure 11–28 needs no labels because it appears in a sales brochure, rather than in a more technical document such as an operating manual. Winner of the 1998 "Best of Show" award in the STC's International Technical Art Competition, this illustration received praise for its "attention to detail, rendering technique, [and] accuracy of construction." Called a multiturn electric actuator, the device is used to open and close valves, mostly in municipal facilities.

As noted earlier, many if not most drawings are produced with the assistance of computers, and Figure 11–28 provides a good example. In correspondence with the author, the illustrator listed the following steps he completed to produce his award-winning drawing:

> *First step:*   I usually get CAD drawings of separate parts, but in this case I was given an actual part. I took the actuator apart, measured and drew each part in 2-D.
>
> *Second step:* I projected each part in an isometric view, placing each part in position relative to [the] other[s]. I did the 2-D and 3-D projection in Adobe Illustrator. [At this point] all of the parts are unpainted with a black outline.
>
> *Third step:*   I brought each part into Adobe Photoshop and painted it, adding color, shadows, and highlights.
>
> *Last step:*   I merged all of the parts in Photoshop using different levels of transparency and shading.
>
> (email correspondence on 11/14/98 to W.S. Pfeiffer from Michael Llewellyn)

This list provides a snapshot of the process of producing one expert drawing. Yet note that the tools used by the illustrator are available to many writers. As technology creates more graphics possibilities, the challenge for you will be to combine sound principles of design with appropriate technical tools.

MISUSE OF GRAPHICS

Technology has revolutionized the world of graphics by placing sophisticated tools in the hands of many writers. Yet this largely positive event has its dark side. You will see many graphics that—in spite of their slickness—distort data and misinform the reader. The previous sections of this chapter have established principles and guidelines to help writers avoid such distortion and misinformation. This last section shows what can happen to graphics when sound design principles are *not* applied.

Description of the Problem

The popular media give a good glimpse into the problem of faulty graphics. One observer has used newspaper reports about the October 19, 1987, stock market plunge as one indication of the problem. Writing in *Aldus Magazine,* Daryl Moen noted that 60 percent of U.S. newspapers included charts and other graphics about the market drop the day after it occurred.* Moen's study revealed that one out of eight had data errors, and one out of three distorted the facts with visual effects. That startling statistic suggests that faulty illustrations are a genuine problem.

Edward R. Tufte analyzes graphics errors in more detail in his excellent work, *The Visual Display of Quantitative Information.* In setting forth his main principles, Tufte notes that "graphical excellence is the well designed presentation of interesting data—a matter of *substance* of *statistics,* and of *design.*" He further contends that graphics must "give to the viewer the greatest number of ideas in the shortest time with the least ink in the smallest space."†

One of Tufte's main criticisms is that charts are often disproportional to the actual differences in the data represented. The next subsection shows some specific ways that this error has worked its way into contemporary graphics.

Examples of Distorted Graphics

There are probably as many ways to distort graphics as there are graphical types. This section gets at the problem of misrepresentation by showing several examples and describing the errors involved. None of the examples commits major errors, yet each one fails to represent the data accurately.

● *Example 1: Faulty Comparisons of Modified Bar Chart*

Figure 11–29 accompanied a newspaper article about changes in mailing costs and service. The problem here is that the chart's decoration—the mailboxes—inhibits rather than promotes clear communication. Although the writer intends to use mailbox symbolism in lieu of precise bars, the height of the mailboxes does not correspond to the *actual* increase in second-class postage rates.

A revised graph should include either (1) mailboxes that correctly approximate the actual differences in second-class rates or (2) a traditional bar chart without the mailboxes.

● *Example 2: "Chartjunk" That Confuses the Reader*

Figure 11–30 concludes a report from a county government to its citizens. Whereas the dollar backdrop is meant to reinforce the topic—that is, the use to which tax funds are put—in fact, it impedes communication. Readers cannot quickly see comparisons. Instead, they must read the entire list below the illustration, mentally rearranging the items into some order.

At the very least, the expenditures should have been placed in sequence, from least to greatest percentage or vice versa. Even with this order, however, one could argue that the dollar bill is a piece of what Edward Tufte calls "chartjunk," which fails to display the data effectively.

---

*Daryl Moen, "Misinformation Graphics," *Aldus Magazine* (January/February 1990), 64.

†Edward R. Tufte, *The Visual Display of Quantitative Information* (Cheshire, Conn.: Graphics Press, 1983), 51.

FIGURE 11–29   Faulty comparisons on modified bar chart
Adapted from *Atlanta Constitution*, 30 Nov. 1989, p. H–1. Used by permission.

Where Your County Taxes are Spent

Management 7%
Courts 24%
Police 18%
Bus service 13%

Community growth 4%
Parks and libraries 18%
Administrative services 11%
Other government 5%

FIGURE 11–30   "Chartjunk" that confuses the reader

● *Example 3: Confusing Pie Charts*

The pie chart in Figure 11–31 (1) omits percentages that should be attached to each of the budgetary expenditures, (2) fails to move in a largest-to-smallest, clockwise sequence, (3) includes too many divisions, many of which are about the same size and thus difficult to distinguish, and (4) introduces a third dimension that adds no value to the graphic.

	FUND NAME	DESCRIPTION	FY '89 BUDGET
A	General Fund	Basic government activities	$112,895,822
B	Transit Fund	Implementation of bus system	10,812,522
C	Fire District Fund	Operation of Fire Department	21,253,523
D	Bond Funds	General obligation bond issue proceeds	11,073,371
E	Road Sales Tax Fund	1% special purpose sales tax for road improvements	116,869,904
F	Water & Pollution Control Fund	Daily water system operation	60,572,506
G	Debt Service Fund	Principal & interest payments for general obligation bonds	8,240,313
H	Water RE&I Fund	Maintenance of existing facilities	27,365,744
I	Solid Waste RE&I Fund	Maintenance of existing facilities	1,111,237
J	Solid Waste Disposal Facilities	Landfill operations	6,003,367
K	Water Construction Fund	Construction of new facilities	110,884,507
L	Other Uses*		18,384,364
		SUB-TOTAL	$505,467,180
		LESS INTERFUND ACTIVITY	− 23,393,042
		TOTAL EXPENDITURES	**$482,074,138**

*Other Uses includes: Community Service Block Grants, Law Library, Claims Fund, Capital Projects, Senior Services, Community Development Block Grant, Grant Fund

In addition to the General Fund, the county budgets a number of other specialized funds. These include the Fire District Fund, Transit Fund, Road Sales Tax Fund, and enterprise funds such as Water and Pollution Control, and Solid Waste Disposal Facilities.

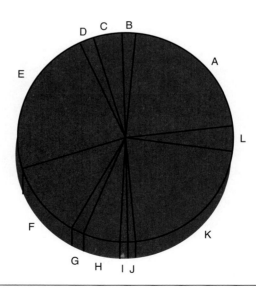

FIGURE 11–31 Confusing pie chart
Adapted from Cobb County 1988–89 Annual Report (Cobb County, GA), 14. Used by permission.

Figure 11–32 attempts the visual strategy of alternating shades, but it only succeeds in overloading the chart with too many small percentage divisions in uncertain order. Moreover, the reader cannot easily see how the pie slices would be grouped under the four headings listed below the chart. A grouped bar chart would have better served the purpose, with "Southeast," "New England," "Mid-Atlantic," and "Other" providing the groupings.

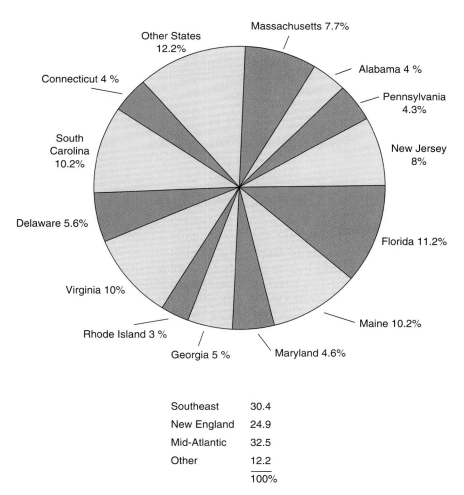

Location of All Ding-Dong Convenience Stores

Southeast	30.4
New England	24.9
Mid-Atlantic	32.5
Other	12.2
	100%

FIGURE 11–32   Confusing pie chart

Communication Challenge

"Massaging McDuff's Annual Report"

Just about everyone considered McDuff's 2001 annual report a boring piece of work. It contained pages of text with only a few tables for visual relief. Wanting to spice up the next annual report, the company hired a graphics firm to create a more appealing design and graphics. GeeWhiz Graphics is completing the 2002 annual report draft for review by McDuff's corporate staff. What follows are highlights of the graphical challenge, along with some questions and comments for discussion.

Beth's Part of the Project

When GeeWhiz got the McDuff job, Beth Chang and Rick Ford were assigned the account. They were to take text and data provided by McDuff's PR department and create a graphically interesting format for readers of the annual report—who are mostly stockholders, McDuff employees, or clients. Beth and Rick have been asked to make data as appealing as possible, especially in light of criticism the company received for its previous annual report. Beth is creating graphics for four pieces of information that McDuff wants emphasized in the report:

1. **International Sales:** McDuff has become a more international firm. Beth was asked to create some graphics that reflect this shift. In 2002 the international offices accounted for 35% of the total $96 million sales. Of that 35%, the Tokyo office was highest with $12 million in sales, Munich was next with $10 million, and the other four international offices shared the rest.
2. **Total Sales:** Over the last 10 years, total sales have gone up steadily. Figures for 1993 through 2002 are, in millions: $65, 70, 73, 74, 80, 83.5, 87, 90, 90.5, 96.
3. **Number of Employees:** Except for one year, when there was a minor layoff to reduce costs, the number of employees has risen over the last five years, as follows: 1,800 (in 1998), 1,950 (in 1999), 1,925 (in 2000), 2,200 (in 2001), 2,500 (in 2002).
4. **Corporate Overhead:** Of the six service areas covered by the corporate office in Baltimore (see pages 54–56 in chapter 2), the corporate budget spends 40% on human resources, 20% on research, 15% on computer operations, 10% on training, 10% on project management, and 5% on marketing. The company wants to emphasize that the employee-related portion of overhead grew since last year—for example, training went from 5% to 10%, and human resources went from 35% to 40%.

Rick's Part of the Project

Beth's colleague at GeeWhiz Graphics, Rick Ford, was given a similar assignment— that is, to create interesting graphical representations of data about the company. He produced three graphics in his initial work on the report:

1. **Bar Chart:** Rick drew a bar chart to reflect the growth of the international offices in the last five years.

2. Pie Chart: To show the investment portfolio of McDuff's retirement plan, Rick produced a pie chart.

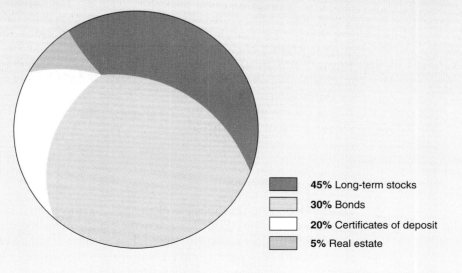

45% Long-term stocks

30% Bonds

20% Certificates of deposit

5% Real estate

Continued

3. **Bar Chart without Bars:** In the past five years, McDuff has made a concerted effort to emphasize preventive medical care and "wellness" programs among its employees. A variation on a bar chart was meant to show this progress in a more visually appealing manner than a conventional bar chart.

| 4,265 | 3,750 | 3,500 | 3,300 | 2,900 |

Sick Days Taken Last 5 Years: McDuff, Inc.

Questions and Comments for Discussion

1. Separate into groups and discuss the kinds of graphics that would be most effective, appropriate, and clear for the four assignments given to Beth Chang. Report the results of your discussion to the entire class.
2. Are there any ethical implications Beth should consider in producing her graphics? If so, what are they?
3. Separate into groups and discuss any ethical or clarity issues suggested by the graphics produced by Rick Ford. Should these graphics be included in McDuff's annual report? If so, why? If not, why not and what would you change? If useful, refer to the ECI (Equal Consideration of Interests) Principle discussed on pages 62–63.

Figure 11–33 negates the value of the pie chart by assuming an oblong shape, rather than a circle. This distortion can make it difficult for the reader to distinguish among sections that are similar in size, such as Sections 1 and 2 in Figure 11–33. The pie chart should be a perfect circle, should have percentages on the circle, and should move in large-to-small sequence from the 12:00 position.

ASSETS OF JONES RETIREMENT FUND

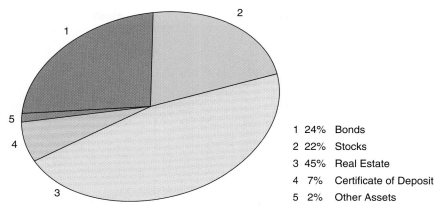

1	24%	Bonds
2	22%	Stocks
3	45%	Real Estate
4	7%	Certificate of Deposit
5	2%	Other Assets

FIGURE 11–33   Confusing pie chart

CHAPTER SUMMARY

More than ever before, readers of technical documents expect good graphics to accompany text, as well as special fonts and color. Graphics (also called "illustrations" or "visual aids") can be in the form of (1) tables (rows and/or columns of data) or (2) figures (a catchall term for all notable illustrations). Both types are used to simplify ideas, reinforce points made in the text, generate interest, and create a universal appeal.

Eight common graphics used in technical writing are tables, pie charts, bar charts, line charts, schedule charts, flowcharts, organization charts, and technical drawings. As detailed in this chapter, you should follow specific guidelines in constructing each type. These basic guidelines apply to all graphics:

1. Determine the purpose of the graphic.
2. Evaluate the accuracy and validity of the data.
3. Refer to all graphics in the text.
4. Think about where to put graphics.
5. Position graphics vertically when possible.
6. Avoid clutter.
7. Provide titles, notes, keys, and source data.

www.prenhall.com/pfeiffer   **Access the Technical Writing Web site by pointing your browser at www.prenhall.com/pfeiffer. Here, you will find additional resources to aid you in learning and applying the information from this chapter.**

ASSIGNMENTS

Your instructor may want you to practice graphics in the context of some of the writing assignments in this textbook, especially in chapters 8, 9, and 10. Here are a few additional exercises.

No. **926**. Energy Production and Consumption, by Major Source: 1960 to 1987

[Btu – British thermal unit. For Btu conversion factors, see text, section 19. See also *Historical Statistics, Colonial times to 1970,* series M 76-92]

YEAR	Total production (quad. Btu)	PERCENT OF PRODUCTION				Total consumption (quad. Btu)	PERCENT OF CONSUMPTION				Consumption/production ratio
		Coal	Petroleum[1]	Natural gas[2]	Other[3]		Coal	Petroleum[1]	Natural gas[2]	Other[3]	
1960	41.5	26.1	36.0	34.0	3.9	43.8	22.5	45.5	28.3	3.8	1.06
1961	42.0	24.9	36.2	34.9	4.0	44.5	21.6	45.5	29.1	3.8	1.06
1962	43.6	25.0	35.6	35.1	4.2	46.5	21.3	45.2	29.5	4.0	1.07
1963	45.9	25.8	34.8	35.4	4.0	48.3	21.5	44.9	29.8	3.7	1.05
1964	47.7	26.2	33.9	35.8	4.0	50.5	21.7	44.2	30.3	3.8	1.06
1965	49.3	26.5	33.5	35.8	4.3	52.7	22.0	44.1	29.9	4.0	1.07
1966	52.2	25.8	33.7	36.4	4.1	55.7	21.8	43.8	30.5	3.8	1.07
1967	55.0	25.1	33.9	36.5	4.4	57.6	20.7	43.9	31.2	4.2	1.05
1968	56.8	24.0	34.0	37.6	4.4	61.0	20.2	44.2	31.5	4.1	1.07
1969	59.1	23.5	33.1	38.7	4.8	64.2	19.3	44.1	32.2	4.4	1.09
1970	62.1	23.5	32.9	38.9	4.7	66.4	18.5	44.4	32.8	4.3	1.07
1971	61.3	21.5	32.7	40.5	5.3	67.9	17.1	45.0	33.1	4.8	1.11
1972	62.4	22.6	32.1	39.7	5.6	71.3	16.9	46.2	31.9	5.0	1.14
1973	62.1	22.5	31.4	39.9	6.2	74.3	17.5	46.9	30.3	5.3	1.20
1974	60.8	23.1	30.5	38.9	7.4	72.5	17.5	46.1	30.0	6.5	1.19
1975	59.9	25.0	29.6	36.8	8.6	70.5	17.9	46.4	28.3	7.4	1.18
1976	59.9	26.1	28.8	36.4	8.6	74.4	18.3	47.3	27.4	7.1	1.24
1977	60.2	26.2	29.0	36.4	8.5	76.3	18.2	48.7	26.1	7.0	1.27
1978	61.1	24.4	30.2	35.6	9.9	78.1	17.6	48.6	25.6	8.1	1.28
1979	63.8	27.5	28.4	35.0	9.1	[4]78.9	19.1	47.1	26.2	7.7	1.24
1980	64.8	28.7	28.2	34.2	8.9	76.0	20.3	45.0	26.8	7.8	1.17
1981	64.4	28.5	28.2	34.2	9.1	74.0	21.5	43.2	26.9	8.4	1.15
1982	63.9	29.2	28.7	32.0	10.2	70.8	21.6	42.7	26.1	9.6	1.11
1983	61.2	28.2	30.1	30.6	11.2	70.5	22.6	42.6	24.6	10.2	1.15
1984	[5]65.8	30.0	28.6	30.7	10.7	74.1	23.0	41.9	25.0	10.0	1.13
1985	64.8	29.8	29.3	29.6	11.3	74.0	23.6	41.8	24.1	10.4	1.14
1986	64.3	30.4	28.6	29.0	12.0	74.3	23.2	43.4	22.5	10.9	1.16
1987	64.6	31.2	27.3	29.5	12.0	76.0	23.7	42.9	22.6	10.8	1.18

[1]Production includes crude oil and lease condensate. Consumption includes domestically produced crude oil, natural gas liquids, and lease condensate, plus imported crude oil and products.   [2]Production includes natural gas liquids; consumption excludes natural gas liquids.   [3]Comprised of hydropower, nuclear power, geothermal energy and other.   [4]Represents peak year for U.S. energy consumption.   [5]Represents peak year for U.S. energy production.
Source: U.S. Energy Information Administration, *Annual Energy Review,* and unpublished data.

FIGURE 11–34   Reference for Assignment 1
Source: U.S. Department of Commerce, Bureau of the Census, *Statistical Abstract of the United States, 1989* (Washington, 1989), 554.

They can be completed either as individual or collaborative projects, depending on directions from your instructor.

1. **Pie, Bar, and Line Charts.** Figure 11–34 shows total energy production and consumption from 1960 through 1987, while also breaking down both into the four categories of coal, petroleum, natural gas, and "other." Use those data to complete the following charts:

   - A pie chart that shows the four groupings of energy consumption in 1987
   - A bar chart that shows the trend in total consumption during these six years: 1960, 1965, 1970, 1975, 1980, and 1985
   - A segmented bar chart that shows the total energy production, and the four percent-of-production subtotals, for 1960 and 1980
   - A single-line chart showing energy production from 1965 through 1975
   - A multiple-line chart that contrasts the coal, petroleum, and natural gas percent of production for any 10-year span on the table

2. **Schedule Charts.** Using any options discussed in this chapter, draw a schedule chart that reflects your work on one of the following:

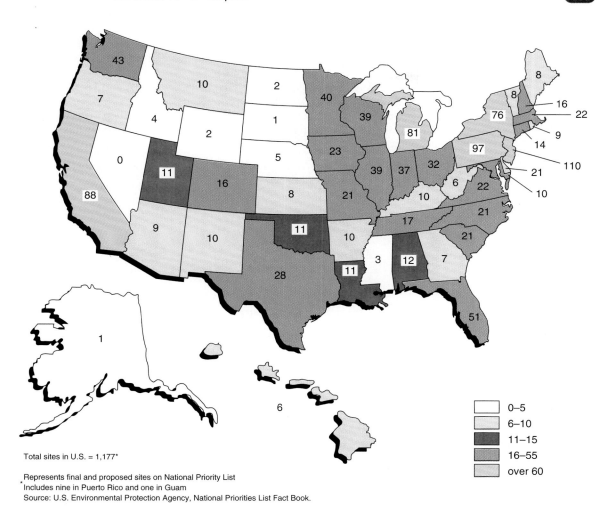

Total sites in U.S. = 1,177*

0–5
6–10
11–15
16–55
over 60

.Represents final and proposed sites on National Priority List
* Includes nine in Puerto Rico and one in Guam
Source: U.S. Environmental Protection Agency, National Priorities List Fact Book.

FIGURE 11–35   Reference for Assignment 6
Source: U.S. Department of Commerce, Bureau of the Census, *Statistical Abstract of the United States, 1989* (Washington, 1989), 220.

- A project at work
- A laboratory course at school
- A lengthy project in a course such as this one

3. **Flowcharts.** Select a process with which you are familiar because of work, school, home, or other interests. Then draw a flowchart that outlines the main activities involved in this process.

4. **Organization Chart.** Select an organization with which you are familiar, or one about which you can find information. Then construct a linear flowchart that would help an outsider understand the management structure of all or part of the organization.

5. **Technical Drawing.** Drawing freehand or using computer-assisted design, produce a simple technical drawing of an object with which you are familiar through work, school, or home use.

6. **Table.** Using the map in Figure 11–35, draw an informal table correlating the five main groupings with the number of states in each.

7. **Misuse of Graphics.** Find three deficient graphics in newspapers, magazines, reports, or other technical documents. Submit copies of the graphics along with a written critique that (1) describes in detail the deficiencies of the graphics and (2) offers suggestions for improving them.

8. **Misuse of Graphics.** Analyze the graphics in Figures 11–36, 11–37, and 11–38. Describe any deficiencies and offer suggestions for improvement.

9. **Ethics Assignment.** Develop a list of practical guidelines that would help writers like yourself avoid ethical errors in creating and using graphics on the job. To create this list, complete the following steps:

   1. Review parts of this chapter that deal with ethical issues, especially the "Misuse of Graphics" and "Communication Challenge" sections.

FIGURE 11–36

FIGURE 11–37

FIGURE 11–38

2. Interview someone who creates or uses graphics frequently, such as a member of the public relations or admissions departments at the college or university that you attend.

3. Review a variety of graphics in diverse media such as popular magazines, textbooks, and the Internet.

Remember to focus on concrete guidelines that could be used during the process of producing a document.

10. **International Communication Assignment.** Examine and report on the graphics preferences of a country other than the United States. For example, you could choose one of the countries in which McDuff has an international office—Venezuela, England, Germany, Kenya, Saudi Arabia, or Japan. Sources of information might include one or more of the following:

1. Books on international communication
2. Internet sites on global cultures
3. Student or work colleagues from the culture being studied

You may be fortunate enough to gain access to documents either written in the culture being studied or written by culturally knowledgeable people for use in the culture. If so, you often can draw useful conclusions from reading such primary sources.

12

Oral Communication

Speeches You Have Heard

General Instructions

Each chapter begins with a Collaboration at Work exercise that applies strategies for working in groups to chapter topics. The exercise assumes you (1) have been divided into groups of about three to six students, (2) will use group time inside or outside of class to complete the case, and (3) will produce an oral or written response. For guidelines about writing in groups, refer to pages 28–34.

Background for Assignment

Even if you have little experience as a public speaker, and even if you have not read this chapter, you already know a lot about what makes a good or bad speech. That's because you have listened to so many presentations in your life, from informal lectures in a classroom to famous speeches by national figures. Every day you see or hear snippets of presentations in the media, so your exposure has been high.

Considering all the "visual input," you probably have developed preferences for certain features in presentations.

Group Assignment

In this exercise, you and your fellow group members will do the following:

1. Share anecdotes about good and bad presentations you have heard, focusing on criteria such as content, organization, delivery, graphics, and gestures.
2. Assemble a first list that includes features of speeches and characteristics of speakers that you consider worthy of modeling.
3. Assemble a second list that includes features of speeches and characteristics of speakers that you think should be avoided.

Be able to provide reasons why these features are positive or negative.

Your career will present you with many opportunities for oral presentations, both formal and informal. At the time they arise, however, you may not consider them to be "opportunities." They may seem to loom on the horizon as stressful obstacles. That response is normal. The purpose of this chapter is to provide the tools that will help oral presentations contribute to your self-esteem and career success. You will find guidelines for preparation and delivery, techniques for dealing with anxiety, and an example of a technical presentation. Also, the chapter addresses the related topic of running effective meetings.

The entire chapter is based on one simple principle: *Almost anyone can become an excellent speaker.* Put aside the myth that competent speakers are born with the talent, that "either they have it or they don't." Certainly some people have more natural talent at thinking on their feet or have a more resonant voice. But success at speaking can come to all speakers, whatever their talent, if they follow the "3 Ps":

Step 1: Prepare carefully
Step 2: Practice often
Step 3: Perform with enthusiasm

These steps form the foundation for all specific guidelines that follow. Before presenting these guidelines, this chapter examines specific ways that formal and informal presentations become part of your professional life.

PRESENTATIONS AND YOUR CAREER

Some oral presentations you will choose to give; others will be "command performances" thrust upon you. Using McDuff, Inc., as a backdrop, the following examples present some realistic situations in which the ability to speak well can lead to success for you and your organization:

- **Getting hired:** As a job applicant with a business degree, you are asked to present several McDuff managers with a 10-minute summary of your education, previous experience, and career goals.
- **Getting customers:** As coordinator of a McDuff proposal team, you have just been informed that McDuff made the "shortlist" of companies bidding on a contract to manage a large construction project. You and your three team members must deliver a 20-minute oral presentation that highlights the written proposal. To be given at the client's office in Grand Rapids, Michigan, in five days, the presentation will begin and end with comments by you, in your role as coordinator. Your three colleagues will each contribute a five-minute talk.
- **Keeping customers:** As a field engineer at McDuff's St. Louis office, you recently submitted a report on your evaluation of a 50-year-old dam in the Ozarks. Now your clients, the commissioners of the county that owns the dam, have asked you to attend their monthly meeting to present an overview of your findings and respond to questions.

- **Contributing to your profession:** As a laboratory supervisor for McDuff, you belong to a professional society that meets yearly to discuss issues in your field. This year you have been asked to deliver a 15-minute presentation on new procedures for testing toxic-waste samples in the laboratory.
- **Contributing to your community:** As an environmental scientist at McDuff's San Francisco office, you have been asked to speak at the quarterly meeting of OceanSave, an activist environmental organization. The group suggests that you speak for half an hour on environmental threats to aquatic life. You accept the invitation because you know that McDuff management encourages such community service.
- **Getting promoted:** As an employee about to be considered for promotion, you will be evaluated on your ability to present information orally. Supervisors will discuss whether they themselves have heard—*or* heard from others about—your effective presentations to colleagues, clients, or community representatives.

As you can see from this list, oral presentations are defined quite broadly. Usually they can be classified according to criteria like these:

1. **Format:** from informal question/answer sessions to formal speeches
2. **Length:** from several-minute overviews to long sessions of an hour or more
3. **Number of presenters:** from solo performances to group presentations
4. **Content:** from a few highlights to detailed coverage

Throughout your career, you will speak to different-sized groups, on diverse topics, and in varied formats. The next two sections provide some common guidelines on preparation, delivery, and graphics.

GUIDELINES FOR PREPARATION AND DELIVERY

The goal of most oral presentations is quite simple: You must present a few basic points, in a fairly brief time, to an interested but usually impatient audience. Simplicity, brevity, and interest are the keys to success. If you deliver what *you* expect when *you* hear a speech, then you will give good presentations yourself.

Although the guidelines here apply to any presentation, they relate best to those that precede or follow a written report, proposal, memo, or letter. Few career presentations are isolated from written work. With this connection in mind, you will note many similarities between the guidelines for good speaking and those for good writing covered in earlier chapters—especially the importance of analyzing the needs of the audience.

● *Presentation Guideline 1: Know Your Listeners*

These features are common to most listeners:

- They cannot "rewind the tape" of your presentation, as opposed to the way they can skip back and forth through the text of a report.

- They are impatient after the first few minutes, particularly if they do not know where a speech is going.
- They will daydream and often need their attention brought back to the matter at hand (expect a 30-second attention span).
- They have heard so many disappointing presentations that they might not have high expectations for yours.

To respond to these realities, you must learn as much as possible about your listeners. For example, you can (1) consider what you already know about your audience, (2) talk with colleagues who have spoken to the same group, and (3) find out which listeners make the decisions.

Most importantly, make sure not to talk over anyone's head. If there are several levels of technical expertise represented by the group, find the lowest common denominator and decrease the technical level of your presentation accordingly. Remember—decision-makers are often the ones without current technical experience. They may want only highlights. Later they can review written documents for details or solicit more technical information during the question-and-answer session after you speak.

● Presentation Guideline 2: Use the Preacher's Maxim

The well-known preacher's maxim goes like this:

> First you tell 'em what you're gonna tell 'em, then you tell 'em, and then you tell 'em what you told 'em.

Why should most speakers follow this plan? Because it gives the speech a simple three-part structure that most listeners can grasp easily. Here is how your speech should be organized (note that it corresponds to the ABC format used throughout this text for writing):

1. Abstract (beginning of presentation): Right at the outset, you should (1) get the listeners' interest (with an anecdote, a statistic, or other technique), (2) state the exact purpose of the speech, and (3) list the main points you will cover. Do not try the patience of your audience with an extended introduction. Use no more than a minute.

Example: "Last year, Jones Engineering had 56 percent more field accidents than the year before. This morning, I'll examine a proposed safety plan that aims to solve this problem. My presentation will focus on three main benefits of the new plan: lower insurance premiums, less lost time from accidents, and better morale among the employees."

2. Body (middle of presentation): Here you discuss the points mentioned briefly in the introduction, in the same order that they were mentioned. Provide the kinds of obvious transitions that help your listeners stay on track.

Example: "The final benefit of the new safety plan will be improved morale among the field-workers at all our job sites. . . ."

3. Conclusion (end of presentation): In the conclusion, review the main ideas covered in the body of the speech and specify actions you want to occur as a result of your presentation.

> Example: "Jones Engineering can benefit from this new safety plan in three main ways. . . . If Jones implements the new plan next month, I believe you will see a dramatic reduction in on-site accidents during the last half of the year."

This simple three-part plan for all presentations gives listeners the handle they need to understand your speech. First, there is a clear "road map" in the introduction so that they know what lies ahead in the rest of the speech. Second, there is an organized pattern in the body, with clear transitions between points. And third, there is a strong finish that brings the audience back full circle to the main thrust of the presentation.

● *Presentation Guideline 3: Stick to a Few Main Points*

Our short-term memory holds limited items. It follows that listeners are most attentive to speeches organized around a few major points. In fact, a good argument can be made for organizing information in groups of *threes* whenever possible. For reasons that are not totally understood, listeners seem to remember groups of three items more than they do any other size groupings—perhaps because:

- The number is simple.
- It parallels the overall three-part structure of most speeches and documents (beginning, middle, end).
- Many good speakers have used triads (Winston Churchill's "Blood, sweat, and tears," Caesar's "I came, I saw, I conquered," etc.).

Whatever the reason, groupings of three will make your speech more memorable to the audience.

● *Presentation Guideline 4: Put Your Outline on Cards, Paper, or Overheads*

The best presentations are "extemporaneous," meaning the speaker shows great familiarity with the material but uses notes for occasional reference. Avoid extremes of (1) reading a speech verbatim, which many listeners consider the ultimate insult, or (2) memorizing a speech, which can make your presentation seem somewhat wooden and artificial.

Ironically, you appear more natural if you refer to notes during a presentation. Such extemporaneous speaking allows you to make last-minute changes in phrasing and emphasis that may improve delivery, rather than locking you into specific phrasing that is memorized or written out word for word.

Depending on your personal preference, you may choose to write speech notes on (1) index cards, (2) a sheet or two of paper, or (3) overhead transparencies. The main advantages and disadvantages of each are in a separate list.

1. **Notes on Cards (3" × 5" or 4" × 6")**

   Advantages

   - Are easy to carry in a shirt pocket, coat, or purse
   - Provide a way to organize points, through ordering of cards
   - Can lead to smooth delivery in that each card contains only one or two points that are easy to view
   - Can be held in one hand, allowing you to move away from lectern while speaking

   Disadvantages

   - Keep you from viewing outline of entire speech
   - Require that you flip through cards repeatedly in speech
   - Can limit use of gestures with hands
   - Can cause confusion if they are not in correct order

2. **Notes on Sheets of Paper**

   Advantages

   - Help you quickly view outline of entire speech
   - Leave your hands free to use gestures
   - Are less obvious than note cards, for no flipping is needed

   Disadvantages

   - Tend to tie you to lectern, where the sheets lie
   - May cause slipups in delivery if you lose your place on the page

3. **Notes on Overhead Transparencies**

   Advantages

   - Introduce variety to audience
   - Give visual reinforcement to audience
   - Can be turned on and off

   Disadvantages

   - Cannot be altered as easily at last minute
   - Must be presented neatly and in parallel style
   - Involve the usual risks that reliance on machinery always introduces into your presentation

---

● *Presentation Guideline 5: Practice, Practice, Practice*

Many speakers prepare a well-organized speech but then fail to add the essential ingredient: practice. Constant practice distinguishes superior presentations from mediocre ones. It also helps to eliminate the nervousness that most speakers feel at one time or another.

In practicing your presentation, make use of four main techniques. They are listed here, from least effective to most effective:

- **Practice before a mirror:** This old-fashioned approach allows you to hear and see yourself in action. The drawback, of course, is that it is difficult to evaluate your own performance while you are speaking. Nevertheless, such run-throughs definitely make you more comfortable with the material.

- **Use of audiotape:** Most presenters have access to a tape player, so this approach is quite practical. The portability of the machines allows you to practice almost anywhere. Although taping a presentation will not improve gestures, it will help you discover and eliminate verbal distractions such as filler words (*uhhhh, um, ya know*).
- **Use of live audience:** Groups of your colleagues, friends, or family—simulating a real audience—can provide the kinds of responses that approximate those of a real audience. In setting up this type of practice session, however, make certain that observers understand the criteria for a good presentation and are prepared to give an honest, forthright critique.
- **Use of videotape:** This practice technique allows you to see and hear yourself as others do. Your careful review of the tape, particularly when done with another qualified observer, can help you identify and eliminate problems with posture, eye contact, vocal patterns, and gestures. At first it can be a chilling experience, but soon you will get over the awkwardness of seeing yourself on film.

● *Presentation Guideline 6: Speak Vigorously and Deliberately*

"Vigorously" means with enthusiasm; "deliberately" means with care, attention, and appropriate emphasis on words and phrases. The importance of this guideline becomes clear when you think back to how you felt during the last speech you heard. At the very least, you expected the speaker to show interest in the subject and to demonstrate enthusiasm. Good information is not enough. You need to arouse the interest of the listeners.

You may wonder, "How much enthusiasm is enough?" The best way to answer this question is to hear or (preferably) watch yourself on tape. Your delivery should incorporate just enough enthusiasm so that it sounds and looks a bit unnatural to you. Few if any listeners ever complain about a speech being too enthusiastic or a speaker being too energetic. But many, many people complain about dull speakers who fail to show that they themselves are excited about the topic. Remember—every presentation is, in a sense, "show time."

● *Presentation Guideline 7: Avoid Filler Words*

Avoiding filler words presents a tremendous challenge to most speakers. When they think about what comes next or encounter a break in the speech, they may tend to fill the gap with filler words and phrases such as these:

uhhhhh . . .

ya know . . .

okay . . .

well . . . uh . . .

like . . .

I mean . . .

umm . . .

These gap-fillers are a bit like spelling errors in written work: Once your listeners find a few, they start looking for more and are distracted from your presentation. To eliminate such distractions, follow these three steps:

Step 1: **Use pauses to your advantage.** Short gaps or pauses inform the listener that you are shifting from one point to another. In signaling a transition, a pause serves to draw attention to the point you make right after the pause. Note how listeners look at you when you pause. Do *not* fill these strategic pauses with filler words.

Step 2: **Practice with tape.** Tape is brutally honest: When you play it back, you will become instantly aware of fillers that occur more than once or twice. Keep a tally sheet of the fillers you use and their frequency. Your goal will be to reduce this frequency with every practice session.

Step 3: **Ask for help from others.** After working with tape machines in Step 2, give your speech to an individual who has been instructed to stop you after each filler. This technique gives immediate reinforcement.

● *Presentation Guideline 8: Use Rhetorical Questions*

Enthusiasm, of course, is your best delivery technique for capturing the attention of the audience. Another technique is the use of rhetorical questions at pivotal points in your presentation.

Rhetorical questions are those you ask to get listeners thinking about a topic, not those that you would expect them to answer out loud. They prod listeners to think about your point and set up an expectation that important information will follow. Also, they break the monotony of standard declarative sentence patterns. For example, here is a rhetorical question used by a computer salesperson in proposing a purchase by one of McDuff's small offices:

> I've discussed the three main advantages that a centralized word processing center would provide your office staff. But is this an approach that you can afford at this point in the company's growth?

Then the speaker would follow the question with remarks supporting the position that the system is affordable.

"What if" scenarios provide another way to introduce rhetorical questions. They gain listeners' attention by having them envision a situation that might occur. For example, a safety engineer could use this kind of rhetorical question in proposing McDuff's asbestos-removal services to a regional bank:

> What if you repossessed a building that contained dangerous levels of asbestos? Do you think that your bank would then be liable for removing all the asbestos?

Again, the question pattern heightens listener interest.

Rhetorical questions do not come naturally. You must make a conscious effort to insert them at points when it is most important to gain or redirect the attention of the audience. Three particularly effective uses follow:

1. **As a grabber at the beginning of a speech:** "Have you ever wondered how you might improve the productivity of your word processing staff?"
2. **As a transition between major points:** "We've seen that centralized word processing can improve the speed of report production, but will it require any additions to your staff?"
3. **As an attention-getter right before your conclusion:** "Now that we've examined the features of centralized word processing, what's the next step you should make at McDuff?"

● *Presentation Guideline 9: Maintain Eye Contact*

Your main goal—always—is to keep listeners interested in what you are saying. This goal requires that you maintain control, using whatever techniques possible to direct the attention of the audience. Frequent eye contact is one good strategy.

The simple truth is that listeners pay closer attention to what you are saying when you look at them. Think how you react when a speaker makes constant eye contact with you. If you are like most people, you feel as if the speaker is speaking to you personally—even if there are 100 people in the audience. Also, you tend to feel more obligated to listen when you know that the speaker's eyes will be meeting yours throughout the presentation. Here are some ways you can make eye contact a natural part of your own strategy for effective oral presentations:

● **With audiences of about 30 or less:** Make regular eye contact with everyone in the room. Be particularly careful not to ignore members of the audience who are seated to your far right and far left (see Figure 12–1). Many speakers tend to focus on the listeners within Section B. Instead, make wide sweeps so that listeners in Sections A and C get equal attention.
● **With large audiences:** There may be too many people or a room too large for you to make individual eye contact with all listeners. In this case, focus on just a few people in all three sections of the audience noted in Figure 12–1. This approach gives the appearance that you are making eye contact with the entire audience.

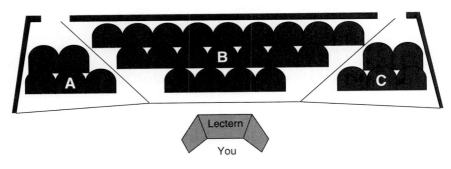

FIGURE 12–1   Audience sections

- **With any size audience:** Occasionally look away from the audience—either to your notes or toward a part of the room where there are no faces looking back. In this way, you avoid the appearance of staring too intensely at your audience. Also, these breaks give you the chance to collect your thoughts or check your notes.

● *Presentation Guideline 10: Use Appropriate Gestures and Posture*

Speaking is only one part of giving a speech; another is adopting appropriate posture and using gestures that will reinforce what you are saying. Note that good speakers are much more than "talking heads" before a lectern. Instead, they:

1. Use their hands and fingers to emphasize major points
2. Stand straight, without leaning on or gripping the lectern
3. Step out from behind the lectern on occasion, to decrease the distance between them and the audience
4. Point toward visuals on screens or charts, without losing eye contact with the audience

The audience will judge you by what you say *and* what they see, a fact that again makes videotaping a crucial part of your preparation. With work on this facet of your presentation, you can avoid problems like keeping your hands constantly in your pockets, rustling change (remove pocket change and keys beforehand), tapping a pencil, scratching nervously, slouching over a lectern, and shifting from foot to foot.

GUIDELINES FOR PRESENTATION GRAPHICS

More than ever before, listeners expect good graphics during oral presentations. Much like gestures, graphics transform the words of your presentation into true communication with the audience. To emphasize the importance of speech graphics, here is an incident from industry once reported to the author:

> Several years ago the competitors for designing a large city's football stadium had been narrowed to three firms. Firm A, a large and respected company, had done some preliminary design work on the project and was expected to get the contract. Firm B, another large and respected firm, was competing fiercely for the job. And Firm C, a small and fairly new company, was considered by all concerned to be a genuine long shot. Yet it had submitted an interesting-enough proposal to be chosen as a finalist.
>
> All three firms were invited to make 10- or 15-minute presentations on their proposals. The presentations by Firms A and B were professional, conventional, and predictable. Firm C, however, took a different and riskier approach. Its presentation was barely 5 minutes long and was given simultaneously with a videotape. As expected, the speech itself stressed the benefits of Firm C's design for the stadium. The accompanying videotape, however, was quite unconventional. It interspersed drawings of Firm C's design with highlights of that city's football team scoring touchdowns, catching passes, and making game-winning tackles.
>
> Firm C's effort to associate the winning football team with its proposed design worked. Shortly after the presentations, the selection committee chose the dark horse, Firm C, to design the new stadium.

The lesson here is *not* that fancy visuals always win a contract. Instead, the point is that innovative graphics, in concert with a solid proposal presentation, can set you apart from the competition. Firm C had a sound stadium design reinforced by an unusual visual display. Granted, it walked the fine line between effective and manipulative graphics. Yet even the fanciest visuals cannot disguise a bad idea to a discerning audience. The Firm C presenters won because they found an effective way to present their proposal. They incorporated graphics into their presentation to reinforce main selling points.

● *Graphics Guideline 1: Discover Listener Preferences*

Some professionals prefer simple speech graphics, such as a conventional flip chart. Others prefer more sophisticated equipment, such as video projectors connected to laptop computers. For example, if your instructor in this course uses presentation graphics, he or she may have chosen overhead transparencies *or* the PowerPoint computer-aided program that is included in the instructional package.

Your listeners are usually willing to indicate their preferences when you call on them. Contact the audience ahead of time and make some inquiries. Also ask for information about the room in which you will be speaking. If possible, request a setting that allows you to make best use of your graphics choice. If you have no control over the setting, then choose graphics that best fit the constraints. Details about lighting, wall space, and chair configuration can greatly influence your selection.

● *Graphics Guideline 2: Think about Graphics Early*

Graphics done as an afterthought usually look "tacked on." Plan graphics while you prepare the text so that the final presentation will seem fluid. This guide-line holds true especially if you rely upon specialists to prepare your visuals. These professionals need some lead time to do their best work. Also, they can often provide helpful insights about how visuals will enhance the presentation—*if* you consult them early enough and *if* you make them a part of your presentation team.

The goal is to use graphics of which you can be proud. Never, never put yourself in the position of having to apologize for the quality of your graphic material. If an illustration is not up to the quality your audience would expect, do *not* use it.

● *Graphics Guideline 3: Keep the Message Simple*

Listeners can be suspicious of overly slick visual effects. Most people prefer the simplicity of simple overhead transparencies and flip charts, for example. If you decide to use more sophisticated techniques, make sure they fit your context. In the previous stadium example, Firm C took a risk by using videotaped football highlights to sell its stadium design. Complex graphics such as videotape can sometimes overshadow the proposal itself. In Firm C's case, however, the tape's purpose was kept quite simple. It simply associated the success of the team with the potential success of the proposed stadium design.

● *Graphics Guideline 4: Make Any Wording Brief and Visible*

Some basic design guidelines apply whether you are using posters, overhead transparencies, or computer-aided graphics such as PowerPoint.

- Use few words, emphasizing just one idea on each frame.
- Use much white space, perhaps as much as 60–70% per frame.
- Use "landscape" format more often than "portrait," especially since it is the preferred default setting for much presentation software.
- Use sans serif large print, from 14–18 pt. minimum for text to 48 pt. for titles.

Your goal should be to create graphics that are easily seen from anywhere in the room and that complement—but do *not* overpower—your presentation.

● *Graphics Guideline 5: Use Colors Carefully*

Colors can add flair to visuals. Follow these simple guidelines to make colors work for you:

- Have a good reason for using color (such as the need to highlight three different bars on a graph with three distinct colors).
- Use only dark, easily seen colors, and be sure that a color contrasts with its background (for example, yellow on white would not work well).
- Use no more than three or four colors in each graphic (to avoid a confused effect).
- For variety, consider using white on a black or dark green background.

● *Graphics Guideline 6: Leave Graphics Up Long Enough*

Because graphics reinforce text, they should be shown only while you address the particular point at hand. For example, reveal a graph just as you are saying, "As you can see from the graph, the projected revenue reaches a peak in 2002." Then pause and leave the graph up a bit longer for the audience to absorb your point.

How long is *too* long? A graphic outlives its usefulness when it remains in sight after you have moved on to another topic. Listeners will continue to study it and ignore what you are now saying. If you use a graphic once and plan to return to it, take it down after its first use and show it again later.

● *Graphics Guideline 7: Avoid Handouts*

Because timing is so important in your use of speech graphics, handouts are usually a bad idea. Readers move through a handout at their own pace, rather than at the pace the speaker might prefer. Thus handouts cause you to lose the attention of your audience. Use them only if (1) no other visual will do, (2) your listener has requested them, or (3) you distribute them as reference material *after* you have finished talking.

● *Graphics Guideline 8: Maintain Eye Contact While Using Graphics*

Do not stare at your visuals while you speak. Maintain control of listeners' responses by looking back and forth from the visual to faces in the audience. To point to the graphic aid, use the hand closest to the visual. Using the opposite hand causes you to cross over your torso, forcing you to turn your neck and head away from the audience.

● *Graphics Guideline 9: Include All Graphics in Your Practice Sessions*

Dry runs before the actual presentation should include every graphic you plan to use, in its final form. This is a good reason to prepare graphics as you prepare text, rather than as an afterthought. Running through a final practice without graphics would be much like doing a dress rehearsal for a play without costumes and props—you would be leaving out parts that require the greatest degree of timing and orchestration. Practicing with graphics helps you improve transitions.

● *Graphics Guideline 10: Use Your Own Equipment*

Murphy's Law always seems to apply when you use another person's audiovisual equipment: Whatever can go wrong, will. For example, a new bulb burns out, there is no extra bulb in the equipment drawer, an extension cord is too short, the screen does not stay down, the client's computer doesn't read your disk—many speakers have experienced these problems and more. Even if the equipment works, it often operates differently from what you are used to. The only sure way to put the odds in your favor is to carry your own equipment and set it up in advance.

However, most of us have to rely on someone else's equipment at least sometimes. Here are a few ways to ward off disaster:

- Find out exactly who will be responsible for providing the equipment and contact that person in advance.
- Have some easy-to-carry backup supplies in your car—an extension cord, an overhead projector bulb, felt-tip markers, and chalk, for example.
- Bring handout versions of your visuals to use as a last resort.

In short, you want to avoid putting yourself in the position of having to apologize. Plan well.

OVERCOMING NERVOUSNESS

The problem of nervousness deserves special mention because it is so common. Virtually everyone who gives speeches feels some degree of nervousness before "the event." An instinctive "fight or flight" response kicks in for the many people who have an absolute dread of presentations. In fact, surveys have determined that most of us rate public speaking at the top of our list of fears, even above sickness and death! Given this common response, this chapter considers the problem and offers suggestions for overcoming it.

Why Do We Fear Presentations?

Most of us feel comfortable with informal conversations, when we can voice our views to friends and indulge in impromptu exchanges. We are used to this type of casual presenting of our ideas. Formal presentations, however, put us into a more structured, more awkward, and thus more tense environment. Despite the fact that we may know the audience is friendly and interested in our success, the formal context triggers nervousness that is sometimes difficult to control.

This nervous response is normal and, to some degree, useful. It gets you "up" for the speech. That adrenaline pumping through your body can generate a degree of enthusiasm that propels the presentation forward and creates a lively performance. Just as veteran actors admit to some nervousness helping to improve their performance, excellent speakers usually can benefit from the same effect.

The problem occurs when nervousness felt before or during a speech becomes so overwhelming that it affects the quality of the presentation. Since sympathy is the last feeling a speaker wants the audience to have, it is worth considering some techniques to combat nervousness.

A Strategy for Staying Calm

As the cliché goes, do not try to eliminate "butterflies" before a presentation—just get them to fly in formation. In other words, it is best to acknowledge that a certain degree of nervousness will always remain. Then go about the business of getting it to work for you. Here are a few suggestions:

● *No Nerves Guideline 1: Know Your Speech*

The most obvious suggestion is also the most important one. If you prepare your speech well, your command of the material will help to conquer any queasiness you feel—particularly at the beginning of the speech, when nervousness is usually at its peak. Be so sure of the material that your listeners will overlook any initial discomfort you may feel.

● *No Nerves Guideline 2: Prepare Yourself Physically*

Your physical well-being before the speech can have a direct bearing on anxiety. More than ever before, most cultures understand the essential connection between mental and physical well-being. This connection suggests you should take these precautions before your presentation:

- **Avoid caffeine or alcohol for at least several hours before you speak.** You do not need the additional jitters brought on by caffeine or the false sense of ease brought on by alcohol.
- **Eat a light, well-balanced meal within a few hours of speaking.** However, do not overdo it—particularly if a meal comes right before your speech. If you are convinced that any eating will increase your anxiety, wait to eat until after speaking.
- **Practice deep-breathing exercises before you speak.** Inhale and exhale slowly, making your body slow down to a pace you can control. If you can control your breathing, you can probably keep the butterflies flying in formation.
- **Exercise normally the same day of the presentation.** A good walk will help invigorate you and reduce nervousness. However, do not wear yourself out by exercising more than you would normally.

● *No Nerves Guideline 3: Picture Yourself Giving a Great Presentation*

Many speakers become nervous because their imaginations are working overtime. They envision the kinds of failure that almost never occur. Instead, speakers should be constantly bombarding their psyches with images of success, not failure. Mentally take yourself through the following steps of the presentation:

- Arriving at the room
- Feeling comfortable at your chair
- Getting encouraging looks from your audience
- Giving an attention-getting introduction
- Presenting your supporting points with clarity and smoothness
- Ending with an effective wrap-up
- Fielding questions with confidence

Sometimes called "imaging," this technique helps to program success into your thinking and control negative feelings that pass through the minds of even the best speakers.

● *No Nerves Guideline 4: Arrange the Room as You Want*

To control your anxiety, assert some control over the physical environment as well. You need everything going for you if you are to feel at ease. Make sure that chairs are arranged to your satisfaction, that the lectern is positioned to your taste, that the lighting is adequate, and so on. These features of the setting can almost always be adjusted if you make the effort to ask. Again, it is a matter of your asserting control to increase your overall confidence.

● *No Nerves Guideline 5: Have a Glass of Water Nearby*

Extreme thirst and a dry throat are physical symptoms of nervousness that can affect delivery. There is nothing to worry about as long as you have water available. Think about this need ahead of time so that you do not have to interrupt your presentation to pour a glass of water.

● *No Nerves Guideline 6: Engage in Casual Banter Before the Speech*

If you have the opportunity, chat with members of the audience before the speech. This ice-breaking technique will reduce your nervousness and help start your relationship with the audience.

● *No Nerves Guideline 7: Remember That You Are the Expert*

As a final "psyching up" exercise before you speak, remind yourself that you have been invited or hired to speak on a topic about which you have useful knowledge. Your listeners want to hear what you have to say and are eager for you to provide useful information to them. So tell yourself, "I'm the expert here!"

● *No Nerves Guideline 8: Do Not Admit Nervousness to the Audience*

No matter how anxious you may feel, never admit it to others. First of all, you do not want listeners to feel sorry for you—that is not an emotion that will lead to a positive critique of your speech. Second, nervousness is almost never apparent to the audience. Your heart may be pounding, your knees may be shaking, and your throat may be dry, but few if any members of the audience can see these symptoms. Why draw attention to the problem by admitting to it? Third, you can best defeat initial anxiety by simply pushing right on through.

● *No Nerves Guideline 9: Slow Down*

Some speakers who feel nervous tend to speed through their presentations. If you have prepared well and practiced the speech on tape, you are not likely to let this happen. Having heard yourself on tape, you will be better able to sense that the pace is too quick. As you speak, constantly remind yourself to maintain an appropriate pace. If you have had this problem before, you might even write "Slow down!" in the margin of your notes.

● *No Nerves Guideline 10: Join a Speaking Organization*

The previous nine guidelines will help reduce your anxiety about a particular speech. To help solve the problem over the long term, however, consider joining an organization like Toastmasters International, which promotes the speaking skills of all its members. Like some other speech organizations, Toastmasters has chapters that meet at many companies and campuses. These meetings provide an excellent, supportive environment in which all members can refine their speaking skills.

EXAMPLE OF MCDUFF ORAL PRESENTATION

This section presents the text and visuals of a short presentation given by Kim Mason, an environmental expert for McDuff's Atlanta office. She has been invited to speak at the monthly lunch meeting of an organization of building owners in the Atlanta region. The agreed-upon topic is the problem of asbestos contamination.

The members of Kim's audience have an obvious interest in the problem: They own buildings that are at risk. Yet they know little more about asbestos than that it is a health issue they must consider when they renovate. Kim's job is to inform them and heighten their awareness. She needs to cover only the highlights, however, because the presentation will be followed by a detailed question-and-answer session. Although some of these owners have been, and will be, clients of McDuff, she has an ethical obligation to avoid promoting McDuff during her presentation.

> Good evening. My name is Kim Mason, and I work for the asbestos-abatement division of McDuff, Inc., in Atlanta. I've been asked to give a short presentation on the problem of asbestos and then to respond to your questions about the importance of removing it from buildings. I'll focus on three main reasons why you, as building owners, should be concerned about the asbestos problem: (TRANSPARENCY 1 [see Figure 12–2])

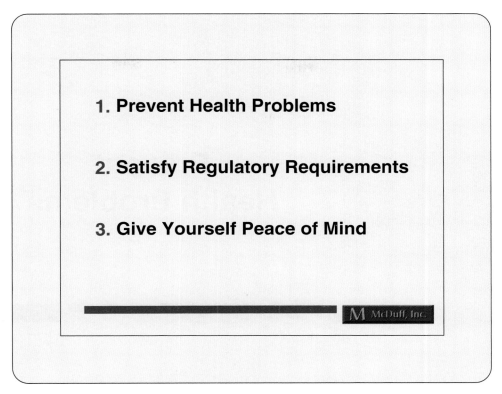

FIGURE 12–2 Transparency 1 for sample presentation

1. To prevent future health problems of your tenants
2. To satisfy regulatory requirements of the government
3. To give yourself peace of mind for the future

Again, my comments will provide just an overview, serving as a basis for the question session that follows in a few minutes. (TRANSPARENCY 2 [see Figure 12–3])

Question: What is the most important reason you need to be concerned about asbestos? Answer: The long-term health of the tenants, workers, and other people in buildings that contain asbestos. Research has clearly linked asbestos with a variety of diseases, including lung cancer, colon cancer, and asbestosis (a debilitating lung disease). Although this connection was first documented in the 1920s, it has only been taken seriously in the last few decades. Unfortunately, by that time asbestos had already been commonly used in many building materials that are part of many structures today.

Here are some of the most common building products containing asbestos. (TRANSPARENCIES 3, 4, and 5 [see Figure 12–4]) As you can see, asbestos was used in materials as varied as floor tiles, pipe wrap, roof felt, and insulation around heating systems. An abundant and naturally occurring mineral, asbestos was fashioned into construction materials through processes such as packing, weaving, and spraying. Its property of heat resistance, as well as its availability, was the main reason for such widespread use.

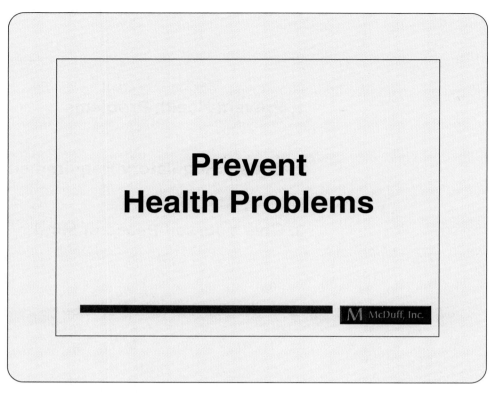

FIGURE 12–3   Transparency 2 for sample presentation

While still embedded in material, asbestos causes no real problems. However, when it deteriorates or is damaged, fibers may become airborne. In this state, they can enter the lungs and cause the health problems mentioned a minute ago. This risk prompted the Environmental Protection Agency in the mid-1970s to ban the use of certain asbestos products in most new construction. But today the decay and renovation of many asbestos-containing building materials may put many of our citizens at risk for years to come. (TRANSPARENCY 6 [see Figure 12–5])

After your concern for occupants' health, what's the next best reason to learn more about asbestos? It's the *law*. Both the Occupational Safety and Health Administration (OSHA) and the Georgia Department of Natural Resources (DNR) require that you follow certain procedures when structures you own could endanger tenants and asbestos-removal workers with contamination. For example, when a structure undergoes renovation that will involve any asbestos-containing material (ACM), the ACM must be removed by following approved engineering procedures. Also, the contaminated refuse must be disposed of in approved landfills. Considering the well-documented potential for health problems related to airborne asbestos, this legislative focus on asbestos contamination makes good sense.

By the way, both OSHA and DNR regulations require removal of asbestos by licensed contractors. These contractors, however, will assume liability only for what they have been

FIGURE 12–4 Transparencies 3–5 for sample presentation

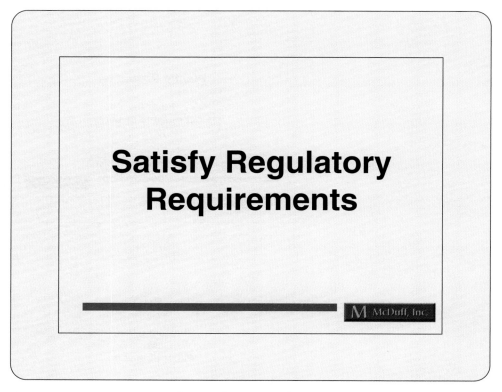

FIGURE 12–5   Transparency 6 for sample presentation

told to remove. They may or may not have credentials and training in health and safety. Therefore, building owners should hire a firm with a professional who will (1) survey the building and present a professional report on the degree of asbestos contamination and (2) monitor the work of the contractor in removing the asbestos. By taking this approach, you as an owner stand a good chance of eliminating all problems with your asbestos.

Yes, it is *your* asbestos. As owner of a building, you also legally own the asbestos associated with that building—*forever*. For example, if a tenant claims to have been exposed to asbestos because of your abatement activity and then brings a lawsuit, you must have documentation showing that you contracted to have the work performed in a "state-of-the-art" manner. If, as recommended, you have hired a qualified monitoring firm and a reputable contracting firm, liability will be focused on the contractor and the monitoring firm—*not* on you. (TRANSPARENCY 7 [see Figure 12–6])

Which brings me to the last reason for concerning yourself with any potential asbestos problem: *peace of mind*. If you examine and then effectively deal with any asbestos contamination that exists in your buildings, you will sleep better at night. For one thing, you will have done your level best to preserve the health of your tenants. For another, as previously noted, you will have shifted any potential liability from you to the professionals you hired to solve the problem—assuming you hired professionals. Your monitoring firm will have continuously documented the contractor's operations and will have pro-

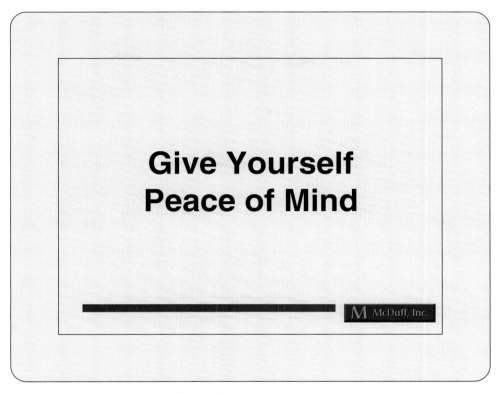

FIGURE 12–6 Transparency 7 for sample presentation

vided you with reports to keep in your files, in the event of later questions by lawyers or regulatory agencies.

In just these few minutes, I have given only highlights about asbestos. It poses a considerable challenge for all of us who own buildings or work in the abatement business. Yet the current diagnostic and cleanup methods are sophisticated enough to suggest that this problem, over time, *will* be solved. Now I would be glad to answer questions.

RUNNING EFFECTIVE MEETINGS

Like formal presentations, meetings are a form of spoken communication that goes hand in hand with written work. Important reports and proposals—even many routine ones—often are followed or preceded by a meeting. For example, you may meet with your colleagues to prepare a group-written report, with your clients to discuss a proposal, or with your department staff to outline recommendations to appear in a yearly report to management. This section will make you a first-class meeting leader by (1) highlighting some common problems with meetings, along with their associated costs to organizations, and (2) describing 10 guidelines for overcoming these problems.

Common Problems with Meetings

Here are six major complaints about meetings held in all types of organizations:

1. They start and end too late.
2. Their purpose is unclear.
3. Not everyone in the meeting really needs to be there.
4. Conversations get off track.
5. Some people dominate while others do not contribute at all.
6. Meetings end with no sense of accomplishment.

As a result of these frustrations, career professionals waste much of their time in poorly run meetings.

Because they waste participants' time, bad meetings also waste a lot of money. To find out what meetings cost an organization, do this rough calculation. Use information about an organization for which you work or for which a friend or family member works.

1. Take the average weekly number of meetings in an office.
2. Multiply that number by the average length of each meeting, in hours.
3. Multiply the result of step 2 by the average number of participants in each meeting.
4. Multiply the result of step 3 by the average hourly salary or billable amount of the participants.

The result, which may surprise you, is the average weekly cost of meetings in the office that you investigated. With these heavy costs in mind, the next section presents some simple guidelines for running good meetings.

Guidelines for Good Meetings

When you choose (or are chosen) to run a meeting, your professional reputation is at stake—let alone the costs just mentioned. Therefore, it is in your own best interests to make sure meetings run well. When you are a meeting participant, you also have an obligation to speak up and help accomplish the goals of the meeting.

The guidelines that follow will help create successful meetings. They fall into three main stages:

Stage 1: Before the meeting (Guidelines 1–4)
Stage 2: During the meeting (Guidelines 5–9)
Stage 3: After the meeting (Guideline 10)

These 10 guidelines apply to *working* meetings—that is, those in which participants use their talents to accomplish specific objectives. Such meetings usually involve a lot of conversation. The guidelines do not apply as well to *informational* meetings wherein a large number of people are assembled only to listen to announcements.

● *Meeting Guideline 1: Involve Only Necessary People*

"Necessary" means those people who, because of their position or knowledge, can contribute to the meeting. Your goal should be a small working group—four to six people is ideal. If others need to know what occurs, send them a copy of the minutes after the meeting.

● *Meeting Guideline 2: Distribute an Agenda Before the Meeting*

To be good participants, most people need to think before the meeting about the objectives of the session. The agenda also gives you, as leader, a way to keep the meeting on schedule. If you are worried about having time to cover the agenda items, consider attaching time limits to each item. That technique helps the meeting leader keep the discussion moving.

● *Meeting Guideline 3: Distribute Readings Before the Meeting*

Jealously guard time at a meeting, making sure to use it for productive discussion. If any member has reading materials that committee members should review as a basis for these discussions, such readings should be handed out ahead of time. Do not use meeting time for reading. Even worse, do not refer to handouts that all members have not had the opportunity to go over.

● *Meeting Guideline 4: Have Only One Meeting Leader*

To prevent confusion, one person should always be in charge. The meeting leader should be able to perform these tasks:

- **Listen carefully** so that all views get a fair hearing
- **Generalize accurately** so that earlier points can be brought back into the discussion when appropriate
- **Give credit to participants** so that they receive reinforcement for their efforts
- **Move toward consensus** so that the meeting does not involve endless discussion

● *Meeting Guideline 5: Start and End on Time*

Nothing deadens a meeting more than a late start, particularly when it is caused by people arriving late. Tardy participants are given no incentive to arrive on time when a meeting leader waits for them. Even worse, prompt members become demoralized by such delays. Latecomers will mend their ways if you make a practice of starting right on time.

It is also important to set an ending time for meetings so that members have a clear view of the time available. Concerning meeting length, most people do their best work in the first hour. After that, productive discussion reaches a point of diminishing returns. If working meetings must last longer than an hour, make sure to build in short breaks and stay on the agenda.

Communication Challenge

"Ethics and the Technical Presentation"

Carlos Santiago had been looking forward to attending his first international conference. The Association of Medical Technology invited him to speak at its annual meeting. Now, however, some of his research presents him with an ethical dilemma. The following sections present background on his invitation and some questions and comments for discussion.

An Opportunity in Caracas

Carlos Santiago works as an equipment development specialist in McDuff's San Francisco office. For two years he has been helping a large medical equipment firm, MedExcel, Inc., improve its C-2000 electrocardiograph. An electrocardiograph records electrical changes that occur during the human heartbeat. One of the most important pieces of equipment in diagnosing heart problems, it is made up of four main parts:

- Electrodes attached to the patient's body
- Electronics that convert and amplify the signal traveling from the electrodes to the computer
- A computer that interprets information provided by the electrodes
- An output screen and printer that provide information to the medical professional

In his work on the C-2000, Carlos (1) developed new electrodes that take extremely accurate readings, (2) designed a panel that is easier to read, and (3) created electronics that convert and amplify signals better than any model he has seen produced by any firm. These major refinements, as well as a host of smaller ones, have created what he and MedExcel believe is the most user-friendly and accurate electrocardiograph on the market.

As a result, Carlos' work has been recognized in the industry. Next week he will fly from San Francisco to Caracas, Venezuela, to speak at the annual meeting of the Association of Medical Technology (AMT). MedExel, which conveyed the invitation to him from the AMT, will be paying all his expenses for the eight-day trip. He has been asked to speak on "Advances in the Electrocardiograph." As you might expect, MedExcel expects that Carlos will showcase the C-2000 as an example of the most current technological advances.

A Late-Breaking Surprise

Everything has come together for Carlos. His client is happy, his branch manager is overjoyed that McDuff's small equipment development group is gaining recognition in a major industry, and a high-visibility professional association is about to showcase his work. The public relations and marketing potential is significant, both for McDuff and MedExcel.

Now the other shoe has dropped. Yesterday, just a week before the conference, Carlos was shocked to learn that Worldwide Medical, the main competitor of MedExcel, just came out with an updated version of its electrocardiograph, the HeartCart 300. From the article he reviewed in a weekly health-care newspaper, the machine is competitive with the C-2000 in accuracy of electrode readings and in electronics. In addition, the panel pictured in the article appears to be a significant improvement over the previous HeartCart model. After reading the article, Carlos went to his computer to find the "Home Page" for Worldwide Medical on the World Wide Web. From the additional information he found there, it appears the HeartCart 300 will be competitive with the C-2000.

Carlos has mixed emotions about this finding. On the one hand, he is glad to see that another firm is investing research dollars in this technology that is crucial to health care throughout the world. On the other, he is understandably concerned that his two years of work may become upstaged by the work of a competing company. For the sake of the investment his client has made in this research, he was hoping that the C-2000 would enjoy sales that would justify the research investment.

Questions and Comments for Discussion

More to the immediate point, Carlos is wondering what effect, if any, this new information should have on his presentation—a presentation that is being underwritten by a $5,000 grant from MedExcel. Here are some questions for you to consider.

1. Do you think Carlos is obligated to revamp his presentation to include information about the HeartCart 300? If your answer is yes, explain exactly how he should proceed. If your answer is no, give your rationale.
2. Should his association with MedExcel have any bearing on his decision?
3. Should the manner by which Carlos received this new information have any effect on the answer to question 1?
4. Give what you think would be the opinion of the Association of Medical Technology in response to question 1.
5. Explain how your answer to question 1 relates to the Equal Consideration of Interests (ECI) ethical principle discussed in chapter 2 (see pages 62–63).
6. Generally, do you see any larger concern here in the relationship between professional associations and private enterprise? Explain your answer.

● *Meeting Guideline 6: Keep Meetings on Track*

By far the biggest challenge for a meeting leader is to encourage open discussion while still moving toward resolution of agenda items. As a leader, you need to be assertive, yet tactful, in your efforts to discourage these three main time-wasters:

- Long-winded digressions by the entire committee
- Domination by one or two outspoken participants
- Interruptions from outside the meeting

● *Meeting Guideline 7: Strive for Consensus*

Consensus means agreement by all those present. Your goal should be to orchestrate a meeting such that all members, after a bit of compromise, feel comfortable with a decision. Such a compromise, when it flows from healthy discussion, is far preferable to a decision generated by voting on alternatives. After all, you are trying to reach a conclusion that everyone helps produce, rather than one only part of the committee embraces.

● *Meeting Guideline 8: Use Visuals*

Graphics help make points more vivid at a meeting. They are especially useful for recording ideas that are being generated rapidly during a discussion. Toward this end, you may want someone from outside the discussion to write important points on a flip chart, a chalkboard, or an overhead transparency.

● *Meeting Guideline 9: End with a Summary*

Before the meeting adjourns, take a few minutes to summarize what items have been discussed and agreed to. This wrap-up gives everyone the opportunity to clarify any point brought up during the meeting. Also, each member will know exactly what actions come next.

● *Meeting Guideline 10: Distribute Minutes Soon*

Write and send out minutes within 48 hours of the meeting. For even routine meetings, it is important that there be a record of the meeting's accomplishments. Also, if any discussion items are particularly controversial, consider having committee members approve minutes with their signature and return them to you, before final distribution.

CHAPTER SUMMARY

The main theme of this chapter is that anyone can become a good speaker by preparing well, practicing often, and giving an energetic performance. This effort pays off richly by helping you deal effectively with employers, customers, and professional colleagues.

Ten guidelines for preparation and delivery will lead to first-class presentations:

1. Know your listeners.
2. Use the preacher's maxim.
3. Stick to a few main points.
4. Put your outline on cards, paper, or overheads.
5. Practice, practice, practice.
6. Speak vigorously and deliberately.
7. Avoid filler words.
8. Use rhetorical questions.
9. Maintain eye contact.
10. Use appropriate gestures and posture.

You should also strive to incorporate illustrations into your speeches by following these 10 guidelines:

1. Discover listener preferences.
2. Think about graphics early.
3. Keep the message simple.
4. Make any wording brief and visible.
5. Use colors carefully.
6. Leave graphics up long enough.
7. Avoid handouts.
8. Maintain eye contact while using graphics.
9. Include all graphics in your practice sessions.
10. Use your own equipment.

A major problem for many speakers is fear of giving presentations. You can control this fear by following these guidelines:

1. Know your speech!
2. Prepare yourself physically.
3. Picture yourself giving a great presentation.
4. Arrange the room as you want.
5. Have a glass of water nearby.
6. Engage in casual banter before the speech.
7. Remember that you are the expert.
8. Do not admit nervousness to the audience.
9. Slow down.
10. Join a speaking organization.

Besides conventional presentations, you often will be required to speak in meetings—as either a leader or a participant. Given the effect that poor meetings have on morale and profits, it is important to develop a strategy for running them well. Here are 10 guidelines:

1. Involve only necessary people.
2. Distribute an agenda before the meeting.
3. Distribute readings before the meeting.

4. Have only one meeting leader.
5. Start and end on time.
6. Keep meetings on track.
7. Strive for consensus.
8. Use visuals.
9. End with a summary.
10. Distribute minutes soon.

Access the Technical Writing Web site by pointing your browser at www.prenhall.com/pfeiffer. Here, you will find additional resources to aid you in learning and applying the information from this chapter.

ASSIGNMENTS

1. **2–3 Minute Presentation Based on McDuff Projects.** Select one of the projects in the chapter 2 color insert. Use information from the project and, if you wish, additional details you invent that could relate to the project. For this presentation, assume you are a McDuff marketing specialist talking to a group of potential clients in a meeting. They may hire McDuff for a similar project and thus want a summary of the job described in the color insert.

2. **2–3 Minute Presentation Based on Your Academic Major.** Give a presentation in which you discuss (a) your major field, (b) reasons for your interest in this major, and (c) specific career paths you may pursue. Assume your audience is a group of students, with undecided majors, who may want to select your major.

3. **5–6 Minute Presentation Based on Short Report.** Select any of the short written assignments in chapters that you have already completed. Prepare a presentation based on the report you have chosen. Assume that your main objective is to present the audience with the major highlights of the written report, which they have all read. Use at least one visual aid.

4. **5–6 Minute Presentation Based on Proposal.** Prepare a presentation based on the proposal assignment at the end of chapter 10. Assume that your audience wants highlights of your written proposal, which they have read.

5. **10–12 Minute Presentation Based on Formal Report.** Prepare a presentation based on any of the long-report assignments at the end of chapter 9. Assume that your audience has read or skimmed the report. Your main objective is to present highlights, along with some important supporting details. Use at least three visual aids.

6. **Group Presentation.** Prepare a group presentation in the size groups indicated by your instructor. It may be related to a collaborative writing assignment in an earlier chapter or it may be done as a separate project. Review the chapter 1 guidelines for collaborative work. Though related to writing, some of these suggestions apply to any group work.

   Your instructor will set time limits for the entire presentation and perhaps for individual presentations. Make sure that your group's members move smoothly from one speech to the next; the individual presentations should work together for a unified effect.

7. **Ethics Assignment.** Suzanne Anthony, a prominent ecologist with McDuff's Atlanta office, has been asked to make a 30-minute speech to a public workshop on environmentalism, sponsored by SprawlStopper, a regional environmental action organization. She agrees to give the talk—for which she will receive an "honorarium" of $500—on her area of expertise: the effects of unplanned growth on biological diversity of plant and

animal species. Suzanne views the talk as a public service and has no knowledge of the sponsoring organization.

A few weeks before the speech, Suzanne's boss, Paul Finn, gets heartburn over his morning coffee as he reads an announcement about Suzanne's speech in the "Community Events" column of the local paper. Just yesterday a large local builder, Action Homes, accepted his proposal for McDuff to complete environmental site assessments on all Action's construction sites for the next three years. Paul is aware of the fact that Action has an on-going court battle with SprawlStopper concerning Action's desire to develop a large site adjacent to a Civil War national park north of Atlanta. Although McDuff is not now involved in the suit, Paul is worried that if Action sees the name of a McDuff scientist associated with an event sponsored by SprawlStopper, Action may have second thoughts about having chosen McDuff for its site work.

So, do you think Paul should say something to Suzanne? If so, *what* should he say? And *why* should he say it? If not, *why* not? If you were in Suzanne's place, how would you respond to a suggestion by Paul that her speech might be inappropriate? Are there any similarities between the situation described here and the one characterized in this chapter's "Communication Challenge"?

8. **International Communication Assignment.** Prepare a group presentation that results from research your group does on the Internet concerning speech communication in a country outside the United States.

   Option A: Retrieve information about one or more businesses and/or careers in a particular country. Once you have split up the group's initial tasks, conduct some of your business by email. Then present the results of your investigation in a panel presentation to the class. For example, your topic could be the computer software industry in England, the tourist industry in Costa Rica, or the textile industry in Malaysia.

   Option B: Retrieve information on subjects related to this chapter—for example, features of public speaking, business presentations, presentation graphics, and meeting management.

Surfing the Turf

General Instructions

Each chapter begins with a Collaboration at Work exercise that applies strategies for working in groups to chapter topics. The exercise assumes you (1) have been divided into groups of about three to six students, (2) will use group time inside or outside of class to complete the case, and (3) will produce an oral or written response. For guidelines about writing in groups, refer to pages 28–34.

Background for Assignment

The Internet has greatly expanded the range of information you can find on a subject and the speed with which it can be found. Yet the Internet also has introduced new challenges. While conducting this type of research, you must do the following:

- Stay focused so that the inevitable distractions of new links and fascinating data do not detract from your main purpose.

- Evaluate the reliability of sources that are quite different from traditional hard-copy sources found in a library.
- Determine the mix of Internet and traditional sources that will provide the best support for your topic.
- Keep good notes so that later you can properly document information that has been secured from the Internet.

To start you thinking about the process of Internet research, this exercise asks you to work with your group on a short project.

Group Assignment

First, agree as a group on a topic about which you want to find information on the Internet. Second, work individually to locate three to five sources of information on the topic (the information itself—not just a list of sources). Third, come back together as a group and discuss the relative value of the sources of the information you found.

*R*esearch does not end with the last college term paper. In fact, your career often will require you to gather technical information from libraries, interviews, questionnaires, the Internet, and other sources. Such on-the-job research will produce documents as diverse as reports, proposals, conference presentations, published papers, newspaper articles, World Wide Web sites, or essays in company magazines. Your professional reputation may depend on your ability to locate information, evaluate it, and use it effectively in everyday research tasks.

This chapter takes you through the research process used in college courses and practiced on the job. Specifically, the chapter has eight main sections:

- Getting started
- Searching online catalogs
- Searching in the library
- Searching the Web
- Using questionnaires and interviews
- Using borrowed information correctly
- Selecting and following a documentation system
- Writing research abstracts

A common thread throughout the chapter is a McDuff case study. Tanya Grant, who works in marketing at McDuff's Atlanta office, is searching for information to complete a project for Rob McDuff, the president of the firm. We'll observe Tanya as she gathers research material.

Like Tanya, later you will have to apply the research process to a specific technical-writing task in college or on the job. It is one thing to read about doing research; it is quite another to dive into your project and work directly with the books, periodicals, electronic databases, Web sites, and other resources in the library and on the Internet. You should seek firsthand research experience as soon as you can.

Finally, remember that the best research writing *smoothly* merges the writer's ideas with supporting data. Such writing should (1) impress the reader with its clarity and simplicity and (2) avoid sounding like a strung-together series of quotations. These two goals present a challenge in research writing.

GETTING STARTED

In chapter 1, you learned about the three phases of any writing project: planning, drafting, and revising. Research can occur in the planning stage, right before you complete an outline, but it can also occur again and again throughout the project. Before starting your research, ask yourself questions like these, to give direction to your work:

- What *questions* must be answered during the research phase?
- What *print, multimedia, and electronic sources* would be most useful?
- What is the nature and extent of information that is needed? Should it be scholarly or popular? Current or historical?
- What are the best strategies and research tools for locating information?
- Whom can you talk to who might have expertise on the topic?
- What are the best criteria for critically evaluating information for reliability, validity, accuracy, timeliness, or point of view or bias?
- What *format* must be used to document material borrowed from sources and what copyright permissions must be acquired to use the information?

Let's take a look at the way Tanya Grant answers these questions during her project.

Tanya has just been given an important task. The company president, Rob McDuff, wants her to examine the feasibility of the company switching to hybrid electric-powered cars in their American offices. McDuff believes the company's success in using these vehicles in their Asian offices is so significant that the company should consider using them in the U.S. offices. Success with hybrid electric cars could help address serious air quality issues and offset escalating petroleum prices. Also, he hopes his firm becomes a major player in refining the technology and fostering its use. In short, this move could serve as a public relation effort, a budgetary control, and a marketing tool for McDuff products.

Rob McDuff told Tanya she should write a report investigating the success and failures of hybrid electric cars. Upper-level management at an upcoming meeting will review the report. She should study the impending tax and other legislation affecting fuel-efficient vehicles. In addition, she should investigate potential of hybrid vehicles. Some of her report will look at start-up costs for switching the vehicle fleet, employee needs, and other in-house matters. The following outline shows how Tanya would probably answer the questions previously noted, as she begins her research:

1. Main question: Should McDuff switch to hybrid electric vehicles?
2. Main types of information needed:

   - What are the advantages and disadvantages of this technology?
   - What is the research saying about the outlook of hybrid electric vehicles?
   - What tax and other legislation is pending at both state and federal levels?
   - Who are the current and potential consumers and what do they think about hybrid electric cars?
   - What are the real costs in maintaining a fleet?
   - What could McDuff gain by the switch? What would they lose?
   - What impact would such a switch have on the competition, potential clients, or employees?

3. Possible sources:

   - Directories (of periodicals, newsletters, newspapers, electronic journals, organizations)

- Indexes and abstracts to journals, periodicals, newspapers
- Bibliographies
- Government documents
- Books
- World Wide Web sites
- Actual copies of environmental magazines or Web pages
- Questionnaires
- Interviews with some potential readers, such as current clients

4. Format for documentation: Tanya will submit a short report to Rob McDuff, documenting her research using the author-date system of citing borrowed information (the same format used by McDuff engineers and scientists in their research reports).

With this basic plan in mind, Tanya can begin her work. Her first decision is *where* to start her research. Choices include the following:

- The public library
- Her corporate library
- The university library
- The Internet—specifically, the World Wide Web, or, more simply, the Web

She eliminates the public library and the McDuff corporate library—the first is too general and the second is too specialized. The remaining options are the Web and the local university library. Tanya will need to *use both* to do a thorough job of research. She knows about the university's extensive print collections in science and technology, their Web-based online catalog, and the electronic databases. In addition, the staff in the reference department and interlibrary loan office will help her identify and track down sources. She also uses the Web regularly for finding business information, news, entertainment, and discussion on just about anything. The two, the Web and the library, work amazingly well together—they intersect, overlap, and complement one another, and each contributes unique sources, provided you know the basic research techniques and tools. A good researcher uses both the Web and the library to take advantage of the strengths, and overcome the weaknesses, of each.

Deciding whether to start in the library or on the Internet can depend on personal factors such as

- Your current knowledge of the topic
- The speed of your Internet connection and computer
- Your experience and skill in using effective search strategies
- Your ability to stay focused
- Your ability to critically evaluate the materials you locate
- Your schedule and the blocks of time you have for research

A planning trip to the library can save you hours of time searching the Web in unfamiliar subject areas. On the other hand, a few hours' preparation online—for example, searching a library's Web-based catalog—can make your trip to the library more productive.

Tanya schedules two days this week and one next week to work in the library and begins her research from her computer in the office. The next section introduces some of the strategies, tools, and basic concepts you need for searching library online catalogs, searching in the library, and searching on the Web.

SEARCHING ONLINE CATALOGS

Both the library and the Internet can seem like intimidating places when you first start a project. Once you learn a few basics, however, you will become comfortable, and even confident about, using these amazing resources. This section includes information on using online library catalogs to locate books, journals, and other resources. We will look at the basics and more advanced techniques that you need for searching not only library catalogs but also most Web search engines and databases.

Books provide well-supported, tested information about a topic, but, by definition, the information is often dated. Even a book just published has information that is one to two years old, given the time it takes to put a book-length manuscript into print. Keep this limitation in mind as you search.

Most college libraries organize their collections by the Library of Congress (LC) system (see Figure 13–1) or by the Dewey decimal system (see Figure 13–2). The library's catalog is a road map to its collection of books, periodicals, and other material; it is an alphabetical list by author, title, and subject. These days, the majority of colleges and university libraries offer sophisticated online catalogs that can be searched at the library or remotely from home or office. The online catalog, unlike the manual card catalog, often has additional features such as keyword and Boolean searching and information about whether a book is available or checked out.

The current trend in university and college libraries is to provide access to the library's catalog through the Internet and usually the Web. Now you can search not only your library's catalog for the best books and periodicals on your topic but also the catalogs of many other libraries around the world (see Figure 13–3). Your library's Web site often provides additional helpful information about services and hours, electronic databases, access to other library catalogs, and links to important Web sites. One of the easiest ways to locate the address for your library's Web site is to telephone the library staff and ask.

The rules for searching online catalogs vary depending on the computer program used by the library. The online catalog's "help screen" is the best guide to search techniques. Here are some general strategies for effective searching.

● *Author or Title Search*

If you know specific authors or titles of potentially useful books, conduct an author or title search to locate the call numbers. Study the catalog entry, especially the subject heading; similar books can be found if you search by subject using these terms.

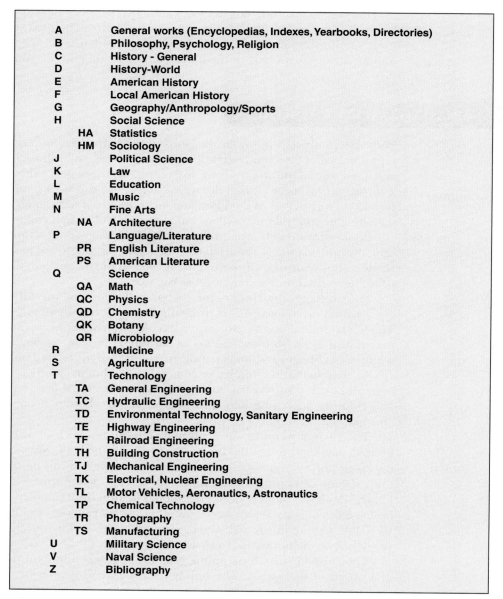

A		General works (Encyclopedias, Indexes, Yearbooks, Directories)
B		Philosophy, Psychology, Religion
C		History - General
D		History-World
E		American History
F		Local American History
G		Geography/Anthropology/Sports
H		Social Science
	HA	Statistics
	HM	Sociology
J		Political Science
K		Law
L		Education
M		Music
N		Fine Arts
	NA	Architecture
P		Language/Literature
	PR	English Literature
	PS	American Literature
Q		Science
	QA	Math
	QC	Physics
	QD	Chemistry
	QK	Botany
	QR	Microbiology
R		Medicine
S		Agriculture
T		Technology
	TA	General Engineering
	TC	Hydraulic Engineering
	TD	Environmental Technology, Sanitary Engineering
	TE	Highway Engineering
	TF	Railroad Engineering
	TH	Building Construction
	TJ	Mechanical Engineering
	TK	Electrical, Nuclear Engineering
	TL	Motor Vehicles, Aeronautics, Astronautics
	TP	Chemical Technology
	TR	Photography
	TS	Manufacturing
U		Military Science
V		Naval Science
Z		Bibliography

FIGURE 13–1   Library of Congress system

● *Subject Search*

Once you know the subject headings assigned to books or journals on your topic, searching by subject can be very efficient. Libraries select the subject headings from a set of books titled *The Library of Congress Subject Headings*. Knowing exactly which

```
                    General Breakdown and Sample Subdivisions

        000    General Works
        100    Philosophy
        200    Religion                    510    Mathematics
        300    Social Science                     511  Arithmetic
        400    Language                           512  Algebra
        500    Science                            513  Geometry
        600    Technology                              513.1  Plane Geometry
        700    The Arts
        800    Literature
        900    History
```

FIGURE 13–2   Dewey decimal system

This page provides links to online catalogs of Georgia libraries. For links to Georgia library web sites, go to the User Communities section in About GALILEO.

NOTE: Most University System of Georgia on-line catalogs are part of the GIL (GALILEO Interconnected Libraries) system and are accessible via web browsers. Other catalogs may require additional software for access.

University System of Georgia Libraries

- Abraham Baldwin Agricultural College GIL@ABAC
- Albany State University GIL@ASU
- Armstrong Atlantic State University GIL@AASU
- Atlanta Metropolitan College GIL@AMC
- Augusta State University GIL@ASU
- Bainbridge College GIL@BC
- Clayton College and State University GIL@CC&SU
- Coastal Georgia Community College GIL@CGCC
- Columbus State University GIL@CSU
- Dalton State College GIL@DSC
- Darton College GIL@DC
- East Georgia College GIL@EGC
- Floyd College GIL@FC
- Fort Valley State University GIL@FVSU
- Gainesville College GIL@GC
- Georgia College & State University GIL@GC&SU
- Georgia Institute of Technology
- Georgia Perimeter College GIL@GPC
- Georgia Southern University GIL@GASOU
- Georgia Southwestern State University GIL@GSW
- Georgia State University GIL@GSU
- Gordon College GIL@GDN
- Gwinnett University Center GIL@GUC
- Kennesaw State University GIL@KSU
- Macon State College GIL@MSC
- Medical College of Georgia GIL@MCG
- Middle Georgia College GIL@MGC
- North Georgia College and State University GIL@NGC&SU
- Savannah State University GIL@SSU
- South Georgia College GIL@SGC
- Southern Polytechnic State University GIL@SPSU
- State University of West Georgia GIL@SUWG
- University of Georgia GIL@UGA
- University of Georgia Law Library (GAVEL)
- Valdosta State University GIL@VSU
- Waycross College GIL@WC

FIGURE 13–3   A list of other catalogs available on one library's Web site
Source: Galileo—Georgia Library Learning Online. Courtesy of the Board of Regents of the University of System Georgia.

subject words to use can be a matter of trial and error, but once found, these headings can serve as powerful tools to gather information on your topic.

● *Keyword Search*

This strategy is probably your best choice, for it allows you to scan through all the fields in a book's library record—author field, title, subject headings, dates, notes, publishers, etc.—to locate books that match your request. Searching by keyword permits use of natural language instead of rigid subject headings. Pay close attention to the catalog's rules for keyword searches; you can often improve your results by limiting searches to particular fields. See Figure 13–4 for a typical keyword search.

> **TIP:** One of the easiest ways to find material on your topic is to switch between keyword and subject searching. For example, begin with a keyword search, selecting the books that match your request, examine the subject headings used to describe those books, and then do a subject search.

Georgia Tech Library Catalog (GTEC) - Basic Search Page
For database status and news, see GTEC News

Need help? Check our Search Tips page. Please logout when you are done. [Logout]

For more options, go to our Advanced Search page.

Search Tip: Do not use stop words in a search. Stop words are A, AN, BY, FOR, IN, OF, ON, THE, TO.
To search a title such as *"IBM Journal of Research and Development"* enter "ibm journal research development."

| hybrid electric cars | Search Everything ▲▼ |

[Start Search] [Clear Form]

No. of Records to display per page: [20 ▲▼]

View Search History: [▲▼]

| about us | architecture library | ask a librarian | contact us | site map | help |

FIGURE 13–4   Results of a typical keyword search
Source: GIL Across Georgia Union Catalog. Used by permission.

Georgia Tech Library Catalog (GTEC)
Please logout when you are done. [Logout]
Search Results for:
S1:
6 Record(s) found. (This page: 1 ~ 6)

[search] [prev. group] [next group]

Results are sorted by Year, newest-oldest.
Want to change sort order?
[Year ‡] [Descending ‡] Sort by in order [Re-Sort]   [Format for printing]

[Select marked records]

1 ☐	**Title:** Forward drive : the race to build "clean" cars for the future / Jim Motavalli. **Author:** Motavalli, Jim. **Call Number:** TL221.15 .M68 2000. **Year:** 2000.
2 ☐	**Title:** Electric vehicles : socio-economic prospects and technological challenges / edited by Robin Cowan, Staffan Hulten. **Author:** **Call Number:** TL220 .E4485X 2000. **Year:** 2000.
3 ☐	**Title:** History of the electric automobile : hybrid electric vehicles / Ernest Henry Wakefield. **Author:** Wakefield, Ernest Henry, 1915-. **Call Number:** TL220 .W343 1998. **Year:** 1998.
4 ☐	**Title:** Evaluation of a Toyota prius hybrid system (THS) [[microform] /] Karl H. Hellman, Maria R. Peralta, Gregory K. Piotrowski. **Author:** Hellman, Karl H. **Call Number:** EP6.2:P93. **Year:** 1998.
5 ☐	**Title:** The household market for **electric** vehicles : testing the **hybrid** household hypothesis-- reflexively designed survey of new-**car**-buying, multip **Author:** Turrentine, Thomas. **Call Number:** TL220 .T88X 1995. **Year:** 1995.
6 ☐	**Title:** **Electric and hybrid** vehicles : selected papers through 1980 / prepared under the auspices of the **Electric** Vehicle Committee, Passenger **Car** Act **Author:** **Call Number:** TL220 .E37 1981. **Year:** 1981.

FIGURE 13–4 continued

503

Georgia Tech Library Catalog (GTEC)
Search Results for:
S2:
Record 3 out of 6

Please logout when you are done. [Logout]

| search | result list | prev. group | next group | first record | last record |
| prev. record | next record |

[Request Document Delivery]

Call Number: **TL220 .W343 1998.**
Main Author: Wakefield, Ernest Henry, 1915-.
Main Title: **History of the electric automobile : hybrid electric vehicles / Ernest Henry Wakefield.**
Pub Info: Warrendale, Pa. : Society of Automotive Engineers, c1998.
Physical Desc: xxii, 332 p. : ill. ; 26 cm.
Notes: Includes bibliographical references and index.
Subject: Automobiles, Electric--History.
Subject: Hybrid electric cars--History.
Record Type: MON.
Language: ENG.
ISSN/ISBN: 0768001250.
Voyager Number: 461227.
OCLC: 38566039.
LCCN: 8003420.

Location and availability:

1) TL220 .W343 1998
   Location 1: Currently checked out--to recall, ask at Circulation desk, 1st floor West

| search | result list | prev. group | next group | first record | last record |
| prev. record | next record |

[Logout] Please logout when you are done.

FIGURE 13-4   continued

● *Advanced Search Techniques*

A library catalog may also include advance strategies such as Boolean searching, positional operators, and truncation. Many times you may not be aware that you are using these tools. Learning to use these techniques is important because they are

---

BOOLEAN SEARCHING

AND: **Example: periodicals AND directories.**
Locates only those records where both terms are present.
Use this to narrow your search and reduce the number of matches.

OR **Example: periodicals OR journals OR zines.**
Locates records in which any one of these terms can appear.
Use this to broaden or enlarge your search.

NOT: **Example: periodicals NOT magazines.**
Eliminates records containing the excluded term.
Use this sparingly to narrow your search.

FIGURE 13–5 Boolean search examples

used in library catalogs, most periodical databases, and Web search engines. Look for the "help screens" in your catalog that describe the advanced search options and practice using them whenever possible; they can save you time and produce excellent results. Here is a brief description of some of the most common.

- A **Boolean search** outlines the relationship of words and phrases using simple AND, OR, NOT statements. See Figure 13–5 for examples.
- **Positional operators** stipulate the relative location of each term within the record. For example, you can often specify that terms must be adjacent or within a certain number of words.
- **Truncation** allows for variant spelling or plurals. For example, in some catalogs entering "wom*n" would retrieve records with either the words "woman" or "women."

Searching other library catalogs can sometimes be as simple as selecting a link from your library's Web site to a library consortium or union catalog of university and college libraries within your region. Searching libraries close to home has the advantage of easier access to their collection whether you visit in person or gain access through your library's interlibrary loan service. If you want to see "what's out there" in larger or more specialized libraries, try searching for library catalogs on the Web or ask the reference librarian if your library provides access to OCLC's *Worldcat*. Following are two Web directories of library catalogs that have been around for some time. (Keep in mind, however, that the nature of the Web is such that they may have disappeared since this book was written.) Try pointing your browser to:

- **Libcat-Metronet** http://www.metronet.lib.mn.us/lc/lcl.cfm
  or
- **Libweb: Library Servers via WWW:** http://sunsite.berkeley.edu/Libweb/

TIP: Keep close track of what you find and *where* you find it so that you don't waste your time searching for books on the shelves of your library when they are actually located elsewhere. Consult the reference department of your library to learn about your library's interlibrary loan service or borrowing privileges at other libraries.

When you use the Web for searching catalogs, remember to evaluate what you find as you search. Begin thinking critically as soon as you start, and work to keep this perspective throughout your research. For example, when looking at a reference to a book, a journal, an article, or a Web site, ask yourself the following:

- What are the author's academic qualifications?
- Who is the publisher and what is its reputation?
- What are the scope and content of the work?
- What are the trends in information on this topic and how does this book, article, journal, or Web site fit in?
- How current is this information?

Finally, use your library's online catalog to find out what research databases are available to you. You may be able to search important research databases such as periodical indexes and abstracts, encyclopedias, dictionaries, directories, and other reference sources (see Figure 13–6). Many of these databases can be searched remotely from your home or office, but some are restricted to in-library searching only. Policies governing who can search, from where, passwords, whether searching is fee-based or free vary widely depending on the contracts between the library and the database vendor. You will learn more about using research databases in the next section, but for now keep in mind that the quality of information you will retrieve from research databases is usually superior to the material you may locate searching the vast World Wide Web. In addition, online catalogs may provide links to recommended, high-quality Web sites you might not otherwise locate. Explore your online options and discuss your needs with the reference staff at your library.

Tanya Grant's McDuff Project

Tanya Grant, for example, rightly thinks that books will not be her main source of information about hybrid electric cars, since the topic has developed relatively recently. Yet she at least wants to see what range of sources the catalog offers. Tanya does a keyword search and determines that the correct subject heading is "hybrid electric vehicles." Her own library's holding are somewhat limited, but one item is worth reviewing. She decides to search the online catalog from another local university with an automotive engineering school. Here she finds a better selection of books, and since her library card gives her borrowing privileges at all state university system libraries, she decides to take a trip to this library. Her final search is in OCLC's Webcat; this comprehensive database leads her to a few more noteworthy titles that she will borrow through her library interlibrary loan department.

SEARCHING IN THE LIBRARY

At some point during your search for secondary resources, you will need to visit your academic library. The library's services and collections of books, journals, microforms, and reference materials, while complex, will support your research and help you locate information. Fortunately, academic and research libraries are organized along similar principles, and the skills you gain from using one library can generally be transferred to other libraries. This section will highlight some of the services and resources you can expect to find as you conduct your research in the library.

All Databases
Arts & Humanities
Business & Economics
Conferences & Proceedings
Earth Sciences
Education
Engineering & Physical Sciences
Full-Text Books
Full-Text Journal/Newspaper Articles
General Indexes
Government Publications
High School
Kid's Stuff
Life Sciences
Medicine & Health Sciences
Newspapers
Public Affairs & Law
Reference Sources
Social Sciences
Spanish / Español

Databases by Password

Ask a Librarian

▶**Engineering and Physical Sciences**
To search a database, click the name below. For database description, click
the ❶ information icon.

Jump to: **A B C E G I L M P R S T**

◀ Prev Group   Next Group ▶

1. **Academic Search Premier (at EBSCOhost)** ❶
   1984- | daily | some full text & images from 1990- | Versión en Español
2. **AccessScience@McGraw-Hill (sci/tech encyclopedia)** ❶
   current file | daily | full text
3. **Applied Science and Technology Index** ❶
   1983- | monthly
4. **Bioengineering Abstracts** ❶
   1993- | monthly | Backup Site (Ohio)
5. **Biotechnology and Bioengineering Set** ❶
   1989- | monthly | Backup Site (Ohio)
6. **Computer and Information Systems Abstracts** ❶
   1981- | monthly | Backup Site (Ohio)
7. **Computer Source: Consumer Edition** ❶
   1984- | daily | full text
8. **Current Contents (scholarly journals)** ❶
   1992- | weekly | will be replaced by **Current Contents Connect**
9. **Current Contents Connect (scholarly journals at ISI)** ❶
   1997- | weekly | will replace **Current Contents** | This database is not yet
   available off campus.
10. **Electronics and Communications Abstracts** ❶
    1981- | monthly | Backup Site (Ohio)

◀ Prev Group   Next Group ▶

Display in groups of: **10** | **20** | **30** | **All**   Your Institution: **Southern Polytechnic
State University**

FIGURE 13–6   Research databases available on one library's Web site
Source: Galileo—Georgia Library Learning Online. Courtesy of the Board of Regents of the University System of Georgia.

Library Services

The services that your library can provide during the research process are many. The
three most important ones are (1) reference and information, (2) interlibrary loan,
and (3) circulation.

1. Reference and Information

Discussing your research topic with a reference librarian can be a very productive first
step. The librarian can recommend reference books; provide instruction on how to
search indexes, abstracts, and electronic databases; and guide you to collections such
as government documents, microform sets, and noteworthy Web sites that you may
not have found on your own. Make the most of your time with the reference staff by

asking specific questions and returning whenever you need more help in locating or using sources.

2. Interlibrary Loan

In the course of research you will identify excellent sources not owned by your library. Fortunately, in most libraries you can arrange to borrow or photocopy these sources through the interlibrary loan service. Be aware that the lending process can take a number of weeks and may require payment for the service. Interlibrary loan and other document delivery services have become increasingly popular as the popularity of computer networks and electronic databases grows.

3. Circulation

The circulation department is responsible for the lending and returning of library materials. In addition, the department may register library users, provide access to reserve materials, recall checked-out books, renew materials to extend the loan, and search for material not on the shelf. Circulation services are generally highly automated with a growing trend toward self-service checkout.

Library Resources

This section includes information on the following resources: books; periodicals; newspapers; company directories; and dictionaries, encyclopedias, and other general references.

● *Resource 1: Books*

As previously discussed, the library catalogs these days are generally automated, and very few traditional wooden card catalogs remain. Books are arranged in an alphanumeric call number order in the library stacks, generally by Library of Congress number or Dewey decimal number (see Figures 13–1 and 13–2). Once you locate the exact book for which you are searching, browse through the books located beside this title. You will likely find other useful and related material. Ask for assistance at the reference desk or circulation desk if you cannot locate the books on your topic.

● *Resource 2: Periodicals*

The *American Heritage Dictionary* defines periodical as a publication issued at "regular intervals." The term encompasses

- Popular magazines that take commercial advertising, such as *Time, Science,* and *National Geographic*
- Professional and scholarly journals such as *IEEE Transactions on Professional Communication*

Most library visitors are familiar with the section that houses current periodicals, either in alphabetical order or by call number. Yet they are less familiar with the part of the library containing back issues. Libraries keep back issues of the periodicals considered most important to its users. They may be in the form of bound volumes, microfilm or microfiche, or full-text electronic versions of periodicals.

Your key to locating information in periodicals is a periodical index or abstract. By looking up your subject in the index, you can find articles that provide the information you need. Some indexes and abstracts like *Readers' Guide to Periodical Literature* deal with popular periodicals. Others, like the *Engineering Index,* deal with a broad range of technical information. Still others, like *Mechanical Engineering Abstracts,* focus on periodicals, books, Web sites, and papers in specialized technical fields. The periodicals covered in the index or abstract are listed in the volumes or in the online information screen, along with the inclusive dates of the issues indexed.

A periodical abstract has the added advantage of providing a brief description of articles so that you can decide whether or not the entire article is worth finding. Abstracts are especially useful when the article being summarized is not available in your library. The abstract can help you decide whether to (1) visit another library, (2) order an article through the interlibrary loan service, or (3) disregard an article altogether.

Most periodical indexes and abstracts are available as electronic databases that can be searched quickly and thoroughly once you know the basics. Your library may subscribe to these sources in a variety of ways—CD-ROMs, locally mounted online files, or, now more commonly, as a Web site linked from the library's online catalog. Many electronic databases provide extensive abstracts or even full-text copies of the articles. Some libraries permit you to search these databases from your home or office, while other libraries, because of license requirements with the database vendors, permit searching within the library only. Still other libraries provide professional search services where, for a fee, the research staff conducts the search for you.

The rules for searching electronic indexes and abstracts vary widely. Each database has unique features and searching requirements. You must invest time and energy to learn these rules to take full advantage of the information the database offers. Start your search by reading the "help screens" and the instructional materials about the database or any support materials that the library provides. You will save yourself time and improve your search results if you understand the basic search strategies and have a grasp of the scope of the database. At a minimum, make sure that you know the rules for printing, emailing, or saving to disc the results of your search before you get too far into your research.

Most of the electronic databases have search strategies similar to what you may have encountered when searching the online catalog for books. The strategies will likely include subject, keyword searching, Boolean search techniques, positional operators, truncation options, and language and date limiting options. The more you practice, the better your searching will become and the more precise your results will be.

TIP: Many libraries offer classes on the basics of online searching. Check with the reference staff to see if your library offers this type of instruction.

Once you have evaluated and narrowed the list of articles on your topic that you wish to read, your next step is to determine which are owned by the library, which need to be requested through interlibrary loan, and which are owned by other nearby libraries. Libraries are often able to flag their holding in the online database. Also database vendors sometimes provide online full-text copies of articles that can be printed immediately or emailed to your home or office. Keep in mind, however, that many excellent periodical articles are *not* available electronically. Try to avoid the trap of arbitrarily limiting your research only to those periodicals with readily available full text. Be aware that the process of getting copies of the article can be tedious and time-consuming; be sure to reserve sufficient time for this important step.

There are hundreds of indexes and abstracts available. Many libraries provide handouts or guides to their print and electronic databases in their online catalogs. Ask the reference staff to help you locate the indexes and abstracts that are most appropriate for your topic. Here are a few of the well-known indexes and abstracting services available in print and/or computer format:

- *ABI/Inform Global* (business and trade journals)
- *Applied Science and Technology Abstracts* (print title: *Applied Science and Technology Index*)
- *Biological Abstracts*
- *Cambridge Scientific Abstracts*
- *Chemical Abstracts*
- *Computer Abstracts*
- *Current Contents*
- *Engineering Index* (electronic title: *EI Tech Index or Compendex*)
- *Findex*
- *General Science Abstracts* (print index: *General Science Index*)
- *GPO Monthly Catalog* (index to government documents)
- *LEXIS-NEXIS Academic Universe*
- *Periodical Abstracts*
- *Psychological Abstracts* (electronic title: *PsycINFO*)
- *Public Affairs Information Service* (electronic title: *PPAIS International*)
- *Science Citation Index, Social Science Citation Index, Arts & Humanities Citation Index* (online through the Web of Science)

Tanya Grant's McDuff Project

Tanya decided to consult a few of the electronic databases recommended by the reference librarian.

1. She conducted a search using *Environmental Engineering Abstracts,* which the library subscribed to electronically through Cambridge Scientific Abstracts. The scope and content of the abstract was exactly what she wanted since it targeted the technological and engineering aspect of hybrid electric vehicles. Because the database was new to her, she spent time learning how to conduct a search and save her results. She limited her search to articles from the last few years. The search not only retrieved useful articles, but also provided links to six high-quality Web sites. Scanning the results, she selected the most promising articles and Web site and emailed a copy to herself and printed a copy of the list to use for locating the periodicals in the library. See Figure 13–7 for Tanya's search.

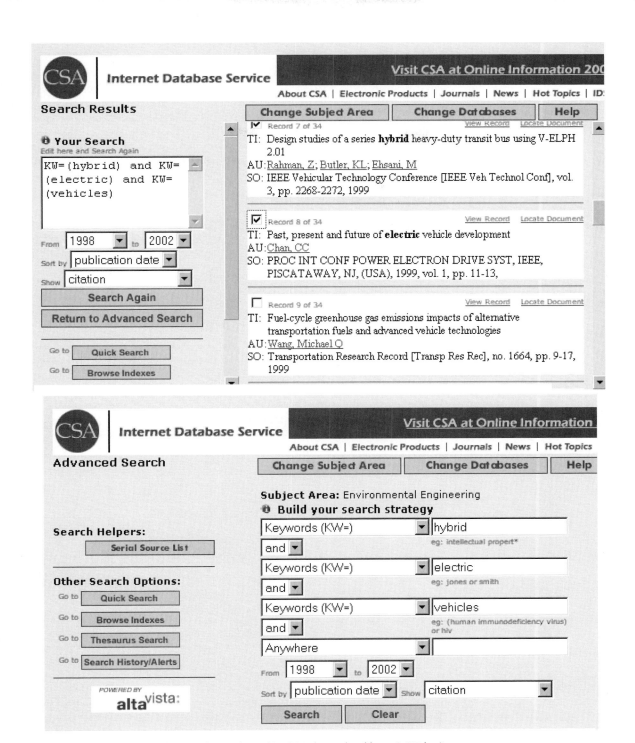

FIGURE 13–7 Results of a search conducted in CSA through a library's Web site

Source: Cambridge Scientific Abstracts. Used by permission.

2. Next she consulted *ABI/Inform Global*, an online database that covers business and management trade journals produced by ProQuest. She was interested in looking at business viewpoint on hybrid vehicles. Her search produced 72 items published since 1999, many of which had full-text copies of the article available for her to read immediately. After sampling a few articles, she flagged those she wanted and emailed them to herself. Tanya decided to redo her search and narrow it to "peer-reviewed" articles only. The nine articles she retrieved in her second search have undergone review and evaluation by experts in the field prior to publishing. These articles will be particularly noteworthy. See Figure 13–8 for Tanya's search.

3. Finally, Tanya consulted *Academic Search Premier*, a comprehensive, general-purpose database created by EBSCO. It covers almost 4,000 periodicals, 2,300 of which are scholarly. Again, she was able to narrow her search to peer-reviewed articles and located some very current and useful information. One of the full-text articles referred to an organization she wanted to investigate further, the Partnership for a New Generation of Vehicles.

● *Resource 3: Newspapers*

If your research topic demands the most current information, newspapers provide an excellent source. One disadvantage is that newspaper information has not "stood the test of time" to the same extent as information in journals and books. Despite this drawback, newspaper articles can give you insight, facts, and opinion on many contemporary issues. Two particularly noteworthy newspapers are the *New York Times* and the *Wall Street Journal.* These well-respected newspapers have a long tradition of high-quality journalism. Both titles are thoroughly indexed from the earliest issues to the present, and many libraries keep microfilm or microfiche back issues for these and other significant national and regional newspapers.

Access to current regional, national, and international newspapers has steadily improved in the last decade. Many newspapers have established elaborate Web sites and commercial database vendors have created vast electronic indexes to hundreds of newspapers, frequently providing full-text and graphics online. Some significant electronic databases for newspapers include *Lexis-Nexis' Academic Universe,* a full-text index to some 5000 publications including newspapers, wire services, legal news, and government publications; *ProQuest's National Newspapers,* an index to some 33 national and regional newspapers; and *Knight-Ridder's Newspaper Source,* with daily feeds and full text for some 139 regional U.S. newspapers. Check your library's online catalog to see if it provides additional links to some of the Web-based news services.

Tanya Grant's McDuff Project
Tanya decided to see what kind of newspaper coverage hybrid electric cars were receiving and try to uncover some of the tax legislation being proposed by each state. Her first search in *ProQuest's National Newspapers* located 327 newspaper articles—some written just the previous week. Reading through a few full-text articles convinced her that this would be her best approach to get the consumer perspective she needed. With her second search, Tanya added the concept "tax" and uncovered 64 articles from major newspapers from around the country describing various tax legislation efforts under way. These articles served as a starting point for studying the complex tax legislation being proposed by various state legislatures. See Figure 13–9 for Tanya's search results.

FIGURE 13–8 Results of a search conducted in *ABI/Inform* database through a library's Web site
Image published with permission of ProQuest Information and Learning Company. Further reproduction is prohibited without permission. Image produced by ProQuest Information and Learning Company. Inquiries may be made to: ProQuest Information and Learning Company, 300 North Zeeb Road, Ann Arbor, MI 48106-1346 USA. Telephone (734) 761-7400; E-mail: info@il.proquest.com; Web-page: www.il.proquest.com.

Save Link   Saves this search as a Durable Link under "Results-Marked List"

289 articles matched your search.

☑ 1. 🖹 📄   <u>Electric Cars May Lose Push In Northeast</u>; *Richard Perez-Pena*; **New York Times**, New York, N.Y.; Nov 12, 2001; Late Edition (East Coast); pg. F.1

☑ 2. 🖹 📄   <u>A Tale of 2 Engines: How Hybrid Cars Tame Emissions</u>; *Matt Lake*; **New York Times**, New York, N.Y.; Nov 8, 2001; Late Edition (East Coast); pg. G.7

☐ 3. 🖹   <u>TRADING UP 2002 MODELS OFFER MORE POWER, MORE STYLE AT NEW ENGLAND INTERNATIONAL AUTO SHOW.; [THIRD Edition]</u>; **Boston Globe**, Boston, Mass.; Nov 4, 2001; pg. J.1

☐ 4. 🖹 📄   <u>Tokyo Show Is Smaller and Subdued, but Still Outlandish</u>; *William Diem*; **New York Times**, New York, N.Y.; Oct 26, 2001; Late Edition (East Coast); pg. F.1

☑ 5. 🖹 📄   <u>Clear Winners Hope to Emerge at Car Challenge; [Home Edition]</u>; *JOHN O'DELL*; **The Los Angeles Times**, Los Angeles, Calif.; Oct 24, 2001; Record edition; pg. G.1

☐ 6. 🖹   <u>Paying at the pump ; Analysts don't see higher gasoline prices affecting what vehicles people buy--yet; [Chicagoland Final , CN Edition]</u>; *Cheryl Jensen Special to the Tribune*; **Chicago Tribune**, Chicago, Ill.; Oct 21, 2001; pg. 1

☐ 7. 🖹 📄   <u>Fuel-Cell Vehicles May Need a Big Push to Get Rolling; Technology * A study says zero-emission cars face major roadblocks, including getting government subsidies, before they'll catch on.; [Home Edition]</u>; *JOHN O'DELL*; **The Los Angeles Times**, Los Angeles, Calif.; Oct 17, 2001; Record edition; pg. G.1

ProQuest®   Southern Polytechnic Library   **?Help**

| Collections | Search Methods ▼ | Topic Finder | Browse Lists ▼ | Results ▼ | Search Guide |

Enter one or more search terms.

Search for: `hybrid`   in the `All Basic Search Fields*` field.

`AND` ▼ `electric`   in the `All Basic Search Fields*` field.

`AND` ▼ `vehicles`   in the `All Basic Search Fields*` field.

`AND` ▼ `Newspapers` ▼   in the **Publication type** field.

`AND` ▼ `All`   in the **Article type** field.

AND Date range: `Current` ▼ `Any Month` ▼ `1999` ▼ to `Any Month` ▼ `2001` ▼

☐ Show results with full text availability only
☐ Show peer reviewed articles
☑ Show total number of articles

Search   Clear

Tips

For the best results, enter the most specific words or phrase.

FIGURE 13–9   Results of a search conducted in the ProQuest Newspapers through a library's Web site

Image published with permission of ProQuest Information and Learning Company. Further reproduction is prohibited without permission. Image produced by ProQuest Information and Learning Company. Inquiries may be made to: ProQuest Information and Learning Company, 300 North Zeeb Road, Ann Arbor, MI 48106-1346 USA. Telephone (734) 761-7400; E-mail: info@il.proquest.com; Web-page: www.il.proquest.com.

514

● *Resource 4: Company Directories*

Often your research needs may require that you find detailed information about specific firms. For example, you could be completing research about a company that may hire you, or you may seek information about companies that compete with your own. There are many options these days for finding company information ranging from traditional printed directories to online and Web-based products. Most companies now produce sophisticated Web sites about their services and products. While not without bias, these can be an excellent source of information. The following is a small sample of some useful directories that are available; ask the reference librarian to recommend others and for assistance in using the online versions of these and other directories.

Compact D/SEC (generally on compact disc or online)

Corp Tech Directory of Technology Companies

D & B Million Dollar Directory

Moody's Industrial Manuals (and other Moody products)

Standard & Poor's Register of Corporations, Directors, and Executives

Thomas Register of American Manufacturers

Ward's Business Directory of U.S. Private and Public Companies

Who's Who in Science and Engineering

● *Resource 5: Dictionaries, Encyclopedias, and other General References*

Sometimes you may need some general information to help you get started on a research project. In this case, you may wish to consult specialized dictionaries, handbooks, or encyclopedias. Most general encyclopedias are available in some electronic format such as CDs or Web-based products. For example, the *Encyclopedia Britannica* (is currently available in print, interactive CD, and online at its own Web site (http://www.britannica.com). There are, however, advantages with using a specialized subject-based encyclopedia or dictionary over a general one in that the articles target a more scholarly audience, assume greater subject expertise, and reference more scholarly materials in their bibliographies. Here is a list of a few specialized dictionaries, handbooks, and encyclopedias you may find in the reference collection:

Blackwell Encyclopedia of Management

Cambridge Encyclopedia of Life Sciences

CRC Handbook of Chemistry and Physics

Dictionary of Business and Economics

Dictionary of the History of Science

Encyclopedia of Associations

Encyclopedia of Business Information Sources

Handbook of Industrial Engineering

International Business Information

McGraw-Hill Encyclopedia of Science and Technology

Van Nostrand's Scientific Encyclopedia

Tanya Grant's McDuff Project

At this point, Tanya had spent many hours examining the library's online catalog, searching in online periodicals indexes and abstracts, and newspaper indexes. She has located books and articles in scholarly and technical periodicals, as well as articles in popular magazines and newspapers. She requested a few promising items not locally owned through interlibrary loans. In the meantime she has plenty to read and begin creating notes. She has a couple of leads to reliable Web sites from the library's online catalog and an organization she wants to research. She has a good start, and plenty of work ahead.

SEARCHING THE WEB

Throughout this chapter you have seen references to the Internet and the World Wide Web. This vast global computer network has changed the way we communicate, market services and products, and collect and distribute information. The World Wide Web is the largest and fastest growing portion of the Internet with its appealing graphic interface that incorporates text, images, and sound, and its ability to move from one Web page to another through hyperlinks. This section will highlight some of the terms and concepts, challenges, and strategies associated with using the Web.

Fundamentals of Web Searching

Mining the Web for useful resources is always challenging and frequently frustrating, but it can yield terrific results. Why is searching such a challenge?

- The Web is huge; it contains tens of millions of documents and is growing at an astounding rate.
- The Web is constantly changing—sites are added, altered, moved, and disappear without warning.
- Search engines and subject directories don't work very well—they retrieve too much, they don't cover the entire Web, the relevancy ranking defies logic, and no two search engines work alike.
- The content of the Web is unregulated; anyone can add anything—fact, fiction, or fiction that looks like fact.
- There is no central index to the Web and few rules for describing Web pages.
- The process of searching, sifting through results, downloading pages, and critically evaluating each Web page is time-consuming.
- The Web is full of distractions that make it difficult to stay focused.

Despite these challenges, the Web offers access to extraordinary resources that often have no print counterpart. Because of the Web's sheer size, a search will usually find something on any topic—possibly something of value or perhaps something useless. Some studies have estimated that scholarly sites represent only 10–20% of the Web, but this number is still significant. Most people agree that the Web's strength lies in its information on current events, business and industry, popular culture, the government, computing, and technology, but all disciplines are represented in some way. Some resources you can find on the Web are

1. Directories of people, business, and organizations
2. Government documents
3. Periodicals, newspapers, and magazines
4. Books
5. Conference proceeding and reports
6. Reference tools like guides, indexes to periodicals, dictionaries
7. An increasing number of "by subscription only" information sources
8. Sound and video clips
9. Illustrations
10. Raw data

● *Using Your Evaluation Skills*

When you search the Web, be prepared to invest time and effort in critically evaluating what you find. Unlike books and articles that undergo a rigorous editing and review process, any Web site can be loaded directly onto the Internet. You will encounter misinformation, grossly biased content, and poor text and graphic design. Evaluate Web sources using the criteria discussed earlier in this chapter. However, since Web sources do not generally follow standard publishing practices, be prepared to invest your valuable research time determining the authority, timeliness, reliability, accuracy, point-of-view, and validity of the source. Once you develop a systematic approach to evaluating sources, you will quickly recognize both the high- and low-quality Web sources. Be particularly alert to the following:

- Obscured authorship: Often a Web designer is credited as the author when in fact an organization or a corporation is the real source.
- Out-of-date information: The Web is littered with abandoned and unmaintained Web sites. A high-quality Web site will prominently display the date.
- Subtle and obvious bias: Many Web sites are elaborate advertisements promoting products, services, causes, or points-of-view. Data manipulation, false arguments, and unsubstantiated opinions are common.
- Poor-quality links: Links from a high-quality Web site usually lead you to other valuable sites; links from a poor-quality site usually lead you to other poor-quality sites. Spending time examining the links will help you determine the quality of the site.
- Flawed style and design: Well-organized and accessible Web sites support the research process. While there are many cases of good research in poorly designed sites, be aware that extracting the information from overly complex sites drains away your research time.

com	Commercial organization (for profit)
edu	Educational institution
gov	Government organization (non-military)
int	International non-profit organizations
mil	Military organization (US)
net	Networking organization
org	Non-profit organization
ca	Canada (country of origin)
uk	United Kingdom (country of origin)

FIGURE 13–10

Spend time evaluating the source up front before you spend time reading the document. If you cannot determine the scope, authority, or date of the Web site, don't use it.

● *Learning the Basics*

The Web is made up of millions of "*Web pages.*" A Web site is a collection of Web pages generally with a "*home page,*" (the introductory first page) and links to additional pages within the site or to other Web sites. A Web page is uniquely identified by a "*URL* or *Uniform Resource Locator*"; this is the key to locating and documenting the Web page. The elements of a URL address are

- Letters that indicate the method of access. For Web sites, *http://* indicates that *hypertext transfer protocol is used.*
- The *address and domain of the computer.* The domain defines the country of origin or the type of computer on which the document is stored.
- The directory paths that will lead to the specific page being referenced.

The URL is generally preceded with *http://* to indicate that *hypertext transfer protocol* will be used. See Figure 13–10 for a list of common domains and examples.

"Web browsers," such as Microsoft Internet Explorer and Netscape, are software applications for viewing Web documents and navigating the Web. Both share similar features: a line for entering URLs, options for creating "bookmarks" (or "favorite" sites), basic navigational features for moving forward and backward, and stopping, and options for setting preferences to customize the browser.

AltaVista	http://www.altavista.com	Keyword
Argus Clearinghouse	http://www.clearinghouse.net	Subject guide plus
AskJeeves	http://www.askjeeves.com	Keyword & subject prompting
BUBL Information Service	http://bubl.ac.uk	Subject guide plus
Dogpile	http://www.dogpile.com	Meta-search engine
Excite	http:www.excite.com	Keyword & subject directory
Google	http://www.google.com/	Keyword results based on links
HotBot	http://hotbot.lycos.com	Keyword & subject directory
Internet Public Library	http://www.ipl.org	Subject guide plus
Lycos	http://www.lycos.com	Keyword & subject directory
NorthernLight	http://www.northernlight.com	Keyword & subject directory
ProFusion	http://www.profusion.com/	Meta-search engine
WebCrawler	http://www.webcrawler.com	Keyword & subject directory
Yahoo!	http://www.yahoo.com	Keyword & subject directory

FIGURE 13–11   Popular search engines and subject guides

Web Search Options

Options for searching the Web include the following:

- Searching by URL when you know the specific address
- Searching by keyword in an index-type "search engine" or "meta-search engine"
- Searching by subject category using a "subject directory"
- Using guides to reviewed and recommended Web sites

Figure 13–11 lists some of the most popular search tools. Keep in mind that Web address changes or improved applications may have appeared since this list was created.

> **TIP:** Competition among search engines is high, and new features and applications appear regularly. To keep up with search engine development and testing, try Search Engine Watch: *http://searchenginewatch.com*

● *Searching by URL: Uniform Resource Locator*

Searching by a URL—Uniform Resource Locator—is the most effective strategy, provided you have complete information and the Web page still exists. URLs are regularly included in books, journals, television and radio broadcasts, and marketing and advertising literature. One good URL may lead you to other well-written and maintained Web sites. Take care when recording and entering a URL—the punctuation, special marks, and abbreviations can be complex and lead to frustrating "cannot locate" messages if entered incorrectly.

● *Searching by Keywords Using Search Engines and Meta-Search Engines*

Hundreds of companies on the Web have created massive catalogs or databases of Web sites and provide keyword "search engines." These companies are actually in the business of selling advertising, and the search engines attract potential customers.

The databases are built without human intervention using computer programs called "robots or spiders" that move throughout the Web, searching for new sites. No single search engine indexes the entire Web, and there is fierce competition among companies for the distinction of having the largest, most current, or most useful database.

The value of a search engine depends on quality, quantity, and currency of the information it retrieves. However, just as important are indexing, speed, basic and advanced search options, and relevancy ranking of the results. The best search engines will provide simple, clear instructions and explain their criteria for building the database and ranking sites. Look for these instructions and pay particular attention to the advanced search features.

Most search engines use a combination of keyword and Boolean searching and often provide options for positional operators and truncation. An identical set of keywords entered into different search engines will produce entirely different information. This inconsistency can be attributed to variations in the search rules, the content, and the characteristics of the database. Experienced searchers recommend using two to four search engines to do a search.

> **TIP:** Master the features of one search engine before moving on to the next.

"Meta-search engines" (or multi-search engines) use the databases of a number of search engines at once to respond to a request. A request is forwarded to a variety of search engines; then the results are collected and displayed. You can save time using a meta-search engine, particularly in narrow, well-defined topics, but you lose the ability to refine searches using the features of individual search engines.

Hundreds of search engines now exist, and a new and better one is always on the way. Second-generation search engines feature *intelligent agents* designed to help refine your question by providing suggestions and alternate lines of inquiry. Other second-generation search engines provide continual updating services using "*push technology*" that store your search profile, run searches, and report results automatically. Some newer search engines retrieve Web sites based on keywords and the number of links to and from that site.

Keeping up with developments in search engines is a challenge. Periodically search for new ones or ask colleagues to recommend one. Keep trying different ones until you find a few that meet your needs.

> **TIP:** Save time searching by creating a file of search engines using the bookmark or favorite site feature on your browser.

● *Searching with Subject Directories*

Searching the Web with a subject directory is similar to browsing in a library; it can be very useful when you are not exactly sure what you are looking for or just want to get a feel for where sources are located. *Yahoo!* is the best-known subject directory. The distinction between a subject directory and a true keyword search engine is disappearing as companies rush to add both features to attract potential customers and advertisers.

The strength of a subject directory lies in how selective it is and how well it classifies Web sites into subject categories. Subject directories are generally compiled by people who review and index sites, although robots are used to retrieve potential additions. Web-site authors can also request their sites be added. Subject directories, unlike robot-driven search engines, do not attempt to be a comprehensive index to the Web. Instead, they try to capture a segment of the Web that appeals to their customers and advertisers. Trying to determine the criteria used to build the subject database is difficult, but generally the goal is to attract customers. The addition of the keyword search feature in subject directories has improved access and eliminated burrowing through multiple layers of information.

● *Using Reviewed and Recommended Subject Guides*

Finding reviewed and recommended guides to Web sites is the most efficient route to high-quality information. One list of recommended sources compiled and maintained by an expert can be all you need to do a thorough job of research. Naturally, the reviewer's reliability and authority, the scope and criteria for what is included, and the timeliness of the list determine its value. The best guides will publish a clear statement of their criteria, background information on their reviewers, and a "last updated" statement. Many lists have been compiled by librarians, scholars, and other subject experts associated with professional organizations or government agencies. You may discover your library's online catalog includes links to such lists. Figure 13–12 shows an example of a guide to recommended Web sites found in an online catalog.

Tanya Grant's Web Search

Tanya began her Web search using a URL one of the McDuff engineers gave her, which led her to a Web site maintained by the U.S. Department of Energy's Office of Transportation Technology. This comprehensive site helped her organize the issues, policies, and research trends, as well as locate articles, reports, and other information sources on the subject. Next, she followed up on a reference to an organization she found mentioned in a journal article. Using the advanced search feature of *Google,* she entered the organization's name as a phrase and located the Web site immediately. It was here that she located a number of useful Canadian and international documents. She spent 3–4 hours reviewing the sites and following up the links.

Tanya spent half an hour searching for a guide to recommended and reviewed Web sites on the topic. First she checked the library's online catalog subject guide to the World Wide Web. While she did find a few useful guides on more general topics, there was nothing exactly on topic. She next checked *BUBL Information Service* and *Argus Clearinghouse* for prepared guides; neither one had anything on target. Finally, she tried *AskJeeves* and had better results. From here she was guided to the Web sites for a number of government agencies, private companies, and universities conducting hybrid vehicle studies.

Academia & Libraries
Area Studies & Ethnic Studies
Arts & Humanities
Business & Economics
Career & Job Information
Consumer Resources
Demographics & Census Data
Education
Electronic Journals & Texts
General & Reference Sources
Government & Politics
News, Media, & Publishing
Sciences & Medicine
Social Sciences
Student Resources, K-12
Technology
Weather & Climate

Special thanks to Eric Griffith and the University of Georgia Libraries, for kindly allowing us to use their collection of web links.

GALILEO | Internet Resources

To browse recommended Web sites for freely available resources, click on a selection from the categories on the left.

View recent additions to Internet Resources.

To access databases purchased or licensed for use by GALILEO participants, click the GALILEO Home Page button below, then choose "Databases."

Search our collection of recommended Internet resources:

◉ All Terms   ○ Any Term

[Search] [Search Tips]

Exploring the Internet

Internet Directories
Browse Internet sites by subject category (including one for kids).

Internet Search Engines
Search the Internet for key words or phrases.

About the Internet

Information about the Internet
User Guides, Definitions, E-mail Addresses.

How to cite Internet Resources
Guidelines for citing Internet sites & articles from GALILEO databases (using APA, MLA formats, etc.).

GALILEO Home Page   Leave a Comment

FIGURE 13–12   Recommended Web sites found in one library's online catalog
Source: Galileo—Georgia Library Learning Online. Courtesy of the Board of Regents of the University System of Georgia.

She came across a Web site for a local research center at a nearby university. She bookmarked the page and made a note to contact the center later that day.

Tanya needed current and specific information on hybrid vehicles and tax incentives. This narrow search worked well using the advanced search features of *Google, Excite,* and particularly, *Hotbot,* which allowed multiple domain limits such as *.gov, .edu,* and *.org.* Just to double-check, Tanya tried *Dogpile,* a meta-search engine; the results were mixed and she found she was spending too much time trying to evaluate sources. Ultimately, since Tanya needed to be confident she had located the most accurate and current information, she returned to her library's online catalogs for a guide to government documents on the Web and was referred to *FirstGov* (www.firstgov.gov). Her search here was uncluttered and produced information that she could confidently use.

Tanya's final search used the popular subject directory *Yahoo!* After a few false starts using the subject tree, she used the keyword feature to locate appropriate sites. Although she found a few new sites, most of them were familiar. This is a sure sign that she had completed her Web research and should move on.

USING QUESTIONNAIRES AND INTERVIEWS

Sometimes, your research project may require collecting firsthand information yourself, such as questionnaires or personal interviews. This section covers both questionnaires or interviews. You will learn how to (1) prepare, send out, and report the results of a questionnaire, and (2) prepare, conduct, and summarize a personal interview.

Tanya Grant's McDuff Project
You will recall that Tanya, who works in Marketing at McDuff, has been asked by the company president to write a report that examines the successes and failures of hybrid electric cars. This report will look at start-up costs for switching the vehicle fleet, tax and other incentives, and the potential of the technology.

Now, before reporting her findings to Rob McDuff, she wants to find out what corporate users of the technology think of its potential. She believes her best approach would be to (1) send a questionnaire to companies that have hybrid vehicle fleets, and (2) personally interview three or four respondents who will help management decide on the company's direction.

Questionnaires

Tanya Grant has the same challenge you would face in developing a questionnaire. Like you, she receives many questionnaires herself. Most of them she tosses in the recycle bin because they don't warrant her time, are too long, or seem confusing. Now that the shoe is on the other foot, she wants to design a questionnaire that attracts the attention of readers and entices them to complete it. To accomplish this feat, she goes through the following three-stage process:

● *Step 1: Preparing the Questionnaire*

Obviously, your questionnaire is useful only if readers complete and return it. You must focus just as much on your readers' needs as you do on your own objectives. Before readers complete a form, they must perceive that (1) it benefits them personally or professionally and (2) it is easy to fill out and return. Keep these two points in mind as you design the form and the cover letter. Here are some specific guidelines for preparing a reader-focused document:

1. Write a precise purpose statement. As in other documents, a one-sentence statement of purpose provides a good lead-in for your cover letter that accompanies the survey (see next section). For example, Tanya Grant has prepared the following purpose statement for her survey concerning hybrid electric vehicles: "The purpose of this survey is to find out ways in which your experience with hybrid cars can benefit others." As obvious as that statement sounds, it will help busy readers who don't have time to wade through long rationales.

2. Limit the number of questions. Every question must serve to draw out information that relates to your purpose statement. For example, Tanya Grant knows her questions must focus on the reader's experience with hybrid electric vehicles. She must resist the temptation to clutter the questionnaire with irrelevant questions on other alternative fuel vehicles such as natural gas or electric cars.

3. Ask mostly objective questions. You need to design your form so that (1) questions are easy to answer and (2) responses are easy to compile. Although open-ended questions yield more detailed information, the answers take time to write and are difficult to analyze. Instead, your goal is breadth, not depth, of response. With the exception of one or two open-ended questions at the end of your questionnaire, reserve long-answer responses for personal interviews you conduct with a select audience. For example, Tanya Grant has decided to include an optional open-ended question at the end of her questionnaire. There she will ask hybrid users to recommend design improvements for hybrid electric vehicles.

Objective questions come in several forms. Four common types are described here, along with examples of each.

- *Either/Or Questions:* Such questions give the reader a choice between two options, such as "yes" or "no." They are useful only when your questions present clear, obvious choices. EXAMPLE: "Do you believe your hybrid vehicles accelerate well in all driving situations?" (followed by "yes" and "no" blocks), or "The hybrid accelerates well in all driving situations."
- *Multiple-Choice Questions:* These questions expand the range of possibilities for the reader to three or more, requiring a longer response time. EXAMPLE: "If you answered 'yes' to the preceding question [a question asking if the hybrid vehicle accelerates well], what is your typical driving terrain: (a) flat, (b) hilly, (c) combination of flat and hilly, (d) mountainous?"
- *Graded-Scale Questions:* By permitting degrees of response, these questions help gauge the relative strength of the reader's opinion. EXAMPLE: "Use of hybrid vehicles has met our day-to-day driving needs: (a) strongly agree, (b) agree, (c) disagree, (d) strongly disagree, or (e) have no opinion."
- *Short-Answer Questions:* Use these questions when the possible short answers are too numerous to list on your form. EXAMPLE: "List the makes of vehicles that your company has purchased in the last five years."

4. Provide clear questions that are easy to answer. Like other forms of technical writing, questionnaires can frustrate readers when individual questions are unclear. Four common problems are (1) bias in phrasing, (2) use of undefined terms, (3) use of more than one variable, and (4) questions that require too much homework. Here are some examples of right and wrong ways to phrase questions, along with a brief comment on each problem:

Biased Question:

Original question: "Are the federal and state government's excessive tax credits for purchasing alternative fueled vehicles affecting your purchasing decision?" (Words like *excessive* reflect a bias in the question, pushing a point of view and thus skewing the response.)

Revised question: "Do you believe that the federal and state tax credits affected your purchasing decision?"

Undefined Technical Terms:

Original question: "Are you familiar with the work of the PNGV on AFV's?" (Your reader may not know that PNGV is short for Partnership for a New Generation of Vehicles or that AFV stands for "alternative fuel vehicles." Thus some "no" answers may be generated by confusion about terminology.)

Revised question: "Are you familiar with the work of the Partnership for a New Generation of Vehicles on alternative fuel vehicles?"

Mixed Variables:

Original question: "Were the dealer's maintenance technicians prompt and thorough in their work?" (There are two questions here, one dealing with promptness and the other with thoroughness.)

Revised question (two separate questions): "Were the dealer's maintenance technicians prompt?" "Were the dealer's maintenance technicians thorough?"

Question That Requires Too Much Homework:

Original question: "What other alternative fuel vehicles has your company researched, tested, or purchased in the last ten years?" (This question asks the reader to conduct research for an accurate answer. If they do not have the time for that research, they may leave the answer blank or provide an inaccurate guess. In either case, you are not getting valid information.)

Revised question: "Has your company tried other alternative fuel vehicles?"

5. Include precise and concise instructions at the top of the form. Your instructions can be in the form of an easy-to-read list of points that start with action verbs, such as the following list:

- Answer Questions 1–20 by checking the correct box.
- Answer Questions 21–30 by completing the sentences in the blanks provided.
- Return the completed form in the envelope provided by October 15, 2002.

Or, if instructions are brief, they can be in the form of a short, action-centered paragraph, like this: "After completing this form, please return it in the enclosed stamped envelope by October 15, 2002."

6. Apply principles of document design. Although you must strive for economy of space when designing a questionnaire, use adequate white space and other design principles to make the document attractive to the eye.

7. Test the questionnaire on a sample audience. Some sort of "user test" is a must for every questionnaire. For example, after completing her survey, Tanya decided to test it on three people:

- A fellow marketing colleague at McDuff, who has conducted several questionnaires for the firm
- A psychologist Tanya knows through a local professional association
- A vehicle fleet manager whom she knows well enough to ask for constructive criticism on the form

Thus her user test will solicit views from people with three quite different perspectives.

● *Step 2: Conducting the Project*

After you have designed a good form, the next task is to distribute it. Following are guidelines for selecting a good sampling of potential respondents, introducing the questionnaire to your audience, and encouraging a quick response from a high percentage of readers.

1. Choose an appropriate audience. Selecting your audience depends on the purpose of your questionnaire. If you manage a 100-employee engineering firm and want to gauge customer satisfaction with recent construction jobs, you might send your questionnaire to all 156 clients you have served in the last two years. Restricting the mailing list would be unnecessary, since you have a small sample.

However, if you are in Tanya Grant's position at McDuff, with a mailing list totaling about 3,200 corporations that have purchased hybrid vehicles in 2000 and 2001, you will need to select a random sample. Tanya's research suggests that she will receive about a 25 percent rate of return on her questionnaires. (Actually, this rate would be quite good for an anonymous questionnaire.) Given that she wants about 200 returned forms, she must send out about 800 questionnaires in expectation of the 25 percent return rate.

With a client list of 3,200, she simply selects every fourth name from the alphabetized list to achieve a random list of 800 names. Note that the selection of client names from an alphabetized list preserves what is essential—that is, the random nature of the process.

Of course, you can create more sophisticated sampling techniques if necessary. For example, let's assume Tanya wants an equal sampling of companies that purchased in each of the two years—2000 (with 1,200 names) and 2001 (with 2,000 names). In other words, she wants to send an equal number of forms to each year's hybrid owners, even though the number of corporate hybrid owners varies from year to year. In this case, first she would select 400 names—or every third name—from the 1,200 alphabetized names for 2000. Then she would select the other 400 names—or every fifth name—from the 2,000 alphabetized names for 2001. As a result, she has done all she can do to equalize the return rate for two years.

This strategy will help you choose the audience for simple questionnaire projects. You may want to consult a specialist in statistics if you face a sophisticated problem in developing an appropriate sampling.

2. Introduce the questionnaire with a clear, concise cover letter. In 15 or 20 seconds, your letter of transmittal must persuade readers that the questionnaire is worth their time. Toward this end, it should include three main sections (which correspond to the letter pattern presented in chapter 7):

- *Opening paragraph:* Here you precisely state the purpose of the questionnaire and perhaps indicate why this reader was selected.
- *Middle paragraph(s):* Here you state the importance of the project. Strive to emphasize ways that it may benefit the reader.
- *Concluding paragraph:* Specify when the questionnaire should be returned, even though this information will be included in the directions on the questionnaire itself.

3. Encourage a quick response. If your questionnaire is not anonymous, you may need to offer an incentive for respondents to submit the form by the due date. For example, you can offer to send them a report of survey results, a complimentary pamphlet or article related to their field, or even something more obviously commercial, when appropriate. Obviously, any incentive must be fitting for the context. Keep in mind also that some experts believe an incentive of any kind introduces a bias to the sample.

If the questionnaire is anonymous or if complimentary gifts are inappropriate or impractical, then you must encourage a quick response simply by making the form as easy as possible to complete. Clear instructions, frequent use of white space, a limited number of questions, and other design features mentioned earlier must be your selling points.

● *Step 3: Reporting the Results*

After you tabulate results of the survey, you must return to the needs of your original audience—the persons who asked you to complete the questionnaire. They expect you to report the results of your work. Described next are the major features of such a report.

First, you want to show your audience that you did a competent job of preparing, distributing, and collecting the questionnaire. Thus the body of your report should give details about your procedures. Appendices may include a sample form, a list of respondents, your schedule, extensive tabulated data, and other supporting information.

Second, you want to reveal the results of the survey. Here is where you need to be especially careful. You should present only those conclusions that flow clearly from data. Choose a tone that is more one of suggesting than declaring. In this way, you give readers the chance to draw their own conclusions and to feel more involved in final decision making. Graphs are an especially useful way to present statistical information (see chapter 11, "Graphics").

Finally, remember that your report and the completed questionnaires may remain on file for later reference by employees who know nothing about your project. Be sure that your document is self-contained. Later readers who uncover the "time capsule" of your project should be able to understand its procedures and significance from the report you have written.

Interviews

Besides questionnaires, interviews are another common way to gather primary research. Often they are conducted after a questionnaire has been completed, as a follow-up activity with selected respondents. Interviews also may be done as primary research independent of questionnaires. In either case, you need to follow some common guidelines to achieve success in the interview. Here are a few basic pointers for preparing, conducting, and recording the results of your interviews:

● *Step 1: Preparing for the Interview*

You should put at least as much effort into planning the interview as you do into conducting it. Good planning will put you at ease *and* show interviewees that you value their time. Specifically, follow these guidelines:

- **Develop a list of specific objectives for the interview.** Know exactly what you want to accomplish so that you can convey this significance to the person you interview.
- **Make clear your main objectives when you make contact for the interview.** This conversation should (1) stress the uniqueness of the person's contribution, (2) put him or her at ease with your goals and the general content of the proposed discussion, and (3) set a starting time and approximate length for the interview. If handled well, this preliminary conversation will serve as a prelude to the interview, giving direction to the next meeting.
- **Prepare an interview outline.** Persons you interview understand your need for written reference during the interview. Indeed, they will expect it of any well-prepared interviewer. A written outline should include (1) a sequential list of topics and subtopics you want to cover and (2) specific questions you plan to ask.
- **Show that you value your interviewee's time.** You can do this first by showing up a few minutes early so that the interview can begin on time. You also show this courtesy by staying on track and ending on time. Never go beyond your promised time limit unless it is absolutely clear that the person being interviewed wants to extend the conversation further than planned.

- *Step 2: Conducting the Interview*

Your interview will be successful if you stay in control of it. Maintaining control has little to do with force of personality, so don't worry if you are not an especially assertive person. Instead, keep control by sticking to your outline and not letting time get away from you. If you find your speakers straying, for example, gently bring them back to the point with another question from your list. Here are additional pointers for conducting the interview:

- **Ask mostly "open" questions.** Open questions require your respondent to say something other than "yes," "no," or other short answers. They are useful to the speaker because they offer an opportunity to clarify an opinion or a fact. They are useful to you because you get the chance to listen to the speaker, to digest information, and to prepare for the next question.

  For example, McDuff's Tanya Grant may ask questions like these: "Could you describe two or three ways in which your expectations for hybrid vehicles have been met? For what purposes is your company currently using its fleet of hybrid vehicles?" or "I've been told that your company has a high commitment to environmental issues in the Denver area. How has purchasing and using hybrid vehicles been part of that commitment?"
- **Ask closed questions when you need to nail down an answer.** For example, Tanya Grant may ask persons she interviews, "Would you be willing to meet with our fleet supervisor to discuss your experience with maintaining hybrid vehicles?" A "yes" or "perhaps" answer will give her an opening for calling this person several months later. A closed question works when commitment is needed.
- **Use summaries throughout the interview.** Brief and frequent summaries serve as important resting points during the conversation. They give you the chance to make sure you understand the answers that have been given. They give your

counterpart the chance to amplify or correct previous comments. For example, Tanya Grant may comment to her speaker, "So, in other words, you are saying that hybrid vehicles make most sense right now for in-city driving where only one or two people share the vehicle." This summary will elicit either a "yes" or a clarification, either of which will help Tanya record the interview accurately.

● *Step 3: Recording the Results*

Throughout the interview you will have taken notes. The actual mechanics of this process may influence the accuracy of your note taking. Here are three possible approaches:

- **Option 1: Number reference:** Using this approach, you will begin the interview with a list of numbered questions on your outline page. Then when you are taking notes, simply list the number of the question, followed by your notes. This approach gives you as much space as you want to write questions, but it does require that you move back and forth between your numbered question list and note page.
- **Option 2: Combined question-and-answer page:** Here you place a major question or two on each page, leaving the rest of the page to record answers to these and related questions that may be discussed. While this strategy requires considerably more paper and separates your prepared list of questions, it does help you focus quickly on each specific question and answer.
- **Option 3: Split page:** Some interviewers prefer to split each page lengthwise, writing questions in the left column and corresponding answers in the right column. Some questions may have been prepared ahead of time, as in Option 2. Others may be written as they are asked. In either case, you have a clear visual break between questions on one side and answers on the other. The advantage over Option 2 is that you have a visual map that shows you your progress during the conversation. Questions and answers are woven together into the fabric of your interview.

USING BORROWED INFORMATION CORRECTLY

Most errors in research papers occur in transferring borrowed information with accuracy and correct form to your report, paper, or other document. This section has two goals: (1) to explain why you must acknowledge sources you have used, and (2) to outline a research process from the point at which you identify possible sources of information to the point at which you have written the first draft.

Avoiding Plagiarism

One basic rule underlies the mechanical steps described in the rest of this chapter:

> With the exception of "common knowledge," you should cite sources for ALL borrowed information used in your final document. This includes quotations, paraphrases, and summaries.

"Common knowledge" is information generally available from basic sources in the field. In the case of Tanya Grant's research project, common knowledge would be a definition of hybrid electric vehicles. When you are uncertain whether or not a piece of borrowed information is common knowledge, go ahead and cite the source. It is better to err on the side of excessive documentation than to leave out a citation and risk a charge of *plagiarism* (the intentional or unintentional use of the ideas of others as your own). Here are three main reasons for documenting sources thoroughly and accurately:

1. **Ethics:** You have an *ethical* obligation to show your reader where your ideas stop and those of another person begin. Otherwise you would be parading the ideas of others as your own.
2. **Law:** You have a *legal* obligation to acknowledge information borrowed from a copyrighted source. In fact, you should seek written permission for the use of borrowed information that is copyrighted when you plan to publish your document or when you are using your document to bring in profit to your firm (as in a proposal or report). If you need more specific information about copyright laws or about the legalities of documentation, see a research librarian.
3. **Courtesy:** You owe readers the *courtesy* of citing sources where they can seek additional information on the subject. Presumably, sources for quotations, paraphrases, and summaries would provide such a reference point.

Certainly some plagiarism occurs when unscrupulous writers intentionally copy the writing of others without acknowledging sources. However, most plagiarism results from sloppy work during the research and writing process. Described here are two common types of unintentional plagiarism. Though the errors are unintentional—that is, the writer did not intend to "cheat"—both result in the inappropriate use of another person's work. That's plagiarism.

Mike Pierson, a supervisor at McDuff's Cleveland office, has been asked to deliver a presentation at an upcoming conference on hybrid electric vehicles. In his last-minute rush to complete the presentation—which will be published in a collection of papers from the meeting—Mike is taking notes from a source in the company library. He hurriedly writes a note card on a source but fails to indicate the source. Later, when he is writing the paper draft, he finds the card and does not know whether it contains information that was borrowed from a source or ideas that came to him during the research process. If he incorporates the passage into his paper without a source, he will have committed plagiarism.

In our second case, Mike transfers a direct quotation from a source onto a note card. He remembers to place source data on the card, but he forgets to include quotation marks. If later he were to incorporate the quotation into his presentation *with* the source citation but *without* quotation marks, he would have plagiarized. Why? Because he would be parading the exact words of another writer as his own paraphrase. The passage would give the appearance of being his own words that are supported by the ideas of another, when in fact the passage would be a direct quote. Again, remember that the test for plagiarism is not one's intent; it is the result.

The next section shows you how to avoid plagiarism by completing the research process carefully. In particular, it focuses on a methodical process that involves

(1) bibliography cards, (2) a rough outline, (3) note cards of three main kinds, (4) a final outline, and (5) drafts.

Following the Research Process

The research process is no different from most other technical tasks. If you carefully attend to this five-step procedure, you can avoid problems at the end.

● *Step 1: Write Complete Bibliography Cards*

Using 3″ × 5″ index cards, you should write a complete bibliography card for each source that may help you later. Called a "working bibliography," this batch of cards becomes the foundation for the rest of your research. (Figure 13–13 shows a bibliography card that Tanya Grant might have collected in her research on hybrid vehicles.)

Errors made at this stage—in transferring information from sources to your cards—can easily work their way into a final document. In writing your cards, therefore, be sure to take these precautions:

1. **Include all information needed for the final-copy citation in your paper.** Use the exact wording for titles and publication information. Common errors are to leave off articles (*a, an, the*) and to abbreviate words in titles, with the writer thinking there will be time later to double-check the original source. In fact, that final check often does not occur, leading to errors in the final citation. Also, for Internet sources you must include the names of online databases, the exact citation for Uniform Resource Locators (URLs), and the date the Internet source was consulted.
2. **Save space at the bottom of the card for a reminder to yourself about the use-fulness of the source.** For example, a notation such as "includes excellent chapter on water resources" may help you later as you begin your research. While

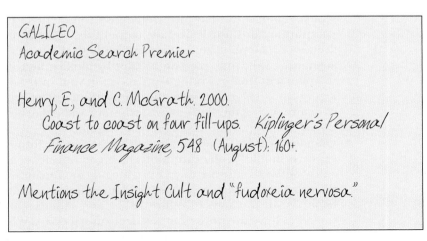

FIGURE 13–13 Sample bibliography card

they are mainly to record source information, bibliography cards can also provide some guidance in the next stage of note taking.

3. **Include information in the exact format as it will appear in your final bibliography, down to the indenting, punctuation, and capitalization.** Tanya Grant's card in Figure 13–13, for example, follows the capitalization guidelines described on pages 535–536 for the author-date system. Again, do not assume that you will have time later to transcribe every card into another format. That time will not be there. Also, using the same format ensures that you will make sure to take down all information needed later for the source page.

● Step 2: Develop a Rough Outline

This outline is *not* the one you will use to write the first draft. Instead, it is essentially a list of topics in the approximate sequence they will be covered in the paper. It serves to direct your writing of note cards during the next step.

● Step 3: Take Careful Notes on Large Note Cards

Most plagiarism results from sloppy note taking. This important stage requires that you attend to detail and follow a rigorous procedure. The procedure suggested here divides note cards into three types: summary, paraphrase, and quotation. Figure 13–14 gives examples of all three types.

- **Summary cards** are written in your words and reduce a good deal of borrowed information to a few sentences. They are best written by reading a section of source material, looking away from the source, and summarizing the passage in your own words. In this way, you can later use any of this information with confidence in your paper, without worrying about the absence of quotation marks. Of course, a summary card may contain a few quoted passages, but the card's main purpose is to reduce considerable source material to a short summary.

- **Paraphrase cards** include a close rephrasing of material from your sources. Unlike summary cards, which condense a considerable amount of information, paraphrase cards usually include more of the original text. Thus they demand even more attention than summary cards to the problem of plagiarism. Like summary cards, they are best written by looking away from the source for a moment and then rewriting the passage in your own words. You can use a few key words from the passage, but do *not* duplicate exact phrasing or sentence structure. Using the "look away" technique will help you avoid creating a paraphrase that too closely resembles the original.

- **Quotation cards** include only words taken directly from the source. Your main concern should be the care with which you transfer sentences from source to card—with absolute accuracy. Even include grammatical or spelling errors that the original source may contain (and, in so doing, use the word *sic*—see Handbook). Also, as shown in Figure 13–14, you should use ellipses (spaced dots) when you leave out words that you deem unnecessary. When using ellipses, however, be sure not to alter the meaning of the passage you are quoting.

QUOTATION CARD   Henry + McGrath

"Although the car [Insight] will never lead the pack at
Indy, you won't fear for your life getting on the freeway...
But note this: The Insight's 365-pound payload capacity
means a couple of hefty passengers can overload [sic]
the car. Honda says that wouldn't damage the vehicle
but would impede performance and mileage."

PARAPHRASE CARD   Henry + McGrath

The Insight won't win any races, but it will get you safely
on the freeway. Its load weight (365 pounds) is poor,
however. Honda says that it wouldn't damage the car
to overload it, but that performance and mileage would
suffer.

SUMMARY CARD   Henry + McGrath

The Insight has modest power but can't carry much
weight.

FIGURE 13–14 Sample note cards on same passage

● *Step 4: Organize Research in an Outline*

With note cards in hand, you are now ready to render order from chaos—to create an outline that flows from the technical research related in your note cards. (See outline suggestions in chapter 1.) Key your note cards to the outline by placing one or both of these items at the top of the card: (1) the wording of the related topic, as it appears on the outline, or (2) the letter and/or number that exists on the outline for the related topic. Then place your cards in the order that you will use them in writing your first draft.

● *Step 5: Write the Draft from the Outline and the Cards*

This step poses the greatest challenge. Here you must incorporate borrowed information with your own ideas to create fluid prose. Your goal should be to demonstrate (a) a smooth transition between your ideas and those you have borrowed and (b) absolute clarity about when borrowed ideas and quotations start and end.

SELECTING AND FOLLOWING A DOCUMENTATION SYSTEM

"Documentation" refers to the mechanical system you use to cite sources from which you borrow information. This section first highlights the main approaches to documentation and then outlines one common approach. For details about a particular documentation system you are using, consult one of the manuals in the list that follows *or* consult the Web site of the organization that publsihes the manual. Pay special attention to new guidelines these manuals may provide for documenting information from online databases and the Internet.

Varieties of Documentation

There are almost as many systems as there are professional organizations. Yet all have the same goal of showing readers the sources from which you gathered information. Here are just a few documentation manuals commonly used in business, industry, and the professions:

The Chicago Manual of Style

Columbia Guide to Online Style

Council of Biology Editors Style Manual

MLA Handbook for Writers of Research Papers

Publication Manual of the American Psychological Association

Style Manual for Engineering Authors and Editors

U.S. Government Printing Office Style Manual

These and other approaches to documentation can be grouped into three main categories, with these features:

1. Footnotes at the bottom of pages, with an alphabetical list of sources at the end of the document
2. Endnotes collected on a page at the end of the document, followed by an alphabetical list of all sources
3. Parenthetical references in the text of the paper, with an alphabetical list of sources at the end of the document

Most professions have adopted some form of the third system—parenthetical documentation. It is the simplest one for writers to use and the easiest one for readers to decipher. Source information is given right within the text, rather than at the bottom of the page or the end of the document. Then a list of works cited occurs at the end of the document. The next section gives guidelines for one such form of parenthetical documentation.

Author–Date System with Pages

Often used in scientific and technical fields, the author-date system closely resembles systems recommended by the Modern Language Association (for the humanities), the American Psychological Association (for behavioral sciences), and the *Chicago Manual of Style* (for a variety of fields). The version described here includes the page number as well as author and date, giving it enough versatility to be used in many professional fields.

Citations within the Text. Follow these six guidelines for acknowledging borrowed information within the text. Remember that you must note *any* quotations, paraphrases, or summaries that are not considered common knowledge.

1. **Immediately after the borrowed information occurs in the text, place the author's last name, publication date, and page number in parentheses.** As always, your main goal is to consider the reader's needs. EXAMPLE: "Aerial photos made at this site in 1965 revealed geomorphic features indicative of fault activity (Spears 1966, 76–77)." Note that the parenthetical citation is placed *inside* the punctuation for the sentence and that there is no comma between author and year.
2. **To show where borrowed information starts, or just to vary your style, you also have the option of placing the author's name at the beginning of the text passage.** In this case, remove the author's name from the parenthetical citation. EXAMPLE: "As Spears has shown, aerial photos made at the site in 1965 revealed geomorphic features indicative of fault activity (1966, 76–77)."
3. **When a source has no author or editor, use a short form of the title.** EXAMPLE: (Faults today 1981, 12). Make sure that the first word of your short title is also the first word under which the title is alphabetized in the list of references.
4. **For works with two authors, list both.** EXAMPLE: (Hobbs and Smith 1989, 52). For works with three or more authors, list the first author's name followed by *et al.*, Latin for "and others." EXAMPLE: (Munson et al. 1967).

5. **When your references contain two or more works with the same author and with the same date, distinguish them from each other by placing letters after the date.** EXAMPLE: (Jones 1987a, 76) versus (Jones 1987b, 28). Then you will make this same distinction in the list of references (Works Cited).

6. **If you need to mention several sources in the same citation, separate them by semicolons.** EXAMPLE: (Barns 1945, 34; Timm 1956, 12).

List of References. Like other forms of parenthetical documentation, the author-date system requires that you place a list of references at the end of the document. This alphabetized listing includes all sources you cited in the document text. However, it does *not* include references you may have consulted but did not cite. View the list as the spot to which readers turn for complete bibliographical information on sources mentioned in the text, in case they want to find the sources themselves in a library. A sample list of references is shown in Figure 13–15. When assembling such a list, follow these basic rules:

1. **Arrange the sources alphabetically according to author.** When there is no author, list a source by its title.

2. **For books, give (a) the author (last name first, followed by initials or by first name and middle initial), (b) year, (c) book title (underlined or in italic with only the first word and proper nouns in initial caps), (d) edition (if any), (e) city of publication (along with state if city is not well known), and (f) publisher.** Note punctuation of examples in Figure 13–15. For additional authors, place names in last-name-last order, after the first author's name.

3. **For articles, give the (a) author, (b) year, (c) article title (no quotation marks, and with only first word and proper nouns in caps), (d) periodical title (underlined or in italics with main words in initial caps), (e) volume number in arabic numbers, and (f) inclusive page numbers on which the article appears.**

Figure 13–15 shows additional details about books and article citations and about other specialized sources.

Online Documentation

Because information found in online sources is also borrowed, it deserves the same level of care in documentation as you apply to print sources. After all, borrowed information is borrowed information, whatever the source.

However, documenting electronic sources presents some special problems, and the rules are not completely settled. Most important and perhaps most obvious, you do not have the assurance with online information—as you do with print sources—that material will stay the same each time it is used. Thus documentation format must include information that clearly notes the date of access and the exact location. Generally, your entry should include the same kind of data used for print sources, with the addition of the following information:

● The name of the database or information service provider you consulted to locate the article, newspaper, electronic source, etc.

- The URL address of the database, information service provider, or specific electronic source
- Date you used the online source

If the electronic source you are citing is also available in print, you must provide a reference to the print format. Take care to present URL addresses completely and accurately. When you must break the address, do so at a logical point.

Book with two or more authors: Note that the second author's name is given in normal order, with a comma between the two authors' names.

Andrews, T.S., and L. Kelly. 1965a. *Geography of central Oregon*. 4th ed. New York: Jones and Caliber Press.

Book by same authors: Note that the blank line indicates another source by same authors. The "b" is used because the same authors published two cited books in the same year. Also, the state abbreviation is used for clarity, since Hiram is a small town.

_____. 1965b. *Geography today*. Hiram, OH.: Pixie Publishers.

Journal article: Note that there are no spaces between the first page number of this journal article and the colon that follows the volume number. Also, there is only a space, but no punctuation, between the journal title and the volume number.

Cranberg, E. V. 1986. Fossils are fun: The life of a geologist in the 1980s. *Geology Issues* 34:233–344.

Article in collection: This entry is for an article that appears in a collection, with a general editor. Note the comma and "ed." (for "edited by") after the collection title. This same format would be used in other cases where you were referring to a piece from a collection, such as a paper in a conference proceedings.

Fandell, C. N., L. Guest, H. M. Smith, and Z. H. Taylor. 1976. Achieving purity in your sampling techniques. In *Geotechnical Engineering Practices*, ed. J. Schwartz, 23–67. Cleveland: Hapsburg Press.

Interview: Refer to yourself in the third person as the "author."

Iris, J. G. 1988. Resident of Summer Hills Subdivision. Interview with the author at site of toxic-waste dump, 23 March.

Newspaper article: This reference includes the day, section, and page number of the article. If no author had been listed, the entry would have begun with the article title.

Mongo, G. P. 1989. Sinkhole psychology: The ground is falling, Chicken Little. *Dayton Gazette,* 12 July, sec. D, 5.

Article from popular magazine: This reference is handled like a journal article except that the date of the particular issue is also included, after the volume number. Note that there is no extra spacing around the parentheses.

Runyon, D. G., and L. P. Moss. 1967. Sinkholes are coming to your area soon. *Timely News* 123 (15 July):65–66.

FIGURE 13–15 Sample list of references

Information from personal email message: This entry concerning a personal email message contains information on both sender and receiver, as well as the date and subject of the message.

Jankowski, S. X. (sjan@spdu.edu). Nature of the Mahikari movement. Personal email to Matthew B. Knob (mknob@aol.com). 22 Dec. 1999.

CD-ROM citation: Reference to a CD-ROM does not require inclusion of the date you consulted the source.

Jasic, H. D. (1972). Japanese religions today [CD-ROM]. Available: USD Quest File Ondisc Item QF 54/4.

Database and Web citations: Note that web citations must include the date that the information was found.

Traeger, C. (n.d.) The will of the Prius. *Carlist.com.* Retrieved 10 August 2001 from the World Wide Web: http://www.carlist.com/newcars/2001/toyotaprius.html

Einstein, P. (1999, October 29). The benefits of Insight. *Professional Engineering,* 12.19, 23. Retrieved August 3, 2001 from GALILEO, All databases, ProQuest: http://www.galileo.peachnet.edu

Glynn, M. (2001, January 19). Automakers will offer "2 for the price of 1": Many major manufacturers are hoping gas-electric hybrids can power future sales. *Seattle Times,* F1. Retrieved August 4, 2001 from GALILEO, All databases, ProQuest: http://www.galileo.peachnet.edu

FIGURE 13–15   continued

See Model 13–1 at the end of this chapter for Tanya Grant's complete memo report that cites research.

WRITING RESEARCH ABSTRACTS

The term *abstract* has been used throughout this book to describe the summary component of any technical document. As the first part of the ABC pattern, it gives decision-makers the most important information they need. However, here we use abstract for a narrower purpose. It is a stand-alone summary that provides readers with a capsule version of a piece of research, such as an article or a book. This section (1) describes the two main types of research abstracts, with examples of each, and (2) gives five guidelines for writing research abstracts.

Types of Abstracts

There are two types of abstracts: informational and descriptive. As the following definitions indicate, informational abstracts include more detail than descriptive abstracts:

Informational Abstract.
- **Format:** This type of abstract includes the major points from the original document.

- **Purpose:** Given their level of detail, informational abstracts give readers enough information to grasp the main findings, conclusions, and recommendations of the original document.
- **Length:** Though longer than descriptive abstracts, informational abstracts are still best kept to one to three paragraphs.
- **Example:** A sentence from such an abstract might read, "The article notes that functional resumes should include a career objective, academic experience, and a list of the applicant's skills." (See corresponding example in definition of a descriptive abstract.)

Descriptive Abstract.

- **Format:** This type of abstract gives only main topics of the document, without supplying supporting details such as findings, conclusions, or recommendations.
- **Purpose:** Given their lack of detail, descriptive abstracts can only help readers decide whether they want to read the original document.
- **Length:** Their lack of detail usually ensures that descriptive abstracts are no more than one paragraph.
- **Example:** A sentence from such an abstract might read, "The article lists the main parts of the functional resume." (See corresponding example in definition of an informational abstract.)

You may wonder when you'll need to write abstracts during your career. First, your boss may ask you to summarize some research, perhaps because he or she lacks your technical background. Second, you may want to collect abstracts as part of your own research project. In either case, you need to write abstracts that accurately reflect the tone and content of the original document.

Assume, for example, that your McDuff supervisor has asked you to read some current research on strategies for negotiating. Later your boss plans to use your abstracts to get an overview of the field and to decide which, if any, of the original full-length documents should be read in full. The examples that follow show both informational and descriptive abstracts of the section of chapter 14 that covers negotiating. Note that the informational abstract actually lists the guidelines contained in the chapter, whereas the descriptive abstract notes only that the article includes the guidelines.

Informational Abstract: "Guidelines for Negotiating"

This article suggests that modern negotiations should replace "I win, you lose" thinking with a "we can both win" attitude. To achieve this change, these six main guidelines are prescribed: (1) think long term, (2) explore many options, (3) find the shared interests, (4) listen carefully, (5) be patient, and (6) do look back. Although this strategy applies to all types of negotiation, this article focuses on a business context. It includes an extended example that involves establishing an appropriate entry salary for a job applicant in computer systems engineering.

Descriptive Abstract: "Guidelines for Negotiating"

This article describes six main guidelines that apply to all types of negotiations. The emphasis is on strategies to be used in the context of business. All the suggestions in the article support the need for a "we both win" attitude in negotiating, rather than an "I win, you lose" approach.

Guidelines for Writing Research Abstracts

The guidelines given here will help you (1) locate the important information in a document written by you or someone else and (2) present it with clarity and precision in an abstract. In every case, you must present a capsule version of the document in language the reader can understand. The ultimate goal is to save the readers time.

● *Abstracting Guideline 1: Highlight the Main Points*

This guideline applies whether you are abstracting a document written by you or someone else. To extract information that will be used in your abstract, follow these steps:

- Find a purpose statement in the first few paragraphs.
- Skim the entire piece quickly, getting a sense of its organization.
- Read the piece more carefully, underlining main points and placing comments in margins.
- Pay special attention to information gained from headings, first sentences of paragraphs, listings, graphics, and beginning and ending sections.

● *Abstracting Guideline 2: Sketch Out an Outline*

From the notes and marginal comments gathered in Abstracting Guideline 1, write a brief outline that contains the main points of the piece. If you are dealing with a well-organized piece of writing, it will be an easy task. If you are not, it will be a challenge. Here is an outline for the negotiation section of chapter 14, as abstracted in the previous examples.

Outline for "Guidelines for Negotiating." Purpose: to provide rules that help readers adopt a "we both win" strategy
- I. Think long-term
  - A. Focus on building mutual trust
  - B. Project the long-term attitude into every part of the negotiation process
- II. Explore many options
  - A. Get away from thinking there are only two choices
  - B. Put diverse options on the table early in the negotiation
- III. Find the shared interests
  - A. Stress points of agreement, rather than conflict
  - B. Use mutual concerns to defuse contentious issues
- IV. Listen carefully
  - A. Ask questions and listen, rather than talk
  - B. Use probing questions to move discussion along in salary discussion
    - 1. Break out of attack/counterattack cycle
    - 2. Uncover motivations

      3. Expose careless logic and unsupported demands

      4. Move both sides closer to objective standards

V. Be patient

    A. Avoid the mistakes that come from hasty decisions

    B. Avoid the bad feeling that results when people feel pressured

VI. Do look back

    A. Keep a journal in which you reflect on your negotiations

    B. Analyze the degree to which you followed the previous five guidelines

● *Abstracting Guideline 3: Begin with a Short Purpose Statement*

Both descriptive and informational abstracts should start with a concise overview sentence. This sentence acquaints the reader with the document's main purpose. Stylistically, it should include an action verb and a clear subject. Here are three options that can be adapted to any abstract:

- The article "Recycle Now!" states that Georgia must intensify its effort to recycle all types of waste.
- In "Recycle Now!" Laurie Hellman claims that Georgia must intensify its effort to recycle all types of waste.
- According to "Recycle Now!" Georgians must intensify their efforts to recycle all types of waste.

● *Abstracting Guideline 4: Maintain a Fluid Style*

One potential hazard of the abstracting process is that you may produce disjointed, awkward paragraphs. You can reduce the possibility of this stylistic flaw by following these steps:

- Write in complete sentences, without deleting articles (*a, an, the*)
- Use transitional words and phrases between sentences
- Follow the natural logic and flow of the original document itself

● *Abstracting Guideline 5: Avoid Technical Terms Readers May Not Know*

Another potential hazard is that the abstract writer, in pursuit of brevity, will use terms unfamiliar to the readers of the abstract. This flaw is especially bothersome to readers who do not have access to the original document. As a general rule, use no technical terms that may be unclear to your intended audience. If a term or two are needed, provide a brief definition in the abstract itself.

    Note, also, that abstracts that might become separated from the original document should include a bibliographical citation (see the previous examples).

Communication Challenge

"To Cite or Not To Cite"

Dan Gibbs works as a benefits and finance specialist at McDuff's corporate office in Baltimore. As the number of McDuff employees has grown, he has received many inquiries about ways to save for retirement. Recently he wrote and distributed a four-page flyer on the topic, using materials from print and online sources. The response was so positive that his boss wants to send the flyer to clients as a "freebie"—both to help clients' employees and to create "good will" in marketing. This use of the flyer has made Dan re-think how he developed the piece. His research process, his results, and some questions for discussion follow.

Background of Retirement Booklet

Unlike Tanya Grant in the hybrid vehicle project described in this chapter, Dan didn't have time or interest in pursuing a full-scale library search about retirement strategies. Besides, he has personnel magazines in the office with data that would support his points. In addition, he is connected to the Internet, a series of loosely connected databases that might provide information directly to him at his office computer.

Like many companies, McDuff has a retirement plan largely in the form of what is called a 401K program. It allows employees to contribute a percentage of their salaries into a tax-deferred retirement account, a portion of which is matched by the employer. Even though McDuff has a generous matching arrangement, many employees do not take full advantage of the program. Therefore, Dan wrote the retirement flyer to remind them that it is never too early to plan for retirement. As it happens, he learned that many U.S. workers are failing to put away enough money for their retirement years.

The Research Process

After outlining his goals for the booklet, Dan began "surfing" through related information on the Internet. He made use of four pieces of information from an *Atlanta-Journal Constitution** article found on an Internet database called "Human

*The database is a fictional one created just for this case. However, the data are real and included in the following source: "A Special Report: The Rocky Road to Retirement," by Hank Ezell, *Atlanta Journal-Constitution,* September 25, 1995, pp. E1, E4–6. The article cites these sources for data listed: the Rand Corporation, National Vital Statistics System, and Ibbotson Associates for item #1; the Social Security Administration for item #2; and Hewitt Associates for item #3.

CHAPTER SUMMARY

This chapter highlights the process of conducting technical research and writing about the results. Much of the information is presented within the context of a research project conducted at McDuff.

Resources Retrieval." Following are three of the themes he stressed, along with related information he used from the newspaper article.

1. *Theme #1: We're living longer past retirement.* The following changes occurred in years of life expected after age 65: for men, 12.8 in 1960, 13.1 in 1970, 14.1 in 1980, and 15.1 in 1990; for women, 15.8 in 1960, 17.0 in 1970, 18.3 in 1980, and 18.9 in 1990.

2. *Theme #2: We cannot depend exclusively on Social Security.* As many more people retire from the "Baby Boom" generation born between 1946 and 1964, fewer and fewer workers paying Social Security are supporting each person getting it. The following numbers are actual and projected number of workers supporting each retiree: 41.9 in 1945, 8.6 in 1955, 4.0 in 1965, 3.2 in 1975, 3.3 in 1985, 3.3 in 1995, 3.1 in 2005, 2.7 in 2015, 2.2 in 2025, and 2.0 in 2035.

3. *Theme #3: We should begin saving when we're young.* If you start at age 25 saving $3,000 yearly in a tax-deferred account, with a 9% annual return, you'll accumulate $166,971 by age 45, $457,685 by age 55, and $1,170,330 by age 65. If you start at age 35, you'll have $48,378 by 45, $166,971 by 55, and $457,685 by 65. If you start at age 45, you'll have $48,378 by 55 and $166,971 by 65.

Questions and Comments for Discussion

1. If you were presenting the data above in a research report, what format would you choose? Why? (See chapter 11.)
2. Considering the four data sets Dan took from the newspaper article/Internet sources, which ones need documentation and which, if any, do not? Explain your answer.
3. Does the fact that the flyer will be sent to clients have any effect on your answer to question #2?
4. Do a "works cited" reference for the data in Theme #1 above. See the footnote to this Communication Challenge for actual source information. Assume that Dan found the piece on the Internet on March 4, 1996.

Before starting your search for information, you need to decide what main question you are trying to answer. Also, think about the types of information you need, the types of sources that would be useful, and the format required for the final document. Once in the library, you have many sources available to you: books, periodicals, newspapers, company directories, general references such as dictionaries and encyclopedias, abstracts—both online and in print. You can also use the Internet, World Wide Web, questionnaires, and interviews.

As you begin to locate sources, follow this five-step research process: (1) write complete bibliography cards, (2) develop a rough outline, (3) take careful notes on large note cards, (4) organize research in an outline, and (5) write the draft from the outline and the cards. For the final paper, choose a documentation system appropriate for your field or organization. Today the preferred approach is to use some form of parenthetical citations, such as the author-date system outlined in this chapter.

Another research skill is writing research abstracts (summaries) of articles, books, or other sources of information. Abstracts can be either descriptive (quite brief) or informational (somewhat more detailed).

 Access the Technical Writing Web site by pointing your browser at www.prenhall.com/pfeiffer. Here, you will find additional resources to aid you in learning and applying the information from this chapter.

ASSIGNMENTS

If your instructor considers it appropriate, use a copy of the Planning Form at the end of the book for completing these assignments.

1. **General Research Paper.** Using a topic approved by your instructor, follow the procedure suggested in this chapter for writing a paper that results from some technical research. Be sure that your topic (1) relates to a technical field in which you have an interest, by virtue of your career or academic experience, and (2) is in a field about which you can find information in nearby libraries.

2. **Research Paper—Your Major Field.** Write a research paper on your major field. Consider some of these questions in arriving at your thesis for the paper: What is the history of your major? What types of jobs do majors in your discipline pursue? Do their job responsibilities change after 5, 10, or 15 years in the field? What kinds of professional organizations exist to support your field?

3. **Research Paper—McDuff.** As a McDuff engineer or scientist, you have been asked to write a research paper for McDuff's upper management. Choose your topic from one of the technical fields listed here. Assume that your readers are gathering information about the topic because they may want to conduct consulting work for companies or government agencies involved in these fields. Focus on advantages and disadvantages associated with the particular technology you choose. Follow the procedure outlined in this chapter.

   - Artificial intelligence
   - Chemical hazards in the home
   - Fiber optics
   - Forestry management
   - Geothermal energy
   - Human-powered vehicles
   - Lignite-coal mining

- Organic farming
- Satellite surveying
- Solar power
- Wind power

4. **Abstract—One Article or Several Articles.**
    Option A. Visit your college library and find a magazine or journal in a technical area, perhaps your major field. Then photocopy a short article (about five pages) that does not already contain a separate abstract or summary at the beginning of the article. Using the guidelines in this chapter, write both an informative and a descriptive abstract for a nontechnical audience. Submit the two abstracts, along with the copy of the article.
    Option B. Follow the instructions in Option A, but use a short article that has been selected or provided by your instructor.
    Option C. Read 3 to 5 current articles in your major field. Write an abstract that summarizes *all* of them on one page.

5. **Questionnaire—Analysis.** Using the guidelines in this chapter for questionnaires, point out problems posed by the following questions:
    a. Is the poor economy affecting your opinion about the current Congress?
    b. Do you think the company's severe morale problem is being caused by excessive layoffs?
    c. Was the response of our salespeople both courteous and efficient?
    d. Of all the computer consultants you have used in the last 15 years, which category most accurately reflects your ranking of our firm: (a) the top 5%, (b) the top 10%, (c) the top 25%, (d) the top 50%, or (e) the bottom 50%?
    e. In choosing your next writing consultant, would you consider seeking the advice of a professional association such as the STC or the CPTSC?
    f. Besides the position just filled, how many job openings at your firm have been handled by Dowry Personnel Services?

6. **Questionnaire—Writing.** Design a brief questionnaire to be completed by students on your campus. Select a topic of general interest, such as the special needs of evening students or the level of satisfaction with certain college facilities or services. Administer the questionnaire to at least 20 individuals (in classes, at the student union, in dormitories, etc.). *After* you analyze the results, write a brief report that summarizes your findings. *Note:* Before completing this exercise, make sure that you gain any necessary approvals of college officials, if required.

7. **Interviews—Simulation and Analysis.** Divide into groups of three or four students, as your instructor directs. Two members of the group will take part in a simulated interview between a placement specialist at your school and a personnel representative from an area company. Assume that the firm might have a number of openings for your school's graduates in the next few years.

    As a group, create some questions that would be useful during the simulated interview. Then have the two members perform the role-playing exercise for 15 to 20 minutes. Finally, as a group, critique the interview according to the suggestions in this chapter and share your findings with the entire class.

8. **Interview—Real-World.** Select a simple research project that would benefit from information gained from an interview. (Your project may or may not be associated with a written assignment in this course.) Using the suggestions in this chapter, conduct the interview with the appropriate official.

9. **Ethics Assignment.** This assignment is best completed as a group exercise.

Assume your group has been chosen to develop a Web-based course in technical communication. Team members are assembling materials on a Web site that can be used by students like you—materials such as (1) guidelines and examples from this book, (2) scholarly articles on communication, (3) newspaper articles and graphics from print and online sources, and (4) examples of technical writing that have been borrowed from various engineering firms.

Your team has been told that, generally speaking, the "fair use" provision of the Copyright Act permits use of limited amounts of photocopied material from copyrighted sources without the need to seek permission from, or provide payment to, the authors—as long as use is related to a not-for-profit organization, such as a college. Your tasks are as follows:

1. Research the Copyright Act to make sure you understand its application to conventional classroom use. If possible, also locate any guidelines that relate to the Internet.
2. Develop a list of some specific borrowed materials your group would want to include on the site for the technical communication course. These materials may fall inside or outside the four general groupings noted previously.
3. Discuss how the medium of the Internet may influence the degree to which the "fair use" provision would be applicable to your Web course. Be specific about the various potential uses of the material.
4. Consult an actual Web-based college course in any field and evaluate the degree to which you think it follows legal and ethical guidelines for usage.
5. Prepare a report on your findings (written or oral, depending on the directions you have been given by your instructor).

10. **International Communication Assignment.** Using interviews, books, periodicals, or the Internet, investigate the degree to which writers in one or more cultures besides your own acknowledge borrowed information in research documents. For example, you might want to seek answers to one or more of the following questions: Do you believe acknowledging the assistance of others is a matter of absolute ethics? Or should such issues be considered relative and therefore influenced by the culture in which they arise? For example, would a culture that highly values teamwork and group consensus take a more lenient attitude toward acknowledging the work of others? These are not simple questions. Think them through carefully.

MEMORANDUM

DATE: August 12, 2002
TO: Rob McDuff, President
FROM: Tanya Grant, Technical Writer
SUBJECT: Hybrid-Electric Vehicle Research and Recommendations

INTRODUCTORY SUMMARY

When you heard that our Tokyo office had purchased some Toyota minivans powered by a combination gasoline-electric engine, you asked me to research hybrid-electric vehicles (HEVs) and recommend whether or not McDuff should use them in its U.S. offices. After making several phone calls, checking useful web sites, and reviewing related books and articles, I recommend McDuff not switch to HEVs at this time. However, we should consider the idea again when HEVs similar to the trucks and SUVs we now use in our U.S. offices are available. This report highlights advantages and disadvantages of HEVs, costs, and facts about the current and future HEV market.

ADVANTAGES

HEVs offer five main advantages: excellent fuel economy, low emissions, long range, conventional fueling, and adequate power.

Excellent Fuel Economy

HEVs are able to approach double the fuel economy of the current conventional car. The subcompact Toyota Prius is rated by the U.S. Environmental Protection Agency (2000) at 52 miles per gallon (mpg) city driving, and at 46 highway. That's right: the Prius gets better mileage in the city, in part because it makes better use of regenerative braking to capture energy for its battery (Garretson, 2001) and has an electric motor-gasoline engine configuration that favors slower stop-and-go traffic (Traeger). The two-seat Honda Insight gets even better mileage, rating 61 mpg city and 68 highway. Because the gasoline engine gets a boost when needed from the electric motor, the engine needn't be large.

Another gas saver is the fact that the gasoline engine shuts down when at a stoplight, to be restarted immediately by the electric motor when you engage the clutch. However, having the engine shut off can be unnerving. One new Insight owner turned the ignition key every time the engine cut out, continuing to do so for two weeks until her husband figured out that the car would start back up by itself (Anderton, 2001).

Low Emissions

In an HEV, the electric motor releases no emissions, and the small gasoline engine recharges the battery and has to work less often, so its emissions are much lower. The California Air Resources Board (2001) rates the Honda Insight as an Ultra-Low Emissions Vehicle (ULEV), which means that the Insight is at least fifty percent cleaner than the average 2001 model car. The Prius does even better, rating as a Super-Ultra-Low Emissions Vehicle (SULEV) because its emissions are ninety percent below the average.

MODEL 13–1
Memo report citing research

continues

Long Range

Because the gasoline engine automatically recharges the electric motor, HEVs can go many hundreds of miles before refueling due to their excellent fuel economy. With fuel tanks of 11.9 and 10.6 gallons, respectively, both the Toyota Prius and the Honda Insight can easily go over 500 miles between gas stations. Henry and McGrath (2000) estimate that the Insight could go across the U.S. on only four tanks of gas.

Conventional Fueling

HEVs run on gasoline, so fueling is no different than for a standard automobile. This feature greatly increases their ease of use for those who may be concerned about making the transition from conventional vehicles.

Adequate Power

For their size, the existing HEVs have decent power. The Toyota Prius can go from 0-60 in 13.0 seconds and achieve a top speed of 99 m.p.h. The Honda Insight has somewhat better power, going to 60 in 10.6 seconds with a top speed of 107 m.p.h. (Road test, 2001). Anderton (2001) admits, however, that the Insight can "lag a bit on hills."

DISADVANTAGES

The three main disadvantages of HEVs are as follows: small selection, low load capacity, and complexity.

Small Selection

Presently, only two HEVs are available in the US: the Honda Insight and the Toyota Prius. Both are small cars. The Insight seats only two, and the Prius supposedly five, but three of them had better be children. The only company vehicles used in our U.S. offices are pickup trucks and SUVs. Currently there are no HEVs on the market that meet McDuff's immediate needs.

Low Load Capacity

The Toyota Prius has the cab room and trunk space of a Corolla, and the Honda Insight has much less. The total load of the Insight is only 365 pounds, so two engineers of 180 pounds each would nearly weigh it to capacity, even without any equipment. The trunk space is so small that Henry and McGrath (2000) complain that a golf bag doesn't even fit. Though the Prius can carry a greater load, it still would not come close to meeting the needs we have for our current fleet vehicles.

Complexity

Although electric motors are relatively simple, gasoline engines are quite complex. Add the fact that the two engines must be linked, and you end up with a very complex system. Einstein (1999) warns that "with two powertrains, hybrids are far more complex than conventional cars and that could mean more down time and higher maintenance costs."

TAX AND LEGISLATIVE INFORMATION

Some information of possible use in your decision concerns tax implication for HEVs and related federal legislation.

MODEL 13–1
continued

Tax Credits

Some state and federal government incentives exist to encourage companies or consumers to buy HEVs.

State

Arizona has eliminated the state sales tax on new hybrid vehicles, which would cut at least $1,000 from the purchase price of a $20,000 Prius. Although no states in which McDuff has offices currently offer tax incentives, Massachusetts is considering legislation similar to Arizona's (Lewis, 2001), which could benefit our Boston office down the road.

Federal

Under an existing U.S. Department of Energy program, HEV purchasers can get a tax credit of 7.5 percent of the vehicle's cost, which would result in not having to pay tax of about $1,500 on a $20,000 Insight or Prius. The credits lessen each year, however, and are scheduled to disappear in 2004 (U.S. Department of Energy).

Currently, the Bush energy plan passed recently by the House and not yet approved by the Senate offers tax incentives of between $1,250 and $5,000 for the purchase of an HEV car or light truck, depending on the battery/gas ratio, the increase in fuel economy, and the projected lifetime fuel savings. If the plan passes the Senate intact, McDuff could gain a $200,000 tax credit if the 80 vehicles in our Atlanta office were switched to HEVs getting medium fuel savings (U.S. House).

Proposed HOV Use Legislation

Though it is not a tax incentive, the current form of the Bush energy plan does propose that states allow solo drivers of HEVs in High Occupancy Vehicle (HOV) lanes (U.S. House).

COST CONSIDERATIONS

There remain important issues regarding HEV purchase and maintenance costs that also may influence your decision.

Purchase

HEVs are more expensive than conventional gasoline vehicles. The Toyota Prius costs $20,000, which is about $4,000 more than the cost of the comparable Toyota Corolla. The even smaller Honda Insight sells for as much as the Prius. Though neither of the current HEVs on the market are cheap, both Honda and Toyota are selling their hybrids at a significant loss, estimated to be between a few thousand dollars and $30,000 per vehicle by industry analysts (Swoboda and Brown, 2001). Neither company is commenting on the cost of production.

Daimler-Chrysler has announced plans to market an HEV Dodge Durango SUV in 2003, and when it does, the company says that it will cost about $3,000 more than the current gasoline model (Glynn, 2001).

Maintenance

Both the Insight and the Prius have standard 3-year/36,000-mile warranties, with 8-year/100,000-mile warranties on the $2,000 batteries. Many mechanics are wary, however, including Tom and Ray Magliozzi of *Cartalk:* "[D]on't delude yourself," they warn. "There's no doubt that buying a vehicle such as this [the Prius] places you squarely among the ranks of the hybrid pioneers" (Toyota Prius, 2001).

MODEL 13–1
continued

continues

Because the technology is so new, it's difficult to find maintenance data. I phoned two fleet mechanics for the city of Denver, which recently used a Federal grant to acquire forty Priuses. Harry Stone, of the Denver Fire Department, says that he's "afraid to touch" the ten HEVs in his fleet until he gets more training. Dan Ramsey of the Denver Public Works Department, which has had thirty of the Priuses for three months now, said that they've had a few problems with the cars. Five had dead batteries, though Ramsey admits that they were sitting on the lot for two months without being started. One Prius had a check-engine light come on, which was fixed by the dealer, who is contracted to do all maintenance. "They're pretty hush-hush about it," Ramsey says of the Toyota dealer's maintenance.

THE HEV MARKET

Manufacturers are not yet sure about the target market for HEVs, which has been characterized as "greenies, techno-geeks, tightwads, urban commuters, hip kids?" (Morgan 2000). What is certain is that production is low and that the size of cars produced runs counter to the current popularity of SUVs.

Low Production

Partly because they're losing money with every Insight and Prius sold, Honda and Toyota have kept the number of these first HEV models small. Only 9,360 were sold in the U.S. last year (Miller and Plungis 2001), and only about 16,000 to 20,000 are expected to sell this year (Trapps 2001; Stutzman 2001). HEVs are responsible for only 0.01 percent of new cars sold in the U.S. While both companies report waiting lists for the cars (Henry and McGrath 2000; Swoboda and Brown 2001), the case seems to be one of small demand and even smaller supply.

Small Cars

Another reason why HEVs have not sold well yet in the U.S. is that both models on the market are small cars, and data indicate that Americans like big vehicles. A survey by the Alliance for Automobile Manufacturers shows that the 50 vehicle models that get 30 mph or better make up only two percent of U.S. auto sales (Bodipo-Memba 2001). "It's not the case of 'If you build it, they will come,' " says Steve Douglas, director of environmental affairs for the Alliance's Sacramento office about smaller, more energy-efficient cars. "We build it, but they haven't come." Meanwhile, SUV sales were up nine percent last year, despite fluctuating gas prices and the fact that the total new car and truck market declined by seven percent. The fuel economy figures for new American cars were the worst since 1980 (Stutzman 2001).

OUTLOOK FOR HEVS

This section describes the near-term outlook for the manufacture of American and Japanese HEVs, as well as long-term prospects in the industry.

Short-Term: American Automakers

Americans may be willing to buy HEVs if they're similar to the big cars, trucks, and SUVs that are currently in demand. The Big Three automakers have been reluctant to rush a small hybrid to a market that isn't demanding it; instead, they've been developing HEV trucks and SUVs similar to the models that U.S. consumers are buying.

MODEL 13–1
continued

Both Ford and Dodge plan to market a hybrid SUV in 2003. The Ford Escape should get about 40 mph and cost less than $25,000 (McDiarmid 2000). Estimates are not yet publicly available for the HEV Dodge Durango. In 2004, General Motors should have out a hybrid SUV that can go from 0-60 in just over seven seconds (GM 2001), the first of its many vehicles planned for their ParidiGM gas-electric system. That year or next, Dodge plans to introduce its Ram Contractor Special, allowing its owner to "take it to a job site and plug in the power tools. And if the power goes out at home, he can just plug his house into the truck's 220-volt system to watch television" (Swoboda and Brown 2001).

Short-Term: Japanese Automakers

Toyota and Honda have been making further HEV plans too. Toyota plans to market a hybrid with a modified HEV power system this fall. The new system is less expensive than that used in the Prius and is more adaptable, so Toyota can use it in pre-existing models. However, it is not as stingy on gas, and it reduces emissions by only fifty percent—forty percent less than the Prius (Tanikawa 2001). Just this year Toyota has been selling an HEV minivan in Japan—the ones that our Tokyo office is using—but it is unclear when the automaker might bring it to the U.S. Honda should have an HEV Civic on the Market by March 2002, hoping to triple its HEV sales in the U.S. (Honda plans 2001).

Long-Term Prospects

Four years ago, engineer Victor Wouk called HEVs "the car of the near future" (1997). It looks as if he was right, though HEV sales are not expected to be overwhelming. The Office for the Study of Automotive Technology at the University of Michigan predicts that only three to four percent of the market will consist of hybrids by 2010 (Morgan 2000). Ford Motor Company Chairman William Clay Ford, Jr., however, believes that hybrids could make up twenty percent of new auto sales by then (McCracken 2001).

In the same speech, Ford also said that fuel cells could "be the predominant automotive power source in 25 years" (McCracken 2001). Nearly all other industry analysts agree, and many believe that fuel cell vehicles, which use hydrogen as fuel and emit only water as a waste product, will be in production by the end of the decade. GM President G. Richard Wagoner, Jr., referring to fuel cells as "the Holy Grail for the pollution-free car," has set a goal for his company to have ten percent of its vehicles using fuel cells by 2010 (Jones 1999). Not to be outdone, Honda and Toyota have both developed fuel cell prototypes. So the HEV may be only a brief chapter in automotive history.

CONCLUSION

In my view, McDuff should not add HEVs to its fleet at present—mainly because the available HEVs are smaller than what the company needs. When larger HEVs come to the market in a few years, however, McDuff will probably want to buy several. Though the vehicles should be more expensive, we probably can recoup the initial expense in saved fuel costs and tax savings. Furthermore, we can continue buying vehicles using our discount arrangement with Ford, as Ford will probably be the first to offer a hybrid SUV. Furthermore, because McDuff is well known for its environmental services, having part of the fleet go "green" in the long term would reinforce our positive image in the minds of customers.

MODEL 13–1
continued

continues

References   6.

Anderton, F. 2001. Green dream comes true. *Los Angeles Times,* 21 June, E1. Retrieved
   August 3, 2001, from GALILEO, All databases, ProQuest: http://www.galileo.peach-
   net.edu

Bodipo-Memba, A. 2001. Incentives appeal to automakers. *Detroit Free Press,* 17 May.
   Retrieved August 6, 2001, from GALILEO, All databases, Newspaper Source:
   http://www.galileo.peachnet.edu

California Air Resources Board 2001. *2001 model year guide to passenger cars*, 20 April.
   Retrieved August 10, 2001, from the World Wide Web: http://www.arb.ca.gov/
   msprog/ccbg/2001pc.htm

Einstein, P. 1999. The benefits of Insight. *Professional Engineering* (29 Oct.): *12.19,* 23.
   Retrieved August 3, 2001, from GALILEO, All databases, ProQuest: http://www.
   galileo.peachnet.edu

Garretson, C. 2001. In an era of gas-gulpers, hybrid cars get by on sip. *The Cincinnati Post,*
   10 May. Retrieved August 10, 2001, from the World Wide Web: http://www.
   cincypost.com/2001/may/10/hybrid051001.html

Glynn, M. 2001. Automakers will offer "2 for the price of 1": Many major manufacturers are
   hoping gas-electric hybrids can power future sales. *Seattle Times,* 19 Jan., F1. Retrieved
   August 4, 2001 from GALILEO, All databases, ProQuest: http://www.galileo.
   peachnet.edu

GM announces hybrid production plans. 2001. *PR Newsline,* 9 Jan. Retrieved August 3,
   2001, from GALILEO, All databases, ProQuest: http://www.galileo.peachnet.edu

Henry, E., and C. McGrath 2000. Coast to coast on four fill-ups. *Kiplinger's Personal Finance Magazine 54.8,* (Aug.): 160. Retrieved August 3, 2001, from GALILEO, All
   databases, Academic Search Premier: http://www.galileo.peachnet.edu

Honda plans to unveil gasoline-electric Civic. 2000. *Toronto Star,* 26 June, BU.12. Retrieved
   August 4, 2001, from GALILEO, All databases, Newspaper Source: http://www.
   galileo.peachnet.edu

Jones, T. 1999. Whose car is greenest? *Forbes* 164.18 (18 Oct.): 60+.

Lewis, R. 2001. Lawmakers eye incentives for cars that save on fuel. *Boston Globe,* 6 June.
   Retrieved August 3, 2001, from GALILEO, All databases, ProQuest: http://www.
   galileo.peachnet.edu

McCracken, J. 2001. GM Promises gas-electric hybrid for 2004 will get 35 m.p.g. *Detroit
   Free Press,* 9 Jan. Retrieved August 3, 2001, from GALILEO, All databases, ProQuest:
   http://www.galileo.peachnet.edu

McDiarmid, H. 2000. Auto representatives in Michigan say demand will drive need for hybrid cars. *Detroit Free Press*, 21 Oct. Retrieved August 3, 2001 from GALILEO, All
   databases, Newspaper Source: http://www.galileo.peachnet.edu

Miller, J. and J. Plungis 2001. GM cuts output 21%: Automaker plans to sell gas-electric
   SUV in 2004. *Detroit News,* 9 Jan., A.1. Retrieved August 3, 2001 from GALILEO, All
   databases, ProQuest: http://www.galileo.peachnet.edu

MODEL 13–1
continued

Morgan, C. 2000. "Green mobiles" are slowly catching on as Honda, Toyota models hit showrooms. *The Miami Herald*, 20 Oct. Retrieved August 6, 2001, from GALILEO, All databases, Newspaper Source: http://www.galileo.peachnet.edu

Road test digest. 2001. *Car and Driver.com*. Retrieved August 10, 2001, from the World Wide Web: http://www.caranddriver.com/xp/Caranddriver/road_test_digest_current.xml

Stutzman, E. 2001. Hybrid owners testify to vehicles' benefits, drawbacks. [Boulder] *Daily Camera*, 4 June. Retrieved August 3, 2001, from GALILEO, All databases, Newspaper Source: http://www.galileo.peachnet.edu

Swoboda, F., and W. Brown 2001. Easing off the gas: Unsure of demand, big 3 move slowly on hybrid cars. *The Washington Post*, 16 May, E1. Retrieved August 3, 2001, from GALILEO, All databases, ProQuest: http://www.galileo.peachnet.edu

Tanikawa, M. 2001. Toyota to introduce another hybrid engine. *New York Times*, 13 June, W.1. Retrieved August 8, 2001, from GALILEO, All databases, ProQuest: http://www.galileo.peachnet.edu

Toyota Prius. 2001. *Cartalk.com*. Retrieved August 19, 2001, from the World Wide Web: http://cartalk.cars.com/Info/Testdrive/Reviews/toyota-prius-2001.html

Traeger, C. (n.d.). The will of the Prius. *Carlist.com*. Retrieved August 10, 2001, from the World Wide Web: http://www.carlist.com/newcars/2001/toyotaprius.html

Trapps, T. 2001. Scientists urge accelerated pace to greater auto fuel efficiency. *Los Angeles Times*, 21 June. Retrieved August 3, 2001, from GALILEO, All databases, ProQuest: http://www.galileo.peachnet.edu

U.S. Department of Energy. (n.d.). *Alternative fuel vehicle fleet buyer's guide*. Retrieved August 9, 2001, from the World Wide Web: http://www.fleets.doe.gov/fleet_tool.cgi?$$,benefits,2

U.S. Environmental Protection Agency 2000. *Model year 2001 fuel economy guide*, 22 Nov. Retrieved August 10, 2001, from the World Wide Web: http://fueleconomy.gov/feg/FEG2001.pdf

U.S. House. 107th Congress, 1st Session. H.R. 4, Securing America's future energy act of 2001. Retrieved August 9, 2001, from the World Wide Web: http://thomas.loc.gov/home/c107bills.html

Wouk, V. 1997. Hybrid electric vehicles. *Scientific American* (Oct.): *277.4*, 70-74.

7.

MODEL 13–1
continued

Planning for Success

General Instructions

Each chapter begins with a Collaboration at Work exercise that applies strategies for working in groups to chapter topics. The exercise assumes you (1) have been divided into groups of about three to six students, (2) will use group time inside or outside of class to complete the case, and (3) will produce an oral or written response. For guidelines about writing in groups, refer to pages 28–34.

Background for Assignment

Some students attend college for its own sake because they love learning. Others may like to learn but mainly view college as a stepping-stone to a career. Because most students are in the second group, they give a good deal of thought to what they will do with their working lives. If they don't, they should. Achieving success in a career starts with establishing a careful plan for getting from here to there. It is a process that involves considerably more than gaining a degree.

Group Assignment

For this exercise, you and your group members will brainstorm about a strategy for getting a particular job or entering a particular career desired by one of your group's members. (You can do the exercise for several careers, if you have time.) Here are the three steps:

1. Share information about the "ideal" jobs of all group members.
2. Choose one position or career that best fits the assignment (that is, one that lends itself to being achieved in an incremental process that is doable).
3. Generate a list of specific steps for getting the position or entering the career (include deadlines and criteria for success at each stage).

If you have time to conduct some research and have access to the Internet, go to *www.monster.com* for assistance in your research.

*I*n applying for a job, you are selling yourself. It is no time for either self-delusion or false modesty. You must first assess your abilities, then find an appropriate job match, and finally persuade a potential employer that you are the right one for the job.

This chapter offers suggestions for landing a job in your profession. You'll find information on these main activities:

- Researching occupations and companies
- Writing job letters and resumes
- Succeeding in job interviews
- Negotiating with current and potential employers

RESEARCHING OCCUPATIONS AND COMPANIES

Before writing a job letter and resume, you may need information about (1) career fields that interest you (if you have not already chosen one), (2) specific companies that hire graduates in your field, and, obviously, (3) specific jobs that are available. Here are some pointers for finding such information—both from traditional sources and from your computer.

● *Do Basic Research in Your College Library or Placement Office*

Libraries and placement centers offer one starting point for getting information about professions. Following are a few well-known handbooks and bibliographies found in reference collections. They either give information about occupations or provide names of other books that supply such information:

Career Choices Encyclopedia: Guide to Entry-Level Jobs

Dictionary of Occupational Titles

Directory of Career Training and Development Programs

Encyclopedia of Business Information Sources

Encyclopedia of Careers and Vocational Guidance

High-Technology Careers

Occupational Outlook Handbook

Professional Careers Sourcebook: An Information Guide for Career Planning

As noted later, these traditional sources are rapidly being replaced by online sources of career information.

● *Interview Someone in Your Field of Interest*

To get the most current information, arrange an interview with someone working in an occupation that interests you. This abundant source of information often goes untapped by college students, who mistakenly think such interviews are difficult to arrange. In fact, usually you can locate people to interview through (1) your college placement office, (2) your college alumni association, or (3) your own network of family and friends. Another possibility is to call a reputable firm in the field and explain that you wish to interview someone in a certain occupation. Make it clear, however, that you are not looking for a job—only information about a profession.

Once you set up the interview, prepare well by listing your questions in a notebook or on a clipboard that you take with you to the interview. This preparation will keep you on track and show persons being interviewed that you value their time and information. Here are some questions to ask:

- How did you prepare for the career or position you now have?
- What college course work or other training was most useful?
- What types of activities fill your typical working day?
- What features of your career do you like the most? The least?
- What personality characteristics are most useful to someone in your career?
- How would you describe the long-term outlook of your field?
- How do you expect your career to develop in the next 5 years, 10 years, or 15 years?
- Do you know any books, periodicals, or online sources that might help me find out more about your field?
- Do you know any individuals who, like you, might permit themselves to be interviewed about their choice of a profession?

Although this interview *may* lead to a discussion about job openings in the interviewer's organization, the main purpose of the conversation is to retrieve information about an occupation.

● *Find Information on Companies in Your Field*

With your focus on a profession, you can begin screening companies that employ people in the field chosen. First, determine the types of information you want to find. Examples might include location, net worth, number of employees, number of workers in your specific field, number of divisions, types of products or services, financial rating, and names and titles of company officers. The following are some sources that might include such information. They can be found in the reference sections of libraries and in college placement centers:

Business Bankings Annual contains ranked lists related to business and industry, such as "Largest Data Communication Companies."

The Career Guide gives overviews of many American companies and includes information such as types of employees hired, training opportunities, and fringe benefits.

Corporate Technology Directory profiles high-tech firms and covers topics such as sales figures, number of employees, locations, and names of executives.

Dun's Million Dollar Directory: America's Leading Public and Private Companies lists information about 160,000 U.S. businesses with net worth over $500,000.

Facts on File Directory of Major Public Corporations gives essential information on 5,700 of the largest U.S. companies listed on major stock exchanges.

Job Choices in Science and Engineering, an annual magazine published by the College Placement Council, includes helpful articles and information about hundreds of companies that hire technical graduates.

Peterson's Business and Management Jobs provides background on employers of business, management, and liberal-arts graduates.

Peterson's Engineering, Science, and Computer Jobs provides background information on employers of technical graduates.

Standard & Poor's Register of Corporations, Directors, and Executives lists names and titles of officials at 55,000 public and private U.S. corporations.

● *Do Intensive Research on a Selected List of Potential Employers*

The previous steps help get you started finding information on occupations and firms. Ultimately, you will develop a selected list of firms that interest you. Your research may have led you to these companies, or your college placement office may have told you that openings exist there. Now you need to conduct an intensive search to learn as much as you can about the firms. Here are a few sources of information, along with the kinds of questions each source will help to answer.

- **Annual reports** (often available at company Web sites and in your library or placement office): How does the firm describe its year's activities to stockholders? What are its products or services?
- **Web sites or media kits** (available from public relations offices): How does the firm portray itself to the public? Is this firm "Internet-savvy"? What can you infer about the firm's corporate culture?
- **Personnel manuals and other policy guidelines:** What are features of the firm's "corporate culture"? How committed is the firm to training? What are the benefits and retirement programs? Where are its branches? What are its customary career paths?
- **Graduates of your college or university now working for the firm:** What sort of reputation does your school have among decision-makers at the firm?
- **Company newsletters and in-house magazines:** How open and informative is the firm's internal communication?
- **Business sections of newspapers and magazines:** What kind of news gets generated about the firm?
- **Professional organizations or associations:** Is the firm active within its profession?
- **Stock reports:** Is the firm making money? How has it done in the past five years?
- **Accrediting agencies or organizations:** How has the firm fared during peer evaluations?
- **Former employees of the company:** Why have people left the firm?
- **Current employees of the company:** What do employees like, or dislike, about the company? Why do they stay?

Other good sources include the Better Business Bureau, Chamber of Commerce, and local newspapers. In other words, you should thoroughly examine an organization from the outside. The information you gather will help you decide where to apply and, if you later receive a job offer, where to begin or continue your career.

● *Use Your Computer to Gather Data*

Today, many applicants go directly to the computer to find information about professions, organizations, graduate schools, and job openings. No doubt between the time this book is written and then published, the names and number of online resources will change dramatically. Generally, here are some of the kinds of information available:

- College and university catalogs
- Web sites for companies, organizations, and schools
- Employment listings from local and national sources
- Online discussion forums involving recent graduates of colleges and universities

One excellent example of a comprehensive job-search Web site, as of this writing, is *www.monster.com*. At this site you can

1. Get tips on writing your resume and searching for jobs
2. Post your resume at the site, in multiple forms for varied employers
3. Peruse lists of job openings (421,471 at one site visit)
4. Receive advice from professionals who are in careers you wish to pursue
5. Read in-depth reports on particular companies
6. Share comments with other job-seekers

In other words, the Internet will help you locate a variety of information during your job search. Moreover, you can use your computer to search for openings and respond to job ads, as mentioned in the next section.

JOB CORRESPONDENCE

Job letters and resumes must grab the attention of busy readers, who may spend only 60 seconds deciding whether to consider you further. This section gives you the tools to write a successful letter and resume. "Successful," of course, means a letter and resume that will get you an interview. After that, your interpersonal skills will help you land the job. The letter and resume only aim to get you to the next step—the personal interview.

Most job letters and resumes still get sent through the mail. However, a growing number of applicants use the Internet to apply for jobs.

For example, online services can place resumes into a bank used by hundreds, perhaps thousands, of companies. The resumes will be scanned with the use of software, which searches for key words that reflect abilities needed for specific jobs and then sends selected resumes to companies. If you use this kind of service, remember one point: When you send credentials into cyberspace, you cannot be sure where they will land. Do not expect the level of confidentiality and security that you have with personal mail.

Whether you use online techniques like email and resume services or stick with the traditional approach, the same basic writing guidelines apply. Your letter, no longer than one page, should be specific about the job you seek and your main selling points. Then the resume—one page or two at most—should simply, specifically, and neatly highlight your background.

Job Letters

A job letter is just another type of sales letter—except that you are "selling" yourself, not a product or service. In preparing to write one, take the point of view of the persons to whom you are writing. What criteria will they use to evaluate your credentials? How much or how little do they want in the letter? What main points will they be hunting for as they scan your resume? Accordingly, this section first examines the needs of these readers and then gives guidelines for you, the writer. Models 14–1 and 14–2 on pages 581–584 include sample job letters and resumes.

The Readers' Needs.   You probably will not know personally the readers of your job letter, so you must think hard about what they may want. Your task is complicated by the fact that often there are several readers of your letter and resume, who may have quite different backgrounds.

Here is one possible scenario:

Step 1: The letter may go first to the personnel office, where a staff member specializing in employment selects letters and resumes that meet the criteria stated in the position announcement. (In some large employers, letters and resumes may even be stored in a computer where they are scanned for key words that relate to specific jobs.)

Step 2: Applications that pass this screening are sent to the department manager who will supervise the employee that is hired. This manager will interview applicants and ultimately hire the employee. The manager may then select a group to be interviewed.

One variation of this process has the personnel department doing an interview as well as screening letters and resumes—before the department manager even hears about any applications. Another variation, as noted earlier, has the employer relying on an online resume service for the initial screening.

Yet sooner or later, a real human being will read your letter and resume. And most readers, whatever their professional background, have these five characteristics in common:

● *Feature 1: They Read Job Letters in Stacks*

Most search-and-screen processes are such that letters get filed until there are many to evaluate. Your reader faces this intimidating pile of paper, from which you want your letter to emerge as victor.

● *Feature 2: They Are Tired*

Some employment specialists may save job letters for their fresher moments, but many people who do the hiring get to job letters at the end of a busy day or at home in the evening. So they have even less patience than usual for flowery wording or hard-to-read print.

● *Feature 3: They Are Impatient*

Your readers expect major points to jump right out at them. In most cases, they will not dig for information that cannot be found quickly.

● *Feature 4: They Become Picky Grammarians*

Readers of all professional and academic backgrounds expect good writing when they read job letters. There is an unspoken assumption that a letter asking for a chance at a career should reflect solid use of the language. Furthermore, it should have no typographical errors. If the letter does contain a typo or grammar error, the reader may wonder about the quality of writing you will produce on the job.

● *Feature 5: They Want Attention-Grabbers but Not Slickness*

You want your letter and resume to stand out without the use of gimmicks. Most readers prefer a tasteful, reserved format that does not draw too much attention to itself. For example, white or off-white stationary is still the standard, along with traditional fonts with lots of white space for easy reading. If you want to attract attention in a professional manner, consider attaching a business card to your letter so that the reader has your name and number handy. Students can design and print business cards with software available in many college computer labs.

Of course, likes and dislikes vary. An advertising director, who works all day with graphics, would probably want a bolder format design than an engineering manager, who works with documents that are less flashy. If you cannot decide, it is best to use a conservative format and style.

The Letter's Organization.   The job-letter guidelines that follow relate to the features mentioned about readers. Your one and only goal is to tantalize the reader enough to want to interview you. That is all. With that goal and the reader's needs in mind, your job letter should follow this ABC format:

ABC FORMAT: *Job Letters*

Abstract

- Apply for a specific job
- Refer to ad, mutual friend, or other source of information about the job
- (Optional) Briefly state how you can meet the main need of your potential employer

Body

- Specify your understanding of the reader's main needs
- Provide main qualifications that satisfy these needs (but only *highlight* points from resume—do NOT simply repeat all resume information)
- Avoid mentioning weak points or deficiencies
- Keep body paragraphs to six or fewer lines
- Use a bulleted or numbered list if it helps draw attention to three or four main points
- Maintain the "you" attitude throughout (see chapter 7)

Conclusion

- Tie the letter together with one main theme or selling point, as you would a sales letter
- Refer to your resume
- Explain how and when the reader can contact you for an interview

This pattern gives you a starting point, but it is not the whole story. There is one feature of application letters that cannot be placed easily in a formula. That feature is style. Work hard with your draft to develop a unity and flow that, by itself, will set you apart from the crowd. Your attention-grabber will engage interest. But the clarity of your prose will keep readers attentive and persuade them that you are an applicant who should be interviewed.

Resumes

Resumes usually accompany application letters. Three points make writing resumes a challenge:

1. **Emphasis:** You should select just a *few major points of emphasis* from your personal and professional life. Avoid the tendency to include college and employment details best left for the interview.
2. **Length:** You usually should use only *one page.* For individuals with extensive experience, a two-page resume is acceptable—if it is arranged evenly over both pages.

3. **Arrangement:** You should arrange information so that it is *pleasing to the eye and easy to scan.* Prospective employers spend less than a minute assessing your application. They may even use computers to scan resumes, taking even less time.

Computers pose a special challenge to you as resume writer, for they fail to "appreciate" some of the elegant variations and innovations sometimes used to get the human reader's attention. If you are writing a resume that may be read by a computer, you may want to (1) use white or very light-colored paper, (2) focus on key words—especially job skills—that might be picked up by the computer scan, and (3) avoid design features that might present obstacles to the scan such as italics, fancy typefaces, and graphics. Actually, you may find that your placement office uses a computer resume program that requires a particular format, effectively removing decisions about style from your consideration.

This section distills the best qualities of many formats into three basic patterns:

1. The chronological resume, which emphasizes employment history,
2. The functional resume, which emphasizes the skills you have developed, and
3. The combined resume, which merges features of both the chronological and functional formats. See the "Experience" section that follows to learn when to use each format. Choose the pattern that best demonstrates your strengths.

The following paragraphs describe the main parts of the resume. The "Experience" section explains the differences between chronological, functional, and combined resumes. Refer to the models on pages 581–590 for resume examples.

Objective.   Personnel directors, other people in the employment cycle, and even computers may sort resumes by the "Objective" statement. Writing a good one is hard work, especially for new graduates, who often just want a chance to start working at a firm at any level. Despite this eagerness to please, do not make the mistake of writing an all-encompassing statement such as "Seeking challenging position in innovative firm in civil-engineering field." Your reader will find such a general statement of little use in sorting your application. It gives the impression that you have not set clear professional goals.

Most objectives should be short, preferably one sentence. Also, they should be detailed enough to show that you have prepared for, and are interested in, a specific career, yet open-ended enough to reflect a degree of flexibility. If you have several quite different career options, you might want to design a different resume for each job description, rather than trying to write a job objective that takes in too much territory.

Note: Some employers prefer that you *not* include an objective. For example, you may be applying for an entry-level job for which an objective would be inappropriate. As always, consider your reader's needs as you make decisions about objectives.

Education.   Whether you follow the objective with the "Education" or "Experience" section depends on the answer to one question: Which topic is most important to the reader? Most recent college graduates lead off with "Education," particularly if the completion of the degree prompted the job search.

This section seems simple at the outset. Obligatory information includes your school, school location, degree, and date of graduation. It is what you include beyond the bare details, however, that most interests employers. Here are some possibilities:

- **Grade point average:** Include it if you are proud of it; do not if it fails to help your case.
- **Honors:** List anything that sets you apart from the crowd—such as dean's list or individual awards in your major department. If you have many, include a separate "Recognitions" heading toward the end of the resume.
- **Minors:** Highlight any minors or degree options, whether they are inside or outside your major field. Employers place value on this specialized training, even if (and sometimes *especially* if) it is outside your major field.
- **Key courses:** When there is room, provide a short list of courses you consider most appropriate for the kind of position you are seeking. Because the employer probably will not look at your transcripts until a later stage of the hiring process, use this brief listing as an attention-grabber.

Experience. This section poses a problem for many applicants just graduating from college. Students often comment that experience is what they are *looking* for, *not* what they have yet. Depending on the amount of work experience you have gained, consider three options for completing this section of the resume: (1) emphasize specific positions you have held (chronological resume), (2) emphasize specific skills you have developed in your experience (functional resume), or (3) emphasize both experience and skills (combined resume).

Option 1: Chronological Format. This option works best if your job experience has led logically toward the job you now seek. Follow these guidelines:

- List relevant full-time or part-time experience, including co-op work, in *reverse* chronological order.
- Be specific about your job responsibilities, while still being brief.
- Be selective if you have had more jobs than can fit on a one-page resume.
- Include nonprofessional tasks (such as working on the campus custodial staff) *if* it will help your case (for example, the employer might want to know that you worked your way through college).
- Remember that if you leave out some jobs, the interview will give you the chance to elaborate upon your work experience.
- Select a readable format, with appropriate white space.
- Use action verbs and lists to emphasize what you did or what you learned at jobs—for example, "Provided telephone support to users of System/23." Use parallel form in each list.

Option 2: Functional Format. This approach works best if (1) you wish to emphasize the skills and strengths you have developed in your career, rather than specific jobs you have had, or (2) you have had "gaps" in your work history, which would be obvious if you used the chronological format. Although it is sometimes used by those whose job experience is not a selling point, this is not always the case.

Sometimes your skills built up over time may be the best argument for your being considered for a position, even if your job experience also is strong. For example, you may have five years' experience in responsible positions at four different retailers. You then decide to write a functional resume focusing on the three skill areas you developed: sales, inventory control, and management.

If you write a functional resume that stresses skills, you may still want to follow this section with a brief employment history (see Option 3). Most potential employers want to know where and when you worked, even though this issue is not a high priority. *Note:* If you decide to leave out the history, bring it with you to the interview on a separate sheet.

Option 3: Combined Format.   The combined format uses features of both chronological and functional formats. This format works best when you want to emphasize the skills you have developed, while still giving limited information on the chronology of your employment.

Models 14–5 and 14–6 on pages 587–590 show two variations of the combined format. In Model 14–5, the experience section looks exactly as it would in a functional resume, with subheadings giving the names of skills. However, the writer adds a brief skeleton work history near the end of the page; he believes the reader will want some sort of chronological work history, even if it is not the writer's strength. Model 14–6 integrates chronological information into the skills section. The positions held may not be prestigious, but together they show that the applicant has considerable experience developing the two sets of skills listed: Editing/Writing and Teaching/Research.

Activities, Recognitions, Interests.   Most resumes use one or two of these headings to provide the reader with additional background information. The choice of which ones, if any, to use depends on what you think will best support your job objective. Here are some possibilities:

- **Activities:** selected items that show your involvement in your college or your community or both.
- **Recognitions:** awards and other specific honors that set you apart from other applicants. (Do not include awards that might appear obscure, meaningless, or dated to the reader, such as most high-school honors.)
- **Interests:** hobbies or other interests that give the reader a brief look at the "other" you.

However you handle these sections, they should be fairly brief and should not detract from the longer, more significant sections described previously.

References.   Your resume opens the door to the job interview and later stages of the job process, when references will be called. There are two main approaches to the reference section of the resume:

1. Writing "Available upon request" at the end of the page
2. Listing names, addresses, and phone numbers at the end of the resume

The first approach assumes that the reader prefers the intermediate step of contacting you before references are sent or solicited. The second approach assumes that the reader prefers to call or write references directly, without having to contact you first. Use the format most commonly used in your field or, most importantly, the one most likely to meet the needs of a particular employer. As always, be ready to tailor your letter and resume each time you put it in the mail.

Your goal is to write an honest resume that emphasizes your good points and minimizes your deficiencies. To repeat a point made at the outset, you want your resume and job letter to open the door for later stages of the application process. Look upon this writing task as your greatest persuasive challenge. Indeed, it is the ultimate sales letter, for what you are selling is the potential you offer to change an organization and, perhaps, the world as well. Considering such heady possibilities, make sure to spend the time necessary to produce first-rate results.

JOB INTERVIEWS

Your job letter and resume have only one purpose: to secure a personal interview by the personnel director or other official who screens applicants for a position.

Much has been written about job interviews. Fortunately, most of the good advice about interviewing goes back to just plain common sense about dealing with people. Following are some suggestions to show you how to prepare for a job interview, perform at your best, and send a follow-up letter.

Preparation

● *Do Your Homework on the Organization*

You have learned how to locate data about specific companies. Once you have been selected for an interview, review whatever information you have already gathered about the employer. Then go one step further by searching for the *most current* information you can find. Your last source may be someone you know at the organization, or a "friend of a friend."

When you don't have personal contacts, use your research skills again. For large firms, locate recent periodical or newspaper articles by consulting general indexes—such as the *Business Periodicals Index, Wall Street Journal Index, Readers' Guide to Periodicals, New York Times Index,* or the index for any newspaper in a large metropolitan area. For smaller firms, consult recent issues of local newspapers for announcements about the company. Being aware of current company issues will demonstrate your initiative and show your interest in the firm.

As noted earlier in the chapter, the company's Web site can also be a good source of current information about an organization.

● *Write Out Answers to the Questions You Consider Likely*

You probably would not take written answers with you to the interview. But writing them out will give you a level of confidence unmatched by candidates who only ponder possible questions that might come their way. This technique resembles the manner in which some people prepare for oral presentations: First they write out a speech, then they commit it to notes, and finally they give an extemporaneous presentation that reflects confidence in themselves and knowledge of the material. This degree of preparation will place you ahead of the competition.

There are few, if any, original questions asked in job interviews. Most interviewers simply select from some standard questions to help them find out more about you and your background. Here are some typical questions, along with tips for responses in parentheses:

1. **Tell me a little about yourself.** (Keep your answer brief and relate it to the position and company—do *not* wander off into unrelated issues, like hobbies, unless asked to do so.)

2. **Why did you choose your college or university?** (Be sure your main reasons relate to academics—for example, the academic standing of the department, the reputation of the faculty, or the job placement statistics in your field.)

3. **What are your strengths?** (Focus on two or three qualities that would directly or indirectly lead to success in the position for which you are applying.)

4. **What are your weaknesses?** (Choose weaknesses that, if viewed from another perspective, could be considered strengths—for example, your perfectionism or overattention to detail.)

5. **Why do you think you would fit into this company?** (Using your research on the firm, cite several points about the company that correspond to your own professional interests—for example, the firm may offer services in three fields that relate to your academic or work experience.)

6. **What jobs have you held?** (Use this question as a way to show that each previous position, no matter how modest, has helped prepare you for this position—for example, part-time employment in a fast-food restaurant developed teamwork and interpersonal skills.)

7. **What are your long-term goals?** (Be ready to give a 5- or 10-year plan that, preferably, fits within the corporate goals and structure of the firm to which you are applying—for example, you may want to move from the position of technical field engineer into the role of a project manager, to develop your management skills.)

8. **What salary range are you considering?** (Avoid discussing salary if you can. Instead, note that you are most interested in criteria such as job satisfaction and professional growth. If pushed, give a salary range that is in line with the research you did on the career field in general and this company in particular; see the last section of this chapter on negotiating.)

9. **Do you like working in groups or prefer working alone?** (Most employers want to know that you have interest and experience in group work—whether in college courses or previous jobs. But they also admire and reward individual accomplishment. In deciding what part of your background to emphasize, consider the "corporate culture" of the organization interviewing you.)

10. **Do you have any questions of me?** (*Always* be ready with questions that reinforce your interest in the organization and your knowledge of the position—for example, "Given the recent opening of your Tucson warehouse, do you plan

other expansions in the Southwest?" or "What types of in-house or off-site training do you offer new engineers who are moving toward project management?" Other questions can concern issues such as (a) benefits, (b) promotions, (c) type of computer network, and (d) travel requirements.)

● *Do Mock Interviews*

You can improve your chances considerably by practicing for job interviews. One of the easiest and best techniques is role-playing. Ask a friend to serve as the interviewer, and give him or her a list of questions from which to choose. Also, inform that person about the company so that he or she can improvise during the session. In this way you will be prepared for the real thing.

You can get additional information about your interviewing abilities by videotaping your role-playing session. Reviewing the videotape will help you highlight (1) questions that pose special problems for you and for which you need further preparation and (2) mannerisms that need correction. This preparation technique is especially useful if you are one of the growing number of applicants who take part in a video interview with a recruiter.

● *Be Physically Prepared for the Interview*

Like oral presentations, job interviews work best when you are physically at your best. Thus all the old standbys apply:

- Get a good night's rest before the interview.
- Avoid caffeine or other stimulants.
- Eat about an hour beforehand so that you are not distracted by hunger pangs during the session.
- Take a brisk walk to dispel nervous energy.

Performance

Good planning is your best assurance of a successful interview. Of course, there are always surprises that may catch you. Remember, however, that most interviewers are seriously interested in your application and want you to succeed. Help them by selling *yourself* and thus giving them a reason to hire you. Here are some guidelines for the interview.

● *Dress Appropriately*

Much has been written on the topic of appropriate attire for interviews. Here are some practical suggestions that are often emphasized:

- Dress conservatively and thus avoid drawing attention to your dress—for example, do not use the interview as an opportunity to break in a garment in the newest style.
- Consider the organization—for example, a brokerage-firm interview may require a dark suit for a man and a tailored suit for a woman, whereas an interview at a construction firm may require less formal attire.

- Avoid excessive jewelry.
- Pay attention to the fine points—for example, wear shined shoes and carry a tasteful briefcase or notebook.

● *Take an Assertive Approach*

Either directly or indirectly, use everything you say to make the case for your hiring. Be positive, direct, and unflappable. Use every question as a springboard to show your capabilities and interest, rather than waiting for point-blank questions about your qualifications. To be sure, the degree to which you assert yourself partly depends on your interpretations of the interviewer's preference and style. Although you do not want to appear "pushy," you should take the right opportunities to sell yourself and your abilities.

● *Use the First Few Minutes to Set the Tone*

What you have heard about first impressions is true: Interviewers draw conclusions quickly. Having given many interviews, they are looking for an applicant who injects vitality into the interview and makes their job easier. Within a minute or two, establish the themes and the tone that will be reinforced throughout the conversation— that is, your relevant background, your promising future, and your eagerness (*not* pushiness). In this sense, the interview subscribes to the Preacher's Maxim mentioned in chapter 12: "First you tell 'em what you're going to tell 'em, then you tell 'em, and then you tell 'em what you told 'em."

● *Maintain Eye Contact While You Speak*

Although you may want to look away occasionally, much of the time your eyes should remain fixed on the person interviewing you. In this way you show interest in what she or he is saying.

If you are being interviewed by several people, make eye contact with *all* of them throughout the interview. No one should feel ignored. You are never quite certain exactly who may be the decision-maker in your case.

● *Be Specific in the Body of the Interview*

In every question you should see the opportunity to say something specific about you and your background. For example, rather than simply stating that your degree program in computer science prepared you for the open position, cite three specific courses and briefly summarize their relevance to the job.

● *Do Not Hesitate*

A job interview is no time to hesitate, unless you are convinced the job is not for you. If the interviewer notes that the position involves 40 percent travel, quickly respond that the prospect of working around the country excites you. The question is this: Do you want the job or not? If you do, then accept the requirements of the position and show excitement about the possibilities. You can always turn down the job if you receive an offer and decide later that some restrictions, like travel, are too demanding.

● *Reinforce Main Points*

The interviewer has no text for the session other than your resume. Therefore, you should drive home main points by injecting short summaries into the conversation. After a five-minute discussion of your recent work experience, take 15 seconds to present a capsule version of relevant employment. Similarly, orchestrate the end of the interview so that you have the chance to summarize your interest in the position and your qualifications. Here is your chance to follow through on the "tell 'em what you told 'em" part of the Preacher's Maxim.

Follow-Up Letters

Follow *every* personal contact with a letter or email—whatever is most appropriate—to the person with whom you spoke. Send it within 24 hours of the interview or meeting so that it immediately reinforces the person's recollection of you. This simple strategy gives you a powerful tool for showing interest in a job.

Follow-up letters abide by the same basic letter pattern discussed in chapter 7. In particular, follow these guidelines:

- Write no more than one page.
- Use a short first paragraph to express appreciation for the interview.
- Use the middle paragraph(s) to (a) reinforce a few reasons why you would be the right choice for the position or (b) express interest in something specific about the organization.
- Use a short last paragraph to restate your interest in the job and to provide a hopeful closing.

See chapter 7 for the various formats appropriate for all types of business letters. Here is sample text of a thank-you letter:

Dear Ms. Ferguson:

I enjoyed meeting with you yesterday about the career possibilities at Klub Kola's district headquarters. The growth that you are experiencing makes Klub an especially exciting company to join.

As I mentioned, my marketing background at Seville College has prepared me for the challenge of working in your new Business Development Department. Several courses last semester focused specifically on sales strategies for consumer goods. In addition, an internship this semester has given me the chance to try out marketing strategies in the context of a local firm.

Again, thank you for the chance to learn about your firm's current success and promising future. I remain very interested in joining the Klub Kola team.

Sincerely,

Marcia B. Mahoney

Marcia B. Mahoney

When your audience might appreciate a less formal response, consider writing your interviewer a personal note instead of a typed letter. This sort of note is most appropriate when you plan a short message.

NEGOTIATING

All of us negotiate every day of our lives. Both on the job and in our personal lives, we constantly find ourselves in give-and-take discussions to negotiate issues as diverse as those that follow:

- Major and minor purchases
- Relationships with spouses and friends
- Performance evaluations—with bosses and with subordinates
- Salaries—with those to whom we report and with those who report to us

Because negotiating will become an important part of your career, it receives attention in this final section. After some brief background information, the chapter focuses on six guidelines that will steer you toward successful negotiations—when you are hired and also at other points in your career. The main example used in this section is a salary negotiation for an entry-level position.

How has the art of negotiating changed recently? In the past, the process was often characterized by words like *trickery, intimidation,* and *manipulation.* In this game's lexicon there were "winners" and "losers" and lots of warlike imagery. Participants, seen as battlefield adversaries, took up extreme positions, defended and attacked each other's flanks, finally agreed reluctantly to some middle ground, and then departed wounded and usually uncertain of who had won the battle.

Today, the trend is away from this war-zone approach with its "I win, you lose" mentality. As a negotiator, you must enter the process searching for common ground for a very practical reason: Long-term relationships are at stake. In later negotiations, you are much more likely to achieve success if the present negotiation helps both parties. This goal—"we both win"—requires a new set of practices at the negotiation table.

Specifically, six guidelines should drive the negotiation process. All of them embody the viewpoint that successful negotiations involve honest communication wherein both parties benefit. Try to weave these six guidelines into the style of negotiating that you develop.

● *Negotiating Guideline 1: Think Long Term*

Enter every negotiation with a long-term strategy for success. You need to establish and nurture a continuing relationship with the person on the other side of the table. Later dealings might depend on mutual understandings and goodwill that result from your first meeting. First impressions *do* count.

How might such long-term thinking apply to actual contract discussions for jobs, especially for your first position after graduating? If you are fortunate enough to be in demand in the job market, you will have the leverage to discuss salary expectations and other benefits during an interview. Such discussions often are characterized by you and the employer sharing details about your expectations and the employer's offer. You should enter such sessions with a realistic idea of what you can command in the marketplace. Neither sell yourself short nor harbor inflated ideas of your worth. Your college or university placement office should be able to provide information about salary ranges and benefit options for graduates in your field and organizations in your region.

Of course, the "real world" of the job hunt is such that the supply of new talent may overshadow the demand. You may be so glad to receive a good offer that you hesitate to jeopardize it by attempting to negotiate. Yet, ironically, you can damage your long-term interests in an organization by being overly timid before accepting an offer. Even if there is little or no room for salary negotiation, you should engage in a wide-ranging discussion that allows you to explore options for your contract and learn about features of the position. This dialogue helps you learn about the organization. It also gives the employer a healthy respect for your ability to ask serious questions about your career.

Whatever your bargaining position, take advantage of the opportunity to discuss features of your job and the organization. Questions like those that follow may yield important information for you *and* show your interest in developing a long-term relationship with the employer:

- What philosophy underlies the firm's approach to management?
- What is the general timetable for career advancement?
- Where will your specific job lead?
- What opportunities exist for company-sponsored training?
- How will you be evaluated and how often?

Employers respect applicants who have done enough homework to ask informed questions about the firm's employment practices. Both parties benefit from a frank, detailed discussion. You get what you need to make an informed decision about the firm; and your potential employer can showcase the organization and observe your ability to ask perceptive questions.

● *Negotiating Guideline 2: Explore Many Options*

The negotiation process sometimes begins with only two options—your salary objective and the employer's offer—with seemingly little room for movement. You can escape this "either/or" trap by working to explore many options in the early stages of contract negotiation. This technique opens both parties to a variety of possible solutions and keeps the discussion rolling.

For example, assume that McDuff recently decided to add a new computer systems engineer to the staff at the corporate office in Baltimore. As a college senior about to graduate with a degree in computer science, you have applied for the job and have had a good first interview. The next week you are called back for a second interview and are offered a job, with a starting salary of $37,000. You are told this firm offer reflects the standard salary for new engineers with no experience. However, your research suggests that entry-level jobs in your field should pay closer to $40,000 a year, a full $3,000 more than the McDuff offer. While this difference concerns you, you have heard good things about the working environment at McDuff and would like to join the firm.

If you immediately were to state your need for a $40,000 starting salary, the negotiation might be thrown into the "either/or" trap that leaves little room for agreement. Instead, you should keep the conversation going by putting additional options on the table and asking open-ended questions (that is, questions that require more than a "yes" or "no" answer). For example, you could temporarily put aside your salary objective and ask how McDuff arrived at the offer figure. While giving the McDuff representative the chance to get facts on the table, this strategy also gives you opportunities to develop and then offer alternatives other than the two salary figures. The discussion might lead to options like these: (1) starting at $37,000 but moving to a higher figure after a successful 90-day trial period, (2) starting at the $37,000 figure but receiving an enhanced stock-option package upon being hired, or (3) starting at $39,000 but giving up the standard $2,000 moving allowance offered to entry-level employees.

The point is that you must be careful to avoid rigidity. Consider possibilities other than the two ideal goals both parties brought to the negotiation table.

● *Negotiating Guideline 3: Find the Shared Interests*

If you succeed in keeping options open during the negotiation process, you will discover points on which you agree. Psychologically, it is to your advantage to draw attention to these points rather than to points of conflict. Finding shared interests helps establish a friendship that, in turn, makes your counterpart more willing to compromise.

Let's go back to the preceding McDuff example. Assume you are continuing to discuss a number of salary options but have reached no agreement. Chances for closure may increase if you temporarily stop discussing salary and instead search for points, however minor, upon which you agree. For example, you could ask about job tasks in the position. When you learn that new engineers spend about 25 percent of their workday writing reports, you comment that your college training included two electives in technical writing, along with a senior-level research report. The McDuff representative praises the extra effort you made to prepare for the communication tasks in a technical profession.

This discussion about writing, though brief, has highlighted information that may have been missed during McDuff's early reviews of your application. The company's interest in good writing overlaps with the extra effort you gave to this discipline in college. That shared interest may motivate the company to offer a salary figure closer to what you desire. At the very least, you will have reinforced the decision McDuff officials made to offer you the job over three other finalists.

● *Negotiating Guideline 4: Listen Carefully*

Despite shared interests, negotiations often return to basic differences. An effective technique at this point is to seek information on the rationale behind your counterpart's views. It furthers the negotiation and, in fact, your own case to ask questions and then listen carefully to the answers coming from the other side of the table.

How are we helped by asking questions? Let's return to the McDuff example. When you are confronted with the salary offer, ask how McDuff arrived at that figure. Your question may uncover what is really behind the offer. Did McDuff recently make similar offers to other applicants? Is McDuff aware of national salary surveys that tend to support your request? Asking such questions benefits both you and the entire negotiation process in four ways:

- You give your counterparts the opportunity to explain their views (thus breaking out of the either/or cycle).
- You discover what motivates them (making it more likely that you will reach consensus).
- You expose careless logic and unsupported demands.
- You move closer to objective standards on which to base negotiations.

From your persistent questioning, careful listening, and occasional responses, information may emerge that would otherwise have remained buried. You may discover, for example, that McDuff is basing its salary offer on data pertaining to another part of the country, where both salaries and costs of living are lower. That would give you the opportunity to argue for a higher starting salary on the basis of regional differences in compensation.

● *Negotiation Guideline 5: Be Patient*

In the old hard-sell negotiations, participants frequently pushed for quick decisions, often to the regret of at least one of the parties. The better approach is to slow down the process. For example, you might want to delay agreement on a final salary figure until a later meeting, giving both you and your counterpart the chance to digest the conversation and consider options.

The main benefit of slowing down the process is to prevent basing decisions on the emotionalism of the moment. When objectivity takes a backseat to emotions in any negotiation—with an applicant, a client, a spouse, or a vendor—it is always best to put on the brakes, for two reasons:

- Good negotiated settlements should stand the test of time. When one party feels pressured, mistakes are made.
- Well-thought-out decisions are more likely to produce better long-term relationships, a major goal of your negotiations.

Negotiating Guideline 6: Do Look Back

Conventional wisdom has it that once you negotiate an agreement, you should not look back to second-guess yourself, since it will only make you less satisfied with what cannot be changed. That kind of thinking assumes that negotiations are spontaneous phenomena that cannot be analyzed, which is not true. If you have conducted your negotiations methodically, you will have much to gain from postmortems—particularly if they are in writing. Keep a negotiation journal to review before every major negotiation starts. Besides reminders, this journal should contain a short summary of previous negotiations. Make these entries immediately after a session ends, being sure to answer these questions:

- What options were explored before a decision was made?
- What shared interests were discovered?

Communication Challenge

"20-Something—Have Degree, Won't Travel"

Zach Lipkowski has been out of Mountain State College for almost a year and is still looking for his first real job. In the meantime, he has become a part-time ski instructor and worked in construction. With the help of a cheap apartment and a loan from his parents, he is managing to cover his bills, but he certainly hasn't used his biology degree from Mountain State. Now he finally has a break, having learned about an opening at McDuff's Denver office. What follows is a description of Zach's background, along with some questions related to his application letter, resume, and job interview.

A Checkered College Career

At Mountain State, Zach majored in biology, a field that had interested him since childhood and that he thought promised a good career. His boyhood hobby of studying poison arrow frogs kindled his interest in the life sciences. It also had led to a field trip to the Costa Rican rain forest in high school, with a Denver biological group. He had even helped a well-known frog expert on a study, providing the Denver researcher with observations about exotic frogs Zach kept as pets at home. So biology was a good fit.

In the first couple years of college, Zach considered many career options. Perhaps he could enter veterinarian or medical school. Or he might do research for industry or the government in environmental science. Another possibility was graduate school in biology toward a teaching career. By the time he was a junior, however, it was clear his grades probably were not high enough to get into medical or vet school, or a Ph.D. program in biology. He graduated with a 2.5 overall and a 2.9 in his major—respectable but not stellar. Some low grades in subjects like math, history, and English had hurt his average.

Despite some mediocre grades, he had the chance to work on an interesting senior project in which he studied "microbe digesters," organisms used to clean up pollution sites by converting toxins into nontoxic substances through the chemical digestive process. These critters are one of the newest environmental remediation techniques. Though only one member of a six-person team, he helped place microbes and then gather data at the site—a paper mill with toxic water. A team member who was a graduate student actually wrote most of the report, but Zach helped draft the summary of fieldwork. His modest part in the project had further inspired his "hands-on" interest in biology.

Zach Meets the Job Market—Twice

Although Zach did not seem destined to inhabit the ivy-covered walls of graduate school, he thought he would have a good shot at entry-level jobs in environmental

Continued

science, and he was right. He applied for many environmental science openings and got two good offers, both in the Midwest.

After much thought he turned down both offers, wanting to stay in Colorado because his mother was ill and, frankly, because he had always wanted to work in Colorado for his career. He was confident that an opportunity would open up as the Denver job market improved. When his mother recovered from her illness, Zach started working in August for a small construction firm as a day laborer, stopping for several months to work at a ski resort. He also took off a month for the trip he had always dreamed about—a fishing and canoeing venture to British Columbia.

By May, a year after graduation, he was regretting not having taken one of the job offers a year earlier. Most of his college friends were working in their fields, even if only at modest jobs, yet he was still doing the same kind of work he had done during summers in college. So Zach decided to reapply for entry-level jobs in environmental consulting firms. These jobs would involve the following main tasks: (1) visiting sites that needed environmental remediation, (2) overseeing lab tests, (3) forming conclusions and recommendations based on these tests, (4) writing reports that summarized activities, and (5) writing proposals and giving presentations to seek new work.

An Offer Brings Self-Assessment

This week Zach's college placement center called him about an opening as an environmental scientist at McDuff's Denver office. He planned to send a job letter and resume and hoped for a chance to interview.

Taking the advice of a placement specialist at his alma mater, Zach evaluated his skills, abilities, and experience before writing his new letter and resume. He jotted down these basic points about his background, some of which were detailed previously:

Positive

- Good work ethic
- Long-time interest in applied biology
- Took four-course sequence in environmental science—two courses short of a minor in the field

- Did you emphasize these shared interests?
- What questions did you ask?
- How did you show that you were listening to responses?

So do look back. Analyze every negotiation to discover what went right and what went wrong during the proceedings. Like other communication skills, such as writing and speaking, the ability to negotiate improves with use. With a few basic guidelines in mind and a journal upon which to reflect, you will discover the power of friendly persuasion.

- Enjoys outdoors and travel
- Worked on microbe project in college, helping to write report

Neutral or Negative

- Did not take professional position after college
- Had mediocre college grade point average
- Focused more on straight biology than on applied environmental science
- Got "C" in required technical writing course
- Had no outstanding extracurricular activities—study, part-time work in the dining hall, and "winding down" seemed to take most time

Questions and Comments for Discussion

Put yourself in Zach's shoes. As he prepares for the job application process, what advice would you give him on the following issues? Refer to the guidelines in the chapter in formulating your answers.

1. This chapter cites specific goals for cover letters. What should Zach emphasize in his letter to McDuff?
2. Zach plans to submit a one-page resume with his letter. What format should he select? Why?
3. Re-read Zach's list of "positive" and "neutral or negative" features. Do you agree with his self-assessment? What features of his background will be of most concern to McDuff? Why? What can Zach do to reduce the impact of negative items in his background during the entire application process?
4. Specifically, how should Zach deal with (a) his mediocre grade point average and (b) the year he has spent doing work unrelated to his intended profession?
5. If his letter and resume interest McDuff, Zach will interview with Ken Pierson, a McDuff project manager who knows very little about biology or microbes. His field is geology. Yet he manages projects that use the talents of a wide range of scientists, engineers, and technicians. What advice would you give Zach to make the best possible impression during the interview?

CHAPTER SUMMARY

This chapter surveys the entire process of searching for jobs, from performing your initial research to negotiating a contract. As a first step in the process, use the Internet and other sources of information to learn about occupations and specific employers that interest you. Second, write letters and resumes that get attention and respond to specific needs of employers. You can choose from chronological, functional, or combined resume formats, using the patterns of organization and style that best highlight your background. Third, prepare carefully for your job interview, especially in anticipating the questions that may be asked. Then perform with confidence. Also, do not forget to send a thank-you letter or email soon after the interview. Finally, use the negotiation phase of the job-search process to begin building a long-term relationship with your future employer, rather than focusing only on short-term gains of your first contract.

Access the Technical Writing Web site by pointing your browser at www.prenhall.com/pfeiffer. Here, you will find additional resources to aid you in learning and applying the information from this chapter.

ASSIGNMENTS

1. **Job Letter and Resume.** Find a job advertisement in the newspaper, on the Internet, or at your college placement office. The ad should match either qualifications you have now *or* those you plan to have after you complete the academic program upon which you are now working. Write a job letter and resume that respond to the ad. Submit the letter, resume, and written advertisement to your instructor.

   If useful for this assignment and if permitted by your instructor, you may fictionalize part of your resume so that it lists a completed degree program and other experience not yet acquired. In this way, the letter and resume will reflect the background you would have if you were applying for the job. Choose the resume format that best fits your credentials.

2. **Job Interview.** Pair up with another classmate for this assignment. First, exchange the letters, resumes, and job ads referred to in assignment 1. Discuss the job ads so that you are familiar with the job being sought by your counterpart, and vice versa. Then perform a role-playing exercise during which you act out the two interviews, one person as applicant and the other as interviewer.

   Option: Include a third member in your group. Have this person serve as a recorder, providing an oral critique of each interview at the end of the exercise. Then collaborate among the three of you in producing a written critique of the role-playing exercise. Specifically, explain what the exercise taught you about the main challenges of the job interview.

3. **Follow-Up Letter.** Write a follow-up letter to the interview that resulted from assignment 2.

4. **Follow-Up Letter—McDuff Projects.** Last month you submitted a job letter and resume for a position with the Barlow Group in Dallas, Texas. Now the firm has written to express interest in your application. It wants to know more about some summer employment you mentioned on your resume—the project you worked on was similar to some of Barlow's projects.

   Assume the summer work in question was as "student-in-training" on one of the projects described in the color insert in chapter 2. Write a letter that briefly describes the project and your participation in it. Use information from the project, along with invented details about the activities you completed as an assistant. Even if your tasks were not especially glorified—manual labor or office support, for example—strive to describe learning experiences that would be meaningful to your reader. You are writing Daniel C. Yates, Barlow Consulting Group, 600 Industrial Way, Dallas, TX 75221.

5. **Negotiation for Entry-Level Job.** As in assignment 2, pair up with another student. Assume that the letters, resumes, interviews, and follow-up letters from the preceding assignments have resulted in a second interview for one of you. (That is, select one of the positions, with one of you acting as applicant and the other as interviewer.)

The topic of this second interview is the position being offered to you. After talking with your team member about the context of this simulated interview, conduct a negotiation session wherein the two of you discuss one or more aspects of the position being offered (salary, benefits, travel schedule, employee orientation, training arrangement, career development, etc.).

6. **Negotiation with McDuff Client.** In this exercise, you and a classmate will simulate a negotiation session between Sharon Gibbon, a McDuff training manager at the Cleveland office, and Bernard Claxton, training director of Cleveland's Mercy Hospital. Study the following details before beginning your 10- or 15-minute discussion.

   Option: Collaborate with your teammate in writing an evaluation of this role-playing exercise. Explain the major obstacles encountered by both Gibbon and Claxton, and describe the techniques attempted by both parties to overcome these obstacles.

   General Background: Gibbon recently submitted a proposal to Claxton, offering to have McDuff conduct three hazardous-waste seminars for the plant staff at Mercy Hospital. Claxton calls Gibbon to say that he wants to go ahead with the seminars, contingent on some final negotiations between the two. Claxton and Gibbon agree to meet in a few days, presumably to iron out a final agreement. Gibbon wants the contract, and Claxton's staff needs the training. Yet the deal won't be sealed until they have their discussion and resolve several issues. Following are their respective points of view.

   Gibbon's Viewpoint: Sharon Gibbon has offered to have McDuff teach five one-day seminars for a fee of $15,000 ($3,000 per seminar). Each seminar will be team-taught by two of McDuff's certified industrial hygienists, Tom Rusher and Susan Sontack. They are expert trainers with much field experience in the identification and safe use of hazardous chemicals and other wastes. The $3,000 course fee is standard for McDuff's hazardous-waste seminars, though the company has on rare occasions given 10 percent discounts for any of the same seminars after the first one for the same client. Gibbon is interested in picking up Mercy Hospital as a client, but she also recognizes that the two instructors she has committed for the seminars may be needed for jobs the company has not yet scheduled. She is leery of cutting fees for Mercy Hospital when there may be other full-fee work right around the corner.

   Claxton's Viewpoint: Claxton knows that the hospital staff must have hazardous-waste training to conform to new county and city regulations, and he has heard from other hospitals that McDuff has the best training in the business. Yet he has real problems spending $15,000 on five seminars. Proposals from other firms were in the $9,000 to $12,000 range for the five seminars, and Claxton's training budget is modest. Though the other firms that submitted proposals did not share McDuff's reputation, they too offered team-taught seminars by certified industrial hygienists. Although Claxton would prefer to hire McDuff and although he knows that good training is worth the money, he is hoping to get Gibbon to lower McDuff's fee when they meet for their negotiation session. He knows that one benefit he can offer McDuff is continued training contracts from the hospital, since the high employee turnover will necessitate frequent training in hazardous waste. In addition, there may be other training opportunities for McDuff at the hospital, once Claxton completes his upcoming needs assessment of the staff training program.

7. **Ethics Assignment.** Searching for employment presents job seekers with some ethical challenges. A few "ethically challenged" individuals paint the portrait in their resume of someone who only remotely resembles the real thing. Certainly lying and deception do occur. Yet most writers simply want to present what they have accomplished and learned in the best possible light. As this chapter suggests, the resume is no time to be overly modest. With the goal of *supportable* self-promotion in mind, evaluate the degree to which the following resume entries are accurate representations of the facts that follow them.

1. Resume Entry: June-September 1997—Served as apprentice reporter for a Detroit area weekly newspaper.

   Reality: Worked for a little over three months as a fact-checker for a group of reporters. Was "let go" when the assistant editor decided to offer the apprentice position to another more promising individual with more journalistic experience.

2. Resume Entry: July 1995—Participated in university-sponsored trip to Germany. Reality: Flew to Germany with two fraternity brothers for a two-day fraternity convention in Munich, after which the three of you toured Bavaria for a week in a rental car.

3. Resume Entry: Summer 1996, 1997, 1998—Worked for Berea Pharmacy as a stock clerk, salesperson, and accountant.

   Reality: Helped off and on with the family business, Berea Pharmacy, during three summers while in college—placing merchandise on shelves, working the cash register, and tallying sales at the end of the day. Your father had regular help so you were able to spend at least half of each summer camping with friends, playing in a softball league, and retaking a couple of college courses.

8. **International Communication Assignment.** This chapter presents an approach to negotiation that attempts to bridge cultural differences; its aim is to be useful in many cultures. Yet no one negotiation strategy can describe the subtleties and nuances negotiations must take on throughout the world. If you are in the position of negotiating with persons from a culture different from your own, you would be wise to do some additional research.

   This assignment requires that you research and report on the negotiation strategies of a culture other than your own. Your instructor will indicate whether your report will be oral or written. Information can be acquired from sources such as the following:

   - Internet sites on other nations and on international communication
   - Friends and colleagues familiar with the culture you have chosen
   - Books and articles on international communication

   Of the many books that might be useful, here are just two: (1) Richard D. Lewis, *When Cultures Collide: Managing Successfully Across Cultures* (London: Nicholas Brealey, 1996), and (2) Terri Morrison, Wayne A. Conway, and Joseph J. Douress, *Dun and Bradstreet's Guide to Doing Business Around the World* (Englewood Cliffs: Prentice Hall, 1997).

201 Edge Drive
Norcross, PA 17001
March 15, 2002

Mr. James Vernon, Personnel Director
McDuff, Inc.
105 Halsey Street
Baltimore, MD 21212

Dear Mr. Vernon:

My academic advisor, Professor Sam Singleton, informed me about an electrical-engineering opening at McDuff, where he worked until last year. I am writing to apply for the job.

I understand that McDuff is making a major effort to build a full-scale equipment development laboratory. That prospect interests me greatly, because of my academic background in electrical engineering. At Northern Tech, I took courses in several subjects that might be useful in the lab's work—for example, microprocessor applications, artificial intelligence, and fiber optics.

Also, related work at Jones Energy & Automation, Inc., has given me experience building and developing new electronics systems. In particular, my work as an assembler taught me the importance of precision and quality control. I'd like the opportunity to apply this knowledge at McDuff.

Personal business will take me to Baltimore April 8-10. Could you meet with me on one of those days to discuss how McDuff might use my skills? Please let me know if an interview would be convenient at that time.

Enclosed is a resume that highlights my credentials. I hope to be talking with you in June.

Sincerely,

Donald Vizano

Donald Vizano

Enclosure: Resume

MODEL 14–1
Job letter (modified block) and chronological resume

continues

Donald Vizano
201 Edge Drive
Norcross, PA 17001
(300) 555-7861
dvizano@nct.edu

OBJECTIVE: A full-time position in electrical engineering, with emphasis on designing new equipment in automation and microprocessing

EDUCATION: 1996-2002 Bachelor of Science in Electrical Engineering (expected June 2002)
Northern College of Technology, Shipley, PA
3.5 GPA (out of 4.0 scale)

Major Courses:

Fiber Optics	Artificial Machine Intelligence
Robotic Systems	Communication Control Systems
Microprocessor Control	Microcomputer Applications
Microcomputer Systems	Digital Control Systems
	Semiconductor Circuits & Devices

Related Courses:

BASIC Programming	FORTRAN
Business Communication	Engineering Economy
Industrial Psychology	Technical Communication

**ACTIVITIES
AND HONORS:**   Institute of Electrical and Electronic Engineering (IEEE)
Dean's List, 8 quarters.

EMPLOYMENT:

1998-2002   Electronic Assembler (part-time)
Jones Energy & Automation, Inc.
Banner, PA

1997-1998   Lab Monitor (part-time)
Computer Services
Northern College of Technology
Shipley, PA

PERSONAL: Willing to travel, fluent in Spanish

REFERENCES: Available upon request

MODEL 14–1
continued

1523 River Lane
Worthville, OH 43804
August 6, 2002

Mr. Willard Yancy
Director, Automotive Systems
XYZ Motor Company, Product Development Division
Charlotte, NC 28202

Dear Mr. Yancy:

Recently I have been researching the leading national companies in automotive computer systems. Your job ad in the July 6 *National Business Employment Weekly* caught my eye because of XYZ's innovations in computer-controlled safety systems. I would like to apply for the automotive computer engineer job.

Your advertisement notes that experience in computer systems for machinery or robotic systems would be a plus. I have had extensive experience in the military with computer systems, ranging from a digital communications computer to an air traffic control training simulator. In addition, my college experience includes courses in computer engineering that have broadened my experience. I am eager to apply what I have learned to your company.

My mechanical knowledge was gained from growing up on my family's dairy farm. After watching and learning from my father, I learned to repair internal combustion engines, diesel engines, and hydraulic systems. Then for five years I managed the entire dairy operation.

With my training and hands-on experience, I believe I can contribute to your company. Please contact me at 614/882-2731 if you wish to arrange an interview.

Sincerely,

James M. Sistrunk

James M. Sistrunk

Enclosure: Resume

MODID 14–2

continues

MODEL 14–2
Job letter (block style) and chronological resume

James M. Sistrunk
1523 River Lane
Worthville, OH 43804
(614) 882-2731

Professional Objective:

To contribute to the research, design, and development of automotive computer control systems

Education:

B.S., Computer Engineering, 1999-present
Columbus College, Columbus, Ohio
Major concentration in Control Systems with minor in Industrial Engineering. Courses included Microcomputer Systems, Digital Control Systems, and several different programming courses.

Computer Repair Technician Certification Training, 1996-1997
U.S. Air Force Technical Training Center, Keesler Air Force Base, Biloxi, MS
General Computer Systems Option with emphasis on mainframe computers. Student leader in charge of processing and orientation for new students from basic training.

Career Development:

Computer Repair Technician, U.S. Air Force, 1996-1999
Secret Clearance

Responsibilities and duties included:

- Repair of computer systems
- Preventative maintenance inspections
- Diagnostics and troubleshooting of equipment

Accomplishments included:

- "Excellent" score during skills evaluation
- Award of an Air Force Specialty Code "5" skill level

Assistant Manager, Spring Farm, Wootan, Ohio, 1990-1995
Responsible for dairy operations on this 500-acre farm. Developed management and technical skills; learned to repair sophisticated farm equipment.

Special Skills:

Macintosh desktop publishing
Microsoft Word
Assembly Language
C++ Programming

References:

Available upon request

MODEL 14–2
continued

456 Cantor Way, #245
Gallop, Minnesota 55002
September 3, 2002

Ms. Judith R. Gonzalez
American Hospital Systems
3023 Center Avenue
Randolf, Minnesota 55440

Dear Ms. Gonzalez:

My placement center recently informed me about the Management Trainee opening with Mercy Hospital. As a business major with experience working in hospitals, I wish to apply for the position.

Your job advertisement notes that you seek candidates with a broad academic background in business and an interest in hospital management. At Central State College, I've taken extensive coursework in three major areas in business: finance, marketing, and personnel management. This broad-based academic curriculum has provided a solid foundation for a wide variety of management tasks at Mercy Hospital.

My summer and part-time employment also matches the needs of your position. While attending Central State, I've worked part-time and summers as an assistant in the Business Office at Grady Hospital. That experience has acquainted me with the basics of business management within the context of a mid-sized hospital, much like Mercy.

The enclosed resume highlights the skills that match your Management Trainee opening. I would like the opportunity to talk with you in person and can be reached at 612-111-1111 for an interview.

Sincerely,

Denise Ware Sanborn

Denise Ware Sanborn

MODEL 14–3
Job letter (modified block) and functional resume

continues

Denise Ware Sanborn
456 Cantor Way, #245
Gallop, Minnesota 55002
612-111-1111

Objective
Entry-level management position in the health care industry. Seek position that includes exposure to a wide variety of management and business-related tasks.

Education
Bachelor of Arts Degree, June 2002
Central State College
Gallop, Minnesota

Major: Business Administration
Grade Point Average: 3.26 of possible 4.0, with 3.56 in all major courses
All college expenses financed by part-time and summer work at Grady Hospital in St. Paul, Minnesota.

Skills and Experience
Finance
Helped with research for three fiscal year budgets
Developed new spreadsheet for monthly budget reports
Wrote accounts payable correspondence
Marketing
Solicited copy from managers for new brochure
Designed and edited new brochure
Participated in team visits to ten area physicians
Personnel
Designed new performance appraisal form for secretarial staff
Interviewed applicants for Maintenance Department jobs
Coordinated annual training program for nursing staff

Awards
1996 Arden Award for best senior project in the Business Administration Department (paper that examined Total Quality Management)

Dean's list for six semesters

References
Academic and work references available upon request.

MODEL 14–3
continued

2389 Jenson Court
Gulfton, MS 39200
(601) 111-1111
February 17, 2002

Mr. Nigel Pierce, Personnel Director
Structural Systems, Inc.
105 Paisley Way
Jackson, MS 39236

Dear Mr. Pierce:

I am writing in response to your ad for a technical representative in the July 13 (Sunday) edition of the *Jackson Journal*. I believe my experience and education make me an excellent candidate for this position.

I am very familiar with your products for the wood construction market. The laminated beams and floor joists your company manufactures were specified by many of the architects I have worked with during my co-op experience at Mississippi College. Work I have done in the residential and small commercial construction industry convinced me of the advantages of your products over nominal lumber.

Enclosed is my resume, which focuses on the skills gained from my co-op work that would transfer to your firm. I look forward to meeting you and discussing my future with your company.

Sincerely,

Todd L. Fisher

Todd L. Fisher

Enclosure: Resume

MODEL 14–4
Job letter (modified block) and functional resume

continues

Todd L. Fisher
2389 Jenson Court
Gulfton, MS 39200
(601) 111-1111
tlfish@ail.com

PROFESSIONAL OBJECTIVE	Use my education in civil engineering and my construction experience to assume a technical advisory position.
EDUCATION	Mississippi College Hart, Mississippi; Bachelor of Science, Civil Engineering Technology June 2001, GPA: 3.00 (out of 4.00)
PROFESSIONAL EXPERIENCE	Financed education by working as co-op student for two Jackson construction firms for 18 months.
Design Skills	Assisted with the layout and design of wall panels for Ridge Development condominium project. Created layout and design for complete roof and floor systems for numerous churches and small commercial projects.
Computer Skills	Introduced computerization to the design offices of a major construction company (HP hardware in HPbasic operating system). Designed trusses on Sun workstations in the UNIX operating system. Operated as the system administrator for the office. Learned DOS operating system and the Windows environment (on IBM hardware and its clones).
Leadership Skills	Instructed new CAD (computer-assisted design) operators on the operation of design software for panel layout and design. Designed and implemented management system for tracking jobs in plant.
INTERESTS	Family, gardening, sailing, travel
REFERENCES	References available upon request.

MODEL 14–4
continued

SUSAN A. MARTIN

PRESENT ADDRESS	**PERMANENT ADDRESS**
540 Wood Drive	30 Avon Place
Bama, CA 90012	Atlas, CA 90000
(901) 666-2222	(901) 555-6074

PROFESSIONAL OBJECTIVE:	Analyze and solve problems involving natural and pollution control systems as an Environmental Scientist.
EDUCATION:	Pierce College, Bama, California Bachelor of Science, Environmental Science May 2000, GPA: 3.15 (out of 4.00) Pleasant Valley College, Barnes, Nevada Associate in Applied Science, Engineering Science May 1998, GPA: 3.15 (out of 4.00)
PROFESSIONAL EXPERIENCE:	
Research Skills:	• Worked as lab assistant in a research project to analyze the effect of acid rain on frog reproduction in Lake Lane. • Designed Pierce College computer program to analyze data on ozone depletion.
Leadership Skills:	• Taught inventory procedures to new employees of Zane's Office Supply. • Helped incoming freshmen and transfer students adjust to Pierce College (as dormitory resident assistant).
Organizational Skills:	• Maintained academic department files as student assistant in Environmental Science Department. • Organized field trips for Pierce College Mountaineering Club.
HONORS AND ACTIVITIES:	Dean's List (five semesters) President of Cycling Club
INTERESTS:	Photography, camping, biking, traveling
EMPLOYMENT HISTORY:	*Dormitory Resident Assistant*, Pierce College, Bama, CA, 1999-2000 *Trainer*, Zane's Office Supply, Bama, CA, 1998-1999
REFERENCES:	References and transcripts available upon request

MODEL 14–5
Combined resume

589

Karen S. Patel
300 Park Drive
Burtingdale, New York 20092

Home: (210) 400-2112   **Messages:** (210) 400-0111

OBJECTIVE   Position as in-house technical writer and as trainer in
                         communication skills

EDUCATION   **Sumpter College, Marist, Vermont**
                         M.S. in Technical Communication, GPA: 4.0
                         December 2000

                         Warren College, Aurora, New York
                         M.A. in English, Cum Laude, June 1997

                         University of Bombay, India
                         B.A. in English, First Class Honors, June 1994

EMPLOYMENT   **Public Relations Office, Sumpter College, 2000-present**
Editing/   Administrative Assistant: Write press releases and conduct
Writing   interviews. Publish news stories in local newspapers and in
                         Sumpter Express. Edit daily campus newsletter.

                         Hawk Newspapers, Albany, New York, 1995-1996
                         Warren College Internship: Covered and reported special
                         events; conducted interviews; assisted with proofreading,
                         layout, headline count. Scanned newspapers for current events;
                         conducted research for stories. Published feature stories.

Teaching/   **Sumpter College, Marist, Vermont, 1999-2000**
Research   Teaching Assistant: Tutored English at the Writing Center,
                         answered "Grammar Hotline" phone questions, edited and
                         critiqued student papers, taught English to non-English
                         speakers, and helped students prepare for Regents exams.

                         Warren College, Aurora, New York, 1996-1997
                         Teaching Assistant: Taught business writing, supervised peer
                         editing and in-class discussions, held student conferences, and
                         graded student papers.

                         Research Assistant: Verified material by checking facts, wrote
                         brief reports related to research, researched information and
                         bibliographies.

COMPUTER SKILLS   Wordperfect, Microsoft Word, Pagemaker, Unix, Excel

REFERENCES   Available upon request

MODEL 14–6
Combined resume

Leslie Highland
997 Simmons Drive
Boise, Idaho 88822

OBJECTIVE: A full-time position in architectural design with emphasis on model-making and renderings for future buildings.

EDUCATION: **Boise Architectural College**
Boise, Idaho
Bachelor of Science
Architectural Engineering Technology
June 2000

Harvard University
Cambridge, Massachusetts
Certificate in Advance Architectural Delineation
August 1994

ACTIVITIES AND HONORS: **Boise Architectural College**
Winner of Senior Design Project
Architectural Engineering Technology

Charter Member of American Society of
Architectural Perspectives

EMPLOYMENT:

1994-2000 **Architectural Designer and Delineator**
Dorsey-Hudson, Architects
Boise, Idaho

1992-1994 **Architectural Designer and Renderer**
Windsor and Associates, Architects
St. Lake, Utah

1989-1992 **Architectural Renderer and Drafter**
Sanders and Associates, Architects
Provo, Utah

1987-1989 **Architectural Drafter**
Brown Engineering
St. Lake, Utah

REFERENCES: References and portfolio available upon request.

MODEL 14–7
Resume with graphics—*not* effective for computer scanning

Chapter 15

Style in Technical Writing

Describing "Style"

General Instructions

Each chapter begins with a Collaboration at Work exercise that applies strategies for working in groups to chapter topics. The exercise assumes you (1) have been divided into groups of about three to six students, (2) will use group time inside or outside of class to complete the case, and (3) will produce an oral or written response. For guidelines about writing in groups, refer to pages 28–34.

Background for Assignment

As this chapter points out, the term "style" refers to the way you choose to express an idea, as opposed to the content of the idea itself. The definition of style that occurs early in this chapter makes it clear that writers adopt particular styles for different

CHAPTER OUTLINE

contexts. For example, following are three passages that express the same idea in three different ways:

1. The results of the experiment strongly suggest to the team conducting the study that the hypothesis is valid.
2. After evaluating the results of the experiment, we concluded that the hypothesis is valid.
3. We believe the experiment worked.

Group Assignment

Describe how the previous three passages convey information differently to the reader. How would you describe the differences in style? When would one be more appropriate to use than another?

*T*his chapter, as well as the Handbook (Appendix C), focuses on the last stage of the writing process—revising. As you may already have discovered, revision sometimes gets short shrift during the rush to finish documents on time. That's a big mistake. Your writing must be clear, concise, and correct if you expect the reader to pay attention to your message. Toward that end, this chapter offers a few basic guidelines on style. The Handbook contains alphabetized entries on grammar, mechanics, and usage.

After defining style and its importance, this chapter gives suggestions for achieving five main stylistic goals:

- Writing clear sentences
- Being concise
- Being accurate in wording
- Using the active voice
- Using nonsexist language

OVERVIEW OF STYLE

This section (1) provides an overview of "style" as it applies to technical writing and (2) defines one particularly important aspect of style—called "tone"—that relates to every guideline in this chapter.

Definition of Style

Just as all writers have distinct personalities, they also display distinct features in their writing. Writing style can be defined in this way:

> **Style:** the features of one's writing that show its individuality—separating it from the writing of others and shaping it to fit the needs of particular situations. Style results from the conscious and subconscious decisions each writer makes in matters like word choice, word order, sentence length, and active and passive voice. These decisions are different from the "right and wrong" matters of grammar and mechanics (see the Handbook). Instead, they comprise choices writers make in deciding how to transmit ideas.

Style is largely a series of personal decisions you make when you write. As noted in chapter 1, however, much writing is being done these days by teams of writers. Collaborative writing requires individual writers to combine their efforts to produce a consensus style, usually a compromise of stylistic preferences of the individuals involved. Thus personal style becomes absorbed into a jointly produced product. Similarly, many companies tend to develop a company style in documents like reports and proposals.

Importance of Tone

Tone is a major component of style and thus deserves special mention here. Through tone you express an *attitude* in your writing—for example, neutral objectivity, on the one hand, or unbridled enthusiasm, on the other. The attitude evident in your tone exerts great influence over the reader. Indeed, it can determine whether your document achieves its objectives. Much like the broader term "style," tone refers to the way you say something rather than what you say.

The following adjectives show a few examples of the types of tone or "attitude" that can be reflected in your writing. Here they are correlated with specific examples of documents:

1. **Casual tone:** Email to three colleagues working with you on a project.
2. **Objective tone:** Formal report to a client in which you present data comparing cost information for replacing the company's computer infrastructure.
3. **Persuasive tone:** Formal proposal to a client in hopes of winning a contract for goods or services.
4. **Enthusiastic tone:** Recommendation letter to a university to accept one of your employees in a master's program.
5. **Serious tone:** Memorandum to employees about the need to reduce the workforce and close an office.
6. **Authoritative tone:** Memo to an employee in which you reprimand him or her for violations of a policy about documenting absences.
7. **Friendly tone:** Letter to long-term clients inviting them to an open house at your new plant location.

Although there are almost as many variations in tone as there are occasions to write documents, one guideline always applies: Be as positive as you can possibly be, considering the context. Negative writing has little place in technical communication. In particular, a condescending or sarcastic tone should be avoided at all costs. It is the kind of writing you will regret. When you stress the positive, you stand the best chance of accomplishing your purpose and gaining the reaction you want from the reader.

Despite the need to make style conform to group or company guidelines, each individual remains the final arbiter of her or his own style in technical writing. Most of us will be our own stylists, even in firms in which in-house editors help "clean up" writing errors. This chapter will help such writers deal with everyday decisions of sentence arrangement, word choice, and the like. But in seeing style as a personal statement, you should not presume that "anything goes." Certain fundamentals are part of all good technical style in the professional world. Let's take a look at these basics.

WRITING CLEAR SENTENCES

Each writer has his or her own approach to sentence style. Yet everyone has the same tools with which to work: words, phrases, and clauses. This section defines some basic terminology in sentence structure. Then it provides simple stylistic guidelines for writing clear sentences.

Sentence Terms

The most important sentence parts are the subject and verb. The *subject* names the person doing the action or the thing being discussed (*He* completed the study/ The *figure* shows that); the *verb* conveys action or state of being (She *visited* the site/ He *was* the manager).

Whether they are subjects, verbs, or other parts of speech, words are used in two main units: phrases and clauses. A *phrase* lacks a subject or verb or both and it thus must always relate to or modify another part of the sentence (She went *to the office*./As *project manager,* he had to write the report). A *clause,* on the other hand, has both a subject and a verb. Either it stands by itself as a *main clause* (*He talked to the group*) or it relies on another part of the sentence for its meaning and is thus a *dependent clause* (*After she left the site,* she went home).

Beyond these basic terms for sentence parts, you also should know the four main types of sentences:

- A *simple sentence* contains one main clause *(He completed his work).*
- A *compound sentence* contains two or more main clauses connected by conjunctions *(He completed his work, but she stayed at the office to begin another job).*
- A *complex sentence* includes one main clause and at least one dependent clause *(After he finished the project, he headed for home).*
- A *compound-complex sentence* contains at least two main clauses and at least one dependent clause *(After they studied the maps, they left the fault line, but they were unable to travel much farther that night).*

Guidelines for Sentence Style

Knowing the basic terms of sentence structure makes it easier to apply stylistic guidelines. Here are a few fundamental ones that form the underpinnings for good technical writing. As you review and edit your own writing or that of others, put these principles into practice.

● *Guideline 1: Place the Main Point Near the Beginning*

One way to satisfy this criterion for good style is to avoid excessive use of the passive voice (see "Using the Active Voice" on pages 604–605). Another way is to avoid lengthy introductory phrases or clauses at the beginnings of sentences. Remember that the reader usually wants the most important information first.

> *Original:* "After reviewing the growth of the Cleveland office, it was decided by the corporate staff that an additional lab should be constructed at the Cleveland location."

>   ***Revision:*** "The corporate staff decided to build a new lab in Cleveland after
>   reviewing the growth of the office there."

● *Guideline 2: Focus on One Main Clause in Each Sentence*

When you string together too many clauses with "and" or "but," you dilute the meaning of your text. However, an occasional compound or compound-complex sentence is acceptable, just for variety.

>   ***Original:*** "The McDuff hiring committee planned to interview Jim Steinway
>   today, but bad weather delayed his plane departure, and the committee had to reschedule the interview for tomorrow."
>
>   ***Revision:*** "The McDuff hiring committee had to change Jim Steinway's interview from today to tomorrow because bad weather delayed his flight."

● *Guideline 3: Vary Sentence Length but Seek an Average Length of 15–20 Words*

Of course, do not inhibit your writing process by counting words while you write. Instead, analyze one of your previous reports to see how you fare. If your sentences are too long, make an effort to shorten them, such as by making two sentences out of one compound sentence connected by an "and" or a "but."

You should also vary the length of sentences. Such variety keeps your reader's attention engaged. Make an effort to place important points in short, emphatic sentences. Reserve longer sentences for supporting main points.

>   ***Original:*** "Our field trip for the project required that we conduct research on
>   Cumberland Island, a national wilderness area off the Georgia Coast,
>   where we observed a number of species that we had not seen on previous field trips. Armadillos were common in the campgrounds,
>   along with raccoons that were so aggressive that they would come out
>   toward the campfire for a handout while we were still eating. We saw
>   the wild horses that are fairly common on the island and were introduced there by explorers centuries ago, as well as a few bobcats that
>   were introduced fairly recently in hopes of checking the expanding
>   population of armadillos."
>
>   ***Revision:*** "Our field trip required that we complete research on Cumberland
>   Island, a wilderness area off the Georgia Coast. There we observed
>   many species we had not seen on previous field trips. Both armadillos and raccoons were common in the campgrounds. Whereas the
>   armadillos were docile, the raccoons were quite aggressive. They
>   would approach the campfire for a handout while we were still eating. We also encountered Cumberland's famous wild horses, introduced centuries ago by explorers. Another interesting sighting was a
>   pair of bobcats. They were brought to the island recently to check
>   the expanding armadillo population."

BEING CONCISE

Some experts believe that careful attention to conciseness would shorten technical documents by 10 to 15 percent. As a result, reports and proposals would take less time to read and cost less to produce. This section on conciseness offers several techniques for reducing verbiage without changing meaning.

● *Guideline 1: Replace Abstract Nouns with Verbs*

Concise writing depends more on verbs than it does on nouns. Sentences that contain abstract nouns, especially ones with more than two syllables, can be shortened by focusing on strong verbs instead. By converting abstract nouns to action verbs, you can eliminate wordiness, as the following sentences illustrate:

Wordy:	"The *acquisition* of the property was accomplished through long and hard negotiations."
Concise:	"The property was *acquired* through long and hard negotiations."
Wordy:	"*Confirmation* of the contract occurred yesterday."
Concise:	"The contract was *confirmed* yesterday."
Wordy:	"*Exploration* of the region had to be effected before the end of the year."
Concise:	"The region had to be *explored* before the end of the year."
Wordy:	"*Replacement* of the transmission was achieved only three hours before the race."
Concise:	"The transmission was *replaced* only three hours before the race."

As the examples show, abstract nouns often end with "-tion" or "-ment" and are often followed by the preposition "of." These words are not always "bad" words; they cause problems only when they replace action verbs from which they are derived. The following examples show some noun phrases along with the preferred verb substitutes:

assessment of	assess
classification of	classify
computation of	compute
delegation of	delegate
development of	develop
disbursement of	disburse
documentation of	document
elimination of	eliminate
establishment of	establish
negotiation of	negotiate
observation of	observe
requirement of	require
verification of	verify

● *Guideline 2: Shorten Wordy Phrases*

Many wordy phrases have become common in business and technical writing. Weighty expressions add unnecessary words and rob prose of clarity. Here are some of the culprits, along with their concise substitutes:

afford an opportunity to	permit
along the lines of	like
an additional	another
at a later date	later
at this point in time	now
by means of	by
come to an end	end
due to the fact that	because
during the course of	during
for the purpose of	for
give consideration to	consider
in advance of	before
in the amount of	of
in the event that	if
in the final analysis	finally
in the proximity of	near
prior to	before
subsequent to	after
with regard to	about

● *Guideline 3: Replace Long Words with Short Ones*

In grade school, most students are taught to experiment with long words. Although this effort helps build vocabularies, it also can lead to a lifelong tendency to use long words when short ones will do. Of course, sometimes you want to use longer words just for variety—for example, using an occasional "approximately" for the preferred "about." As a rule, however, the following long words in the left column routinely should be replaced by the short words in the right column:

advantageous	helpful
alleviate	lessen, lighten
approximately	about
cognizant	aware
commence	start, begin
demonstrate	show

discontinue	end, stop
endeavor	try
finalize	end, complete
implement	carry out
initiate	start, begin
inquire	ask
modification	change
prioritize	rank, rate
procure	buy
terminate	end, fire
transport	move
undertake	try, attempt
utilize	use

● *Guideline 4: Leave Out Clichés*

Clichés are worn-out expressions that add words to your writing. Though they once were fresh phrases, they became clichés when they no longer conveyed their original meaning. You can make writing more concise by replacing clichés with a good adjective or two. Here are some clichés to avoid:

as plain as day

ballpark figure

efficient and effective

few and far between

last but not least

leaps and bounds

needless to say

reinvent the wheel

skyrocketing costs

step in the right direction

● *Guideline 5: Make Writing More Direct by Reading It Aloud*

Much wordiness results from talking around a topic. Sometimes called "circumlocution," this stylistic flaw arises from a tendency to write indirectly. It can be avoided by reading passages aloud. Hearing the sound of the words makes problems of wordiness quite apparent. It helps condense all kinds of inflated language, including the wordy expressions mentioned earlier. Remember, however, that direct writing must also retain a tactful, diplomatic tone when it conveys negative or sensitive information.

Indirect: "We would like to suggest that you consider directing your attention toward completing the project before the commencement of the seasonal monsoon rains in the region of the project area."

Direct: "We suggest you complete the project before the monsoons begin."

Indirect: "At the close of the last phase of the project, a bill for your services should be expedited to our central office for payment."

Direct: "After the project ends, please send your bill immediately to our central office."

Indirect: "It is possible that the well-water samples collected during our investigation of the well on the site of the subdivision could possibly contain some chemicals in concentrations higher than is allowable according to the state laws now in effect."

Direct: "Our samples from the subdivision's well might contain chemical concentrations beyond those permitted by the state."

● *Guideline 6: Avoid "There Are," "It Is," and Similar Constructions*

"There are" and "it is" should not be substituted for concrete subjects and action verbs, which are preferable in good writing. Such constructions delay the delivery of information about who or what is doing something. They tend to make your writing lifeless and abstract. Avoid them by creating (1) main subjects that are concrete nouns and (2) main verbs that are action words. Note that the following revised passages give readers a clear idea of who is doing what in the subject and verb positions.

Original: "There are many McDuff projects that could be considered for design awards."

Revision: "Many McDuff projects could be considered for design awards."

Original: "It is clear to the hiring committee that writing skills are an important criterion for every technical position."

Revision: "The hiring committee believes that writing skills are an important criterion for every technical position."

Original: "There were 15 people who attended the meeting at the client's office in Charlotte."

Revision: "Fifteen people attended the meeting at the client's office in Charlotte."

● *Guideline 7: Cut Out Extra Words*

This guideline covers all wordiness errors not mentioned earlier. You need to keep a vigilant eye for *any* extra words or redundant phrasing. Sometimes the problem comes in the form of needless connecting words, like *to be* or *that*. Other times it appears as redundant points—that is, those that have been made earlier in a sentence, paragraph, or section and do not need repeating.

Delete extra words when their use (1) does not add a necessary transition between ideas or (2) does not provide new information to the reader. (One important exception is the intentional repetition of main points for emphasis, as in repeating important conclusions in different parts of a report.) The examples in Figure 15–1 display a variety of wordy or redundant writing, with corrections made in longhand.

Example 1: Preparing the client's final bill involves ~~the~~ checking ~~of~~ all
   ^project^ ~~^~~invoices ~~for the project.~~

Example 2: The report examined what the McDuff project manager
   considered ~~to be~~ a technically acceptable risk.

Example 3: During ~~the course of~~ its field work, the McDuff team will ~~be~~
   ~~engaged in the process of~~ reviewing all ~~of the~~ notes ~~that have~~
   ~~been~~ accumulated in previous studies.

Example 4: ~~Because of his position~~ as head of ~~the~~ ^McDuff^ public relations group~~, at~~
   ~~McDuff,~~ he planned ~~such that he would be able~~ to attend the
   meeting.

Example 5: She believed ~~that the~~ recruiting ~~of~~ more minorities for the
   technical staff is essential.

Example 6: The department must determine its ~~aims and~~ goals so that
   they can be included in ~~the~~ ^McDuff's 2002^ annual strategic plan ~~produced by~~
   ~~McDuff for the year of 2002.~~

Example 7: Most McDuff managers ~~generally~~ agree that all ~~of the~~
   company~~'s~~ employees ~~at all the offices~~ deserve ~~at least~~ some
   ~~degree of~~ training each year ~~that they work for the firm.~~

FIGURE 15–1   Editing of wordy or redundant writing

BEING ACCURATE IN WORDING

Good technical writing also demands accuracy in phrasing. Technical professionals place their reputations and financial futures on the line with every document that goes out the door. That fact shows the importance of taking your time on editing that deals with the accuracy of phrasing. Accuracy often demands more words, not fewer. The main rule is this:

> Never sacrifice clarity for conciseness.

Careful writing helps to limit liability that your organization may incur. Your goal is very simple: Make sure words convey the meaning you intend—no more, no less. Here are some basic guidelines to follow:

● *Guideline 1: Distinguish Facts from Opinions*

In practice, this guideline means you must identify opinions and judgments as such by using phrases like "we recommend," "we believe," "we suggest," or "in our opinion." EXAMPLE: "In our opinion, spread footings would be an acceptable foundation for the building you plan at the site." If you want to avoid repetitious use of such phrases, group your opinions into listings or report sections. Thus a single lead-in can show the reader that opinions, not facts, are forthcoming. EXAMPLE: "On the basis of our site visit and our experience at similar sites, we believe that (1) _____ , (2) _____ , and (3) _____ ."

● *Guideline 2: Include Obvious Qualifying Statements When Needed*

This guideline does not mean you have to be overly defensive in every part of the report. It does mean that you must be wary of possible misinterpretations. EXAMPLE: "Our summary of soil conditions is based only on information obtained during a brief visit to the site. We did not drill any soil borings."

● *Guideline 3: Use Absolute Words Carefully*

Avoid words that convey an absolute meaning or that convey a stronger meaning than you intend. One notable example is "minimize," which means to reduce to the lowest possible level or amount. If a report claims that a piece of equipment will "minimize" breakdowns on the assembly line, the passage could be interpreted as an absolute commitment. The reader could consider any breakdown at all to be a violation of the report's implications. If instead the writer had used the verb "limit" or "reduce," the wording would have been more accurate and less open to misunderstanding.

USING THE ACTIVE VOICE

Striving to use the active voice can greatly improve your technical writing style. This section defines the active and passive voices and then gives examples of each. It also lists some practical guidelines for using both voices.

What Do Active and Passive Mean?

Active-voice sentences emphasize the person (or thing) performing the action—that is, somebody (or something) does something ("Matt completed the field study yesterday"). Passive-voice sentences emphasize the recipient of the action itself—that is, something is being done to something by somebody ("The field study was completed [by Matt] yesterday"). Here are some other examples of the same thoughts being expressed in first the active and then the passive voice:

- EXAMPLES: Active-Voice Sentences:

    1. "We *reviewed* aerial photographs in our initial assessment of possible fault activity at the site."
    2. "The study *revealed* that three underground storage tanks had leaked unleaded gasoline into the soil."
    3. "We *recommend* that you use a minimum concrete thickness of 6 in. for residential subdivision streets."

- EXAMPLES: Passive-Voice Sentences:

    1. "Aerial photographs *were reviewed* [by us] in our initial assessment of possible fault activity at the site."
    2. "The fact that three underground storage tanks had been leaking unleaded gasoline into the soil *was revealed* in the study."
    3. "*It is recommended* that you use a minimum concrete thickness of 6 in. for residential subdivision streets."

Just reading through these examples gives the sense that passive constructions are wordier than active ones. Also, passives tend to leave out the person or thing doing the action. Although occasionally this impersonal approach is appropriate, the reader can become frustrated by writing that fails to say who or what is doing something.

When Should Actives and Passives Be Used?

Both the active and passive voices have a place in your writing. Knowing when to use each is the key. Here are a few guidelines that will help:

- *Use the active voice when you want to:*

    1. Emphasize who is responsible for an action ("*We recommend* that you consider . . .")
    2. Stress the name of a company, whether yours or the reader's ("*PineBluff Contracting has expressed* interest in receiving bids to perform work at . . .")

3. Rewrite a top-heavy sentence so that the person or thing doing the action is up front (*"Figure 1 shows* the approximate locations of . . .")

4. Pare down the verbiage in your writing, since the active voice is usually a shorter construction

● *Use the passive voice when you want to:*

1. Emphasize the receiver of the action or the action itself rather than the person performing the action (*"Samples will be sent* directly from the site to our laboratory in Sacramento")

2. Avoid the kind of egocentric tone that results from repetitious use of "I," "we," and the name of your company (*"The project will be directed* by two programmers from our Boston office")

3. Break the monotony of writing that relies too heavily on active-voice sentences

Although the passive voice has its place, it is far too common in business and technical writing. This stylistic error results from the common misperception that passive writing is more objective. In fact, excessive use of the passive voice only makes writing more tedious to read. In modern business and technical writing, strive to use the active voice.

USING NONSEXIST LANGUAGE

Language usually *follows* changes in culture, rather than *anticipating* such changes. An example is today's shift away from sexist language in business and technical writing—indeed, in all writing and speaking. The change reflects the increasing number of women entering previously male-dominated professions such as engineering, management, medicine, and law. It also reflects the fact that many men have taken previously female-dominated positions as nurses and flight attendants.

This section on style defines sexist and nonsexist language. Then it suggests ways to avoid using gender-offensive language in your writing.

Sexism and Language

Sexist language is the use of wording, especially masculine pronouns like "he" or "him," to represent positions or individuals who could be either men or women. For many years, it was perfectly appropriate to use "he," "his," "him," or other masculine words in sentences such as these:

● "The operations specialist should check page 5 of his manual before flipping the switch."

● "Every physician was asked to renew his membership in the medical association before next month."

● "Each new student at the military academy was asked to leave most of his personal possessions in the front hallway of the administration building."

The masculine pronoun was understood to represent any person—male or female. Such usage came under attack for several reasons:

1. As already mentioned, the entry of many more women into male-dominated professions has called attention to the inappropriate generic use of masculine pronouns.
2. Many people believe the use of masculine pronouns in a context that could include both genders constrains women from achieving equal status in the professions and in the culture. That is, the use of masculine pronouns encourages sexism in society as a whole.

Either point supplies a good enough reason to avoid sexist language. Many women in positions of responsibility may read your on-the-job writing. If you fail to rid your writing of sexism, you risk drawing attention toward sexist language and away from your ideas. Common sense argues for following some basic style techniques to avoid this problem.

Techniques for Nonsexist Language

This section offers techniques for shifting from sexist to nonsexist language. Not all these strategies will suit your taste in writing style; use the ones that work for you.

● *Technique 1: Avoid Personal Pronouns Altogether*

One easy way to avoid sexist language is to delete or replace unnecessary pronouns:

Example:

Sexist Language: "During *his* first day on the job, any new employee in the toxic-waste laboratory must report to the company doctor for *his* employment physical."

Nonsexist Language: "During *the* first day on the job, each new employee in the toxic-waste laboratory must report to the company doctor for *a* physical."

● *Technique 2: Use Plural Pronouns Instead of Singular*

In most contexts you can shift from singular to plural pronouns without altering meaning. The plural usage avoids the problem of using masculine pronouns.

Example:

Sexist Language: "*Each* geologist should submit *his* time sheet by noon on the Thursday before checks are issued."

Nonsexist Language: "*All* geologists should submit *their* time sheets on the Thursday before checks are issued."

Interestingly, you may encounter sexist language that uses generic female pronouns inappropriately. For example, "Each nurse should make every effort to complete *her* rounds each hour." As in the preceding case, a shift to plural pronouns is appropriate: "Nurses should make every effort to complete *their* rounds each hour."

● *Technique 3: Alternate Masculine and Feminine Pronouns*

Writers who prefer singular pronouns can avoid sexist use by alternating "he" and "him" with "she" and "her." Using this technique usually avoids the unsettling practice of switching pronoun use within too brief a passage, such as a paragraph or page. Instead, writers may switch every few pages, or every section or chapter.

Although this technique is not yet in common usage, its appeal is growing. It gives writers the linguistic flexibility to continue to use masculine and feminine pronouns in a generic fashion. However, one problem is that the alternating use of masculine and feminine pronouns tends to draw attention to itself. Also, the writer must work to balance the use of masculine and feminine pronouns, in a sense to give "equal treatment."

● *Technique 4: Use Forms Like "He or She," "Hers or His," and "Him or Her"*

This solution requires the writer to include pronouns for both genders.

Example:

> *Sexist Language:*   "The president made it clear that each McDuff branch manager will be responsible for the balance sheet of *his* respective office."
>
> *Nonsexist Language:*   "The president made it clear that each McDuff branch manager will be responsible for the balance sheet of *his or her* respective office."

This stylistic correction of sexist language may bother some readers. They feel that the doublet structure, "her or his," is wordy and awkward. Many readers are bothered even more by the slash formations "he/she," "his/her," and "her/him." Avoid this usage.

● *Technique 5: Shift to Second-Person Pronouns*

Consider shifting to the use of "you" and "your," words without any sexual bias. This technique is effective only with documents in which it is appropriate to use an instructions-related "command" tone associated with the use of "you."

Example:

> *Sexist Language:*   "After selecting *her* insurance option in the benefit plan, each new nurse should submit *her* paperwork to the Human Resources Department."
>
> *Nonsexist Language:*   "Submit *your* paperwork to the Human Resources Department after selecting *your* insurance option in the benefit plan."

● *Technique 6: Be Especially Careful of Titles and Letter Salutations*

When you do not know how a woman you are writing to prefers to be addressed, use "Ms." Even better, call the person's employer and ask if the recipient goes by "Miss,"

Communication Challenge

"An Editorial Adjustment"

McDuff, Inc., has hired a technical writer/editor at its Cleveland office, the smallest branch in the company. The office finally generates enough reports and proposals to justify the addition, and Evelyn Tobin started the job a month ago. Some of the Cleveland employees, who were comfortable with the old system, are now having trouble adjusting to having an editor. What follows is some background on the hiring of Evelyn, the changes that she is making in office writing, and some questions and comments for discussion.

Winds of Change

For years the staff at McDuff's Cleveland office handled all its own writing and editing. Managers, engineers, scientists, accountants, trainers, and others had to draft and edit their own copy. Because they could depend on no one else to help, they gave great attention to the process and prided themselves on the quality of their writing. With the aid of several good secretaries, who often corrected grammar while they typed, the documents produced seemed adequate.

The growth of the office, however, increased the number and complexity of the reports and proposals that went out the door. The quality of editing began to decline. Those who observed the trend tied it to the following changes:

- Each writer simply had a higher volume of reports and proposals to complete, to keep the office competitive with similar firms.
- A new mobility in the workforce meant that fewer employees had received on-the-job training from old-timers at the office. In fact, over half the positions requiring a college degree had been filled in the last three years.
- This new workforce came from many different academic backgrounds and from other firms, making it harder than it used to be to impose a set "style" at the office.
- The experienced secretaries, who had been expert editors, had now retired. Most of the replacements were excellent keyboard operators but did not have the same editorial skills as their predecessors. They assumed their job was to type exactly what writers gave them.

The branch manager had observed these changes. Perhaps the last straw came when one long-time client returned a report with corrections made in red ink, along with this note: "You guys used to turn out good reports. What's happened?" With that embarrassment, the branch manager quickly hired an in-house technical writer/editor.

New Editor Takes Charge

When Evelyn Tobin started work a month ago, she met with all the staff to discuss her duties. At that meeting, there was general agreement that Evelyn would (1) provide writing advice, (2) perform a style edit for some reports, (3) be the lead writer for key proposals, (4) help with training in the office, and (5) do a quick grammar edit on as many reports as she had time to review.

With a B.S. in Technical Communication and two years of editing experience with a government agency, Evelyn was used to simplifying writing that was confusing, convoluted, or too technical. Although she had not worked with technical firms like McDuff, she assumed all her experience would translate to the new job. As it happened, most of her initial work involved style edits of reports that were to be sent to a mixed readership—some readers had a technical background, but others did not. Following are several changes Evelyn made in the reports, along with the original passages:

1. *Original:* The purpose of the new well is to allow Tank, Inc., to perform monthly water-level monitoring at three locations at the oil field so that the results can be sent to the Water Quality Control Board.

   Evelyn's Revision: The new well will allow Tank, Inc., to monitor water levels at the oil field. Then the data will be sent to the Water Quality Control Board.

2. *Original:* During the drilling of the boring, some soil sampling was performed by our technicians for the purpose of determining the exact location of the water table at the site.

   Evelyn's Revision: While drilling the boring, the technicians sampled soils to locate the water table.

3. *Original:* This letter proposal has been prepared by us for use by whatever attorney you select so that he can present a ballpark figure of costs to the college governing board.

   Evelyn's Revision: We prepared this proposal for whatever attorney you select. Then she can present a cost estimate to the college governing board.

4. *Original:* At this point in time, it is our belief that you should give equal consideration to both alternatives, for both can afford you the opportunity to complete expansion of the office complex prior to summer.

   Evelyn's Revision: At this point, we believe you should consider both alternatives. Either one will allow you to expand the office by next summer.

5. *Original:* There are a total of two ways we are recommending that you consider changing the plans in order to minimize the chance for earthquake damage.

   Evelyn's Revision: We recommend two changes to reduce the chance for earthquake damage.

Continued

Questions and Comments for Discussion

1. Study the before and after versions carefully.

   - Explain the rationale you think Evelyn would have for each of the changes she made.
   - Given the audience for the documents, was she right to make the changes?
   - Can you see changes in content that the original writer may find unacceptable?
   - Are there any cases where you need more context surrounding the passage to provide adequate answers to the two previous questions? Explain.

2. Would your answers to any of the previous questions change if the only audience for the report had been a group of technical experts?

3. Are there any alternative revisions that you think would be as effective as, or more effective than, the revisions Evelyn made?

4. Suggest what you think would be the best way for Evelyn to convey her revisions to the writer. Would this method change or stay the same as she gains more experience at the office?

5. One employee has come to Evelyn for advice on some grammar-checking software. He noted that his software stopped at every passive-voice sentence and suggested an active-voice replacement, and he wondered if he should always make the change. If you were in Evelyn's position, what answer would you give?

6. Discuss the effect that hiring a technical writer/editor might have on an office like McDuff-Cleveland. That is, how might the change affect the "corporate culture" of a such a company, where the professional staff spends from 25% to 50% of its time writing and editing documents?

7. If you were working at the Cleveland office, how would you feel about having your documents reviewed for style? for grammar?

"Mrs.," "Ms.," or some other title. (When calling, also check on the correct spelling of the person's name and her current job title.) Receptionists and secretaries expect to receive such inquiries.

When you do not know who will read your letter, never use "Dear Sir" or "Gentlemen" as a generic greeting. Such a mistake may offend women reading the letter and may even cost you some business. "Dear Sir or Madam" is also inappropriate. It shows you do not know your audience, and it includes the archaic form "madam." Instead, call the organization for the name of a particular person to whom you can direct your letter. If you must write to a group of people, replace the generic greeting with an "Attention" line that denotes the name of the group.

Examples:

Sexist Language: Dear Miss Finnegan: [to a single woman for whom you can determine no title preference]

Nonsexist Language: Dear Ms. Finnegan:
Sexist Language: Dear Sir: or Gentlemen:
Nonsexist Language: Attention: Admissions Committee

No doubt the coming years will bring additional suggestions for solving the problem of sexist language. Whatever the culture finally settles on, it is clear that good technical writing style will no longer tolerate the use of such language.

CHAPTER SUMMARY

Style is an important part of technical writing. During the editing process, writers make the kinds of changes that place their personal stamp on a document. Style can also be shaped (1) by a group, in that writing done collaboratively can acquire features of its diverse contributors, or (2) by an organization, in that an organization may require writers to adopt a particular writing style. Yet the decision-making process of individual writers remains the most important influence on the style of technical documents.

This chapter offers five basic suggestions for achieving good technical writing style. First, sentences should be clear, with main ideas at the beginning and with one main clause in most sentences. Although sentences should average only 15 to 20 words, you should vary sentence patterns in every document. Second, technical writing should be concise. You can achieve this goal by reading prose aloud as you rewrite and edit. Third, wording should be accurate. Fourth, the active voice should be dominant, though the passive voice also has a place in good technical writing. And fifth, the language of technical documents should be free of sexual bias.

 Access the Technical Writing Web site by pointing your browser at www.prenhall.com/pfeiffer. Here, you will find additional resources to aid you in learning and applying the information from this chapter.

ASSIGNMENTS

1. **Conciseness—Abstract Words.** Make the following sentences more concise by replacing abstract nouns with verbs. Other minor changes in wording may be necessary.
   a. Verification of the agreement was indicated by the signing of the contract by members of the McDuff corporate staff.
   b. The inspectors indicated that observation of the site occurred on July 16, 1996.
   c. Negotiation of the final contract was to happen on the day after their arrival.
   d. After three hours of discussion, the branch managers agreed that establishment of a new McDuff mission statement should take place in the next fiscal year.
   e. Assessment of the firm's progress will happen during the annual meeting of the McDuff Board of Directors.
   f. The entire company agreed that classification of employees according to level of education was inappropriate.
   g. Documentation of the results of the lab test appeared in the final report.

    h. Unlike the previous year, this year the disbursement of stock dividends will occur after the annual meeting.

    i. In analyzing the managerial style of the manager, the outside evaluators determined that delegation of authority appeared to be a problem for her.

    j. The financial statement showed that computation of the annual revenues had been done properly.

2. **Conciseness—Wordy Phrases and Long Words.** Condense the following sentences by replacing long phrases and words with shorter substitutes.

    a. In the final analysis, we decided to place the new pumping station in the proximity of the old one.

    b. Prior to commencing the project, they met to prioritize their objectives.

    c. Endeavoring to complete the study on time, Sheila transported the supplies immediately from the field location to the McDuff lab.

    d. During the course of his career, he planned to utilize the experience he had gained in the ambulance business.

    e. His work with the firm terminated due to the fact that he took a job with another competing firm.

    f. In the event that two clients need a crew in Austin next week, we can give consideration to using the same crew for both projects and lowering travel costs for both clients.

    g. Rob McDuff was not cognizant of the fact that younger employees felt differently than older employees about the expansion of their office building.

    h. To implement the Phoenix asbestos project, we made adjustments in the workload of two engineers so that they could be available to undertake the project in Phoenix.

    i. Subsequent to the announcement he made, he held a news conference for approximately one hour of time.

    j. At this point in time, she had every hope that her annual bonus would afford her family the opportunity to take an additional family vacation.

3. **Conciseness—Clichés and "There Are/It Is" Constructions.** Rewrite the following sentences by eliminating clichés and the wordy constructions "there are" and "it is."

    a. They all agreed that the issue had been discussed repeatedly for the last 10 years; thus they did not want to reinvent the wheel during the current study.

    b. There are many examples of skyrocketing equipment costs affecting the final budget for a project.

    c. It is a fact that most employees at McDuff believe the company has taken a step in the right direction by adding international offices.

    d. Needless to say, it is clear that Karen is looking forward to the three-week vacation.

    e. She explained to her staff that it was as plain as day that they would have to decrease their labor costs.

    f. The prospective client asked for a ballpark figure of the project costs.

    g. Last but not least, there was the issue of quality control that he wanted to emphasize in his speech.

    h. In these modern times today, there are new approaches that college graduates should take to the job search.

    i. Susan ended the meeting by concluding that there were a number of mutually agreeable solutions that could be explored so that the new departments in conflict could peacefully coexist.

    j. It is a fact that our boss ended the meeting about a loss of profits by noting that we are all in the same boat.

4. **Sentence Clarity.** Improve the clarity of the following sentences by changing sentence structures or by splitting long sentences into several shorter ones.

    a. Therefore, to collect a sample from above the water table, and thus to follow the directions provided by the client, the initial boring was abandoned and the drill rig was repositioned about two feet away and a new boring was drilled.

    b. After capping the soil sample ring with PVC end caps and then notifying all members of the project team, we placed it in a cooler for storage on-site and transportation later to a chemical analytical laboratory.

    c. Based on the geotechnical data obtained from the subsurface exploration program, the results of the percolation testing, and the planned plumbing fixtures, the feasibility of installing a leachfield-type on-site sewage-disposal system was evaluated.

    d. Percolation test #1 was performed approximately 40 feet east of the existing pump house and percolation test #2 was performed near the base of the slope approximately 65 feet west of the pump house, and then the results were submitted to the builder.

    e. We appreciate the opportunity to provide our services on this project and look forward to continuing our relationship with XYZ Trading and Transportation Company when we begin the Zanter Project with your Finance Department next spring.

    f. All of the earth materials encountered in our exploration can be used for trench backfill above manhole and pipe bedding, provided they are free of organic material, debris, and other deleterious materials, and they are screened to remove particles greater than six inches in diameter.

    g. This study was conducted to identify, to the extent possible, based on available information from the city files and the criteria described in our proposal of June 18, 1998, whether activities near the site may have involved the use, storage, disposal, or release of hazardous or potentially hazardous substances to the environment.

    h. The properties consist of approximately 5,000 acres, including those parcels of Heron Ranch owned by American Axis Insurance Company, the unsold Jones Ranch parcels, the village commercial area, the mobile home subdivisions, two condominium complexes, a contractor's storage area, an RV storage area, a sales office, a gatehouse, open space parcels, and the undeveloped areas for future Buildings 1666, 1503, 1990, and 1910.

    i. Having already requested permits for the construction of the bathhouse, medical center, maintenance building, boat dock, swimming pool, community building, and an addition to the community building, we still need to apply for the storeroom permit.

    j. A report dated May 25, 1998, for the ABC Corporation confirmed that the updated business plan had been completed the previous month, but a new plan had to be submitted by May 25, 1999.

5. **Active and Passive Voice Verbs.** Make changes in active and passive voice verbs, where appropriate. Refer to the guidelines in the chapter. Be able to supply a rationale for any change you make.

    a. It was recommended by the personnel committee that you consider changing the requirements for promotion.

    b. No formal report about assets was reported by the corporation before it announced the merger.

    c. The graphs showing the differences in depreciation and interest and the net loss on the investment are shown in Appendix A.

    d. It has been noted by the Department of Environmental Services that the laundry business was storing toxic chemicals in an unsafe location.

   e. The samples from the Scottish Highlands will be sent to McDuff's engineering lab in London.

   f. The violation of ethical guidelines was reported by the commissioner to the president of the association.

   g. No complete equipment inventory has been made by McDuff's Boston office.

   h. It was concluded by the employee committee that McDuff's retirement program needed to be revised.

   i. Dirt brought to the site should be evaluated by the engineer on-site before it is placed in the foundation.

   j. Due to presence of a good deal of sand at the location, excavations are anticipated by us to be relatively unstable.

6. **Sexist Language.** Revise the following sentences to eliminate sexist language.

   a. The department decided to advertise for a department chairman in three national newspapers.

   b. Although each manager was responsible for his own budget, some managers obviously had better accounting skills than others.

   c. The company policy manual states that each secretary should submit her time card twice a month.

   d. If an hourly worker misses no work for sickness during a calendar year, he will receive a $500 bonus at year's end.

   e. Each flight attendant is required to meet special work standards as long as she is employed by an international airline.

   f. Typically, a new engineer at McDuff receives his first promotion after about a year.

   g. Every worker wonders whether he is saving enough for retirement.

   h. If a pilot senses danger, she should abort the takeoff.

   i. Upon arriving at the site, a McDuff scientist should make immediate contact with his client representative.

   j. [greeting section of a letter] Gentlemen:

7. **Advanced Exercise—Conciseness.** The following sentences contain more words than necessary. Rewrite each passage more concisely, without changing the meaning. If appropriate, make two sentences out of one.

   a. The disbursement of the funds from the estate will occur on the day that the proceedings concerning the estate are finalized in court.

   b. During the course of the project that we conducted for Acme Pipe, several members of our project team were in the unfortunate position of having to perform their fieldwork at the same time that torrential rains hit the area, totaling three inches of rain in one afternoon.

   c. At a later date we plan to begin the process of prioritizing our responsibilities on the project so that we will have a clear idea of which activities deserve the most attention from the project personnel.

   d. Needless to say, we do not plan to add our participation to the project if we conclude that the skyrocketing costs of the project will prohibit our earning what could be considered to be a fair profit from the venture.

   e. The government at this point in time plans to discontinue its testing of every item but will undertake to implement testing again in approximately five months.

   f. Hazerd, Inc., will endeavor to finalize the modifications of the blueprints for a ballpark figure of about $850.

g. For us to supply the additional supplies that the client wishes to procure from us, the client will have to initiate a change order that permits additional funds to be transferred into the project account.

h. Upon further analysis of the many and varied options that we are cognizant of at this time, it is our opinion that the long-term interests of our firm would be best served by reducing the size of the production staff by 300 workers.

i. Prior to the implementation of the state law with regard to the use of asbestos as a building material, it was common practice to utilize this naturally occurring mineral in all kinds of facilities, some of which became health hazards subsequently.

j. In the event that we are given permission to undertake the research, be sure to make certain to perform an efficient and effective search of available literature in a research facility so that we do not end up, in the final analysis, reinventing the wheel with regard to knowledge of superconductors.

8. **Advanced Exercise—General Style Rules.** Revise the following sentences by applying all guidelines mentioned in this chapter. When you change passive verbs to active, you may need to make some assumptions about the agent of the action, since the sentences are taken out of context.

   a. Based on our review of the available records, conversations with the various agencies involved, including the Fire Department and the Police Department, and a thorough survey of the site where the spill occurred, it was determined that the site contained chemicals that were hazardous to human health.

   b. After seven hours at the negotiation table, the union representatives and management decided that the issues they were discussing could not be resolved that evening, so they met the next day at the hotel complex, at which point they agreed on a new contract that would increase job security and benefits.

   c. It is recommended by us that your mainframe computer system be replaced immediately by a newer model.

   d. After the study was completed by the research team and the results were published in the company newsletter the following month, the president decided to call a meeting of all senior-level managers to discuss strategies for addressing problems highlighted by the research team.

   e. Our project activities can be generally described in this way. The samples were retrieved from the site and then were transported to the testing lab in the containers made especially for this project, and at the lab they were tested to determine their soil properties; the data were analyzed by all the members of the team before findings and conclusions were arrived at.

   f. First the old asbestos tile was removed. Then the black adhesive was scraped off. Later the floor was sanded smooth. The wood arrived shortly. Then the floor was installed.

   g. The figures on the firm's profit margins in July and August, along with sales commissions for the last six months of the previous year and the top 10 salespersons in the firm, are included in the Appendix.

   h. It was suggested by the team that the company needs to invest in modern equipment.

   i. It is the opinion of this writer that the company's health plan is adequate.

   j. Shortly after the last change in leadership, and during the time that the board of directors was expressing strong views about the direction that the company was taking, it became clear to me and other members of the senior staff that the company was in trouble.

    k. Each manager should complete and submit his monthly report by the second Tuesday of every month.

    l. After completing our engineering analysis, it is clear that metal fatigue caused the structure to fail.

    m. Upon hearing the captain's signal, each flight attendant should complete her checklist of preflight procedures.

    n. Our weed-spraying procedure will have minimal impact on shrubbery that surrounds the building site.

    o. It was reported today from the corporate headquarters that the health-care plan has been approved by the president.

9. **Editing Paper of Classmate.** For this assignment, exchange papers with a member of your class. Use either the draft of a current assignment *or* a paper that was completed earlier in the term. Edit your classmate's work in accordance with this chapter's guidelines on style. Then explain your changes to the writer.

10. **Editing Sample Memo.** Using the guidelines in this chapter, edit the following memorandum. The assignment can be completed individually, or in teams as a group-editing project.

DATE: January 12, 2002
TO: All Employees of Denver Branch
FROM: Leonard Schwartz, Branch Manager
SUBJECT: New Loss-Prevention System

As you may have recently heard, lately we received news from the corporate headquarters of the company that it would be in the best interest of the entire company to pay more attention to matters of preventing accidents and any other safety-related measures that affect the workplace, including both office and field activities related to all types of jobs that we complete. Every single employee in each office at every branch needs to be ever mindful in this regard so that he is most efficient and effective in the daily performance of his everyday tasks that relate to his job responsibilities such that safety is always of paramount concern.

    With this goal of safety ever present in our minds, I believe the bottom line of the emphasis on safety could be considered to be the training that each of us receives in his first, initial weeks on the job as well as the training provided on a regular basis throughout each year of our employment with McDuff, so that we are always aware of how to operate in a safe manner. The training vehicle gives the company the mechanism to provide each of you with the means to become aware of the elements of safety that relate to the specific needs and requirements of your own particular job. Therefore, at this point in time I have come to the conclusion in the process of contemplating the relevance of the new corporate emphasis on safety to our particular branch that we need, as a branch, to give much greater scrutiny and analysis to the way we can prevent accidents and emphasize the concern of safety at every stage of our operation for every employee. Toward this end, I have asked the training coordinator, Kendra Jones, to assemble a written training program that will involve every single employee and that can be implemented beginning no later than June of this year. When the plan has been written and approved at the various levels within the office, I will conduct a meeting with every department in order to emphasize the major and minor components of this upcoming safety program.

It is my great pleasure to announce to all of you that effective in the next month (February) I will give a monthly safety award of $100 to the individual branch employee at any level of the branch who comes up with the best, most useful suggestion related to safety in any part of the branch activities. Today I will take the action of placing a suggestion box on the wall of the lunchroom so that all of you will have easy access to a way to get your suggestions for safety into the pipeline and to be considered. As an attachment to the memo you are now reading from me I have provided you with a copy of the form that you are to use in making any suggestions that are then to be placed in the suggestion box. On the last day of each month that we work, the box will be emptied of the completed forms for that month, and before the end of the following week a winner will be selected by me for the previous month's suggestion program and an announcement will be placed by me to that effect on the bulletin board in the company workroom.

If you have any questions in regard to the corporate safety program as it affects our branch or about the suggestion program that is being implemented here at the Denver office at McDuff, please do not hesitate to make your comments known either in memorandum form or by way of telephonic response to this memorandum.

11. **Ethics Assignment.** Reread this chapter's "Communication Challenge," concerning Evelyn Tobin, the new editor at McDuff's Cleveland office. Now assume that you, as an electrical engineer at Evelyn's office, have asked for her help in preparing an article for publication in a professional journal. As you hand her the article, you are quick to add you have long-standing problems organizing information and editing well. Two days later the draft appears in your mailbox looking like your first graded paper in English 101 in college. Evelyn has even provided a suggested outline for reorganizing the entire piece. Upon reading her comments and reviewing the outline, you find that you agree with almost all her suggestions. You follow her suggestions and proceed to meet with her several times and show her three more drafts, including the final that she edits and proofs.

Feeling that she has done more on your article than she would normally do as part of her job responsibilities, Evelyn diplomatically asks how you plan to acknowledge her work on the final published article. How would you respond to her? Would you list her on the title page as co-author, would you mention her in a footnote as an editor, or would you adopt some other approach? Explain the rationale you would give Evelyn after telling her your decision. What are the main ethical considerations in making the decision?

12. **International Communication Assignment.** One major problem with international communication occurs when product instructions are written (or translated) by individuals who do not have enough familiarity with the language being used. The problem can be solved by choosing writers or translators who ARE in fact native speakers and writers. For this assignment, locate a set of instructions written in English with stylistic errors that would not have been made by a native speaker/writer. Point out these errors and suggest appropriate revisions.

Appendix A

Online Technical Documents

*J*uanita Ramirez felt a little uncomfortable as she looked around at the other attendees in the meeting. She counted the vice presidents and directors and noted that she was a bit out of her usual level. As McDuff's training manager, she had dealt with most of these people in one-on-one meetings, but this was her first time in a meeting at a peer level with all of them. She listened and took notes as the discussion wandered around the problems of distributing information about policies, procedures, and company events throughout McDuff's domestic and international offices, not to mention keeping employees and customers informed on technical updates and project schedules. Her notes looked like a diagram of a chemical molecule as she tried to keep track of the complex relationships among departments and knowledge bases that were emerging from the discussion. She was wondering what the ultimate point of the meeting was when the president let the first shoe drop.

"I'm convinced we need to take our company's documentation online. It's the only way we can keep up-to-date information available to our employees and customers worldwide." Juanita was wondering how he planned to make that happen when he dropped the other shoe, "And Juanita, I'm going to ask you to spearhead this project since you handle most of the technical documentation already."

All Juanita could do was wonder what she had gotten into and hope her skills in working with paper documents would help her with online documents.

Like McDuff, more and more companies are putting information online, and technical communicators like Juanita are often asked to take the lead. They soon learn that their background in designing and writing technical documents on paper is useful, but that online documents have some special twists. This appendix focuses on the ways online documents differ from paper documents and offers additional guidelines for working with online documents.

WHAT IS ONLINE DOCUMENTATION?

Many documents come under the general category of "online documentation," but they all have two points in common:

1. Computers are used to access and read them.
2. Their primary purpose is to *locate and display information* (as opposed to software programs whose primary purpose is to *process data and display the results*).

Since online documents exist on computers, they allow the reader to interact with the document in a way that could never be done with paper. This interaction is so predominant that many developers of online documentation do not even use the word "reader"—they prefer the term "user" instead. Whatever we call them, readers or users, people generally interact with online documents in the following ways:

- Inquiries
- Responses
- Navigation

1. Inquiries—readers can go to search engines or online indexes to enter keywords or names of topics about which they need information.

2. Responses—many online documents present questions to the readers, and the reader responds by clicking on predetermined choices or by typing (or even speaking) their answers into the computer. We even label certain types of screens used in software applications as "dialog boxes" since they enable the reader and computer to exchange information.

3. Navigation—readers can influence how information is presented to them (and in what order) through devices such as "browse" buttons, "previous" and "next" buttons, and, most commonly, "hypertext" or "hypermedia" links. These are words or graphics that take the reader to a different topic if the reader clicks on them with the mouse pointer.

A common feature of much online documentation is its use of multimedia. Just as paper documents can include media other than text, such as photographs or illustrations, online documents also can include multimedia elements, such as sound, video, animation, and graphics. These more active media elements help give online documents a heightened sense of interaction. However, when overused (as they often are) they provide a heightened potential for distraction.

- ● *Online Guideline 1: Use multimedia to support the needs of the reader and the purpose of the document—not to show off.*

Multimedia can be useful for the following objectives:

1. Attract the reader's attention. But once the reader's attention is engaged, these same multimedia elements can become distractions. Some particularly annoying examples include music that plays when the reader first comes into a document but which can't be turned off, and highly animated graphics that stay on the screen while the reader is trying to concentrate on text. In one usability test, readers covered one such graphic with their hands while trying to read the rest of the page. Some simple design solutions to these problems include the following:

- Make it easy for the reader to "turn off" sound and animated elements.
- Provide links or navigation that takes the reader to a calmer place to actually read in-depth text. In other words, first get their attention. Then move to a more business-like environment within your online document.

2. Illustrate processes. Many online documents include video or animation that demonstrates what the text describes. Online encyclopedias, for example, make extensive use of media elements that may show such diverse topics as how elephants look and sound and how the heart beats. Good designers give much control to the reader in starting, stopping, and replaying these elements.

3. Provide multiple channels for complementary information. Online documentation for software products often includes an animated "demonstration" of a procedure being done while a text box or even a narrator's voice explains the background.

These examples illustrate how media can be used correctly. They also illustrate the power of online documents as effective communications.

Types of Online Documents

In general, online technical documents can be categorized as follows:

- Help files
- Web-based documents
- Online books
- Computer-based training

Help Files. "Help files" are documents that are integrated with software products they support. Help files tend to focus on procedural information (e.g., "how to do" topics) and structural information (e.g., "what parts do what" topics). Many "help" documents make effective use of "pop-ups," small message screens that appear over the current page when a link is clicked. Pop-ups are commonly used for definitions and as labels for elements in a graphic.

Web-based Documents.   These are networked documents that are accessed and read through special software called "browsers." Web-based documents are distributed over three kinds of networks:

1. **Intranets**—in-house networks that can only be accessed by employees. Companies use Intranets for in-house technical documentation, for departmental news, and for publication of policies and procedures.
2. **Extranets**—networks that are physically accessible by anyone around the world but that need a password to be viewed. Companies often use extranets to provide customers with technical information they do not want accessed by the general public.
3. **Internet**—the worldwide network that allows public access to anyone. "Homepages" are common documents found on the Internet. Companies use these documents to inform the general public about their business, job opportunities, and products. Many companies even sell their products directly over the Internet.

Online Books.   These large documents, such as manuals or catalogs, are organized much like their paper counterparts but distributed in compact electronic packages, such as diskettes or CD-ROMs. It is becoming quite common for companies to send their product manuals on CD-ROM and have the customer print out desired sections they want in "hard copy" (paper version).

Computer-based Training.   "Computer-based-training" (CBT) is specialized, often interactive, online documentation that is generally meant to take the place of classroom instruction. CBT programs are usually produced by trained instructional designers.

Authoring Tools and Delivery Programs

With each of the above types of documents, we need to differentiate between authoring tools and delivery programs.

An "authoring tool" is a software program that lets the writer of online documentation write the text, incorporate the media, and create the user interface, including the interactive devices and links. Common authoring tools for "help" files are RoboHelp, Doc-to-Help, and ForeHelp, all of which help develop "help" files that run in Windows-based applications. Authoring tools for online books include Lotus SmartText and Adobe Acrobat. CBT authoring tools include Asymetrix Toolbook and MacroMedia Authorware. Web pages can be produced with simple text editors like NotePad, but writers often prefer graphical user interface tools like FrontPage and PageMill.

All these tools seek to provide an easy-to-use interface that lets designers of online documents generate their documents without having to understand a programming language or an extensive set of codes. Many of these tools, like the "help" authoring tools and the CBT tools, compile the final product into a compact, easily distributed file. Depending on the product, some of these files need delivery programs for the reader to view the file.

A "delivery program" is a software program that lets the reader view files produced by the online documentation developer. The important consideration is that some of these programs are free and some charge the user or the distributor. These costs are called "run-time" costs. In determining what tool to use to develop online documentation, consider whether or not there will be run-time costs for the reader to view the documentation. Common delivery systems that are free are the WinHelp engine that runs "help" files in Windows (it comes with the Windows operating system), Adobe's viewer called Acrobat (downloaded from the manufacturer's Web site), and common Internet browsers like Netscape and Internet Explorer (built into most computers).

HOW DOES READER ANALYSIS CHANGE?

Up to now, for technical documents, we have analyzed our readers along two dimensions:

1. Technical expertise
2. Decision-making level

With online documentation we have the additional dimension of online literacy. "Online literacy" refers to how well the reader knows how to use online documentation.

● *Online Guideline 2: Assume your readers know very little about how to operate your online document. Keep your navigation methods simple. If necessary, instruct them how to use advanced functions.*

Surveys and usability tests show that, at best, readers generally are familiar with only superficial details of reading online documents. Many don't know how to use search features and indexes, most don't know how to manipulate help screens so that the instructions stay visible or how to put the instructions in a particular location on a screen, and many Internet users don't know how to use advanced features like frames or "plug-ins."

When technical communicators decide to develop online documents, they learn many things about how those documents work but neglect to take into account that their readers still lack these skills. Therefore, much useful information gets hidden because readers don't know the tricks to get to it.

HOW DOES DOCUMENT ORGANIZATION CHANGE?

This book has emphasized the use of an ABC format, where documents have a definite beginning, middle, and end. Even though readers don't always read documents that way, they rely on this consistent structure to find things. For example, if you wanted to read an overview about a product you just bought, you'd expect it to be toward the front of the manual. Readers thumb through documents scanning headers, headings, and illustrations until they find the topic of interest. If all else fails, they know there is a table of contents in the front and an index in the back. In short, readers know a lot about how paper documents are structured and rely heavily on that knowledge when looking for information.

Online documents, however, have been described as documents with a middle but no beginning or end. Even when they have a definite structure, that structure is not as physically evident to the reader as it would be in a paper document.

- ***Online Guideline 3: Give readers clues about where they are in an online document and give them enough information to make navigation decisions.***

We talk about readers being "lost in hyperspace," which happens when a reader takes a series of hypertext links and ends up lost, with no idea how to get back. For example, take the case of a student whose teacher sends her to a Web page to learn about Ethiopia. While on the page, she clicks on a link to a discussion about Ethiopian culture and takes a link called "indigenous recipes." From there she takes a link to "other spicy foods," then to "Cajun recipes," then to "New Orleans," and then to "Bourbon Street." The teacher walks by and scolds her for not doing the assignment, and the student is unable to explain how or why she got there, although every click seemed somehow related to where she was at the time. Worse yet, she might have seen something of interest along the way; however, since she doesn't really remember how she got there, she's not sure she'll ever be able to find it again.

As designers of online documents, there are techniques you can use to keep the reader from getting lost:

- **Provide an online table of contents.** This strategy lets the reader get a sense of structure and navigate by topics of interest, making it easy to get back to the table of contents from every topic screen in the document. Good homepages on a Web site can serve the same purpose.
- **Keep links to a minimum.** You need to be stingy with hypertext links. Links should relate to the topic in a specific way. Provide a link only if it is likely that someone will need or want that additional information. For example, in an instruction called "How to print a file," some designers might provide a link to a topic called "How to rename a file," since they both deal with files. It is unlikely, though, that someone who needs the instructions for printing is also looking for instructions about renaming. Since many readers take a link whether they think they need it or not, many will take the link and begin the process of getting lost in hyperspace.

● **Phrase hypertext links so the reader knows where they go.** A link that merely says <u>"Requirements"</u> is not as effective as one that says, "Click <u>here</u> to see a checklist of hardware and software requirements." (Also, links positioned in the middle of a sentence are easier to see than stand-alone words used as links.)

● *Online Guideline 4: Keep topics short.*

Online documents are not as easy to read as paper documents, so you should keep topics as short as possible. If the topic must be long, give the reader the option to print it out.

Also, when readers go to an online document, they are usually looking for very specific information. Readers generally do not scroll at all or only scroll about two screen lengths looking for information, before giving up. Short topics make information easier to find.

HOW DOES PAGE LAYOUT CHANGE?

With online documents there are no pages, but you do have to worry about how the screen is laid out and where text and pictures are arranged. Since the screen area is smaller than most pages, online documents are stingier with white space. Yet white space is still important. A good rule of thumb is to limit "passive" white space, such as margins, but keep "active" white space, such as space that shows topical organization.

Headings are still important in online documents, but the reader is not as likely to see them. Therefore, many long online documents include navigation bars with the document's headings so that the reader can scan the topics and go directly to one of interest.

● *Online Guideline 5: Put the most important information at the top of a topic.*

Journalists talk about which news stories get placed "above the fold," that is, the top half of the newspaper. Readers focus on what's easy to find, and 80 percent of the articles that get read in a newspaper are "above the fold." The same principle applies to online documentation and what we could call the "scroll line." Information that the reader has to scroll to see will not be read as often as information that first appears in the screen's visible area. Like smart journalists, put your most important information "above the scroll line."

SUMMARY

Although writing technical documents for online is similar to writing them for paper, the wise developer will pay attention to the special capabilities and risks of online documents. On the positive side, they provide a heightened interaction that can engage the reader, they can present information in a wider array of media, and they can be updated and redistributed much more quickly than paper documents. On the downside, the reader must learn new skills in order to use online documents, media can distract rather than enhance, and producing online documents requires more time and higher technical skills than their paper counterparts.

Writers of online documents must be part computer programmer, graphic designer, and writer, and they must have a clear view of the readers' needs and capabilities. Because of the greater potential of online documentation and its higher technical demand, online documents are a natural field for technology students who enjoy technical writing.

Appendix B

Web Site Design

INTRODUCTION

John Ayers works in Technical Publications, a group in McDuff's Research and Training Department. His job is to issue and update Technical Bulletins to McDuff's field offices. These bulletins update employees on technical news, regulations, best practices, and guidelines from the staff researchers. One morning, his boss, Juanita Ramirez, walked into his cubicle.

"I've got good news and bad news," she said as she waved a document in the air.

"Let's hear it," John answered, a bit insincerely.

"The good news is that our latest field survey gives your Technical Bulletins very high marks from the folks who read them."

627

"And the bad?" John asked, feeling a little relieved.

"Only fifteen percent of our field employees have ever seen one. Forty-two percent didn't even know we had such a thing."

John was not altogether surprised. Although each field office was supposed to maintain a three-ring binder with these bulletins sorted and updated, in reality this rarely happened. Even when these notebooks were in good shape, many employees worked remotely or telecommuted and did not have ready access to the notebook in their branch offices.

John started to rebut, "Well, you can lead a horse to water . . ."

Juanita interrupted. "The survey says that the field employees would like to get at these documents over the Internet. I want you to create a Web page on our intranet site and use it to distribute the Technical Bulletins. That includes creating an archive of all the bulletins we've already released." Juanita looked over her glasses at John with her patented "you're the technical writer" look.

"Not a problem," John muttered, hoping his answer sounded more confident than he felt.

John was certainly comfortable *using* the Web, and he and some classmates had even done a Web presentation for one of their technical writing classes in college. The topic had been "Surfing the Internet," and John and his friends had downloaded lots of animated surfer graphics from the Internet. The class had gotten a big kick out of it. This project sounded different, however, putting technical information online for an international engineering company. John couldn't help but wonder how his experience with writing technical bulletins would merge with his experience designing a "cool" Web site. Somehow he felt that animated surfers and hula girls weren't what Juanita had in mind.

No one can become an expert Web site designer overnight, but the design principles for effective Web site design are not that different from those for good page design in print. The secret is to start with pages that have simple functionality but incorporate sound design principles. This appendix offers an understanding of the underlying technology and foundational principles of information design. In particular, you will find discussions of the following topics:

- How documents work on the Web
- Elements of Web site design
- Principles of hypertext links
- Conversion of documents to Web pages
- Usability and accessibility

The first half of this appendix suggests specific guidelines for designing Web sites—guidelines that should augment the ones given throughout this book—and then provides some practical instructions to help get you started. If you get interested in building complex or highly interactive Web pages, you can read more detailed "how-to" books to help you. For information concerning online documentation in general—as opposed to the specific application discussed here—refer to Appendix A in this book.

HOW DOCUMENTS WORK ON THE WEB

The Internet has become a dominant publication medium for technical communicators for many of the reasons Juanita and John were discovering. Paper documents are costly to distribute, difficult to maintain, and sometimes impossible to get to. Many companies are finding the Internet to be a ready-to-use distribution tool for putting timely information in the hands of their employees, business partners, and customers. More and more, these companies are "Web-enabling" many of their traditional transactions, communications included. This expansion of communication media changes the ways that technical communicators must think about designing and writing technical documents. This section explains how the Web works to distribute documents through "client/server" technology and how a set of tags called "HTML" makes it easy for writers to Web-enable their documents.

Client/Server Network

The Internet, commonly called the World Wide Web (or just "the Web"), is really a large client/server network. *Servers* are computers that hold files or applications that can be accessed and used by *client* computers on the same network. A *network* is a system of computers and connections (wire or wireless) along with software protocols that regulate how member-computers communicate. The Internet is so widespread because it uses the world's telephone and cable systems as its network backbone.

Servers, also called *hosts,* store Web pages as text and graphics files. Each server has a unique name or address called a URL (Universal Resource Locator). When you enter an address into your Internet browser, such as *http://www.yahoo.com,* you are entering the name of a specific server. You can also ask for a specific file on that server, which is what you are doing when you type information after the ".com". For example, the URL http://www.acme.com/products/turbo.html is requesting a file named "turbo.html" from a directory named "products" on the server named "acme.com". (When you don't enter a specific file name, the server sends you whatever file has been set up to be the default file.)

All this works because of special software on the client computer. The most important is the *browser.* A browser is a software application that reads Web files and displays them on the computer's monitor. Because the browser has the programming needed to display Web pages, Web files can be kept small. They only need to carry content and simple instructions about document elements. The "heavy lifting," such as formatting and managing the visual display and media players, is done by the browser, reducing the amount of computer instructions that need to be transmitted over the network. Two well-known browsers are *Internet Explorer* and *Netscape.*

The second important software application on the client is the Internet Service Provider (ISP) software. The ISP formats data into the right protocols to send and receive it over the Internet. The ISP also provides the connectivity needed to get onto the Web. A well-known ISP is *America On Line.* All one needs, then, to be on the Internet is a computer with a browser (which comes as part of the initial software),

Network

A **network** is a system of computers and connections (wire or wireless) along with software protocols that regulate how member computers communicate.

FIGURE B–1   Client/server model of an Internet transaction

the ISP software (which is purchased and installed via CDROM), and a MODEM (usually built into the computer) for transmitting and receiving data over phone lines. Figure B–1 illustrates a basic exchange of a request and the resultant file over the Internet.

HTML: The Basic Language of the Web

Web pages are easy to write because they follow a rather simple set of rules called HTML. HTML stands for Hypertext Markup Language. (*Hypertext* is text in one file that serves as a link to another file.) Remember, the browser does the hard work; all the writer of a Web page has to do is "mark up" the text so that the browser knows what the parts of the document are. *Marking up* means identifying the structural elements of the document, such as headings, lists, tables, etc. The markups are usually called *tags*. The markup can also specify links to other files. The browser is preprogrammed to display those elements. For example, the marked-up text shown in Figure B–2 would be displayed by the browser as shown in Figure B–3. Table B–1 explains the HTML used in Figure B–2.

Perhaps the hardest thing for writers to get used to in Web publishing is that they lose a lot of control over how documents look to the reader. If the reader resized the browser window, for example, the text would break at different places. Also in the

```
<HTML>
    <HEAD>
        <TITLE>Glossary: Network</TITLE>
    </HEAD>
    <BODY>
        <H3>Network</H3>
        A <B>network</B> is a system of computers and
        connections (wire or wireless) along with
        software protocols that regulate how member
        computers communicate.<P>
        <IMG SRC="CS model.jpg">
    </BODY>
</HTML>
```

FIGURE B–2 Sample HTML text

FIGURE B–3 HTML file displayed in browser.
Internet Explorer is either a trademark or trademarks of Microsoft Corporation in the
United States and/or other countries.

TABLE B–1 Explanation of HTML used in Figure B–2

HTML	Explanation
<HTML>	Tells the browser that this is an HTML file.
<HEAD>	Identifies the start of header information, i.e., information about the Web page but not displayed as part of the Web page.
<TITLE>Glossary: Network</TITLE>	Sets the text that displays in the title bar of the browser. You will notice that many HTML tags occur in pairs with the format <*tag*>...</*tag*>. This simply tells the browser to apply that tag's attributes to whatever text comes between the two tags.
</HEAD>	Tells the browser that the header information is done.
<BODY>	Identifies the start of the page content. This is the information that the browser will display on the monitor.
<H3>Network</H3>	Creates a level 3 heading that says "Network." The browser will format it according to its rules for displaying level 3 headings.
A <B>network</B> is a system of computers and connections (wire or wireless) along with software protocols that regulate how member computers communicate.<P>	This is the body of the document. Notice that the word "network" has been tagged as boldface. The <p> at the end signals a paragraph break.
<IMG SRC="CS model.jpg">	This line inserts the illustration in file "CS model.jpg". This is the same file used for Figure B–1 in this appendix. We chose a file format that works well in both print and Web documents so we could use the same illustration in either document. This is called "repurposing" content. Modern technical writers anticipate this and choose designs and formats that can be repurposed easily in both print and online media. (In general, use graphic files that end in .jpg or .gif.)
</BODY> </HTML>	Ends the body of the document and ends the HTML file.

example, the bottom of the illustration with the Client PC callouts is *below the scroll line*, i.e., the reader would have to scroll down to see it. For this reason, Web documents need to be designed to be more flexible than their paper counterparts. For one thing, don't force line wraps so that text looks good on your monitor; on another monitor with another browser, these might show up as awkward breaks in the middle of a line.

Although many Web page authors code their HTML pages by hand, most writers use authoring tools like *FrontPage* or *DreamWeaver* to speed up the process. Later this appendix will discuss how you can even use *MS Word* to build HTML documents.

The following two principles summarize the design constraints placed on technical communicators by the technology of the Internet:

Web Design Principle 1: When designing print documents, choose file formats and layouts that repurpose easily to the Web.

Web Design Principle 2: Realize that the final document can look different on different browsers. Keep the design flexible.

ELEMENTS OF WEB SITE DESIGN

Elements of Design

John Ayers was right in suspecting that designing professional Web sites is a bit different than designing a school project. It involves complex tasks of gathering and managing content, and then presenting that content in ways that are easily accessed and pleasing to look at. Designing a Web site at a professional level, therefore, involves three elements of design:

- Architectural design
- Information design
- Visual design

Architectural design is a systematic approach that *organizes* a body of knowledge or a collection of related business activities (e.g., order processing, inventory control, etc.) into components and defines relationships among those components. Architectural design involves the following key activities:

1. Cataloging the inherent structures of the *content domain* (the stuff to be known) or charting the major tasks that can be done on a site
2. Analyzing the information requirements of the end user

Information design focuses on the *presentation of information* in ways so that its components can be easily perceived and its content easily processed. It goes hand in hand with architectural design and adds the additional requirement of understanding how people process written and visual information.

Visual design is the systematic use of media (typography, graphics, animation, and video) as well as layout to deliver content and to allow the necessary interaction between user and computer to navigate, submit queries, and retrieve information.

The outcome of an integrated architecture/information/visual design approach is a coherent and cohesive **structure** that can be easily perceived, navigated, and accessed by end users to retrieve information or perform tasks needed for particular objectives.

Web Site Architecture

In Web site design, the basic unit of construction is the *Web page*. The defining attribute that groups information on a Web page is called that page's *topic*. Web pages can be grouped into larger structures called *categories*. The ensuing architectural structure is called a *Web site*. Figure B–4 shows some of the elements that might be included in an intranet Web site for McDuff, Inc. In this case, the architecture is modeled around the organizational structure of McDuff, Inc. The designer is hoping that the readers' knowledge of the company's structure will help them envision and navigate the site's structure. In a sense, the designer is using the company structure as a metaphor for the Web site's architecture.

FIGURE B–4   Information structures constituting a Web site's architecture

Information Design Elements of a Web Page

From the perspective of information design, a Web page's structure should communicate the structure of that page's information as well as the architectural structure of the Web site.

A typical Web page's structure can include the following information and architectural elements:

- Key content
- Category links
- Topical links
- Orientation indicators
- Cross-references
- Opportunistic links
- Regulatory links

Figure B–5 shows a sample grid structure that designates specific screen locations for the various information design elements, and Table B–2 describes each element. Although the content is given the largest area, note how much of the user interface is dedicated to displaying the overall architecture. This is a common structure, but it is by no means the only structure.

Not all Web pages have or need all of these elements. A very simple Web page might be completely dedicated to content. Generally, the deeper into a Web site the user gets, the fewer of these elements may be required. It is generally desirable, however, to have designated areas for content and category links on all pages.

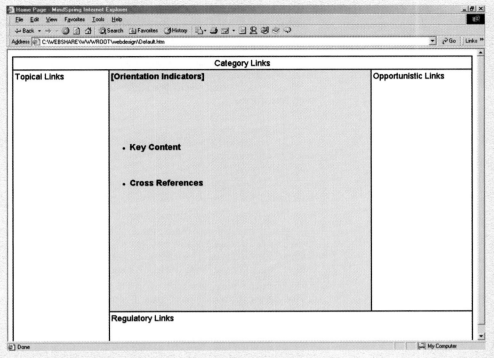

FIGURE B–5 Typical grid structure for a complex Web site
Internet Explorer is either a trademark or trademarks of Microsoft Corporation in the United States and/or other countries.

Visual Design

Lastly, the look and feel of the Web page must support the architectural and information design decisions you make. In addition, it should create a "look and feel" that are consistent with the purpose or tone of the Web site.

Figure B–6 shows the McDuff, Inc., intranet home page, which has been laid out using the grid structure shown in Figure B–5. A marbled background sets the categorical links apart from the content area (and reinforces the McDuff theme of "construction"). The graphic of the man reading a newspaper visually communicates the purpose of that column. Few colors are used, which makes the hypertext links in blue stand out more. The tone is conservative, reflecting an appropriate image for an engineering consulting company. Imagine the same page with a black background, animated cartoon graphics, and lots of different font styles. That could look interesting, even attractive, but it would not look like McDuff, Inc.

Architectural, Informational, and Visual Design Guidelines

Organization structure and look is important in print documents too, but print documents have the luxury of being easily hefted or quickly thumbed through. A Web

TABLE B–2 Web page elements

Element	Description
Key Content	The information or transaction that delivers the value of the page. Content can be the *target information* the user is seeking, or it can be *meta-information*, e.g., additional navigational choices and explanations to help the user focus in on the target information.
Cross-references	Links the user might take based upon associations with specific content elements of the current page. For example, a link to information about Georgia on a page about Atlanta would be a cross-reference. The item being cross-referenced may reside on the same page or on a different page. Many cross-references link to other Web sites.
Category links	Links to the broad categories of information or tasks **across** the Web site (and/or other categories at the same level of hierarchy as the current category).
Topical links	Links to other topics or transactions **within** the current category.
Orientation indicators	Signposts that tell the users where they are within the current architecture. On a home page or a small site, this may be as simple as just the title of the page. On complex sites, it can be a technique called *breadcrumbs,* which gives the exact path the reader has taken to get to this page.
Opportunistic links	Links to topics that are only loosely related to the current topic or to the supposed task of the user on a given Web page. For example, a banner ad for fishing equipment on a Web site about fishing resorts would be an opportunistic link.
Regulatory links	Links to topics that have no relation to content, e.g., privacy statements, links to vendors or sponsors, etc. Their presence is usually dictated in general across the Web site rather than driven by informational content of a given page.

"book" can be seen only one page at a time. Web designs, therefore, need to draw on the users' experience or imaginations to convey structure, not only for the current page being viewed, but for the entire site as well. To meet these additional requirements, here is a special set of design principles:

Web Design Principle 3: Choose an organizational structure that reflects a logical grouping of the content, and use a scheme (or metaphor) the reader will find familiar.

Web Design Principle 4: Make page design emphasize the key content while still showing the overall architecture of the Web site.

Web Design Principle 5: Be sure visual elements support the informational design and create an appropriate "look and feel" based on the purpose of the Web site.

HYPERTEXT LINKS

Hypertext links consist of a target (either text or graphics) that the reader can click on with his or her mouse and that then causes the browser to move to a different location in the file, move to a new file, or initiate some other kind of action (such as downloading a file). It is this ability to link files that gives the Internet its Web-like

FIGURE B–6   The McDuff intranet home page
Internet Explorer is either a trademark or trademarks of Microsoft Corporation in the
United States and/or other countries.

quality. It can be a source of robust information design, but it can also get readers
lost or confused, a condition known as "lost in cyberspace."

Hypertext Tags

The HTML tag that creates a link and its target is quite simple. The following
markup,

<A HREF="techbul.html">Technical Bulletins</A>

would make the text "Technical Publications" a *hot spot,* i.e., text that the reader could
click on. Clicking on that text would cause the browser to open the file "techbul.html".
The browser automatically underlines hot spots and displays them in color. By default,
most browsers display a hypertext link (hot spot) in blue, but change the color to purple if the reader takes the link. That makes it easier for readers to know what links they
have already visited and which ones they have not. This is an important navigational
aid, one that helps the reader get around the site without accidentally revisiting sites
(going around in cyber-circles, so to speak).

Many Web authors prefer to use graphics to be the hypertext links, either in the form of pictures or as word graphics to make their pages look more interesting. If you do this, you must make it obvious that something is a link. Graphics that look like buttons can work well. Many readers, however, do not realize that word graphics are links and, therefore, do not click on them. If you use graphical links, do something to make them "look clickable."

Keeping Readers Oriented

One reason that readers get lost taking hypertext links is that they do not have a clear idea of where a link will take them in the first place. This ambiguity causes them to bounce around the site and sometimes even to accidentally leave the site. Imagine a link labeled "Rob McDuff" on the home page. A reader might take this link expecting to find a biographical sketch about Rob McDuff. If so, that reader would be disoriented if instead it opened an email link to Rob. The same would be true of a reader who took that link thinking it would send an email, and instead, found a biographical sketch. The solution is to word hypertext links so that they give a clear preview of what type of content the reader will find. (Notice that on the home page in Figure B–6, the email link to Rob McDuff is clearly communicated in the working of the link and is supported by an appropriate graphic.)

Figure B–7 shows an interior page (a page inside the Web site) of the McDuff intranet. All of the links are obvious, with perhaps the exception of one. The McDuff logo in the upper left-hand corner is a hyperlink to the Web site's home page. The use of a logo in this way is a well-established Web convention, and savvy users would be disappointed if this site did not use it. Even so, note that the designer has included a text link right below it that also takes the reader to the home page. Robust designs provide alternative text links to support graphic links.

The page in Figure B–7 also provides some signposts to help orient the reader to where he or she is in the Web site. For one, visited links have changed color, helping the reader know which links he or she has already taken. Also, notice the breadcrumbs above the title "Technical Bulletins" that say, "Home> Research & Training> Technical Bulletins." This technique tells the reader what path he or she has taken to get here.

Because readers can get lost more easily in online documents than in print documents, use the following design principles for dealing with hypertext links:

Web Design Principle 6: Make all hypertext links visually obvious to the reader.

Web Design Principle 7: Have links preview the type of content the reader will find.

Web Design Principle 8: Provide signposts that help readers know where they are in the online document.

CONVERSION OF DOCUMENTS TO WEB PAGES

You may come across the guideline that Web "pages" should be short, usually no longer than two or three screen lengths. This is typically true, but in the case of distributing technical documents over the Internet, chopped-up versions can be awkward to print, and the need to continually click to the next section can get annoying

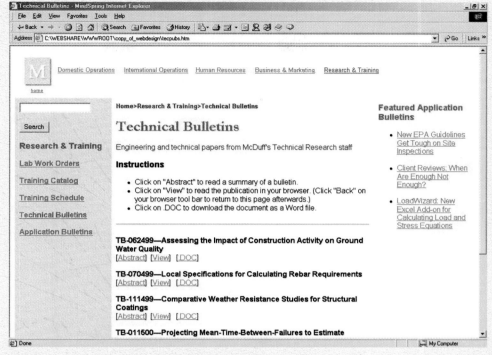

FIGURE B–7 Interior page in the McDuff intranet
Internet Explorer is either a Trademark or Trademarks of Microsoft Corporation in the United States and/or other countries.

for many readers. Still, readers like to get to information quickly, whether in print or on line. There is an easy way to convert large documents (articles or chapters) into Web pages that are easily skimmed on line. This section shows you how to do this with documents created in Microsoft Word. For these documents to be effective online documents, however, you must know how to apply style tags and insert a Table of Contents in MS Word. Then you must know how to convert that document to a Web page (HTML document). This section covers each of these requirements.

Style Tags

Style tags in MS Word are a lot like HTML tags—they describe the structure of content in the document. To make long documents more readable online, use Heading style tags in your document. To do this in MS Word, select View and then Toolbars, and then make sure that Formatting is selected. (See Figure B–8.)

Once you have turned on the Formatting toolbar, you can use the style tag drop-down menu to format your headings as Heading 1, Heading 2, or Heading 3. Merely put your cursor where you want the heading, click on the drop-down style menu, and then select the level heading you want. (See Figure B–9.)

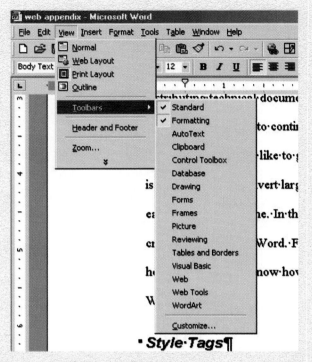

FIGURE B–8   Turning on the Formatting toolbar
Word is either a trademark or trademarks of Microsoft Corporation in the United States and/or other countries.

Inserting a Table of Contents

If you insert a Table of Contents at the front of your MS Word document, it will generate a set of hypertext links automatically when you convert that document to a Web page. To insert the table, put your cursor at the front of the document, select Insert, and then select Index and Tables. A dialog box similar to the one in Figure B–10 appears. Click the tab labeled "Table of Contents" and click the "OK" button.

Converting to a Web Page

Once you have tagged all of your headings with style tags and inserted a Table of Contents, you are ready to convert the document to a Web page. Click File and then select "Save as Web Page" (some versions may say "Save as HTML"). If your document contained graphics, these are converted also and saved as separate files. The Web page looks like the one shown in Figure B–11. Notice that the headings in the Table of Contents are now hypertext links. Clicking on one will jump the reader to that specific section.

If you get into the habit of using style tags in your documents and inserting Tables of Contents (both of which are good practices for paper documents), converting those documents later to Web pages becomes quite simple.

FIGURE B–9 Style tag drop-down menu
Word is either a trademark or trademarks of Microsoft Corporation in the United States and/or other countries.

FIGURE B–10 Table of Contents dialog box
Word is either a trademark or trademarks of Microsoft Corporation in the United States and/or other countries.

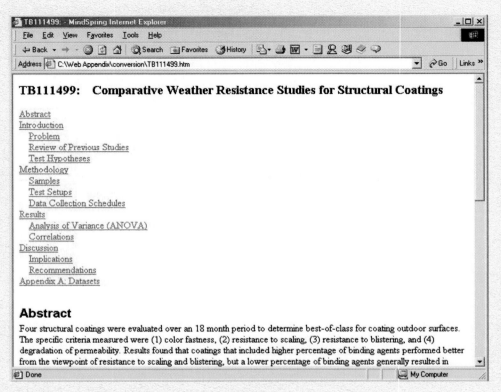

FIGURE B–11   Table of Contents as hypertext links
Internet Explorer is either a trademark or trademarks of Microsoft Corporation in the United States and/or other countries.

USABILITY AND ACCESSIBILITY

With Web-based documents, *usability* generally refers to how easily the reader can navigate and find the information he or she is looking for. It can also include other online tasks, such as book marking or printing the document. *Accessibility* refers to how easily the reader can get to the document, and we generally use the term to describe how well an online document accommodates users with special needs, such as visually impaired readers or readers with diminished motor skills.

Web designers generally cannot tell if their sites are intuitive because they have too much "insider" information. For example, the McDuff logo in Figure B–6 probably looks intuitively clickable as a link back to the home page to the designer. But this is because the designer was already familiar with that particular convention and also because the designer made it be a link. Most designers overestimate how much their users know about the Internet and overestimate the intuitiveness of their designs. The answer to this problem is *usability testing*—seeing the Web site through the eyes of real users.

Usability Testing

A usability test has a user perform a set of tasks in a controlled environment where the user's actions and comments can be observed. A rigorous test uses a cross-functional team as the observation team rather than relying on the observations of a single evaluator. For example, a team can be comprised of the following functions:

- Technical Writer
- Web Designer
- Graphics Specialist
- Subject Matter Expert
- Web Site Sponsor

The important thing is to have someone other than just you observe the users. Others will see things that you miss or will place importance on problems you might dismiss.

A usability test has the following three events:

1. Planning meeting
2. User sessions
3. Action meeting

Planning Meeting.   In the planning meeting the team agrees on the user profile(s) and the list of critical tasks they will have the users do during the test.

The user profile describes the computer expertise, the content knowledge, and the range of experience and management levels the team wants to see represented. If users will vary over a wide range, say from novice Internet users to very experienced users, or from general readers to very technical readers, make sure that at least one of each type is used. Generally, four to seven users give you enough information to make your Web site usable.

The team identifies tasks that are done early with the document (such as getting to the right section in the Web site itself) or which they feel are particularly important to the success of the Web site (such as finding specific information or being able to book mark it). Someone must then take these tasks and convert them into *scenarios,* i.e., high-level descriptions used to instruct the test participants during the test.

Writing Scenarios.   How successful a usability test will be depends on how well the test scenarios are designed. Good scenarios have the following characteristics:

- They place the task in a user context.
- They establish a goal to be accomplished.
- They have a clear ending point.

Figure B–12 shows a sample scenario that could be used to test the McDuff intranet.

A common mistake in writing scenarios is overdocumenting the procedures in the scenarios. Scenarios should put the tasks into a user context, but not give information about how to do the task. If possible, avoid using terminology in the scenario that appears on the user interface itself. For example, the scenario in Figure B–12 would find out if users know the term "rebar" since that is the term used on the

You are helping to set job specifications for a project in Nevada, and you want to know the Nevada building code for using concrete reinforcement bars. Someone has told you there might be a document on the company intranet that gives the codes for each state.

1. Go to the McDuff employee intranet site.
2. See if you can find the information about concrete reinforcement bars for Nevada.
3. Think out loud.
4. Call the facilitator at Ext. 323 when you have found the information.

FIGURE B–12 Sample scenario for McDuff intranet

FIGURE B–13 Typical usability lab

intranet site itself. It would also find out if employees even know how to get onto the intranet site. Lastly, it would find out how users would search for such a document and how easily they can find specific information within it.

User Sessions. Figure B–13 illustrates a typical commercial usability lab. The key features are (1) the user room where the participant works with the product and (2) the observation room connected with a one-way mirror through which the observation team watches the user working with the product. Cameras and microphones in the user room enable the team to record the sessions. You can certainly conduct usability tests in less formal settings, but try to keep the observers out of the user's direct line of sight.

During the user session, the facilitator gives the scenarios to the user one at a time and leaves the user alone in the room while he or she works with the Web site. Users are instructed to "think out loud" while they work through each scenario. While the team watches, the user tries to do the task, thinking aloud about how he or she is making sense of the user interface in the context of the task. One member of the team keeps a running computer log of the entire session.

At the end of each scenario, the facilitator debriefs the user to gather any additional information that might be useful. A typical user session lasts one and a half hours and covers seven to ten different scenarios. For the most part, the same scenarios are used for all the users.

After each user session, the team members meet and compare their individual observations. They rely on the computer log sheet one of the members kept during the session and their own notes. If required, they can refer back to the videotape for clarification. The facilitator records the team's observations on flip-chart sheets that are then posted on the walls. The team avoids having specific discussions about solutions at this time.

Action Meeting.　The action meeting is held one or two days after the completion of the user sessions so that the events are still fresh in the team's mind. During the action meeting, the team collaboratively devises solutions to the issues they observed. Each remedy is assigned to someone on the team and classified into one of the following priority levels:

1. Critical—must fix before release.
2. Important—all efforts will be made to fix before release, but release will not be held up.
3. Useful—change in a later release.

Accessibility

The Americans with Disabilities Act (ADA) requires equal access to computer systems for users who are physically disabled. Many disabled people use the Internet as an important channel to the world. Many of them use special programs or devices known as "user agents" that let them access the computer. For example, visually impaired users have special software that "reads" text out loud to them or converts text to Braille. As a novice designer, you may not know enough HTML to implement all of the following guidelines, but you should insist on them when professional developers are working with you.

General.　In general, design for device independent access. *Device independent access* means that the user may interact with the Web site with a preferred input (or output) device: mouse, keyboard, voice, head wand, or other. For example, if a control can only be activated with a mouse or other pointing device, someone who is using the page without sight, with voice input, keyboard, or some other nonpointing input device will not be able to use that control.

Provide text equivalents for graphics used as links to make it possible for users to interact with them without a pointing device (for example, the word "Home" under a graphical logo link to the home page).

Motor Skill Impairment.   In general, adhere closely to the general guideline for maintaining device independent access. In addition, avoid the need for precise mouse positioning.

- Provide large target areas for link triggers.
- Provide buffer spacing between active controls.
- Where active elements have to be close to each other, keep controls large and use rollovers or similar techniques to emphasize which element has the focus.
- Keep critical navigation links above the scroll line (so users can browse the site more easily).

Visual Impairment.   Depending on the severity of their impairment, visually impaired users rely either on special agents that "read" text and convert it to voiced narration or Braille, or they rely on monitor settings or supplementary software programs that increase the size of the presentation. The most important guideline is to avoid techniques or devices that "work against" these remedies.

- Include ALT tags with all graphics. ALT tags provide a text equivalent for users who do not view the graphic. For example, if a blind reader navigated over the graphic below, the audio reader would say, "model of a client server."

  <IMG SRC="CS Model.jpg" ALT="Model of a Client Server">

- Specify font sizes in percentages rather than absolute values:

  H1 { font-size: 200% }

- If using Cascading Style Sheets, avoid using the "Important" tag to override local browser settings that affect size and color of fonts.
- Test the Web site in a text-only browser and make sure that all critical information and navigation options are delivered.

The *Web Accessibility Initiative* has an excellent Web site with lots of useful information on how to implement Web accessibility techniques at *http://www.w3.org/WAI/*

SUMMARY

Effective use of the Internet for technical communication is more of a function of good architectural, information, and visual design than it is a function of fancy programming techniques. Remember that users are more easily confused and get lost more easily in Web documents than they do in print documents, and follow the design guidelines laid out in this appendix. Follow up on your design by testing it with real users to find out how they make sense of the design (which is often in ways totally unpredicted by the designer). Lastly, don't lock disabled readers out of your Web sites. Be sensitive to the special ways they access the Internet.

Appendix C

Handbook

This handbook includes entries on the basics of writing. Here you will find three main types of information:

1. **Grammar:** the rules by which we edit sentence elements. Examples include rules for the placement of punctuation, the agreement of subjects and verbs, and the placement of modifiers.
2. **Mechanics:** the rules by which we make final proofreading changes. Examples include the rules for abbreviations and the use of numbers. This handbook also includes a list of commonly misspelled words.
3. **Usage:** information on the correct use of particular words, especially pairs of words that are often confused. Examples include problem words like "affect/effect," "complement/compliment," and "who/whom."

Another editing concern, technical style, is the topic of chapter 15. In that chapter are guidelines for sentence structure, conciseness, accuracy of wording, active and passive voice, and nonsexist language. Together, chapter 15 and this handbook will help you turn unedited drafts into final revised documents.

This handbook is presented in alphabetized fashion for easy reference during the editing process. A table of contents follows. Grammar and mechanics entries are in caps; usage entries are in lowercase. Several exercises follow the entries.

A/An

These two words are different forms of the same article. *A* occurs before words that start with consonants or consonant sounds. EXAMPLES:

- a three-pronged plug
- a once-in-a-lifetime job ("once" begins with the consonant sound of *w*)
- a historic moment (many speakers and some writers mistakenly use "an" before "historic")

An occurs before words that begin with vowels or vowel sounds. EXAMPLES:

- an eager new employee
- an hour before closing

A lot/Alot

The correct form is the two-word phrase *a lot*. Though acceptable in informal discourse, "a lot" usually should be replaced by more formal diction in technical writing. EXAMPLE: They retrieved many [*not* a lot of] soil samples from the construction site.

Abbreviations

Technical writing uses many abbreviations. Without this shorthand form, you would end up writing much longer reports and proposals—without any additional content. Follow these seven basic rules in your use of abbreviations, paying special attention to the first three:

● *Rule 1: Do Not Use Abbreviations When Confusion May Result*

When you want to use a term just once or twice and you are not certain your readers will understand an abbreviation, write out the term rather than abbreviating it. EXAMPLE: "They were required to remove creosote from the site, according to the directive from the Environmental Protection Agency." Even though "EPA" is the accepted abbreviation for this government agency, you should write out the name in full *if* you are using the term only once to an audience that may not understand it.

● *Rule 2: Use Parentheses for Clarity*

When you use a term *more* than twice and are not certain that your readers will understand it, write out the term the first time it is used and place the abbreviation in parentheses. Then use the abbreviation in the rest of the document. In long reports or proposals, however, you may need to repeat the full term in key places. EXAMPLE: "According to the directive from the Environmental Protection Agency (EPA), they were required to remove the creosote from the construction site. Furthermore, the directive indicated that the builders could expect to be visited by EPA inspectors every other week."

● *Rule 3: Include a Glossary When There Are Many Abbreviations*

When your document contains many abbreviations that may not be understood by all readers, include a well-marked glossary at the beginning or end of the document. A glossary simply collects all the terms and abbreviations and places them in one location, for easy reference.

● *Rule 4: Use Abbreviations for Units of Measure*

Most technical documents use abbreviations for units of measure. Do not include a period unless the abbreviation could be confused with a word. EXAMPLES: mi, ft, oz, gal., in., and lb. Note that units-of-measurement abbreviations have the same form for both singular and plural amounts. EXAMPLES: 1/2 in., 1 in., 5 in.

● *Rule 5: Avoid Spacing and Periods*

Avoid internal spacing and internal periods in most abbreviations that contain all capital letters. EXAMPLES: ASTM, EPA, ASEE. Exceptions include professional titles and degrees such as P.E., B.S., and B.A.

● *Rule 6: Be Careful with Company Names*

Abbreviate a company or other organizational name only when you are sure that officials from the organization consider the abbreviation appropriate. IBM (for the company) and UCLA (for the university) are examples of commonly accepted organizational abbreviations. When in doubt, follow Rule 2—write the name in full the first time it is used, followed by the abbreviation in parentheses.

● *Rule 7: Use These Common Abbreviations*

These common abbreviations are appropriate for most writing in your technical or business career. They are placed into three main categories of measurements, locations, and titles:

Measurements: Use these abbreviations only when you place numbers before the measurement.

ac:	alternating current	K:	Kelvin
amp:	ampere	ke:	kinetic energy
bbl:	barrel	kg:	kilogram
Btu:	British thermal unit	km:	kilometer
bu:	bushel	kw:	kilowatt
C:	Celsius	kwh:	kilowatt-hour
cal:	calorie	l:	liter
cc:	cubic centimeter	lb:	pound
circ:	circumference	lin:	linear
cm:	centimeter	lm:	lumen
cos:	cosine	log.:	logarithm
cot:	cotangent	m:	meter
cps:	cycles per second	mm:	millimeter
cu ft:	cubic feet	min:	minute
db:	decibel	oz:	ounce
dc:	direct current	ppm:	parts per million
dm:	decimeter	psf:	pounds per square foot
doz *or* dz:	dozen	psi:	pounds per square inch
F:	Fahrenheit	pt:	pint
f:	farad	qt:	quart
fbm:	foot board measure	rev:	revolution
fig.:	figure	rpm:	revolutions per minute
fl oz:	fluid ounce	sec:	second
FM:	frequency modulation	sq:	square
fp:	foot pound	sq ft:	square foot (feet)
ft:	foot (feet)	T:	ton
g:	gram	tan.:	tangent
gal.:	gallon	v:	volt
gpm:	gallons per minute	va:	volt-ampere
hp:	horsepower	w:	watt
hr:	hour	wk:	week
Hz:	hertz	wl:	wavelength
in.:	inch	yd:	yard
j:	joule	yr:	year

Locations: Use these common abbreviations for addresses (on envelopes and letters, for example), but write out the words in full in other contexts.

AL:	Alabama		NM:	New Mexico
AK:	Alaska		NY:	New York
AS:	American Samoa		NC:	North Carolina
AZ:	Arizona		ND:	North Dakota
AR:	Arkansas		OH:	Ohio
CA:	California		OK:	Oklahoma
CZ:	Canal Zone		OR:	Oregon
CO:	Colorado		PA:	Pennsylvania
CT:	Connecticut		PR:	Puerto Rico
DE:	Delaware		RI:	Rhode Island
DC:	District of Columbia		SC:	South Carolina
FL:	Florida		SD:	South Dakota
GA:	Georgia		TN:	Tennessee
GU:	Guam		TX:	Texas
HI:	Hawaii		UT:	Utah
ID:	Idaho		VT:	Vermont
IL:	Illinois		VI:	Virgin Islands
IN:	Indiana		VA:	Virginia
IA:	Iowa		WA:	Washington
KS:	Kansas		WV:	West Virginia
KY:	Kentucky		WI:	Wisconsin
LA:	Louisiana		WY:	Wyoming
ME:	Maine		Alta.:	Alberta
MD:	Maryland		B.C.:	British Columbia
MA:	Massachusetts		Man.:	Manitoba
MI:	Michigan		N.B.:	New Brunswick
MN:	Minnesota		Nfld.:	Newfoundland
MS:	Mississippi		N.W.T.:	Northwest Territories
MO:	Missouri		N.S.:	Nova Scotia
MT:	Montana		Ont.:	Ontario
NE:	Nebraska		P.E.I.:	Prince Edward Island
NV:	Nevada		P.Q.:	Quebec
NH:	New Hampshire		Sask.:	Saskatchewan
NJ:	New Jersey		Yuk.:	Yukon

Titles: Some of these abbreviations go before the name (such as Dr., Ms., and Messrs.), while others go after the name (such as college degrees, Jr., and Sr.).

Atty.:	Attorney
B.A.:	Bachelor of Arts
B.S.:	Bachelor of Science
D.D.:	Doctor of Divinity
Dr.:	Doctor (used mainly with medical and dental degrees but also with other doctorates)
Drs.:	plural of Dr.
D.V.M.:	Doctor of Veterinary Medicine
Hon.:	Honorable
Jr.:	Junior
LL.D.:	Doctor of Laws
M.A.:	Master of Arts
M.S.:	Master of Science
M.D.:	Doctor of Medicine
Messrs.:	Plural of Mr.
Mr.:	Mister
Mrs.:	used to designate married, widowed, or divorced women
Ms.:	used increasingly for all women, especially when one is uncertain about a woman's marital status
Ph.D.:	Doctor of Philosophy
Sr.:	Senior

Accept/Except

These words have different meanings and often are different parts of speech. *Accept* is a verb that means "to receive." *Except* is a preposition or verb and means "to make an exception or special case of." EXAMPLES:

- I *accepted* the service award from my office manager.
- Everyone *except* Jonah attended the marine science lecture.
- The company president *excepted* me from the meeting because I had an important sales call to make the same day.

Advice/Advise/Inform

Advice is a noun that means "suggestions or recommendation." *Advise* is a verb that means "to suggest or recommend." Do not use the verb *advise* as a substitute for *inform,* which means simply "to provide information." EXAMPLES:

- The consultant gave us *advice* on starting a new retirement plan for our employees.

- She *advised* us that a 401(k) plan would be useful for all our employees.
- She *informed* [not *advised*] her clients that they would receive her final report by March 15.

Affect/Effect

These two words generate untold grief among many writers. The key to using them correctly is remembering two simple sentences: (1) *affect* with an "a" is a verb meaning "to influence"; (2) *effect* with an "e" is a noun meaning "result." There are some exceptions, however, such as the following: "Effect" can be a verb that means "to bring about," as in "He effected considerable change when he became a manager." EXAMPLES:

- His progressive leadership greatly *affected* the company's future.
- One *effect* of securing the large government contract was the hiring of several more accountants.
- The president's belief in the future of microcomputers *effected* change in the company's approach to office management. (For a less wordy alternative, substitute "changed" for "effected change in.")

Agree to/Agree with

In correct usage, *agree to* means that you have *consented to* an arrangement, an offer, a proposal, etc. *Agree with* is less constraining and only suggests that you are *in harmony with* a certain statement, idea, person, etc. EXAMPLES:

- Representatives from McDuff *agreed to* alter the contract to reflect the new scope of work.
- We *agree with* you that more study may be needed before the nuclear power plant is built.

All Right/Alright

All right is an acceptable spelling; *alright* is not. *All right* is an adjective that means "acceptable," an exclamation that means "outstanding," or a phrase that means "correct." EXAMPLES:

- Sharon suggested that the advertising copy was *all right* for now but that she would want changes next month.
- Upon seeing his article in print, Zach exclaimed *"All right!"*
- The five classmates were *all right* in their response to the trick questions on the quiz.

All Together/Altogether

All together is used when items or people are being considered in a group or are working in concert. *Altogether* is a synonym for "utterly" or "completely." EXAMPLES:

- The three firms were *all together* in their support of the agency's plan.
- There were *altogether* too many pedestrians walking near the dangerous intersection.

Allusion/Illusion/Delusion/Elusion

These similar sounding words have distinct meanings. Here is a summary of the differences:

1. **Allusion:** a noun meaning "reference," as in your making an allusion to your vacation in a speech. The related verb is *allude*.
2. **Illusion:** a noun meaning "misunderstanding or false perception." It can be physical (as in seeing a mirage) or mental (as in having the false impression that your hair is not thinning when it is).
3. **Delusion:** a noun meaning a belief based on self-deception. Unlike *illusion*, the word conveys a much stronger sense that someone is out of touch with reality, as in having "delusions of grandeur." The related verb is *delude*.
4. **Elusion:** a noun meaning "the act of escaping or avoiding." The more common form is the verb, *elude*, meaning "to escape or avoid."

EXAMPLES:

- His report included an *allusion* to the upcoming visit by the government agency in charge of accreditation.
- She harbored an *illusion* that she was certain to receive the promotion. In fact, her supervisor preferred another department member with more experience.
- He had *delusions* that he soon would become company president, even though he started just last week in the mailroom.
- The main point of the report *eluded* him because there was no executive summary.

Already/All Ready

All ready is a phrase that means "everyone is prepared," whereas *already* is an adverb that means something is finished or completed. EXAMPLES:

- They were *all ready* for the presentation to the client.
- George had *already* arrived at the office before the rest of his proposal team members had even left their homes.

Alternately/Alternatively

Because many readers are aware of the distinction between these two words, any misuse can cause embarrassment or even misunderstanding. Follow these guidelines for correct use.

Alternately. As a derivative of "alternate," *alternately* is best reserved for events or actions that occur "in turns." EXAMPLE: While digging the trench, he used a backhoe and a hand shovel *alternately* throughout the day.

Alternatively. A derivative of "alternative," *alternatively* should be used in contexts where two or more choices are being considered. EXAMPLE: We suggest that you use deep foundations at the site. *Alternatively,* you could consider spread footings that were carefully installed.

Amount/Number

Amount is used in reference to items that *cannot* be counted, whereas *number* is used to indicate items that *can* be counted. EXAMPLES:

- In the last year, we have greatly increased the *amount* of computer paper ordered for the Boston office.
- The last year has seen a huge increase in the *number* [*not* amount] of boxes of computer paper ordered for the Boston office.

And/Or

This awkward expression probably has its origins in legal writing. It means that there are three separate options to be considered: the item before *and/or*, the item after *and/or*, or both items.

Avoid *and/or* because readers may find it confusing, visually awkward, or both. Instead, replace it with the structure used in the previous sentence. That is, write "A, B, or both," *not* "A and/or B." EXAMPLE:

> The management trainee was permitted to select two seminars from the areas of computer hardware, communication skills, or both [not *computer hardware and/or communication skills*].

Anticipate/Expect

These two words are *not* synonyms. In fact, their meanings are distinctly different. *Anticipate* is used when you mean to suggest or state that steps have been taken beforehand to prepare for a situation. *Expect* only means you consider something likely to occur. EXAMPLES:

- *Anticipating* that the contract will be successfully negotiated, Jones Engineering is hiring three new hydrologists.
- We *expect* [*not* anticipate] that you will encounter semicohesive and cohesive soils in your excavations at the Park Avenue site.

Apt/Liable/Likely

Maintain the distinctions in these three similar words.

1. *Apt* is an adjective that means "appropriate," "suitable," or "has an aptitude for."
2. *Liable* is an adjective that means "legally obligated" or "subject to."
3. *Likely* is either an adjective that means "probable" or "promising" or an adverb that means "probably." As an adverb, it should be preceded by a qualifier such as *quite*.

EXAMPLES:

- The successful advertising campaign showed that she could select an *apt* phrase for selling products.

- Jonathan is *apt* at running good meetings. He always hands out an agenda and always ends on time.
- The contract makes clear who is *liable* for any on-site damage.
- Completing the warehouse without an inspection will make the contractor *liable* to lawsuits from the owner.
- A *likely* result of the investigation will be a change in the law. [*likely* as an adjective]
- The investigation will quite *likely* result in a change in the law. [*likely* as an adverb]

Assure/Ensure/Insure

Assure is a verb that can mean "to promise." It is used in reference to people, as in "We want to assure you that our crews will strive to complete the project on time." In fact, *assure* and its derivatives (like "assurance") should be used with care in technical contexts, for these words can be viewed as a guarantee.

The synonyms *ensure* and *insure* are verbs meaning "to make certain." Like *assure*, they imply a level of certainty that is not always appropriate in engineering or the sciences. When their use is deemed appropriate, the preferred word is *ensure;* reserve *insure* for sentences in which the context is insurance. EXAMPLES:

- Be *assured* that our representatives will be on-site to answer questions that the subcontractor may have.
- To *ensure* that the project stays within schedule, we are building in 10 extra days for bad weather. (An alternative: So that the project stays within schedule, we are building in 10 extra days for bad weather.)

Augment/Supplement

Augment is a verb that means to increase in size, weight, number, or importance. *Supplement* is either (1) a verb that means "to add to" something to make it complete or to make up for a deficiency or (2) a noun that means "the thing that has been added." EXAMPLES:

- The power company supervisor decided to *augment* the line crews in five counties.
- He *supplemented* the audit report by adding the three accounting statements.
- The three accounting *supplements* helped support the conclusions of the audit report.

Awhile/A While

Though similar in meaning, this pair is used differently. *Awhile* means "for a short time." Because "for" is already a part of its definition, it cannot be preceded by the preposition "for." The noun *while,* however, can be preceded by the two words "for a," giving it essentially the same meaning as *awhile*. EXAMPLES:

- Kirk waited *awhile* before trying to restart the generator.
- Kirk waited *for a while* before trying to restart the generator.

Balance/Remainder/Rest

Balance should be used as a synonym for *remainder* only in the context of financial affairs. *Remainder* and *rest* are synonyms to be used in other nonfinancial contexts. EXAMPLES:

- The account had a *balance* of $500, which was enough to avoid a service charge.
- The *remainder* [or *rest*, but not *balance*] of the day will be spent on training in oral presentations for proposals.
- During the *rest* [not *balance*] of the session, we learned about the new office equipment.

Because/Since

Maintain the distinction between these two words. *Because* establishes a cause-effect relationship, whereas *since* is associated with time. EXAMPLES:

- *Because* he left at 3 p.m., he was able to avoid rush hour.
- *Since* last week, her manufacturing team completed 3,000 units.

Between/Among

The distinction between these two words has become somewhat blurred. However, many readers still prefer to see *between* used with reference to only two items, reserving *among* for three or more items. EXAMPLES:

- The agreement was just *between* my supervisor and me. No one else in the group knew about it.
- The proposal was circulated *among* all members of the writing team.
- *Among* Sallie, Todd, and Fran, there was little agreement about the long-term benefits of the project.

Bi-/Semi-/Biannual/Biennial

The prefixes *bi* and *semi* can cause confusion. Generally, *bi* means "every two years, months, weeks, etc.," whereas *semi* means "twice a year, month, week, etc." Yet many readers get confused by the difference, especially when they are confronted with a notable exception such as *biannual* (which means twice a year) and *biennial* (which means every two years).

Your goal, as always, is clarity for the reader. Therefore, it is best to write out meanings in clear prose, rather than relying on prefixes that may not be understood. EXAMPLES:

- We get paid twice a month [preferable to *semimonthly* or *biweekly*].
- The part-time editor submits articles every other month [preferable to *bimonthly*].
- We hold a company social gathering twice a year [preferable to *biannually* or *semi-annually*].
- The auditor inspects our safety files every two years [preferable to *biennially*].

Capital/Capitol

Capital is a noun whose main meanings are (1) a city or town that is a government center, (2) wealth or resources, or (3) net worth of a business or the investment that has been made in the business by owners. *Capital* can also be an adjective meaning (1) "excellent," (2) "primary," or (3) "related to the death penalty." Finally, *capital* can be a noun or an adjective referring to uppercase letters.

Capitol is a noun or an adjective that refers to a building where a legislature meets. With a capital letter, it refers exclusively to the building in Washington, D.C., where the U.S. Congress meets. EXAMPLES:

- The *capital* of Pickens County is Jasper, Georgia.
- Our family *capital* was reduced by the tornado and hurricane.
- She had invested significant *capital* in the carpet factory.
- Their proposal contained some *capital* ideas that would open new opportunities for our firm.
- In some countries, armed robbery is a *capital* offense.
- The students visited the *capitol* building in Atlanta. Next year they will visit the *Capitol* in Washington, D.C., where they will meet several members of Congress.

Capitalization

As a rule, you should capitalize *specific* names of people, places, and things—sometimes called "proper nouns." For example, capitalize specific streets, towns, trademarks, geological eras, planets, groups of stars, days of the week, months of the year, names of organizations, holidays, and colleges. However, remember that excessive capitalization—as in titles of positions in a company—is inappropriate in technical writing and can appear somewhat pompous.

The following rules cover some frequent uses of capitals:

1. Major words in titles of books and articles. Only capitalize prepositions and articles when they appear as the first word in titles. EXAMPLES:

   - *For Whom the Bell Tolls*
   - *In Search of Excellence*
   - *The Power of Positive Thinking*

2. Names of places and geographical locations. EXAMPLES:

   - Washington Monument
   - Cleveland Stadium
   - Dallas, Texas
   - Cobb County

3. Names of aircraft and ships. EXAMPLES:

   - *Air Force One*
   - SS *Arizona*
   - *Nina, Pinta,* and *Santa Maria*

4. Names of specific departments and offices within an organization. EXAMPLES:

- Humanities Department
- Personnel Department
- International Division

5. Political, corporate, and other titles that come before names. EXAMPLES:

- Chancellor Hairston
- Councilwoman Jones
- Professor Gainesberg
- Congressman Buffett

Note, however, that general practice does not call for capitalizing most titles when they are used by themselves or when they follow a person's name. EXAMPLES:

- Jane Cannon, a professor in the Business Department.
- Rob McDuff, president of McDuff, Inc.
- Chris Presley, secretary of the Oil Rig Division.

Center On/Revolve Around

The key to using these phrases correctly is to think about their literal meaning. For example, you center *on* (not around) a goal, just as you would center on a target with a gun or bow and arrow. Likewise, your hobbies revolve *around* your early interest in water sports, just as the planets revolve around the sun in our solar system. EXAMPLES:

- All her selling points in the proposal *centered on* the need for greater productivity in the factory.
- At the latest annual meeting, some stockholders argued that most of the company's recent projects *revolved around* the CEO's interest in attracting attention from the media.

Cite/Site/Sight

1. *Cite* is a verb meaning "to quote as an example, authority, or proof." It can also mean "to commend" or "to bring before a court of law" (as in receiving a traffic ticket).
2. *Site* usually is a noun that means "a particular location." It can also be a verb that means "to place at a location," as with a new school being sited by the town square, but this usage is not preferred. Instead use a more conventional verb, such as *built*.
3. *Sight* is a noun meaning "the act of seeing" or "something that is seen." Or it can be a verb meaning "to see or observe."

EXAMPLES:

- We *cited* a famous geologist in our report on the earthquake.
- Rene was *cited* during the ceremony for her exemplary service to the city of Roswell.

- The officer will *cite* the party-goers for disturbing the peace.
- Although five possible dorm *sites* were considered last year, the college administrators decided to build [preferred over *site*] the dorm at a different location.
- The *sight* of the flock of whooping cranes excited the visitors.
- Yesterday we *sighted* five whooping cranes at the marsh.

Complement/Compliment

Both words can be nouns and verbs, and both have adjective forms (complementary, complimentary).

Complement.   This word is used as a noun to mean "that which has made something whole or complete," as a verb to mean "to make whole, to make complete," or as an adjective. You may find it easier to remember the word by recalling its mathematical definition: Two complementary angles must always equal 90 degrees. EXAMPLES:

- As noun: The *complement* of five technicians brought our crew strength up to 100 percent.
- As verb: The firm in Canada served to *complement* ours in that together we won a joint contract.
- As adjective: Seeing that project manager and her secretary work so well together made clear their *complementary* relationship in getting the office work done.

Compliment.   This word is used as a noun to mean "an act of praise, flattery, or admiration," as a verb to mean "to praise, to flatter," or as an adjective to mean "related to praise or flattery, or without charge." EXAMPLES:

- As noun: He appreciated the verbal *compliments,* but he also hoped they would result in a substantial raise.
- As verb: Howard *complimented* the crew for finishing the job on time and within budget.
- As adjective: We were fortunate to receive several *complimentary* copies of the new software from the publisher.

Compose/Comprise

These are both acceptable words, with an inverse relationship to each other. *Compose* means "to make up or be included in," whereas *comprise* means "to include or consist of." The easiest way to remember this relationship is to memorize one sentence: "The parts compose the whole, but the whole comprises the parts." One more point to remember: The common phrase "is comprised of" is a substandard, unacceptable replacement for "comprise" or "is composed of." Careful writers do not use it. EXAMPLES:

- Seven quite discrete layers *compose* the soils that were uncovered at the site.
- The borings revealed a stratigraphy that *comprises* [*not* is comprised of] seven quite discrete layers.

Consul/Council/Counsel

These three words can be distinguished by meaning and, in part, by their use within a sentence.

Consul: A noun meaning an official of a country who is sent to represent that country's interests in a foreign land.

Council: A noun meaning an official group or committee.

Counsel: A noun meaning an adviser or advice given OR a verb meaning to produce advice.

EXAMPLES:

- (consul) The Brazilian *consul* met with consular officials from three other countries.
- (council) The Human Resources *Council* of our company recommended a new retirement plan to the company president.
- (counsel—as noun) After the tragedy, they received legal *counsel* from their family attorney and spiritual *counsel* from their minister.
- (counsel—as verb) As a communications specialist, Roberta helps to *counsel* employees who are involved in various types of disputes.

Continuous/Continual

The technical accuracy of some reports may depend on your understanding of the difference between these two words. *Continuous* and *continuously* should be used in reference to uninterrupted, unceasing activities. However, "continual" and "continually" should be used with activities that are intermittent, or repeated at intervals. If you think your reader may not understand the difference, you should either (1) use synonyms that will be clearer (such as "uninterrupted" for *continuous,* and "intermittent" for *continual*) or (2) define each word at the point you first use it in the document. EXAMPLES:

- We *continually* checked the water pressure for three hours before the equipment arrived, while also using the time to set up the next day's tests.
- Because it rained *continuously* from 10:00 A.M. until noon, we were unable to move our equipment onto the utility easement.

Criterion/Criteria

Coming from the Latin language, *criterion* and *criteria* are the singular and plural forms of a word that means "rationale or reasons for selecting a person, place, thing, or idea." A common error is to use *criteria* as both a singular and plural form, but such misuse disregards a distinction recognized by many readers. Maintain the distinction in your writing. EXAMPLES:

- Among all the qualifications we established for the new position, the most important *criterion* for success is good communication skills.
- She had to satisfy many *criteria* before being accepted into the honorary society of her profession.

Data/Datum

Coming as it does from the Latin, the word *data* is the plural form of *datum*. Although many writers now accept *data* as singular or plural, traditionalists in the technical and scientific community still consider *data* exclusively a plural form. Therefore, you should maintain the plural usage. EXAMPLES:

- These *data* show that there is a strong case for building the dam at the other location.
- This particular *datum* shows that we need to reconsider recommendations put forth in the original report.

If you consider the traditional singular form of *datum* to be awkward, use substitutes such as "This item in the data shows" or "One of the data shows that." Singular subjects like "one" or "item" allow you to keep your original meaning without using the word *datum*.

Definite/Definitive

Though similar in meaning, these words have slightly different contexts. *Definite* refers to that which is precise, explicit, or final. *Definitive* has the more restrictive meaning of "authoritative" or "final." EXAMPLES:

- It is now *definite* that he will be assigned to the London office for six months.
- He received the *definitive* study on the effect of the oil spill on the marine ecology.

Discrete/Discreet/Discretion

The adjective *discrete* suggests something that is separate, or something that is made up of many separate parts. The adjective *discreet* is associated with actions that require caution, modesty, or reserve. The noun *discretion* refers to the quality of being "discreet" or the freedom a person has to act on her or his own. EXAMPLES:

- The orientation program at McDuff includes a writing seminar, which is a *discrete* training unit offered for one full day.
- The orientation program at McDuff includes five *discrete* units.
- As a counselor in McDuff's Human Resources Office, Sharon was *discreet* in her handling of personal information about employees.
- Every employee in the Human Resources Office was instructed to show *discretion* in handling personal information about employees.
- By starting a flextime program, McDuff, Inc., will give employees a good deal of *discretion* in selecting the time to start and end their workday.

Disinterested/Uninterested

In contemporary business use, these words have quite different meanings. Because errors can cause confusion for the reader, make sure not to use the words as synonyms. *Disinterested* means "without prejudice or bias," whereas *uninterested* means "showing no interest." EXAMPLES:

- The agency sought a *disinterested* observer who had no stake in the outcome of the trial.
- They spent several days talking to officials from Iceland, but they still remain *uninterested* in performing work in that country.

Due to/Because of

Besides irritating those who expect proper English, mixing these two phrases can also cause confusion. *Due to* is an adjective phrase meaning "attributable to" and almost always follows a "to be" verb (such as "is," "was," or "were"). It should not be used in place of prepositional phrases such as "because of," "owing to," or "as a result of." EXAMPLES:

- The cracked walls were *due to* the lack of proper foundation fill being used during construction.
- We won the contract *because of* [not *due to*] our thorough understanding of the client's needs.

Each Other/One Another

Each other occurs in contexts that include only two persons, whereas *one another* occurs in contexts that include three or more persons. EXAMPLES:

- Shana and Katie worked closely with *each other* during the project.
- All six members of the team converted with *one another* regularly through email.

e.g./i.e.

The abbreviation *e.g.* means "for example," whereas *i.e.* means "that is." These two Latin abbreviations are often confused, a fact that should give you pause before using them. Many writers prefer to write them out, rather than risk confusion on the part of the reader. EXAMPLES:

- During the trip, he visited 12 cities where McDuff is considering opening offices—*e.g.,* [or, preferably, *for example*] Kansas City, New Orleans, and Seattle.
- A spot along the Zayante Fault was the earthquake's epicenter—*i.e.,* [or, preferably, *that is*] the focal point for seismic activity.

English as a Second Language (ESL)

Technical writing challenges native English speakers and non-native speakers alike. The purpose of this section is to present a basic description of three grammatical forms: articles, prepositions, and verbs. These forms may require more intense consideration from international students when they complete technical writing assignments. Each issue is described using the ease-of-operation section from the memo in Model 3–1 on page 96. The passage, descriptions, and charts work together to show how these grammar issues function collectively to create meaning.

Ease of Operation—Article Usage

The AIM 500 is so easy to operate that **a** novice can learn to transmit **a** document to another location in about two minutes. Here's **the** basic procedure:

1. Press **the** button marked TEL on **the** face of **the** fax machine. You then hear **a** dial tone.
2. Press **the** telephone number of **the** person receiving **the** fax on **the** number pad on **the** face of **the** machine.
3. Lay **the** document face down on **the** tray at **the** back of **the** machine.

At this point, just wait for **the** document to be transmitted—about 18 seconds per page to transmit. **The** fax machine will even signal **the** user with **a** beep and **a** message on its LCD display when **the** document has been transmitted. Other more advanced operations are equally simple to use and require little training. Provided with **the** machine are two different charts that illustrate **the** machine's main functions.

The size of **the** AIM 500 makes it easy to set up almost anywhere in **an** office. **The** dimensions are 13 inches in width, 15 inches in length, and 9.5 inches in height. **The** narrow width, in particular, allows **the** machine to fit on most desks, file cabinets, or shelves.

Articles. Articles are one of the most difficult forms of English grammar for non-native English speakers, mainly because some language systems do not use them. Thus speakers of particular languages may have to work hard to incorporate the English article system into their language proficiency.

The English articles include "a," "an," and "the."

- "A" and "an" express indefinite meaning when they refer to nouns or pronouns or pronouns that are not specific. The writer believes the reader does not know the noun or pronoun.
- "The" expresses definite meaning when it refers to a specific noun or pronoun. The writer believes the reader knows the specific noun or pronoun.

ESL writers choose the correct article only when they (1) know the context or meaning, (2) determine whether they share information about the noun with the reader, and (3) consider the type of noun following the article.

The ease-of-operation passage includes 31 articles that represent two types—definite and indefinite. When a writer and a reader share knowledge of a noun, the definite article should be used. On 25 occasions the articles in the passage suggest the writer and reader share some knowledge of a count noun. Count nouns are nouns that can be counted (pen, cloud, memo). Examples of non-count nouns are sugar, air, and beef.

For example, the memo writer and the memo recipient share knowledge of the particular model fax machine—the AIM 500. Thus, "the" is definite when it refers to "the fax machine" in the memo. Notice, however, that "document" becomes definite only after the second time it is mentioned ("Lay the document facedown. . . ."). In the first reference to "document," "a" document refers to a document about which the writer and reader share no knowledge. The memo writer cannot know which document the reader will fax. Only in the second reference do the writer and reader know the document to be the one the reader will fax.

The indefinite article "a" occurs five times, while "an" occurs once. Each occurrence signals a singular count noun. The reader and the writer share no knowledge of the nouns that follow the "a" or "an," so an indefinite article is appropriate. "A" precedes nouns beginning with consonant sounds. "An" precedes nouns beginning

Articles from "Ease of Operation" Excerpt			
Article	Noun	Type	Comment
The	ATM 500	definite	first mention—shared knowledge
a	novice	indefinite	first mention—no shared knowledge
a	document	indefinite	first mention—no shared knowledge
the	basic procedure	definite	
the	button	definite	
the	face	definite	
the	fax machine	definite	first mention without proper name, with reader/writer shared knowledge
a	dial tone	indefinite	first mention—no shared knowledge
the	telephone number	definite	
the	person	definite	
the	fax	definite	
the	number pad	definite	
the	face	definite	
the	machine	definite	
the	document	definite	
the	tray	definite	
the	back	definite	
the	machine	definite	
the	document	definite	second mention
the	fax machine	definite	
the	user	definite	
a	beep	indefinite	first mention—no shared knowledge
a	message	indefinite	first mention—no shared knowledge
the	document	definite	
the	machine	definite	
the	machine's main functions	definite	
The	size	definite	
the	AIM 500	definite	
an	office	indefinite	first mention—preceding vowel sound—no shared knowledge
The	dimensions	definite	
The	narrow width	definite	
the	machine	definite	

with vowel sounds. Indefinite articles seldom precede non-count nouns unless a non-count functions as a modifier (a beef shortage).

Definite and indefinite articles are used more frequently than other articles; however, other articles do exist. The "generic" article refers to classes or groups of people, objects, and ideas. If the fax machine is thought of in a general sense, the meaning changes. For example, "the fax machine increased office productivity by 33%." "The" now has a generic meaning representing fax machines in general. The same generic meaning can apply to the plural noun, but such generic use requires no article. "Fax machines increased office productivity by 33%." "The" in this instance is a generic article.

Ease of Operation—Verb Usage

The AIM 500 **is** so easy to operate that a novice **can learn** to transmit a document to another location in about two minutes. Here**'s** the basic procedure:

1. **Press** the button marked TEL on the face of the fax machine. You then **hear** a dial tone.
2. **Press** the telephone number of the person receiving the fax on the number pad on the face of the machine.
3. **Lay** the document face down on the tray at the back of the machine.

At this point, just **wait** for the document to be transmitted—about 18 seconds per page to transmit. The fax machine **will** even **signal** the user with a beep and a message on its LCD display when the document **has been transmitted.** Other more advanced operations **are** equally simple to use and **require** little training. **Provided** with the machine **are** two different charts that **illustrate** the machine's main functions.

The size of the AIM 500 **makes** it easy to set up almost anywhere in an office. The dimensions **are** 13 inches in width, 15 inches in length, and 9.5 inches in height. The narrow width, in particular, **allows** the machine to fit on most desks, file cabinets, or shelves.

Verbs. Verbs express time in three ways—simple present, simple past, and future. "Wait," "waited," and "will wait" and "lay" ("to put"), "laid", and "will lay" are examples of simple present, simple past, and future tense verbs. Unfortunately, the English verb system is more complicated than that. Verbs express more than time.

Verbs in the English language system appear as either regular or irregular forms. Regular verbs follow a predictable pattern.

- The form of the simple present tense verbs ("walk") changes to the simple past tense with the addition of "ed" ("walked").
- The past tense "walked" changes to the perfect tense with the addition of an auxiliary (helping) word to the simple past tense form. For example, "I have walked" is present perfect, and "I had walked" is past perfect.

Irregular verbs do not follow a predictable pattern.

- Most importantly, the past tense is not created by adding "ed." The simple present tense of "lay" ("to put") changes completely in the simple past ("laid").
- Like regular verbs, the past tense ("laid") changes to the perfect tense with the addition of an auxiliary verb.

Regular verbs

present	past	future
learn	learned	will learn
wait	waited	will wait
press	pressed	will press
signal	signaled	will signal
require	required	will require
provide	provided	will provide
illustrate	illustrated	will illustrate
allow	allowed	will allow
present perfect	**past perfect**	**future perfect**
have learned	had learned	will have learned
have waited	had waited	will have waited
have pressed	had pressed	will have pressed
have signaled	had signaled	will have signaled
have required	had required	will have required
have provided	had provided	will have provided
have illustrated	had illustrated	will have illustrated
have allowed	had allowed	will have allowed

Irregular verbs

present	past	future
is	was	will be
are	were	will be
hear	heard	will hear
do	did	will do
get	got	will get
see	saw	will see
write	wrote	will write
speak	spoke	will speak
present perfect	**past perfect**	**future perfect**
have been (he, she, it)	had been	will have been
have been (they)	had been	will have been
have heard	had heard	will have heard
have done	had done	will have done
have gotten	had gotten	will have gotten
have seen	had seen	will have seen
have written	had written	will have written
have spoken	had spoken	will have spoken

"I have laid the document facedown on the tray" is present perfect, and "I had laid the document face down on the tray" is past perfect.

Let's examine four specific verb forms in the "Ease of Operation" passage.

1. "Is" represents a being or linking verb in the passage. Being verbs suggest an aspect of an experience or being (existence); for example, "He is still here," and "The fax is broken." Linking verbs connect a subject to a complement (completer); for example, "The fax machine is inexpensive."

2. "Can learn" is the present tense verb "learn" preceded by a modal. Modals assist verbs to convey meaning. "Can" suggests ability or possibility. Other modals and their meanings appear below.

will	would	could	shall	should	might	must
scientific fact possibility determination	hypothetical	hypothetical	formal will	expectation obligation	possibility	necessity

3. "Here's" shows a linking verb ("is") connected to its complement ("here"). The sentence in its usual order—subject first followed by the verb—would appear as, "The basic procedure is here." Article—adjective—noun—linking verb—complement.

4. "Press," "Lay," and "wait (for)" share at least four common traits: present tense, singular number, action to transitive, and understood subject of "you." Though "you" does not appear in the text, the procedure clearly instructs the person operating the fax machine—"you." Action or transitive verbs express movement, activity, and momentum, and may take objects. Objects answer the questions who? what? to whom? or for whom? in relation to transitive verbs. For example, "Press the button," "Hear a dial tone," "Press the telephone number," "Lay the document face down." Press what? Hear what? Lay what?

Ease of Operation—Preposition Usage

The AIM 500 is so easy to operate that a novice can learn to transmit a document **to** another location **in** about two minutes. Here's the basic procedure:

1. Press the button marked TEL **on** the face **of** the fax machine. You then hear a dial tone.
2. Press the telephone number **of** the person receiving the fax **on** the number pad **on** the face **of** the machine.
3. Lay the document face down **on** the tray **at** the back **of** the machine.

At this point, just wait **for** the document to be transmitted—**about** 18 seconds **per** page to transmit. The fax machine will even signal the user **with** a beep and a message **on** its LCD display when the document has been transmitted. Other more advanced operations are equally simple to use and require little training. Provided **with** the machine are two different charts that illustrate the machine's main functions.

The size **of** the AIM 500 makes it easy to set up almost anywhere **in** an office. The dimensions are 13 inches **in** width, 15 inches **in** length, and 9.5 inches **in** height. The narrow width, **in** particular, allows the machine to fit **on** most desks, file cabinets, or shelves.

Verbs from "Ease of Operation" Excerpt			
Verb	**Tense**	**Number**	**Other details**
is	present	singular	linking/being (is, was, been)
can learn	present	singular	"can" is a modal auxiliary implying "possibility"
Here's (is)	present	singular	linking/being
Press	present	singular	understood "you" as subject
hear	present	singular	action/transitive
Press	present	singular	understood "you" as subject
Lay	present	singular	irregular (lay, laid, laid) singular—understood "you" as subject
wait	present	singular	understood "you" as subject
will signal	future	singular	action to happen or condition to experience
has been transmitted	present perfect	singular	passive voice—action that began in the past and continues to the present
are	present	plural	linking/being
require	present	plural	action/transitive
Provided are	present perfect	plural	passive voice—action that began in the past and continues to the present
illustrate	present	plural	action/transitive
makes	present	singular	action/transitive
are	present	plural	linking/being
allows	present	singular	action/transitive

Prepositions.   Prepositions are words that become a part of a phrase composed of the preposition, a noun or pronoun, and any modifiers. Notice the relationships expressed within the prepositional phrases and the ways they affect meaning in the sentences. In the "Ease of Operation" passage, about half the prepositional phrases function as adverbs noting place or time. The other half function as adjectives.

Place	**or**	**Location**	**Time**
at		on	before
in		above	after
below		around	since
beneath		out	during
over		underneath	
within		under	
outside		near	
into		inside	

One important exception is a preposition that connects to a verb to make a "prepositional" verb, "wait for." Another interesting quality of prepositions is that

Prepositions from "Ease of Operation" Excerpt		
Preposition	**Noun Phrase**	**Comment**
to	another location	direction toward
in	(about) two minutes	approximation of time
on	the face	position
of	the fax machine	originating at or from
of	the person	associated with
on	the number pad	position
on	the face	position
of	the machine	originating at
on	the tray	position
at	the back	position of
of	the machine	originating at
At	this point	on or near the time
for	the document	indication of object of desire
about	18 seconds	adverb = approximation
per	page	for every
with	a beep and a message	accompanying
on	its LCD display	position
with	the machine	accompanying
of	the AIM 500	originating at or from
in	an office	within the area
in	width	with reference to
in	length	with reference to
in	height	with reference to
in	particular	with reference to
on	most desks, file cabinets, or shelves	position

sometimes more than one can be used to express similar meaning. In the "Ease of Operation" passage, for example, both "on" the tray and "at" the back indicate position. Another way to state the same information is "on" the tray "on" the back.

Farther/Further

Though similar in meaning, these two words are used differently. *Farther* refers to actual physical distance, whereas *further* refers to non-physical distance or can mean "additional." EXAMPLES:

- The overhead projector was moved *farther* from the screen so that the print would be easier to see.
- *Farther* up the old lumber road, they found footprints of an unidentified mammal.
- As he read *further* along in the report, he began to understand the complexity of the project.
- She gave *further* instructions after they arrived at the site.

Fewer/Less

The adjective *fewer* is used before items that can be counted, whereas the adjective *less* is used before mass quantities. When errors occur, they usually result from *less* being used with countable items, as in this *incorrect* sentence: "We can complete the job with less men at the site." EXAMPLES:

- The newly certified industrial hygienist signed with us because the other firm in which he was interested offered *fewer* [not *less*] benefits.
- There was *less* sand in the sample taken from 15 ft than in the one taken from 10 ft.

Flammable/Inflammable/Nonflammable

Given the importance of these words in avoiding injury and death, make sure to use them correctly—especially in instructions. *Flammable* means "capable of burning quickly" and is acceptable usage. *Inflammable* has the same meaning, but it is *not* acceptable usage for this reason: Some readers confuse it with "nonflammable." The word *nonflammable*, then, means "not capable of burning" and is accepted usage. EXAMPLES:

- They marked the package *flammable* because its contents could be easily ignited by a spark. (Note that *flammable* is preferred here over its synonym, *inflammable*.)
- The foreman felt comfortable placing the crates near the heating unit, since all the crates' contents were *nonflammable*.

Former/Latter

These two words direct the reader's attention to previous points or items. *Former* refers to that which came first, whereas *latter* refers to that which came last. Note that the words are used together only with *two* items or points—*not* with three or more. Also, you should know that some readers may prefer you avoid *former* and *latter* altogether, for the construction may force them to look back to previous sentences to understand your meaning. The second example below gives an alternative.

- (with former/latter) The airline's machinists and flight attendants went on strike yesterday. The *former* left work in the morning, whereas the *latter* left work in the afternoon.
- (without former/latter) The airline's machinists and flight attendants went on strike yesterday. The machinists left work in the morning, whereas the flight attendants left work in the afternoon.

Fortuitous/Fortunate

The word *fortuitous* is an adjective that refers to an unexpected action, without regard to whether it is desirable or not. The word *fortunate* is an adjective that indicates an action that is clearly desired. The common usage error with this pair is the wrong assumption that *fortuitous* events must also be *fortunate*. EXAMPLES:

- Seeing McDuff's London manager at the conference was quite *fortuitous*, since I had not been told that he also was attending.

- It was indeed *fortunate* that I encountered the London manager, for it gave us the chance to talk about an upcoming project involving both our offices.

Generally/Typically/Usually

Words like these can be useful qualifiers in your reports. They indicate to the reader that what you have stated is often, but not always, the case. Make certain to place these adverb modifiers as close as possible to the words they modify. In the first example here, it would be inaccurate to write "were typically sampled," in that the adverb modifies the entire verb phrase "were sampled." EXAMPLES:

- Cohesionless soils *typically* were sampled by driving a 2-in. diameter, split-barrel sampler. (Active-voice alternative: *Typically*, we sampled cohesionless soils by driving a 2-in. diameter, split-barrel sampler.)
- For projects like the one you propose, the technician *usually* will clean the equipment before returning to the office.
- It is *generally* known that sites for dumping waste should be equipped with appropriate liners.

Good/Well

Though similar in meaning, *good* is used as an adjective and *well* is used as an adverb. A common usage error occurs when writers use the adjective when the adverb is required. EXAMPLES:

- It is *good* practice to submit three-year plans on time.
- He did *well* to complete the three-year plan on time, considering the many reports he had to finish that same week.

Imply/Infer

Remember that the person doing the speaking or writing implies, whereas the person hearing or reading the words infers. In other words, the word *imply* requires an active role; the word *infer* requires a passive role. When you *imply* a point, your words suggest rather than state a point. When you *infer* a point, you form a conclusion or deduce meaning from someone else's words or actions. EXAMPLES:

- The contracts officer *implied* that there would be stiff competition for that $20 million waste-treatment project.
- We *inferred* from her remarks that any firm hoping to secure the work must have completed similar projects recently.

Its/It's

These words are often confused. You can avoid error by remembering that *it's* with the apostrophe is used *only* as a contraction for "it is" or "it has." The other form—*its*—is a possessive pronoun. EXAMPLES:

- Because of the rain, *it's* [or *it is*] going to be difficult to move the equipment to the site.
- *It's* [or *it has*] been a long time since we submitted the proposal.
- The company completed *its* part of the agreement on time.

Lay/Lie

These two verbs are troublesome, and you need to know some basic grammar to use them correctly.

1. *Lay* means "to place." It is a transitive verb; thus it takes a direct object to which it conveys action. ("She laid down the printout before starting the meeting.") Its main forms are *lay* (present), *laid* (past), *laid* (past participle), and *laying* (present participle).
2. *Lie* means "to be in a reclining position." It is an intransitive verb; thus it does not take a direct object. ("In some countries, it is acceptable for workers to lie down for a midday nap.") Its main forms are *lie* (present), *lay* (past), *lain* (past principle), and *lying* (present participle).

If you want to use these words with confidence, remember the transitive/intransitive distinction and memorize the principal parts. EXAMPLES:

- (lay) I will *lay* the notebook on the lab desk before noon.
- (lay) I have *laid* the notebook there before.
- (lay) I was *laying* the notebook down when the phone rang.
- (lie) The watchdog *lies* motionless at the warehouse gate.
- (lie) The dog *lay* there yesterday too.
- (lie) The dog has *lain* there for three hours today and no doubt will be *lying* there when I return from lunch.

Lead/Led

Lead is either a noun that names the metallic element or a verb that means "to direct or show the way." *Led* is only a verb form, the past tense of the verb *lead*. EXAMPLES:

- The company bought rights to mine *lead* on the land.
- They chose a new president to *lead* the firm into the twenty-first century.
- They were *led* to believe that salary raises would be high this year.

Like/As

These two words are different parts of speech and thus are used differently in sentences. *Like* is a preposition and therefore is followed by an object—not an entire clause. *As* is a conjunction and thus is followed by a group of words that includes a verb. *As if* and *as though* are related conjunctions. EXAMPLES:

- Gary looks *like* his father.
- Managers *like* John will be promoted quickly.

- If Teresa writes this report *as* she wrote the last one, our clients will be pleased.
- Our proposals are brief, *as* they should be.
- Our branch manager talks *as though* [or *as if*] the merger will take place soon.

Loose/Lose

Loose, which rhymes with "goose," is an adjective that means "unfastened, flexible, or unconfined." *Lose,* which rhymes with "ooze," is a verb that means "to misplace." EXAMPLES:

- The power failure was linked to a *loose* connection at the switchbox.
- Because of poor service, the photocopy machine company may *lose* its contract with McDuff's San Francisco office.

Modifiers: Dangling and Misplaced

This section includes guidelines for avoiding the most common modification errors—dangling modifiers and misplaced modifiers. But first we need to define the term "modifier." Words, phrases, and even dependent clauses can serve as modifiers. They serve to qualify, or add meaning to, other elements in the sentence. For our purposes here, the most important point is that modifiers need to be clearly connected to what they modify.

Modification errors occur most often with verbal phrases. A phrase is a group of words that lacks either a subject or a predicate. The term "verbal" refers to (1) gerunds (*-ing* form of verbs used as nouns, such as "He likes skiing"), (2) participles (*-ing* form of verbs used as adjectives, such as "Skiing down the hill, he lost a glove"), or (3) infinitives (the word "to" plus the verb root, such as "To attend the opera was his favorite pastime"). Now let's look at the two main modification errors.

Dangling Modifiers.   When a verbal phrase "dangles," the sentence in which it is used contains no specific word for the phrase to modify. As a result, the meaning of the sentence can be confusing to the reader. For example, "In designing the foundation, several alternatives were discussed." It is not at all clear exactly who is doing the "designing." The phrase dangles because it does not modify a specific word. The modifier does not dangle in this version of the sentence: "In designing the foundation, we discussed several alternatives."

Misplaced Modifiers.   When a verbal phrase is misplaced, it may appear to refer to a word that it, in fact, does not modify. EXAMPLE: "Floating peacefully near the oil rig, we saw two humpback whales." Obviously, the whales are doing the floating, and the rig workers are doing the seeing here. Yet because the verbal phrase is placed at the beginning of the sentence, rather than at the end immediately after the word it modifies, the sentence presents some momentary confusion.

Misplaced modifiers can lead to confusion about the agent of action in technical tasks. EXAMPLE: "Before beginning to dig the observation trenches, we recommend that the contractors submit their proposed excavation program for our review." On quick reading, the reader is not certain about who will be "beginning to dig"—the contractors or the "we" in the sentence. The answer is the contractors. Thus a correct placement of the modifier would be "We recommend the following: Before the contractors begin digging observation trenches, they should submit their proposed excavation for our review."

Solving Modifier Problems.   At best, dangling and misplaced modifiers produce a momentary misreading by the audience. At worst, they can lead to confusion that results in disgruntled readers, lost customers, or liability problems. To prevent modification problems, place all verbal phrases—indeed, all modifiers—as close as possible to the word they modify. If you spot a modification error while you are editing, correct it in one of two ways:

1.   Leave the modifier as it is and rework the rest of the sentence. Thus you would change "Using an angle of friction of 20 degrees and a vertical weight of 300 tons, the sliding resistance would be . . ." to the following: "Using an angle of friction of 20 degrees and a vertical weight of 300 tons, we computed a sliding resistance of. . . ."

2.   Rephrase the modifier as a complete clause. Thus you would change the previous original sentence to "If the angle of friction is 20 degrees and the vertical weight is 300 tons, the sliding resistance should be. . . ."

In either case, your goal is to link the modifier clearly and smoothly with the word or phrase it modifies.

Number of/Total of

These two phrases can take singular or plural verbs, depending on the context. Here are two simple rules for correct usage:

1. If the phrase is preceded by *the,* it takes a singular verb because emphasis is placed on the group.
2. If the phrase is preceded by *a,* it takes a plural verb because emphasis is placed on the many individual items.

EXAMPLES:

- *The number of* projects going over budget *has* decreased dramatically.
- *The total of* 90 lawyers *believes* the courtroom guidelines should be changed.
- *A number* of projects *have* stayed within budget recently.
- *A total of* 90 lawyers *believe* the courtroom guidelines should be changed.

Numbers

Like rules for abbreviations, those for numbers vary from profession to profession and even from company to company. Most technical writing subscribes to the approach that numbers are best expressed in figures (45) rather than words (forty-five). Note that this style may differ from that used in other types of writing, such as this textbook.

Unless the preferences of a particular reader suggest that you do otherwise, follow these common rules for use of numbers in writing your technical documents:

● *Rule 1: Follow the 10-or-Over Rule*

In general, use figures for numbers of 10 or more, words for numbers below 10. EXAMPLES: three technicians at the site/15 reports submitted last month/one rig contracted for the job.

● *Rule 2: Do Not Start Sentences with Figures*

Begin sentences with the word form of numbers, not with figures. EXAMPLE: "Forty-five containers were shipped back to the lab."

● *Rule 3: Use Figures as Modifiers*

Whether above or below 10, numbers are usually expressed as figures when used as modifiers with units of measurement, time, and money, especially when these units are abbreviated. EXAMPLES: 4 in., 7 hr, 17 ft, $5 per hr. Exceptions can be made when the unit is not abbreviated. EXAMPLE: five years.

● *Rule 4: Use Figures in a Group of Mixed Numbers*

Use only figures when the numbers grouped together in a passage (usually *one* sentence) are both above and below 10. EXAMPLE: "For that project they assembled 15 samplers, 4 rigs, and 25 containers." In other words, this rule argues for consistency within a writing unit.

● *Rule 5: Use the Figure Form in Illustration Titles*

Use the numeric form when labeling specific tables and figures in your reports. EXAMPLES: Figure 3, Table 14-B.

● *Rule 6: Be Careful with Fractions*

Express fractions as words when they stand alone, but as figures when they are used as a modifier or are joined to whole numbers. EXAMPLE: "We have completed two-thirds of the project using the 2 1/2 in. pipe."

● *Rule 7: Use Figures and Words with Numbers in Succession*

When two numbers appear in succession in the same unit, write the first as a word and the second as a figure. EXAMPLE: "We found fifteen 2-ft pieces of pipe in the machinery."

● *Rule 8: Only Rarely Use Numbers in Parentheses*

Except in legal documents, avoid the practice of placing figures in parentheses after their word equivalents. EXAMPLE: "The second party will send the first party forty-five (45) barrels on or before the first of each month." Note that the parenthetical amount is placed immediately after the figure, not after the unit of measurement.

● *Rule 9: Use Figures with Dollars*

Use figures with all dollar amounts, with the exception of the context noted in Rule 8. Avoid cents columns unless exactness to the penny is necessary.

● *Rule 10: Use Commas in Four-Digit Figures*

To prevent possible misreading, use commas in figures of four digits or more. EXAMPLES: 15,000; 1,247; 6,003.

● *Rule 11: Use Words for Ordinals*

Usually spell out the ordinal form of numbers. EXAMPLE: "The government informed all parties of the first, second, and third [not 1st, 2nd, and 3rd] choices in the design competition." A notable exception is tables and figures, where space limitations could argue for the abbreviated form.

Oral/Verbal

Oral refers to words that are spoken, as in "oral presentation." The term *verbal* refers to spoken or written language. To prevent confusion, avoid the word *verbal* and instead specify your meaning with the words "oral" and "written." EXAMPLES:

- In its international operations, McDuff, Inc., has learned that some countries still rely upon *oral* [not *verbal*] contracts.
- Their *oral* agreement last month was followed by a *written* [not *verbal*] contract this month.

Parts of Speech

This term refers to the eight main groups of words in English grammar. A word's placement in one of these groups is based upon its function within the sentence.

Noun. Words in this group name persons, places, objects, or ideas. The two major categories are (1) proper nouns and (2) common nouns. Proper nouns name specific persons, places, objects, or ideas, and they are capitalized. EXAMPLES: Cleveland; Mississippi River; McDuff, Inc.; Student Government Association; Susan Jones; Existentialism. Common nouns name general groups of persons, places, objects, and ideas, and they are not capitalized. EXAMPLES: trucks, farmers, engineers, assembly lines, philosophy.

Verb. A verb expresses action or state of being. Verbs give movement to sentences and form the core of meaning in your writing. EXAMPLES: explore, grasp, write, develop, is, has.

Pronoun. A pronoun is a substitute for a noun. Some sample pronoun categories include (1) personal pronouns (I, we, you, she, he), (2) relative pronouns (who, whom, that, which), (3) reflexive and intensive pronouns (myself, yourself, itself),

(4) demonstrative pronouns (this, that, these, those), and (5) indefinite pronouns (all, any, each, anyone).

Adjective. An adjective modifies a noun. EXAMPLES: horizontal, stationary, green, large, simple.

Adverb. An adverb modifies a verb, an adjective, another adverb, or a whole statement. EXAMPLES: soon, generally, well, very, too, greatly.

Preposition. A preposition shows the relationship between a noun or pronoun (the object of a preposition) and another element of the sentence. Forming a prepositional phrase, the preposition and its object can reveal relationships such as location ("They went *over the hill*"), time ("He left *after the meeting*"), and direction ("She walked *toward the office*").

Conjunction. A conjunction is a connecting word that links words, phrases, or clauses. EXAMPLES: and, but, for, nor, although, after, because, since.

Interjection. As an expression of emotion, an interjection can stand alone ("Look out!") or can be inserted into another sentence.

Passed/Past

Passed is the past tense of the verb *pass*, whereas *past* is an adjective, preposition, or noun that means "previous" or "beyond" or "a time before the present." EXAMPLES:

- He *passed* the survey marker on his way to the construction site.
- The *past* president attended last night's meeting. [adjective]
- He worked *past* midnight on the project. [preposition]
- In the distant *past*, the valley was a tribal hunting ground. [noun]

Per

Coming from the Latin, this word should be reserved for business and technical expressions that involve statistics or measurement—such as "per annum" or "per mile." It should *not* be used as a stuffy substitute for "in accordance with." EXAMPLES:

- Her *per diem* travel allowance of $90 covered hotels and motels.
- During the oil crisis years ago, gasoline prices increased by over 50 cents *per gallon.*
- *As you requested* [not *per your request*], we have enclosed brochures on our products.

Per Cent/Percent/Percentage

Per cent and *percent* have basically the same usage and are used with exact numbers. The one word *percent* is preferred. Even more common in technical writing, however, is the use of the percent sign (%) after numbers. The word *percentage* is only used to express general amounts, not exact numbers. EXAMPLES:

- After completing a marketing survey, McDuff, Inc., discovered that 83 *percent* [or 83%] of its current clients have hired McDuff for previous projects.
- A large *percentage* of the defects can be linked to the loss of two experienced quality-control inspectors.

Practical/Practicable

Though close in meaning, these two words have quite different implications. *Practical* refers to an action that is known to be effective. *Practicable* refers to an action that can be accomplished or put into practice, without regard for its effectiveness or practicality. EXAMPLES:

- His *practical* solution to the underemployment problem led to a 30% increase in employment last year.
- The department head presented a *practicable* response, for it already had been put into practice in another branch.

Principal/Principle

When these two words are misused, the careful reader will notice. Keep them straight by remembering this simple distinction: *Principle* is always a noun that means "basic truth, belief, or theorem." EXAMPLE: "He believed in the principle of free speech." *Principal* can be either a noun or an adjective and has three basic uses:

- **As a noun meaning "head official" or "person who plays a major role."** EXAMPLE: We asked that a *principal* in the firm sign the contract.
- **As a noun meaning "the main portion of a financial account upon which interest is paid."** EXAMPLE: If we deposit $5,000 in *principal*, we will earn 9 percent interest.
- **As an adjective meaning "main or primary."** EXAMPLE: We believe that the *principal* reason for contamination at the site is the leaky underground storage tank.

Pronouns: Agreement and Reference

A pronoun is a word that replaces a noun, which is called the "antecedent" of the pronoun. EXAMPLES: this, it, he, she, they. Pronouns, as such, provide you with a useful strategy for varying your style by avoiding repetition of nouns. Here are some rules to prevent pronoun errors:

● *Rule 1: Make Pronouns Agree with Antecedents*

Check every pronoun to make certain it agrees with its antecedent in number. That is, both noun and pronoun must be singular, or both must be plural. Of special concern are the pronouns "it" and "they." EXAMPLES:

- Change "McDuff, Inc., plans to complete their Argentina project next month" to this sentence: "McDuff, Inc., plans to complete its Argentina project next month."

- Change "The committee released their recommendations to all departments" to this sentence: "The committee released its recommendations to all departments."

● *Rule 2: Be Clear About the Antecedent of Every Pronoun*

There must be no question about what noun a pronoun replaces. Any confusion about the antecedent of a pronoun can change the entire meaning of a sentence. To avoid such reference problems, you may need to rewrite a sentence or even use a noun rather than a pronoun. Do whatever is necessary to prevent misunderstanding by your reader. EXAMPLE: Change "The gas filters for these tanks are so dirty that they should not be used" to this sentence: "These filters are so dirty that they should not be used."

● *Rule 3: Avoid Using "This" as the Subject Unless a Noun Follows It*

A common stylistic error is the vague use of "this," especially as the subject of a sentence. Sometimes the reference is not clear at all; sometimes the reference may be clear after several readings. In almost all cases, however, the use of "this" as a pronoun reflects poor technical style and tends to make the reader want to ask, "This what?" Instead, make the subject of your sentences concrete, either by adding a noun after the "this" or by recasting the sentence. EXAMPLE: Change "He talked constantly about the project to be completed at the Olympics. This made his office-mates irritable" to the following: "His constant talk about the Olympics project irritated his office-mates."

Punctuation: General

Commas.   Most writers struggle with commas, so you are not alone. The problem is basically threefold. First, the teaching of punctuation has been approached in different, and sometimes quite contradictory, ways. Second, comma rules themselves are subject to various interpretations. And third, problems with comma placement often mask more fundamental problems with the structure of a sentence itself.

You need to start by knowing the basic rules of comma use. The rules that follow are fairly simple. If you learn them now, you will save yourself a good deal of time later in that you will not be constantly questioning usage. In other words, the main benefit of learning the basics of comma use is increased confidence in your own ability to handle the mechanics of editing. (If you do not understand some of the grammatical terms that follow, such as "compound sentence," refer to the section on sentence structure.)

● *Rule 1: Commas in a Series*

Use commas to separate words, phrases, and short clauses written in a series of three or more items. EXAMPLE: "The samples contained gray sand, sandy clay, and silty sand." According to current U.S. usage, a comma always comes before the "and" in a series. (In the United Kingdom, the comma is left out.)

● *Rule 2: Commas in Compound Sentences*

Use a comma before the conjunction that joins main clauses in a compound sentence. EXAMPLE: "We completed the drilling at the Smith Industries location, and then we grouted the holes with Sakrete." The comma is needed here because it separates two complete clauses, each with its own subject and verb ("we completed" and "we grouted"). If the second "we" had been deleted, there would be only one clause containing one subject and two verbs ("we completed and grouted"). Thus no comma would be needed. Of course, it may be that a sentence following this comma rule is far too long; do not use the rule to string together intolerably long sentences.

● *Rule 3: Commas with Nonessential Modifiers*

Set off nonessential modifiers with commas—at either the beginning, middle, or end of sentences. Nonessential modifiers are usually phrases that add more information to a sentence, rather than greatly changing its meaning. When you speak, often there is a pause between this kind of modifier and the main part of the sentence, giving you a clue that a comma break is needed. EXAMPLE: "The report, which we submitted three weeks ago, indicated that the company would not be responsible for transporting hazardous wastes." But—"The report that we submitted three weeks ago indicated that the company would not be responsible for transporting hazardous wastes." The first example includes a nonessential modifier, would be spoken with pauses, and therefore uses separating commas. The second example includes an *essential* modifier, would be spoken *without* pauses, and therefore includes *no* separating commas.

● *Rule 4: Commas with Adjectives in a Series*

Use a comma to separate two or more adjectives that modify the same noun. To help you decide if adjectives modify the same noun, use this test: If you can reverse their positions and still retain the same meaning, then the adjectives modify the same word and should be separated by a comma. EXAMPLE: "Jason opened the two containers in a clean, well-lighted place."

● *Rule 5: Commas with Introductory Elements*

Use a comma after introductory phrases or clauses of about five words or more. EXAMPLE: "After completing the topographic survey of the area, the crew returned to headquarters for its weekly project meeting." Commas like the one after "area" help readers separate secondary or modifying points from your main idea, which of course should be in the main clause. Without these commas, there may be difficulty reading such sentences properly.

● *Rule 6: Commas in Dates, Titles, Etc.*

Abide by the conventions of comma usage in punctuating dates, titles, geographic place names, and addresses. EXAMPLES:

- "May 3, 2003, is the projected date of completion." (But note the change in the "military" form of dates: "We will complete the project on 3 May 2003.")
- "John F. Dunwoody, Ph.D., has been hired to assist on the project."
- "McDuff, Inc., has been selected for the project."
- "He listed Dayton, Ohio, as his permanent residence."

Note the need for commas after the year "2003," the title "Ph.D.," the designation "Inc.," and the state name "Ohio." Also note that if the day had not been in the first example, there would be *no* comma between the month and year and no comma after the year.

Semicolons. The semicolon is easy to use if you remember that it, like a period, indicates the end of a complete thought. Its most frequent use is in situations where grammar rules would allow you to use a period but where your stylistic preference is for a less abrupt connector. EXAMPLE: "Five engineers left the convention hotel after dinner; only two returned by midnight."

One of the most common punctuation errors, the comma splice, occurs when a comma is used instead of a semicolon or period in compound sentences connected by words such as "however," "therefore," "thus," and "then." When you see that these connectors separate two main clauses, make sure either to use a semicolon or to start a new sentence. EXAMPLE: "We made it to the project site by the agreed-upon time; however, [or ". . . time. However, . . ."] the rain forced us to stay in our trucks for two hours."

As noted in the "Lists" entry, there is another instance in which you might use semicolons. Place them after the items in a list when you are treating the list like a sentence and when any one of the items contains internal commas.

Colons. As mentioned in the "Lists" entry, you should place a colon immediately after the last word in the lead-in before a formal list of bulleted or numbered items. EXAMPLE: "Our field study involved these three steps:" or "In our field study we were asked to:" The colon may come after a complete clause, as in the first example, or it may split a grammatical construction, as in the second example. However, it is preferable to use a complete clause before a formal list.

The colon can also be used in sentences in which you want a formal break before a point of clarification or elaboration. EXAMPLE: "They were interested in just one result: quality construction." In addition, use the colon in sentences in which you want a formal break before a series that is not part of a listing. EXAMPLE: "They agreed to perform all on-site work required in these four cities: Houston, Austin, Laredo, and Abilene." But note that there is no colon before a sentence series without a break in thought. EXAMPLE: "They agreed to perform all the on-site work required in Houston, Austin, Laredo, and Abilene."

Apostrophes. The apostrophe can be used for contractions, for some plurals, and for possessives. Only the latter two uses cause confusion. Use an apostrophe to indicate the plural form of a word as a word. EXAMPLE: "That redundant paragraph contained seven *area's* and three *factor's* in only five sentences." Although some writers also use apostrophes to form the plurals of numbers and full-cap abbreviations, the current tendency is to include only the "s." EXAMPLES: 7s, ABCs, PCBs, P.E.s.

As for possessives, you probably already know that the grammar rules seem to vary, depending on the reference book you are reading. Here are some simple guidelines:

● *Possessive Rule 1*

Form the possessive of multisyllabic nouns that end in "s" by adding just an apostrophe, whether the nouns are singular or plural. EXAMPLES: actress' costume, genius' test score, the three technicians' samples, Jesus' parables, the companies' joint project.

● *Possessive Rule 2*

Form the possessive of one-syllable, singular nouns ending in "s" or an "s" sound by adding an apostrophe plus "s." EXAMPLES: Hoss's horse, Tex's song, the boss's progress report.

● *Possessive Rule 3*

Form the possessive of all plural nouns ending in "s" or an "s" sound by adding just an apostrophe. EXAMPLES: the cars' engines, the ducks' flight path, the trees' roots.

● *Possessive Rule 4*

Form the possessive of all singular and plural nouns not ending in "s" by adding an apostrophe plus "s." EXAMPLES: the man's hat, the men's team, the company's policy.

● *Possessive Rule 5*

Form the possessive of paired nouns by first determining whether there is joint ownership or individual ownership. For joint ownership, make only the last noun possessive. For individual ownership, make both nouns possessive. EXAMPLE: "Susan and Terry's project was entered in the science fair; but Tom's and Scott's projects were not."

Quotation Marks. In technical writing, you may want to use this form of punctuation to draw attention to particular words, to indicate passages taken directly from another source, or to enclose the titles of short documents such as reports or book chapters. The rule to remember is this: Periods and commas go inside quotation marks; semicolons and colons go outside quotation marks.

Parentheses. Use parentheses carefully, since long parenthetical expressions can cause the reader to lose the train of thought. This form of punctuation can be used when you (1) place an abbreviation after a complete term, (2) add a brief explanation within the text, or (3) include reference citations within the document text (as explained in chapter 13). The period goes after the close parenthesis when the parenthetical information is part of the sentence, as in the previous sentence. (However, it goes inside the close parenthesis when the parenthetical information forms its own sentence, as in the sentence you are reading.)

Brackets. Use a pair of brackets for these purposes: (1) to set off parenthetical material already contained within another parenthetical statement and (2) to draw attention to a comment you are making within a quoted passage. EXAMPLE: "Two McDuff studies have shown that the Colony Dam is up to safety standards. (See Figure 4-3 [Dam Safety Record] for a complete record of our findings.) In addition, the county engineer has a letter on file that will give further assurance to prospective homeowners on the lake. His letter notes that 'After finishing my three-month study [he completed the study in July 1993], I conclude that the Colony Dam meets all safety standards set by the county and state governments.' "

Hyphens. The hyphen is used to form certain word compounds in English. Although the rules for its use sometimes seem to change from handbook to handbook, those that follow are the most common:

● *Hyphen Rule 1*

Use hyphens with compound numerals. EXAMPLE: twenty-one through ninety-nine.

● *Hyphen Rule 2*

Use hyphens with most compounds that begin with "self." EXAMPLES: self-defense, self-image, self-pity. Other "self" compounds, like "selfhood" and "selfsame," are written as unhyphenated words.

● *Hyphen Rule 3*

Use hyphens with group modifiers when they precede the noun but not when they follow the noun. EXAMPLES: A well-organized paper, a paper that was well organized, twentieth-century geotechnical technology, bluish-gray shale, fire-tested material, thin-bedded limestone.

However, remember that when the first word of the modifier is an adverb ending in "-ly," place no hyphen between the words. EXAMPLES: carefully drawn plate, frightfully ignorant teacher.

● *Hyphen Rule 4*

Place hyphens between prefixes and root words in the following cases: (a) between a prefix and a proper name (ex-Republican, pre-Sputnik); (b) between some prefixes that end with a vowel and root words beginning with a vowel, particularly if the use of a hyphen would prevent an odd spelling (semi-independent, re-enter, re-elect); and (c) between a prefix and a root when the hyphen helps to prevent confusion (re-sent, not resent; re-form, not reform; re-cover, not recover).

Punctuation: Lists

As noted in chapter 4 ("Page Design"), listings draw attention to parallel pieces of information whose importance would be harder to grasp in paragraph format. In other words, employ lists as an attention-getting strategy. Following are some general pointers for punctuating lists. (See pages 117–118 in chapter 4 for other rules for lists.)

You have three main options for punctuating a listing. The common denominators for all three are that you (1) always place a colon after the last word of the lead-in and (2) always capitalize the first letter of the first word of each listed item.

Option A: Place no punctuation after listed items. This style is appropriate when the list includes only short phrases. More and more writers are choosing this option, as opposed to Option B. EXAMPLE:

"In this study, we will develop recommendations that address these six concerns in your project:

- Site preparation
- Foundation design
- Sanitary-sewer design
- Storm-sewer design
- Geologic surface faulting
- Projections for regional land subsidence."

Option B: Treat the list like a sentence series. In this case, you place commas or semicolons between items and a period at the end of the series. Whether you choose Option A or B largely depends on your own style or that of your employer. EXAMPLE:

"In this study, we developed recommendations that dealt with four topics:

- Site preparation,
- Foundation design,
- Sewer construction, and
- Geologic faulting."

Note that this option requires you to place an "and" after the comma that appears before the last item. Another variation of Option B occurs when you have internal commas within one or more of the items. In this case, you need to change the commas that follow the listed items into semicolons. Yet you still keep the "and" before the last item. EXAMPLE:

"Last month we completed environmental assessments at three locations:

- A gas refinery in Dallas, Texas;
- The site of a former chemical plant in Little Rock, Arkansas; and
- A waste pit outside of Baton Rouge, Louisiana."

Option C: Treat each item like a separate sentence. When items in a list are complete sentences, you may want to punctuate each one like a separate sentence, placing a period at the end of each. You *must* choose this option when one or more of your listed items contain more than one sentence. EXAMPLE:

"The main conclusions of our preliminary assessment are summarized here:

- At five of the six borehole locations, petroleum hydrocarbons were detected at concentrations greater than a background concentration of 10 mg/kg.
- No PCB concentrations were detected in the subsurface soils we analyzed. We will continue the testing, as discussed in our proposal.

- Sampling and testing should be restarted three weeks from the date of this report."

Regrettably/Regretfully

Regrettably means "unfortunately," whereas *regretfully* means "with regret." When you are unsure of which word to use, substitute the above definitions to determine correct usage. EXAMPLES:

- *Regrettably,* the team members omitted their resumes from the proposal.
- Hank submitted his resume to the investment firm, but, *regrettably,* he forgot to include a cover letter.
- I *regretfully* climbed on the plane to return home from Hawaii.

Respectively

Some good writers may use *respectively* to connect sets of related information. Yet such usage creates extra work for readers by making them reread previous passages. It is best to avoid *respectively* by rewriting the sentence, as shown in the several following options. EXAMPLES:

Original: Appendices A, G, H, and R contain the topographical maps for Sites 6, 7, 8, and 10, respectively.

Revision—Option 1: Appendix A contains the topographical map for Site 6; Appendix G contains the map for Site 7; Appendix H contains the map for Site 8; and Appendix R contains the map for Site 10.

Revision—Option 2: Appendix A contains the topographical map for Site 6; Appendix G for Site 7; Appendix H for Site 8; and Appendix R for Site 10.

Revision—Option 3: Topographical maps are contained in the appendices, as shown in the following list:

Appendix	Site
A	6
G	7
H	8
R	10

Set/Sit

Like *lie* and *lay*, these two verbs are distinguished by form and use. Here are the basic differences:

1. *Set* means "to place in a particular spot" or "to adjust." It is a transitive verb and thus takes a direct object to which it conveys action. Its main parts are *set* (present), *set* (past tense), *set* (past participle), and *setting* (present participle).

2. *Sit* means "to be seated." It is usually an intransitive verb and thus does not take a direct object. Its main parts are *sit* (present), *sat* (past), *sat* (past participle), and *sitting* (present participle). It can be transitive when used casually as a direction to be seated. ("Sit yourself down and take a break.")

EXAMPLES:

- He *set* the computer on the table yesterday.
- While *setting* the computer, on the table, he sprained his back.
- The technician had *set* the thermostat at 75 degrees.
- She plans to *sit* exactly where she sat last year.
- While *sitting* at her desk, she saw the computer.

Sic

Latin for "thus," this word is most often used when a quoted passage contains an error or other point that might be questioned by the reader. Inserted within brackets, *sic* shows the reader that the error was included in the original passage—and that it was not introduced by you. EXAMPLE: The customer's letter to our sales department claimed that "there are too [sic] or three main flaws in the product."

Spelling

All writers find at least some words difficult to spell, and some writers have major problems with spelling. Automatic spell-checking software helps solve the problem. Yet you still need to remain vigilant during the proofreading stage. One or more misspelled words in an otherwise well-written document may cause readers to question professionalism in other areas.

This entry includes a list of commonly misspelled words. However, you should keep your own list of words you most frequently have trouble spelling. Like most writers, you probably have a relatively short list of words that give you repeated difficulty.

absence	analysis	changeable	definitely
accessible	analyze	channel	dependable
accommodate	arctic	column	descend
accumulate	athlete	commitment	dilemma
accustomed	athletic	committee	disappear
achievement	awful	compatible	disappoint
acknowledgment	basically	compelled	disaster
acquaintance	believable	conscience	disastrous
admittance	benefited	conscientious	efficient
advisable	bulletin	conscious	eligible
aisle	calendar	controlled	embarrass
allotting	career	convenient	endurance

environment	ingenious	misspelled	questionnaire
equipment	initially	mortgage	recession
equipped	initiative	movable	reference
essential	insistence	necessary	safety
exaggerate	interfered	noticeable	similar
existence	interference	nuisance	sincerely
experience	interrupt	numerous	specifically
familiar	irrelevant	occasionally	subtle
favorite	judgment	occurred	temperament
February	knowledge	occurrence	temperature
foreign	later	omission	thorough
foresee	latter	pamphlet	tolerance
forfeit	liable	parallel	transferred
forty	liaison	pastime	truly
fourth	library	peculiar	undoubtedly
genius	lightning	possess	unmistakably
government	likely	practically	until
guarantee	loneliness	preference	useful
guidance	maintenance	preferred	usually
handicapped	manageable	privilege	valuable
harass	maneuver	profession	various
height	mathematics	professor	vehicle
illogical	medieval	pronunciation	wholly
incidentally	mileage	publicly	writing
independence	miscellaneous	quantity	written
indispensable			

Stationary/Stationery

Stationary means "fixed" or "unchanging," whereas *stationery* refers to paper and envelopes used in writing or typing letters. EXAMPLES:

- To perform the test correctly, one of the workers had to remain *stationary* while the other one moved around the job site.
- When she began her own business, Julie purchased *stationery* with her new logo on each envelop and piece of paper.

Subject-Verb Agreement

Subject-verb agreement errors are quite common in technical writing. They occur when writers fail to make the subject of a clause agree in number with the verb.

EXAMPLE: "The nature of the diverse geological deposits are explained in the report." (The verb should be "is," since the singular subject is "nature.")

Writers who tend to make these errors should devote special attention to them. Specifically, isolate the subjects and verbs of all the clauses in a document and make certain that they agree. Here are seven specific rules for making subjects agree with verbs:

● Rule 1: Subjects Connected by "And" Take Plural Verbs

This rule applies to two or more words or phrases that, together, form one subject phrase. EXAMPLE: "The site preparation section and the foundation design portion of the report are to be written by the same person."

● Rule 2: Verbs After "Either/Or" Agree with the Nearest Subject

Subject words connected by "either" or "or" confuse many writers, but the rule is very clear. Your verb choice depends on the subject nearest the verb. EXAMPLE: "He told his group that neither the three reports nor the proposal was to be sent to the client that week."

● Rule 3: Verbs Agree with the Subject, Not with the Subjective Complement

Sometimes called a predicate noun or adjective, a subjective complement renames the subject and occurs after verbs such as "is," "was," "are," and "were." EXAMPLE: "The theme of our proposal is our successful projects in that region of the state." But the same rule would permit this usage: "Successful projects in that part of the state are the theme we intend to emphasize in the proposal."

● Rule 4: Prepositional Phrases Do Not Affect Matters of Agreement

"As long as," "in addition to," "as well as," and "along with" are prepositions, not conjunctions. A verb agrees with its subject, not with the object of a prepositional phrase. EXAMPLE: "The manager of human resources, along with the personnel director, is supposed to meet with the three applicants."

● Rule 5: Collective Nouns Usually Take Singular Verbs

Collective nouns have singular form but usually refer to a group of persons or things (for example, "team," "committee," or "crew"). When a collective noun refers to a group as a whole, use a singular verb. EXAMPLE: "The project crew was ready to complete the assignment." Occasionally, a collective noun refers to the members of the group acting in their separate capacities. In this case, either use a plural verb or, to avoid awkwardness, reword the sentence. EXAMPLE: "The crew were not in agreement about the site locations." Or, "Members of the crew were not in agreement about the site locations."

● Rule 6: Foreign Plurals Usually Take Plural Verbs

Although usage is gradually changing, most careful writers still use plural verbs with "data," "strata," "phenomena," "media," and other irregular plurals. EXAMPLE:

"The data he asked for in the request for proposal are incorporated into the three tables."

● *Rule 7: Indefinite Pronouns Like "Each" and "Anyone" Take Singular Verbs*

Writers often fail to follow this rule when they make the verb agree with the object of a prepositional phrase, instead of with the subject. EXAMPLE: "Each of the committee members are ready to adjourn" (incorrect). "Each of the committee members is ready to adjourn" (correct).

To/Too/Two

To is part of the infinitive verb form *or* is a preposition in a prepositional phrase. *Too* is an adverb that suggests an excessive amount *or* that means "also." *Two* is a noun or an adjective that stands for the numeral "2." EXAMPLES:

- He volunteered *to* go [infinitive verb] *to* Alaska [prepositional phrase] *to* work [another infinitive verb form] on the project.
- Stephanie explained that the proposed hazardous-waste dump would pose *too* many risks *to* the water supply. Scott made this point, *too*.

Utilize/Use

Utilize is simply a long form for the preferred verb "use." Although some verbs that end in "-ize" are useful words, most are simply wordy substitutes for shorter forms. As some writing teachers say, "Why use 'utilize' when you can use 'use'?"

Which/That

Which is used to introduce nonrestrictive clauses, which are defined as clauses not essential to meaning (as in this sentence). Note that such clauses require a comma before the *which* and a slight pause in speech. *That* is used to introduce restrictive clauses that are essential to the meaning of the sentence (as in this sentence). Note that such clauses have no comma before the *that* and are read without a pause. *Which* and *that* can produce different meanings, as the following examples show:

- Our benefits package, *which* is the best in our industry, includes several options for medical care.
- The benefits package *that* our firm provides includes several options for medical care.
- My daughter's school, *which* is in Cobb County, has an excellent math program.
- The school *that* my daughter attends is in Cobb County and has an excellent math program.

Note that the above examples with *that* might be considered wordy by some readers. Indeed, such sentences often can be made more concise by deleting the *that* introducing the restrictive clause. However, delete *that* only if you can do so without creating an awkward, choppy sentence.

Who/Whom

These two words give writers (and speakers) fits, but the importance of their correct use probably has been exaggerated. If you want to be one who uses them properly, remember this basic point: *Who* is a subjective form that can only be used in the subject slot of a clause; *whom* is an objective form that can only be used as a direct object or other nonsubject noun form of a sentence. EXAMPLES:

- The man *who* you said called me yesterday is a good customer of the firm. (The clause "who . . . called me yesterday" modifies "man." Within this clause, *who* is the subject of the verb "called." Note that the subject role of *who* is not affected by the two words "you said," which interrupt the clause.)
- They could not remember the name of the person *whom* they interviewed. (The clause "whom they interviewed" modifies "person." Within this clause, *whom* is the direct object of the verb "interviewed.")

Who's/Whose

Who's is a contraction that replaces *who is,* whereas *whose* is a possessive adjective. EXAMPLES:

- *Who's* planning to attend the annual meeting?
- Susan is the manager *who's* responsible for training.
- *Whose* budget includes training?
- Susan is the manager *whose* budget includes training.

Your/You're

Your is an adjective that shows ownership, whereas *you're* is a contraction for *you are.* EXAMPLES:

- *Your* office will be remodeled next week.
- *You're* responsible for giving performance appraisals.

EXERCISE 1: GRAMMAR AND MECHANICS

The following passages contain a variety of grammatical and mechanical errors covered in the handbook. The major focus is punctuation. Rewrite each passage.

1. Some concerns regarding plumbing design are mentioned in our report, however, no unusual design problems are expected.
2. An estimate of the total charges for an audit and for three site visits are based on our standard fee schedules.
3. The drill bit was efficient cheap and available.

4. The plan unless we have completely misjudged it, will increase sales markedly.

5. Our proposal contains design information for these two parts of the project; Phase 1 (evaluating the 3 computers) and Phase 2 (installing the computer selected).

6. If conditions require the use of all-terrain equipment to reach the construction locations, this will increase the cost of the project slightly.

7. An asbestos survey was beyond the scope of this project, if you want one, we would be happy to submit a proposal.

8. Jones-Simon Company, the owners of the new building, were informed of the problem with the foundation.

9. Also provided is the number and type of tests to be given at the office.

10. Calculating the standard usages by the current purchase order prices result in a downward adjustment of $.065.

11. Data showing the standard uses of the steel, including allowances for scrap, waste and end pieces of the tube rolls, are included for your convenience at the end of this report in Table 7.

12. This equipment has not been in operation for 3 months, and therefore, its condition could not be determined by a quick visual inspection.

13. Arthur Jones Manager of the Atlanta branch wrote that three proposals had been accepted.

14. The generator that broke yesterday has been shipped to Tampa already by Harry Thompson.

15. The first computer lasted eight years the second two years.

16. He wants one thing out of their work speed.

17. On 25 September 1993 the papers were signed.

18. On March 23 1993 the proposal was accepted.

19. The meeting was held in Columbus the Capital of Ohio.

20. McDuff, Inc. completed its Indonesia project in record time.

21. He decided to write for the brochure then he changed his mind.

22. Interest by the Kettering Hospital staff in the development of a master plan for the new building wings have been expressed.

23. However much he wants to work for Gasion engineering he will turn the job down if he has to move to another state.

24. 35 computer scientists attended the convention, but only eleven of them were from private industry.

25. Working at a high salary gives him some satisfaction still he would like more emotional satisfaction from his job.

26. His handwriting is almost unreadable therefore his secretary asked him to dictate letters.

27. Any major city especially one that is as large as Chicago is bound to have problems with mass transit.

28. He ended his speech by citing the company motto; "Quality first, last, and always".

29. Houston situated on the Gulf of Mexico is an important international port.

30. The word "effect" is in that student's opinion a difficult one to use.

31. All persons who showed up for the retirement party, told stories about their association with Charlie over the years.

32. The data that was included in the study seems inconclusive.
33. My colleague John handled the presentation for me.
34. Before he arrived failure seemed certain.
35. While evaluating the quality of her job performance a study was made of her writing skills by her supervisor.
36. I shall contribute to the fund for I feel that the cause is worthwhile.
37. James visited the site however he found little work finished.
38. There are three stages cutting grinding and polishing.
39. The three stages are cutting grinding and polishing.
40. Writers occasionally create awkward verbs "prioritize" and "terminate" for example.
41. Either the project engineers or the consulting chemist are planning to visit with the client next week.
42. Besides Gerry Dave worked on the Peru project.
43. The corporation made a large unexpected gift to the university.
44. The reason for his early retirement are the financial incentives given by his employer.
45. Profit, safety and innovation are the factors that affect the design of many foundations.
46. No later than May 1995 the building will be finished.
47. Each of the committee members complete a review of the file submitted by the applicant.
48. The team completed their collaborative writing project on schedule.
49. Both the personnel officers and the one member of the quality team is going to attend the conference in Fargo.
50. He presented a well organized presentation but unfortunately the other speakers on the panel were not well-prepared.

EXERCISE 2: USAGE

For each of the following passages, select the correct word or phrase from the choices within the parentheses. Be ready to explain the rationale for your choice.

1. John (implied, inferred) in his report that TransAm Oil should reject the bid.
2. Before leaving on vacation, the company president left instructions for the manner in which responsibilities should be split (among, between) the three vice presidents.
3. Harold became (uninterested, disinterested) in the accounting problem after working on it for 18 straight hours.
4. A large (percent, percentage) of the tellers is dissatisfied with the revised work schedule.
5. The typist responded that he would make (less, fewer) errors if the partner would spell words correctly in the draft.
6. From her reading of the annual report, Ms. Jones (inferred, implied) that the company might expand its operations.
7. The president's decision concerning flextime will be (effected, affected) by the many conversations he is having with employees about scheduling difficulties.

8. His (principal, principle) concern was that the loan's interest and (principle, principal) remain under $500.
9. Throughout the day, his concentration was interrupted (continuously, continually) by phone calls.
10. He jogged (continuously, continually) for 20 minutes.
11. Five thousand books (compose, comprise) his personal library.
12. The clients (who, whom) he considered most important received Christmas gifts from the company.
13. The company decided to expand (its, it's) operations in the hope that (its, it's) the right time to do so.
14. The (nonflammable, flammable, inflammable) liquids were kept in a separate room, because of their danger.
15. They waited for (awhile, a while) before calling the subcontractor.
16. Caution should be taken to (ensure, insure) that the alarm system will not go off accidentally.
17. The new floors (are comprised of, are composed of, comprise) a thick concrete mixture.
18. He (expects, anticipates) that 15 new employees will be hired this year.
19. The main office offered to (augment, supplement) the annual operating budget of the Boston office with an additional $100,000 in funds.
20. It was (all together, altogether) too late to make changes in the proposal.
21. The arbitrator made sure that both parties (agreed to, agreed with) the terms and conditions of the contract before it was submitted to the board.
22. Option 1 calls for complete removal of the asbestos. (Alternately, Alternatively), Option 2 would only require that the asbestos material be thoroughly covered.
23. They had not considered the (amount, number) of cement blocks needed for the new addition.
24. (Due to, Because of) the change in weather, they had to reschedule the trip to the project site.
25. The health inspector found (too, to) many violations in that room (to, too).
26. They claimed that the old equipment (used, utilized) too much fuel.
27. Gone are the days when a major construction job gets started with a handshake and (a verbal, an oral) agreement.
28. The complex project has 18 (discreet, discrete) phases; each part deals with confidential information that must be handled (discretely, discreetly).
29. He was (definitive, definite) about the fact that he would not be able to complete the proposal by next Tuesday.
30. He usually received (complementary, complimentary) samples from his main suppliers.
31. To (lose, loose) a client for whom they had worked so hard was devastating.
32. It was (fortunate, fortuitous) he was there at the exact moment the customer needed to order a year's worth of supplies, for the sales commission was huge.
33. Among all the information on the graph, he located the one (data, datum) that shows the price of tuna on the Seattle market at 5 P.M. on August 7.
34. Each (principle, principal) of the corporation was required to buy stock.
35. He returned to the office to (assure, ensure) that the safe was locked.

Index

PLANNING FORM

NAME: _____ ASSIGNMENT: _____

I. Purpose: Answer each question in one or two sentences.

  A. Why are you writing this document?_____

_____

  B. What response do you want from readers? _____

_____

II. Reader Matrix: Fill in names and positions of people who may read the document.

	Decision-Makers	Advisers	Receivers
Managers	_____ _____ _____	_____ _____ _____	_____ _____ _____
Experts	_____ _____ _____	_____ _____ _____	_____ _____ _____
Operators	_____ _____ _____	_____ _____ _____	_____ _____ _____
General Readers	_____ _____ _____	_____ _____ _____	_____ _____ _____

III. Information on Individual Readers: Answer these questions about selected members of your audience. Attach additional sheets as is necessary.

  1. What is this reader's technical or educational background?

_____

_____

  2. What main question does this person need answered?

_____

_____

  3. What main action do you want this person to take?

_____

_____

  4. What features of this person's personality might affect his or her reading?

_____

_____

  5. What features does this person prefer in

    Format?_____

    Style?_____

    Organization? _____

IV. Outline: Attach an outline (topic) to use in drafting the **BODY** of this document.